Second Edition

EMERGENCY RESPONSE HANDBOOK for CHEMICAL and BIOLOGICAL AGENTS and WEAPONS

Second Edition

EMERGENCY RESPONSE HANDBOOK for CHEMICAL and BIOLOGICAL AGENTS and WEAPONS

JOHN R. CASHMAN

CRC Press
Taylor & Francis Group
Boca Raton London New York

CRC Press is an imprint of the
Taylor & Francis Group, an **informa** business

CRC Press
Taylor & Francis Group
6000 Broken Sound Parkway NW, Suite 300
Boca Raton, FL 33487-2742

© 2008 by Taylor & Francis Group, LLC
CRC Press is an imprint of Taylor & Francis Group, an Informa business

No claim to original U.S. Government works
Printed in the United States of America on acid-free paper
10 9 8 7 6 5 4 3 2 1

International Standard Book Number-13: 978-1-4200-5265-7 (Hardcover)

Library of Congress Cataloging-in-Publication Data

Cashman, John R.
    Emergency response handbook for chemical and biological agents and weapons / author, John R. Cashman. -- 2nd ed.
        p. cm.
    "A CRC title."
    Rev. ed. of: Emergency response to chemical and biological agents / John R. Cashman. 2000.
    Includes bibliographical references and index.
    ISBN 978-1-4200-5265-7 (hardback : alk. paper)
    1. Hazardous substances--Accidents--Handbooks, manuals, etc. 2. Biological weapons--Safety measures--Handbooks, manuals, etc. 3. Chemical weapons--Safety measures--Handbooks, manuals, etc. 4. Emergency medical services--Handbooks, manuals, etc. I. Cashman, John R. Emergency response to chemical and biological agents. II. Title.

T55.3.H3C377 2008
628.9'2--dc22   2008002764

Visit the Taylor & Francis Web site at
http://www.taylorandfrancis.com

and the CRC Press Web site at
http://www.crcpress.com

Dedication

Both firefighters and statesmen lost a wonderful friend and a constant worker for the establishment of instrumentation and methodology that would provide improvement for meaningful operating procedures of the fire service. Chief John Eversole died on May 20, 2007. John was a giant in thought, word, and deed; and the fire service in the United States lost a gentle giant when he died. Chief John Eversole was a hazardous materials coordinator for the Chicago, Illinois Fire Department.

He was a member of the department for thirty years and worked on some of the busiest engines, trucks, hook-and-ladders, and squad companies in the western part of the city. Eversole was a member of the National Fire Protection Association standard committee that produced national hazardous materials NFPA-471, NFPA-472, and NFPA-473. He was also chairman of the International Association of Fire Chiefs' hazardous materials committee.

He was a leader on the street, and in the committee rooms where the nitty-gritty of hazardous materials response in the United States is thrashed out. Chief John Eversole was approachable to all persons and eloquent in expressing his thoughts. He also had a definite command presence on an incident scene gained through long command experience. John, we are going to miss you.

Contents

About the Author

John R. Cashman, AA, BA, MPA, has been writing about hazardous materials response and control for thirty years. He began his career covering events such as road racing and motocross in the summer, and snowmobile racing in the winter, in his off hours while an employee of the state of Vermont. He became a full-time writer of non-fiction in 1978. To date, he has over 250 magazine articles and six books to his credit. In addition, he has published *Hazardous Materials Newsletter* since April of 1980 (Haznews@msn.com).

Introduction

Definition of topic: An in-depth and complete training manual for first emergency responders and other secondary responders at all levels to incidents involving chemical and biological agents.

This valuable new reference book provides a comprehensive guide that thoroughly and expertly covers the fundamental practices and advanced tactics of emergency response to chemical and biological agents: nerve agents, vesicants, pulmonary agents, cyanogen agents, biological substances (bacteria, rickettsiae, chlamydia, viruses, and toxins), antidotes, pretreatments, vaccines, detectors, decontamination techniques, and medical treatment. In the United States, emergency response to chemical and biological incidents will be done by local hazardous materials response teams (HMRT) and first responders who will be assisted by the federal government. This includes firefighters, emergency medical service personnel, emergency management officials, police and sheriff's agencies, civil defense employees, nurses/physicians and other hospital personnel, military, health regulatory, commercial response contractors and cleanup forces, industrial Haz-Mat teams and fire brigades, county/local/state/federal governments, the Red Cross, the Salvation Army, consultants, National Guard, Community colleges who train firefighters and EMS personnel, FBI, Coast Guard Haz-Mat Team, OSHA, FEMA, State Police, COBRA Teams, EPA, coroners, and suppliers who provide cleanup and manpower services to scenes of destruction. Since September 11, 2001, the U.S. government has spent vast sums of money for protective equipment and training to deal with chemical and biological agents out of fear of terrorism, industrial accidents, misuse, and criminal activities.

This book focuses on actual response techniques and offers advice to first responders and other government agencies for dealing with chemical and biological agents and weapons. Chapter 1 deals with the worst railroad wreck in the last thirty years in which chlorine, a poison gas in World War I, killed nine persons, sickened 554 others (75 of whom were admitted to hospitals), and caused the evacuation from their homes of 5,400 people for fourteen days. Other chapters focus on the killings at Columbine, Colorado, and biological/chemical agents that can wreak havoc and death on the population. Agents included are: Anthrax, Botulism, Brucellosis, Glanders and Melioidosis, Plague, Q Fever, Ricin, S.E.B. (Staphylococcal Enterotoxin B), Smallpox, Trichothecene Mycotoxins T-2, Tularemia, Venezuelan Equine Encephalitis (V.E.E.), and Viral Hemorrhagic Fevers (V.H.F.), among them: Crimean Congo Fever, Ebola Fever, Lassa Fever, Rift Valley Fever, and other VHF fevers.

Chapters focused on biological agents are presented in the following format: Agent (Introduction), Classification (Broad), Duration of Illness, Probable Form of Dissemination, Detection in the Field, Infective Dose (Aerosol), Sign and Symptoms, Incubation Time, Diagnosis, Differential Diagnosis, Vaccine Efficacy, Persistency, Personal Protection, Routes of Entry to the Body, Transmissible from Person-To-Person, Duration of Illness, Potential Ability to Kill, Symptoms & Effects, Defensive Measures, Vaccines, Drugs Available, Decontamination, Specific Ability to Kill, and Characteristics.

Each biological agent chapter has a Response on Scene by First Responders section that includes Caution, Field First Aid, Drugs, Antibiotics, Medical Management, Fire, Personal Protection, Spill/Leak Control, Symptoms, and Vaccines. Also, spread over other chapters, will include basic duties of various first responders including Fire Departments, Emergency Medical Services, Law Enforcement, and Hazardous Materials Response Teams.

Chemical agents considered include Arsenical Vesicants (including Ethyldichloroarsine, Methyldichloroarsine, and Phenyldichloroarsine), Arsine, Cyanogen Chloride, Diphosgene, Phosgene, Distilled Mustard, Hydrogen Cyanide, Lewisite, Mustard Lewisite, Nerve Agent GF, Nerve Agent Sarin, Nerve Agent Soman, Nerve Agent Tabun, Nerve Agent VX, and Nitrogen Mustards (including HN-1, HN-2, and HN-3). These chemical agents and weapons will have basic information as follows: Formula, Vapor Density, Vapor Pressure, Molecular Weight, Liquid Density, Volatility, Medium Lethal Dose, Physical State, Odor, Freezing/Melting Point, Action Weight, Physiological Action, Required Level of Protection, Decontamination, Detection in the Field, Use, CAS Registry Number, RTECS Number, and LD50 (oral). Chemical agents will also have a mirror-image of the "Response on Scene by First Responders" questions addressed above for biological agents (i.e., Caution, Field First Aid, Drugs, etc.).

Our federal government has changed completely since the first attacks in 2001 in its manner of providing state governments with federal money for personal protective equipment, assistance, antibiotics, vaccines, grants, medical programs, detection devices, drugs and defensive measures. As a case in point, South Carolina received $90 million dollars in federal money to fight terrorism.

The Chairman of the National Intelligence Council released a report in Washington entitled, "The Global Infectious Disease Threat and Its Implications for the United States." This report deals with warnings of growing possibilities for American citizens to come down with infections that run rampant in other parts of the world, since the United States is a sizable hub for world travelers, immigration and commerce. Also, we have a high percentage of American military service personnel serving in all sections of the world. The Asian continent has seen steady increases in infectious diseases such as the spread of HIV and AIDS. In addition, an estimated thirty diseases, unknown in the past, have appeared globally since the early 1970s. These diseases include Hepatitis C, Nipah virus that is encephalitis-related, and Ebola hemorrhagic fever, and other diseases that so far remain incurable. In like manner, infectious diseases such as malaria, cholera, and tuberculosis have rejoined our nation since the 1970s. Also, terrorist use of biological agents has increased; there were 140 anthrax hoaxes the United States in the late 1990s and, most recently, actual anthrax attacks in 2001. The report also gives a warning that most infectious diseases originate in other countries and are brought into the United States by travelers, immigrants, imported animals, foodstuffs, and our military troops who have served in far corners of the earth.

The response to chemical and biological warfare weapons, terrorism attacks, and even influenza mini-epidemics can require an immense number of federal/state/county/city/private workers and medical personnel ranging from physicians to registered nurses to emergency medical technicians and paramedics. As an example, the writer has actually seen an incident where thirty-one agencies from all levels of government responded with an untold number of persons who were assigned at the site.

Any terrorist incident that uses chemical or biological agents is basically a hazardous materials incident. Hazardous Materials Response Teams (HMRT) and first responders to such incidents use the following principles and characteristics. They deal with abatement, action checklists, after-action reports, antibiotics, antidotes, biological agents, blister agents, blood agents, breakthrough time, as well as briefings and critiques. CAMEO (Computer Aided Management of Emergency Operations), case histories, chemical agents, chemistry of Haz-Mat, command post operation, compatibility, computers, and containment are well known to such responders. Containers, contingency planning, databases, decision making, decontamination, detectors, dispersants, disposal, emergency response plans, and evacuation are "must know" information for HMRT. Size-up & evaluation, exposures, funding/cost recovery, hazard analysis, incident command system, incident vigilance & discipline, industrial agents, leak/fire/spill control, and "lessons learned" are daily facts of life. They consider levels of incidents, levels of protection, manuals, material safety data sheets (MSDS), medical surveillance, monitoring, and national and local contingency plans among their standard operating procedures. HMRT members know that nerve agents, neutralizers, no-fight situations, offloading/transfer, patching/plugging, perimeter control can help them *or* hurt them. They study personnel safety, pH, physics, post-fire residue, protection, reactivity, recon, and resource materials to be better prepared. They also understand responder liability, scene management, secondary emergencies, S.O.P.(s) & protocols, sorbent materials, staging areas, standards, storage, tactics, team concept, testing methods, toxicology, training, triage, vaccine, vectors, vehicles, weapons of mass destruction (WMD), and zonal delineation.

We are speaking here of "emergency responders" of all possible types and talents including firefighters, police officers, emergency medical technicians and paramedics, ambulance crews, decontamination crews both in the field and at hospitals and clinics, emergency management officials, military troops, health regulatory personnel, commercial response contractors, engineers and construction employees, industry HMRT and fire brigades, county/local/state/federal government responders, consultants and specialists, specialty teams and mortuary workers. First responders and other interested persons who deal with hazardous materials response and control, chemical and biological agent and weapon control, and the ever-growing number of people who are concerned about the steady growth of terrorism in the United States; are looking for hard facts, rather than theories.

The average lethal chemical agents in storage today are thousands of times *less* lethal, by weight, than equivalent amounts of biological warfare agents; because of their very high toxicity, the lethal biological agent dose can be far smaller than that required from chemical agents.

Chemical and biological weapons (CBW) have long been called "the poor man's atomic bomb," but they are actually weapons of mass destruction that once could be afforded only by a few powerful and industrialized nations; however, during the twenty-first century, a proliferation of technology has now made them readily available to second and third rate powers, as well as terrorists and one man or woman acting alone

Terrorists have never had so many options to inflict death, injury, and destruction on a large scale as they do today.

September 11, 2001, A Targeted Nation:
What Terrorists Can Accomplish

Now, we finally know what terrorism is; the unfathomable suddenly became credible. First responders at the local, county, and state levels have been training for Domestic Prepared-ness defense for at least eight years, but when put to the test, they found they were badly at odds with the reality of terrorism. They had been trained to defend against NBC (Nuclear, Biological, and Chemical) agents, WMD, mass casualty incidents, and secondary explo-sives. The federal government's definition of WMD states, "Any explosive, incendiary, or poison gas, bomb, grenade, rocket having a propellant charge of more than four ounces, missile having an explosive or incendiary charge of more than one quarter ounce, mine or device similar to the above; poison gas; any weapon involving a disease organism; or any weapon that is designed to release radiation or radioactivity at a level dangerous to human life." It never mentions commercial airliners with a minimum passenger load to allow take-over carrying a maximum load of aviation fuel and used as missiles. Our domestic citizens at this time find themselves living in a completely changed world. There were a good many dead at the World Trade Center, but very few injured who could be treated and saved. First responders drilling with mass casualty scenarios before September 11, 2001, expected casualties, not piles of dead bodies. What the hijackers created in New York City was a high death/low treatment scenario put into action—terrible, ghastly, and deadly action. The actual scenario played-out proved that the terrorists were eminently successful in reaching their goals. We were totally unprepared for their attacks or level of skill used to organize an attack to kill an estimated 6,000 Americans and others who came from eighty different countries around the world.

The "missiles" that hit the World Trade Center, a world-wide symbol of financial trade in the Wall Street area of New York City, destroyed much more than the military missiles the United States sent against training camps in the desert set up by Osama bin Laden in the past. Although the World Trade Center required seven years to construct and cost $1.5 billion in current dollars when it had its grand opening in 1973, a structure designed to withstand the damage done if the largest airliner available in 1973 would crash into the buildings, it could *not* stand the crash of two airliners, one into each tower of Building 1 (the first tower to be hit at 8:45a.m.) and Building 2 (the second tower to be hit just after 9 a.m., the tower with the observation deck 110 stories in the air).

Nineteen Terrorists from the Middle East

Mohamed Atta, Flight 11, Boston to L.A., American Airlines, Boeing 767, the first plane to hit the World Trade Center towers.

Marwan Al Shehhi , United Flight 175, Boston to L.A., Boeing 767, the second plane to hit the World Trade Center towers.

Hani Hanjour, American Airlines Flight 77, Washington to L.A., Boeing 757, crashed into the Pentagon just outside Washington, D.C.

Wail Alshehri, American Airlines Flight 11.

Waleed M. Alshehri, American Airlines Flight 11.

Abdulaziz Alomani, American Airlines Flight 11.

Ziad Jarrahi, United Airlines Flight 93, Newark to San Francisco, crashed in
   Pennsylvania.
Khalid Al-Midhar, American Airlines Flight 77.
Majed Moqed, American Airlines Flight 77.
Nawaq Alhamzi, American Airlines Flight 77.
Salem Alhamzi, American Airlines Flight 77.
Satam Al-Sugami, American Airlines Flight 11.
Fayez Ahmed, United Airlines Flight 175.
Ahmed Alghamdi, United Airlines Flight 175.
Hamza Alghamdi, United Airlines Flight 175.
Mohald Alshehri, United Airlines Flight 175.
Saeed Alghamdi, United Airlines Flight 93.
Ahmed Alhaznawi, United Airlines Flight 93.
Ahmed Alnami, United Airlines Flight 93.

Although there were some secondary explosions cited just after impact, causing inferences that the hijackers may have hid explosives in the towers before-hand in preparation
for the airliner crashes to come, it is more likely that these later explosions were caused by
the effect of an estimated 35,000 gallons of jet fuel carried by two Boeing 767 aircraft, as
well as structural damage to critical support structures. Building 2 fell first about 10 a.m.,
and Building 1 imploded and fell at roughly 10:30 a.m. A total of seven buildings at the
World Trade Center collapsed and two were damaged.

At least five Kamikaze-like pilots using 14 "soldiers" as their crews hijacked four commercial airlines' planes with great ease and flew two craft into separate towers of the World
Trade Center in New York City, one into the Pentagon in Washington, D.C., and crashed
the remaining plane in a Pennsylvania meadow after possibly heading back toward Washington, D.C. The incident was described as, "an act of war by madmen" by one commenta-
tor, but this writer does not agree they were madmen. They were exceptionally well trained,
equipped, and motivated by a religious cause they expected would lead them to their prom-
ised land. Their carefully laid plans went off without a hitch, and resulted in an estimated
death count of over six thousand U. S. citizens, responders, and foreigner visitors. In the
material that follows, readers will be led through the attack and introduced to nineteen
young men, both pilots and "soldiers" who were essential for the attack but probably did
not plan, organize, or finance the venture. They proved what terrorists from the Middle
East could do to the most powerful country in the world by studying our weaknesses,
strengths, immigration laws, communications systems, and goodwill. They provided, in
turn, evidence to us of disastrous weaknesses in our antiterrorist programs and domestic
defense systems that we have spent billions of U.S. dollars on over the last seven years. Such
weaknesses need to be corrected at once, or they can be utilized again.

American Airlines Flight 11, a Boeing 767 en-route to Los Angeles, California with
ninety-two passengers aboard, left Logan International Airport in Boston, Massachusetts,
lifted into the air and headed west. Before the passengers could even get accustomed to
the early morning trip, the hijackers took over the plane by using knives and box cutters
to stab two flight attendants and a business-class passenger before bursting into the pilots'
compartment and taking over the plane with their own pilot trained in the United States.
Immediately after taking over the cockpit, the aircraft changed direction and began to lose

altitude. The hijackers aboard Flight 11 were named, or had selected names for this attack, Mohamed Atta, Abdulaziz Alomari, Waleed M. Alshehri, Wail Alshehri, and Satam Al Sugami. On September 11, 2001, Mohamed Atta and Abdulaziz Alomai had their photograph taken by a twenty-four-hour security camera as they exited the screening area at Portland, Maine, International Jetport. By 8:10 a.m., they were in control of a Boeing 767 "missile" loaded with 20,000 gallons of aviation fuel headed toward the World Trade Center on lower Manhattan Island. The airliner would write a new page in the history books of the United States. At 8:45 a.m., the hijacker pilot prepared to meet his maker and crashed his missile into the north tower of the World Trade Center. At first, bystanders thought it must be a freak accident, like the plane that crashed into the Empire State Building in 1947. Employees went into the hallways or sought solace from friends nearby, but it was reported that a public address system announcement told them it was all right to return to their offices. Some did not believe this seemingly official word, and started down the stairs. Some who tried to take the elevators down to the lobby of this massive building were burned by fireballs in the elevator shafts. The wonder is that so many survivors were actually able to make it down from as high as the eightieth floor by trudging down the stairs while they watched firefighters, emergency medical technicians, and police officers fight their way up on the same stair wells.

Another Boeing 767 missile, United Flight 175, left Logan Airport in Boston at 8:14 a.m. en-route to Los Angeles, California with fifty passengers and nine crew members. It was taken over in minutes, and later hit the south side of the south tower just after 9 a.m. American Airline Flight 77 using a Boeing 757 took off from Dulles International Airport located in Washington, D.C., a little after 8 a.m. carrying fifty-eight passengers and six crew members bound for Los Angeles, California. After a roundabout flight, Flight 77 hit the southwest side of the five-sided Pentagon at about 9:20 a.m. A Boeing 757, United Airlines Flight 93, took off from Newark, New Jersey, a few seconds after eight in the morning with thirty-eight passengers and seven crew members. This flight ultimately crashed in Shanksville, Pennsylvania, at 10:00 a.m., apparently after a battle between passengers, who had learned by cell phones calls of the attacks on the World Trade Center and the hijackers. The Boeing 767 and the Boeing 757 are reportedly so similar in performance and characteristics that pilots of such aircraft can fly one if they have been trained in the other.

The American Dream came crashing to earth when the World Trade Center's north and south towers collapsed. Domestic terrorism by foreign troops, and the tremendous loss of life involved, led to a new awakening in American citizens of fear, uncertainty, revulsion, anger, awareness, and dedication. The Federal Aviation Administration shut down airports and ordered all flights airborne at the time to land at the closest, feasible airport. The City of New York closed all bridges and tunnels, causing a great walking migration of Manhattan workers to their homes in other locations. Buildings were evacuated near the collapsed Trade Center, as well as the United Nations buildings facing the East River. The New York Stock Exchange, the American Stock Exchange, and NASDAQ had to close. In Washington, D.C., just about everything was either evacuated, shut-down, or guarded after American Airline Flight 77 hit the Pentagon. U. S. borders with both Canada and Mexico were placed on highest alert. Two aircraft carriers and five smaller ships were alerted at Norfolk, Virginia, and ordered out to protect East Coast cities. By mid-afternoon, Urban Search and Rescue Teams and trained canines were alerted around the country from as far away as Sacramento, California, by the Federal Emergency Management Agency who supports a total of twenty-six such teams for area emergency response. Americans, and viewers around the world, began watching television hour-after-hour with rapt attention. They could not tear themselves away from TV sets and radios. Do you remember where you were on September 11, 2001? Do you remember what your first thought was at that time? Was your first thought, 'How could this happen in the World of the Free?' This answer should be provided by our federal government to all citizens in due time, but the writer will try to provide some early idea of Who, What, Why, How and How Much. The reader should be aware that the material that follows introducing the hijackers should be read with caution. First of all, some of these men were probably "sleepers" who laid low in the United States for months, even years, before being ordered to accept this assignment. With just a few mistakes, they were able to hide themselves and their mission — once they were informed what it was — until the mission was successfully completed. Some, responding to orders, may never have revealed the real names they used in their home countries; they may have had driver's licenses, receipts, addresses, cell phone/Mail Boxes Inc. bills, and other matter used for personal identification that provide totally false information regarding correct name, address, and status. The names used in this document are the names and ages that investigators came to know them by on or subsequent to September 11, 2001. The names may have come from stolen identities, complete fictionalization, or use of a deceased person's actual identity.

Mohamed Atta, who probably was the pilot of American Airlines Flight 11 that crashed into the World Trade Center towers first, reportedly thirty-three of age, was born in the United Arab Emirates, but had an Egyptian driver's license. After obtaining a bachelor's degree in 1990 in Cairo, Egypt, he studied urban affairs for eight years at the Technical University in Hamburg, Germany, and reportedly was connected to an Islamic fundamentalist group according to German investigators. Atta said he was a "cousin" to Marwan Al-Shehhi, one of hijackers who died in United Airlines Flight 175 that also left from Boston. Whatever the form of their relationship, they were always together in both Germany and the United States until their death in separate planes on September 11, 2001. Atta came from wealthy and successful parents and siblings. His parents lived in a nice apartment that had a broad view of Cairo, and they had another home for holidays by the Mediterranean Sea. His father was a well-known lawyer and both his sisters had gone to advanced schooling and had been awarded Ph.D. Mohamed Atta was never a poor young boy and never lacked a good education. He wanted for nothing, yet until he became a terrorist he never was able to develop a constant ideal and a vision. He was a bit intense and derided other persons. His parents tried to get Atta to marry a number of times, but their plans did not work out. After a final attempt by his father to arrange a marriage, Atta demurred that he had to go back to Germany to study for his Ph.D., but then left for Florida to enter flight school in quest of a higher goal. In Germany, Mohamed had joined an Islamic fundamentalist cell that planned and organized terrorist actions on American targets where he maintained close relationships with his superiors, whoever they may have been. He would disappear from school for long periods of time, but at Huffman Aviation in Venice, Florida, and while practicing in rental planes and taking flight simulation training at SimCenter Inc. in Opalocka, he was all business and attitude, some good and some bad. An FAA flight examiner tested both Mohamed Atta and' Marwan Al-Shehhi. The examiner tested each man separately in several flights in a twin-engine Piper Seneca and sent his findings to the Federal Aviation Administration. Then, the examiner issued both terrorists temporary pilot certificates. The soon-to-be suicide pilots then moved to a rented apartment in Hollywood, Florida. While in Florida, the two terrorists changed their stories several times, saying they were from Saudi Arabia, Pakistan, or Germany.

Mohamed Atta entered the United States in May of 2000, a man with a short temper who had a suicide assignment and a lot of luck. If his difficult personality got in his way, it did not betray him. He left few signs that he would be involved in terrorism in the United States. Apparently, he was the leader of the cell that took over American Airlines Flight 11, and he might have of had some control over the United Airlines Flight 175 as well. On May 20, 2001, Mohamed applied for and received a Florida state driver's license even though he had owned and driven a red Pontiac Grand Am since July of 2000, and moved into the Tara Gardens condominiums in Coral Springs, Florida. One week before getting his Florida driver's license, he had been stopped by a Broward County "County Mounty" for driving without a license, and ordered to appear in court on May 28, 2001. When he did not appear in court, a bench warrant was issued for his arrest. That is, one of the subleaders of the Boston hijackers was supposedly being sought by police since May 28, 2001. when he crashed into the World Trade Center. At times, the terrorists living in Florida changed their living quarters, thus changing their address record every few weeks or months. However, they seemed to have a lot of meetings no matter where they were living. They kept in touch with others in various terrorist cells or their suppliers by visiting libraries or readily available Internet facilities. Federal investigators have retrieved hundreds of e-mail written in both English and Arabic from as far back as forty days before the actual attack. They also have a list of the addresses where the e-mails were sent throughout the country and abroad which could possibly be extremely helpful in identifying contacts, assistants, supporters, suppliers and payment routes.

Mohamed Atta took a trip to Spain from Miami Airport the first week of July 2001, and returned to Atlanta, Georgia, on July 19, a period for the trip of roughly twelve days. Marwan Al-Shehhi traveled from Amsterdam, Holland, in early May, but it is not clear where else he may have traveled. While enrolled at Huffman Aviation, Atta was warned by school officials to change his attitude or get out. During his stays in the United States, Atta lived in Venice, Coral Springs, and Hollywood, Florida. A rather handsome man, Atta is presently judged one of the main operatives in this conspiracy but not the ringleader in the total effort, not the guy behind the dream but willingly cooperative in the overall effort where he was only a part. Atta was certainly well trained and resourceful, a true believer! His training included learning how to fit into American society without making waves. Like Marwan Al-Shehhi, he went through pilot training at Huffman Aviation in Venice, Florida and continued on towards his goal by taking a couple of three-hour stints of air flight simulations in a Boeing 727 airliner at SimCenter Inc. located at Opalaka, Florida. Late in August of 2001, Atta filled out an American Airlines Frequent Flyer form, although he knew well that his mission was a suicide mission with no free airline mileage involved.

Marwan Al-Shehhi, who died in the second airliner to hit the World Trade Center, was twenty-three years old and lived in the United Arab Emirates. While in the United States taking flight training with Atta, he lived in Venice and Nokomis, Florida. Al-Shehhi spent a year studying electrical engineering at the Technical University in Hamburg, Germany. At the time he lived in Venice, he shared an apartment with Atta. They also lived together while in Germany.

Hani Hanjour died aboard American Airline Flight 77 that left Dulles Airport bound for Los Angeles, California, but was taken over and executed a large circle and ended up crashing into the Pentagon just outside the Capital at 9:40 a.m. Hani may have lived in

Phoenix, Arizona ,and San Diego, California. He had received a commercial pilot's license in 1999. At that time, he stated his home address as a post office box in Saudi Arabia. Hani Hanjour had attended CRM Airline Training Center in Scottsdale, Arizona in 1996, and in December, 1997. Due to personality problems, he tended to be argumentative; Hanjour tried to receive a certificate as a qualified pilot two times but failed each time.

Wail Alshehi, twenty-eight years old, also a pilot, died on American Airlines Flight 11. Some of the "sleepers" among the terrorists had been in the United States for years. An example could be Waleed Alshehri, a Saudi national who received a United States Social Security card in 1994. Another example could be Hani Hanjour, the pilot who crashed into the Pentagon, lived in Arizona where he received pilot training for five years before September 11, 2001. Alshehri, while in the United States, may have lived in Hollywood, Florida, and Newton, Massachusetts.

Waleed M. Alshehri, twenty-five years of age, was also a qualified pilot on American Airlines Flight 11. He graduated from Embry-Riddle Aeronautical University located in Daytona Beach, Florida, in 1997 with a Bachelor's Degree in aeronautical science and the university's commercial pilot license degree. He had a four-year scholarship paid for by the Saudi Arabian government. At Embry-Riddle, students represent more than one hundred nations, with many students being from the Middle East. Waleed also had a commercial pilot's license, permitting Waleed to fly multi-engine planes. While at Embly-Riddle, he lived at 1690 Dunn Avenue in Daytona Beach. He paid about $290 a week for a room at the budget Homing Inn placed along U.S. Route 1 in Boynton Beach, Florida. At one time or another, Satam Al-Suqami, and Wail Al-Shehri also claimed the Homing Inn as their home address when acquiring Florida state photograph I.D. cards. These three men had late evening meetings with Mohamed Atta and Marwan Al-Shehri at a nearby Denny's restaurant.

Abdulaziz Alomari who boarded American Airlines Flight 11 in Boston, was thirty-eight years of age and was rated as a private pilot and flight engineer by the Federal Aviation Administration. For a time, he lived in Vero Beach, Florida, accompanied by his wife and four children. Abdul told the owner of the apartment in Vero Beach he would be out of the house by the end of August, 2001, but actually stayed there until September 3, 2001. Alomari appeared outside the Virginia Department of Motor Vehicles located in Arlington on August 2, 2001, where he located a go-between inside a nearby parking lot to help him to obtain a Virginia identification card. Abdulaziz really needed some type of "good" identification to fulfill his mission. In the parking lot was a man named Herbert Villalobes who was happy to meet his needs. They got a form filled out and had the form signed by a notary public. Villalobes signed the form as "Oscar Diaz" and certified that Abdulaziz Alomari lived in Virginia. At least five of the nineteen hijackers received state of Virginia I.D. cards during the month before the attack on the World Trade Center and the Pentagon. Alomari told the owner he was going home, but made his way to Boston. At an earlier time, he listed his previous employers as Saudi Flight Ops which handles aircraft maintenance for Saudi Arabian Airlines at Kennedy Airport in New York.

Ziad Jarrahi was aboard United Airlines Flight 93 where the passengers, who knew about the crashes into the World Trade Center from what others passengers on this flight had learned from calling their loved ones on cell phones, apparently attempted to rush the hijackers when the plane crashed into a meadow in Pennsylvania. The Federal government has a Hamburg, Germany, pilot's listing in the name of "Ziad Jarrah." At the U.S. Fitness Center in Dania, Florida, there is a record for club member # 5887, Ziad Jarrah, who had a membership for two months from May 7 to July 7, 2001. Ziad Samir Jarrahi did actually take self-defense lessons in Dania. Of Lebanese extraction, Jarrahi was mainly interested in martial arts instruction while in Dania. He seemed to want to learn "street-fighting" techniques like karate and how to control a bigger man with strictly his hands.

Others not identified as "pilots," are treated below. These may include non-military "soldiers" who surfaced in the United States for this single suicide assignment.

Khalid Al-Midhar was aboard American Airlines Flight 77, a Boeing 757 from Dulles Airport. It was reported that he possibly had lived in San Diego, California, and New York, and that he had a B-1 visa good for one-year for business dealings. In December 1999, Khalid Al-Midhar and Nawaf al-Hazmi (see below) were photographed by the security service of Malasia at Kuala Lumpur alongside a man named Tawfiq bin Atash, better known by his terrorist moniker of "Khallad." Khallad was a member of Osama bin Laden's terrorist group, Al Qaeda who had been involved in planning the attack on the *U.S.S. Cole* in Yemen harbor. In mid-August of 2001, both Midhar and Hazmi were already in the United States living undercover preparing for some type of terrorism incident, and Federal Bureau of Investigation agents were looking for them without success. After they landed at Los Angeles National Airport, carrying Saudi passports sometime in the year 2000, Al-Hamzi rented a room in San Diego from September to December, 2000, with Al-Midhar sharing the room. On September 11, 2001, they both were in the airliner that crashed into the Pentagon just outside of Washington, D.C.

Majed Moqed was listed as a passenger on American Airlines Flight 77.

Nawaf Al-Hamzi on Flight 77, while in the United States may have lived in Fort Lee and Wayne, New Jersey and San Diego. Al-Hamzi and Khalid Al-Midhar purchased airline tickets in their own names on American Airlines Flight 77 that crashed into the Pentagon.

Salem Alhamzi on American Airlines Flight 77, while in the United States may have lived in Fort Lee and Wayne, New Jersey.

Satam Al Suqami on Flight 11, stated he was from the United Arab Emirates. He was positively identified as one of the Flight 11 hijackers only after his passport was found during a foot-by-foot search of the debris and rubble from the World Trade Center done by first responders.

Sayez Ahmed on United Airlines Flight 175, while in the United States may have lived in Delray Beach, Florida.

Almed Alghamdi was on United Airlines Flight 175. While in the United States, he was said to have lived in Vienna, Virginia. and may have lived in Delray Beach, Florida.

Hamza Alghamdi was also on Flight 175, and while in the United States he lived at the Delray Beach Racquet Club (Florida) located at 755 Dotteral Road with a few other people. Since this condominium complex hosts the Rod Laver Tennis Academy, they had to get the "look" worn by others around them so they often carried tennis rackets and carried small bags supposedly for storage of their tennis shoes and street clothes. The three men left the racquet club sometime around September 9, 2001. Hamza was only twenty years old, and looked about seventeen or eighteen.

Mohald Alshehi was also on Flight 175, and while in the United States may have lived in Delray Beach, Florida.

Saeed Alghamdi was on board United Flight 93, and while in the United States may have lived in Delray Beach, Florida.

Ahmed Alhaznawi is listed as twenty years old and was aboard United Airline Flight 93. While in the United States, he probably lived in Delray Beach, Florida.

Ahmed Alnami was also on United Airlines Flight 93 and while in the United States may have lived in Delray Beach, Florida.

The hijackers had been trained and counseled to keep them from attracting attention in American society. Particularly for the "sleepers" who stayed in the U.S. for longer periods of time, a guide was produced, one that could easily be duplicated at any copy or print shop, that instructed the men to shave any full-face beards, live in newly constructed areas of town where people do not know one, and drink alcohol, and chase women. They are encouraged to be "social," open, and friendly with people — apparently, this was a tough requirement for the hijackers who had other thoughts on their minds. They are also taught to get and use a cell phone, go to a ballgame, coach a soccer team, and wear a proper "American" hairstyle. They learned how not to use traditional Muslim greetings — a "Peace be with you" or a "May Allah reward you" — could ruin all their planning; not to cause trouble in their adopted neighborhoods, and not to park in "No Parking" zones. The men from the Middle East had absolutely no trouble in securing state driver's licenses or other seemingly acceptable photo identification, obtaining fake addresses at various Mailboxes Inc. facilities, or buying airline tickets with various VISA cards or even buying one-way tickets for cash — supposedly a warning sign for airlines and travel agents. They freely rented cars with questionable identification, rented a variety of apartments and/or rooms, and paid their ever-present U.S. dollars for bar bills. Whether or not they chased women is open to question. They were free spending without having gainful employment, used cell phones with abandon, and sent e-mail at library facilities to confer or report to others. The lived quietly in a very successful fashion, *almost* like citizens of the United States.

"Sleepers" or undercover members probably had training similar to that expressed in the so-called "Al Qaeda Handbook" (actual title in English, "Military Studies in the Jihad Against the Tyrants") U.S. officials referred to in the recent trial of suspects in the 1993 bombing of the same World Trade Center in lower Manhattan. This handbook contains a number of directions on how to set-up clandestine meeting locations. It covers methods of operations, such as "The matters of arming and financing should not be known by anyone except the commander. The apartments should not be rented under real names. They should undergo all security measures related to the military organization' camps. Prior to executing an operation, falsified documents should be prepared for the participating individuals. The documents relating to the operation should be hidden in a secure place and burned immediately after the operation, and traces of the fire should be removed. The means of communication between the operation and participating brothers should be established. Reliable transportation means must be made available. It is essential that, prior to the operation, these means are checked and properly maintained."

The nineteen hijackers even got very explicit instructions when they began their trip to the Magic Kingdom. Below are excerpts from such instructions the terrorists received before their final flight. The language below is a translation from Arabic to English, and provides excerpts from copies of the messages that were found in three different places.

"On Their Last night: 1. Pledge of allegiance. 2. Review the plan carefully. 3. Reading a verse from Quran that calls for endurance (try to train your self and understand the meaning of this verse). 4. Reading a verse from Quran, which says obey God, his Prophet, and don't hesitate so

that you will fail. 5. Night prayers and call to God for help. 6. Reading from Quran. 7. Purify your heart and forget the world, because the time to play has passed and be ready for Day of Judgement. 8. Be happy, because the distance between this life and your joyful new life is short. 9. Close your eyes and remember that you have been done injustice, but will eventually triumph. 10. Remember God's saying, You were waiting for death before you met it. And, how many small groups were able to triumph over large ones? 11. Remember to pray for yourself and your brothers. 12. Drive out the evil. Check your weapon. 14. Tidy up you clothes. 15. Pray the morning prayer with your brothers."

"Stage 2: When the taxi takes you to the airport. When you arrive, say your prayers. And smile, be satisfied, because God is with the faithful and He is guarding you, although you don't feel it. Say your prayers, God made us triumph ... Try not to have others watch you while you are uttering your prayers. Don't be confused, and don't be nervous. Look cheerful and satisfied, because you are doing a job which is loved by God, and you will end your days in heavens where you will join the virgins ..."

Stage 3: When you board the plane and before you step in, read your prayer and repeat the same prayers we mention before, when you take your seat. Read this verse from Quran: 'When you meet a group, be steadfast and remember God, you will be triumphant.' Give a priority to interests of the group and the job. Don't take revenge for yourself and make everything for God. Apply the rules of the prisoners of war. Take them prisoner and kill them as God said; No prophet can have prisoners of war. If everything goes well, each one of you will touch the shoulder of his brother. At airport, at plane and at cabin, remember that what you are doing stands for God and don't confuse your brother. But encourage them and remind them of the saying of God. Open your chest welcoming death in the path of God and utter your prayer seconds before you go to your target. Let your last words be, There is no God but God and Mohammed is his messenger. Then, God willing, you will be in heavens. When you see the infidels, remember that the enemies of Islam were in the thousands, but the faithful were victorious."

The excerpts above will give both Americans and citizens of the world an idea of what they are up against in the War on Terrorism. We are now engaged in a Holy War as far as our enemies are concerned, and it may be a very dirty war. World War I was called The Great War, World War II was called The Good War. What do we call this new war? We don't even know who our enemies are.

Using their own names in some instances, the hijackers left a paper trail of receipts. This would be a plus for them since their identification would make them heroes for all time in the Middle East. Planning for the hijackings may have begun as early as 1996, while reconnaissance of the Logan International Airport/Massport in Boston, probably commenced six months before the actual attack. At least five of the ten terrorists who took over two airliners operating out of Logan International Airport received extensive flight training in United States aviation training facilities.

The terrorists did not work at any jobs besides setting up the most ruinous and deadly terrorist event to yet take place in the United States, yet they had spent an estimated $500,000 for travel around the world, rental cars, fake addresses, martial arts books and tapes, apartments, motel rooms, tuition at pilot training schools, and the high price of Business Class airline travel. However, shortly before the end of their story, they sent a package to a contact abroad. Investigators theorize that the package contained additional money that they were returning. Did the CIA, the FBI, the Department of State, the Department of Justice, and all politicians fail to protect American citizens from the worst terrorist attack to yet hit the United States? Historians will answer "Yes," while many of us will demure since we ourselves *never* really had any idea that the strongest country in the entire world could ever be victimized by this particular type of attack. People may be thinking CIA and the FBI managers should start working at their jobs a bit harder and do away with the constant embarrassment their employees have caused by leaking secret information over recent years. There were previous intelligence reports, in hind-sight, that should have given a clue to those in authority as to what was going on. The terrorists did make some mistakes that could have botched their entire assignment, but nobody reported their mistakes, or if they did, no one in authority did anything about such a report. After the World Trade Center collapsed, some intelligence reports seemed far more significant than before. That is one problem that will cure itself in no time. The American people will now know what terrorists can do, and how well they can do it in the United States in spite of all the money that has gone into Homeland Defense, WMD, and NBC training. As one example, one intelligence report concerned itself about a member of Osama bin Laden's family who had supposedly been warned by a contact to leave Saudi Arabia for a safer location, possibly Afghanistan, before a certain deadline. In addition, bin Laden had a well-known desire to shift his attacks on U.S.-related targets outside the United States to inside the country; that is, to domestic terrorism. Still the people, and the authorities, had no warning at all. Hundreds of firefighters, police officers, and emergency medical technicians died in the World Trade Center because they had insufficient knowledge about the real situation or the method of construction of the towers.

Osama bin Laden is the big, bad bogey-man who has fear and loathing etched into the faces of all Americans, yet his many relatives love the United States; it is estimated that forty of them were living here on September 11, 2001. Most of them left the country within two days, and the United States just let them go rather than taking them into custody as material witnesses. The family is all-important in Saudi Arabian culture as it is in Arab and Muslim populations around the world. Osama secrets himself in the mountains and poverty of Afghanistan, although it is not well known to the American public is that in Boston two of his uncles and one of his sons are wealthy businessmen; Harvard University was given $2,000,000 by the bin Laden family.

Perhaps our government does not really understand that they have a religious war on their hands, rather than a political war. Religious wars have been continuous over the centuries until one side is triumphant and the other side is deceased, whereas political wars come and go as one demagogue follows another and neither one has enough troops to kill everybody on the other side. Religious wars are vastly different than political wars — a lesson that we must learn. Terrorism is directly related to the news media. Terrorism cannot exist without media attention. A high body count of innocent people is most important , but only if the media is there to take pictures and do news reports. Religious terrorists are much more fearful than political terrorists. They know no bounds to destruction. They will not be stopped until they reach their goal, whatever their goal may be.

"A fact that is particularly surprising is that some of the few fathers among the terrorists in New York City and Washington, D.C. brought their families and children with them to the United States. The families went shopping in the malls, the children attended local schools, played with the neighborhood kids, and ate a lot of pizza and other treats. All in all, they appeared to lead ordinary "American" lives, although the fathers among them meant to cause the destruction of America and American citizens.

On September 6, 2001, a white Mitsubishi automobile driven by one or more of the two hijacking groups that were to start their attacks over Boston entered the garage at Logan International Airport and remained there for almost two hours according to surveillance photos taken at the garage. In addition, the white Mitsubishi made two visits on September 9 — two days before the attacks that changed the world — lasting about forty-five minutes for the first visit and over an hour for the second. September 10 saw the same Mitsubishi back at the Logan garage for about one-half hour. The Mitsubishi made its last journey to the Logan garage the morning of the September 11.

Mohamed Atta and Marwan Shehhi felt confident enough to let down a bit at Smukums restaurant in Hollywood, Florida, the Friday before their scheduled meeting with death. They had cocktails (five drinks of vodka for one and rum and coke for the other), and, when the waitress presented the bill for $48 because she was going off shift and wanted to get her tip, Atta took it as an affront to his pride. He became incensed, thinking he was being billed early because restaurant employees thought he did not have any money. Atta reportedly flashed a stash of $50 and $100 bills, and said something about being an American Airlines pilot (which he would be, at least until he died crashing his plane into the World Trade Center four days later) and asked, "Why would I have any problem paying a $48 bar bill?"

In late August of 2001, Waleed M. Alshehri and Wail Ahshehri, who may have been bothers, used a Visa card to purchase tickets on American Airlines Flight 11, and Mohald Alshehri and Fayez Ahmed bought seats on United Airlines Flight 175. A couple of days later, Ahmed Alghamdi and Hamza Algham arranged for seats on United Airlines Flight 175 as well; while Mohamed Atta and Abdulaziz Alomari used a Visa card to secure two seats together in the Business Class section of American Airlines Flight 11. Satam Al-Suqami purchased his ticket with cash. There were absolutely no problems caused by the purchase of these tickets, and nobody asked any questions. On September 10, Mohamed Atta and Abdulaziz Alomari rented a Nissan Altima in Boston, drove north to Portland, Maine, and stayed one night at the Comfort Inn near the Portland International Jetport. Another two terrorists on the same evening probably took their rest at the Milner Hotel in downtown Boston. Ahmed Alghamdi and Hamza Alghamdi had a room at Days Hotel in Brighton, near the Massachusetts Turnpike in the Watertown area, and Waleed Ashehri and Wail Alshehri probably stayed at the Park Inn Hotel in Newton, Massachusetts. Except for Portland all these locations are within easy driving distance from Boston's Logan International Airport. Why Atta and Alomari went to Portland is a mystery. The trip must have been very important to them, but why? Did they have to pick-up a person or persons who entered the United States from Canada, or did they just want to keep the number of terrorists hiding in the Boston area that evening to a minimum number. Mohamed Atta and Abdulaziz Alomani were late in arriving from Portland via a U.S. Air commuter flight the morning of September 11.,and Atta's suitcases did not reach the death plane in time for take-off. From the luggage, and a white Mitsubishi automobile that had visited Logan airport a number of times during the prior week, investigators found two videotapes apparently made by Intelligent Television & Video of cockpit simulation videos depicting commercial jet flight with one video showing in-flight training for both the Boeing 757-200 and the Boeing 747-400. These videos were similar to the Microsoft Flight Simulator 2000 software that enable a commercial jet pilot or flying buff to do simulated flying of a number of different planes including the Boeing 737-400 and the Boeing 777-300. Also included was a training booklet for using flight simulators for both the Boeing 757 and the Boeing 767, the planes actually used to bring down the World Trade Center and to attack the Pentagon.

Student and Other Visas

On September 11, 2001, six of nineteen hijackers were recognized by airline passenger pro-filing systems and received special handling. Two more were recognized because of problems with their identification, while another one was keyed for special evaluation because he was traveling with a man who had questionable personal identification. That is, nine hijackers of nineteen (47.4 %) were recognized as needing special evaluation and scrutiny as a result of a profiling test. The nine had their checked baggage checked for explosives (all passengers have their carry-on baggage and themselves checked). They took over the four planes with box cutters and small knives, which we not considered weapons on September 11, 2001. Two of the hijackers were also on a Federal Bureau of Investigation (FBI) watch list of potential terrorists, but nobody told the airlines of this fact.

> On Monday, March 11, 2002 — six months to the day of the September 11 terrorist attack on America — the Huffman Aviation flight school (Venice, Florida) received student visa approval forms for Mohamed Atta and Marwan Al-Shehhi from the U.S. Immigration and Naturalization Service, a part of the United States Justice Department, permitting the pair to live and study at a flight school six months after they had flown two separate commercial airliners into the World Trade Center killing upwards of 3,000 people. Is that horrific mistake justice, or incompetence for the more than 3,000 America citizens who died in this single attack or incompetence?

Student and regular visas allow entry into the Land of the Free with no checks and balances, where seemingly no one seems to be held accountable. Terrorists have used the freedom allowed by our immigration laws to kill United States' citizens and visitors from other counties. The reader should be aware that the United States has no reliable, national system for controlling or even monitoring immigrants who overstay a student or regular visa. It seems that many immigrants who have a U.S. visa just melt into American society. They have very little chance of being caught;, our present laws and routines almost guarantee that none will be caught. By the simple fact that they have an U.S. visa, they then have same legal protections that American citizens have. North America, in both the United States and Canada, is entirely the land of the free sought out by everybody in the entire world. The number of illegal immigrants within U.S. borders tallies at about 11,000,000. As of January of 2002, four months after the September 11, 2001, tragedy, 314,000 illegal immigrants were denied the right to remain in the United States for crimes or actions, and have been sentenced to be deported, yet the federal government has no information at all as to where in the country they are. A Taliban official could come to the United States with a student or regular U.S. visa, answer a few questions correctly at the U.S. embassy in whatever country he called his home, and, like other terrorists who recently used this gimmick, he would never have to attend a single class. He would be completely free to go underground and set up any program he wanted to conduct against the freedom of the United States or any country in the entire world. The situation, at present, is entirely out-of-hand.

There is some talk among politicians in the Bush administration about broad measures to restrict and control immigration of aliens who have no love or respect for the United States, and have a record to prove it. The central question seems to be, will our federal government respond with security arrangements to keep the terrorists out *and* be accountable for the success of such methods. Is our federal government actually aware that a whole new world began on September 11, 2001? Can the United States handle the challenge of terrorism? In the past, U.S. immigration policy has been a sieve. Most of the nineteen terrorists who hijacked four airliners on September 11 entered the United States **legally** on student, business, or tourist visas. They were not citizens of the United States, but because they has U.S. visas **they had the same rights and protections as United States citizens.**

As an example, Hani Hanjour was a pilot traveling as a passenger aboard American Airline Flight 77 that left Dulles Airport in the early morning on September 11. Hanjour later appeared to be a pilot, trained in the United States, who flew Flight 77 into the Pentagon. Hani Hanjour used a student visa to enter the U.S. in a legal manner. *Time* magazine (October 1, Vol. 158, No. 15) reported, "He had been accepted for an intensive English course at Holy Names College in Oakland, California. When classes actually began, Hani did not show." He resurfaced in American society on September 11, 2001, after learning how to fly commercial airliners in San Diego, California and in Maryland. Mohamed Atta was allowed to enter the United States when immigration authorities noted he had an application for a student visa pending.

Almost anyone can get a visa to visit the United States. Fifteen out of nineteen hijackers received their visas legally in Saudi Arabia. Unrelated as far as is known, a Saudi citizen, Abdulla Noman, worked for the U.S. Commerce Department issuing visas in Jeddal, Saudi Arabia. Noman was arrested in early November of 2001 for selling U.S. visas in Las Vegas, Nevada. A government witness stated to the FBI that he had paid a little over $3,000 to Noman for an U.S. visa in 1998. A wealthy person can actually buy an U.S. visa in a number of cases, and it is all perfectly legal. Our immigration laws are totally out of sync relative to the safety of American citizens.

Does the United States believe in complete tolerance to border control problems, or national security in the face of repeated terrorist attacks? Our government's top priority is, or should be, to protect the rights of American citizens. Some legislative moves have been made by the administration to control and contain student visas, but educators are furiously fighting such measures for a very good reason: although foreign students account for 3.4 percent of total enrollments in 2000, they paid 7.9 percent of tuition and fees. In 2000, foreign students brought $11 billion of outside money into the United States. Students from middle-eastern countries are only a small percentage of foreign students attending school in the United States. A much higher percentage of students from China, India, Japan, South Korea, Taiwan and Canada existed in 2001.

January 6, 2005: A Gas Attack on Home Ground

You have to go back to February 24, 1978, for an incident that brought hazardous materials to the attention of the national media like the 2005 incident in Graniteville, South Carolina. Hazardous materials spattered upon the public consciousness of our nation with incredible force on February 24, 1978, when a single jumbo tank car carrying 27,871 gallons of liquefied petroleum gas (propane) ruptured with a Hiroshima-like fireball in downtown Waverly, Tennessee, killing sixteen persons and leaving scores more to stumble through the business district, skin dripping from their bodies like runny Saran-wrap. Two days earlier, a high-carbon wheel on a gondola car, its incompatible brake shoes overheated by a handbrake negligently left in the applied position, broke seven miles outside of Waverly allowing the damaged wheel-truck to bounce across cross-ties through deserted countryside and finally derail twenty-four cars in downtown Waverly. Within its 25/32-inch-thick steel envelope, tank car UTLX-83013 carried roughly twice the weight and three times the volume of compressed, flammable gas permitted in a single tank car prior to the late 1950s. The benign-appearing yet massive bomb became an attraction, a curiosity luring townspeople into the area. When the liquid propane leaked from that fragile container forty hours after the derailment, it would instantly expand 270 times to highly flammable gas waiting for an ignition source, and the story of Waverly would be etched in 1,700 degree heat.

In Waverly, the hazardous material was propane; in Graniteville, South Carolina on January 6, 2005, the hazardous material was a chemical agent named chlorine, one of the chief toxic Weapons used during World War I. Nine persons died needlessly from a First World War chemical agent. It was not the first rail horror to haunt Graniteville. On November 10, 2004, in Graniteville, five workers at a textile mill driving home after working all night were killed at a railroad crossing. A Norfolk Southern train engine pulling two cars hit their car doing 45 M.P.H. in a stretch of track with a maximum speed allowed of 49 M.P.H. Three automobiles tried to beat the train to the crossing, according to witnesses; but only two won the race. The third car was struck by the train engine. All five occupants in the car were killed.

In the 2005 incident, a freight train operated by Norfolk Southern in Graniteville, with three tank cars of chlorine plus two cars loaded with other hazardous materials, ran a switch that crashed the moving train into an unoccupied train on a siding at a textile plant. Nine people were killed by inhaling chlorine gas. The crash happened at about 2:30 a.m., while about 400 people were working the night shift at Avondale Mill, a textile plant. At least 511 persons were examined by emergency departments after exposure to chlorine gas, 69 persons were admitted to seven hospitals throughout the area, and 18 persons were treated in doctors' offices. In addition to nine dead, at least eight people were in critical condition. The dead bodies in the textile mill amounted to five, plus a very unlucky truck driver from Quebec, Canada, who was apparently sleeping in his truck on mill property, and never work up. The body of a man was found in his home. The engineer of the occupied train either died in the train wreck or from ingestion of chlorine. Another body was found later. Officials ordered all 5,400 people within one mile of the incident to evacuate the scene. Employees in the textile mill right on scene had to run for their lives and evade the green haze of chlorine gas that stayed near the ground since chlorine is heavier than air. One of three chlorine tank cars was leaking, and a second one was apparently questionable. When the chlorine liquid leaked from the one car, it expanded at a 460-to-1 ratio and became a chlorine gas.

Many Graniteville residents were directed towards decontamination stations located at the University of South Carolina at Aiken, and Midland Valley High School. There they had their contaminated clothes confiscated, and after they went through a bath with soap and water they were treated with oxygen if necessary. Roughly twelve persons decided to avoid the evacuation order and stay at home, after initially being told to do just that. A dusk to dawn curfew was put into place within two miles of the wreck site in response to a feeling the cool night air would cause them chlorine to settle close to the ground. Two days after the wreck, about forty tons of crushed lime was dropped near the leaking car to start neutralizing the chlorine. Norfolk Southern paid for hotel rooms and given $100 Wal-Mart gift cards or checks to people who were forced to evacuate. Investigators were trying to determine why the manned locomotive carrying chlorine in three tank cars and two other hazardous materials left the main track and hit a Norfolk Southern train parked in a siding. There seemed to be some suspicion that the crew who had parked the train a few hours earlier had left the switch in the wrong position.

As of January 14, 2005—eight days after the wreck—twenty-two patients remain hospitalized. The Aiken County Sheriff's Office holds warrants for the arrests of fifteen individuals who changed the address on their driver's license to a Graniteville address and later were reimbursed by Norfolk Southern Railway for expenses. The Graniteville-Vaucluse-Warrenville Fire Department has relocated its headquarters to the former Community Services building, and the fire service and Haz-Mat returned today to an all-volunteer force. Animal control officers reunited 287 pets with their owners, and two dead dogs were sent to Columbia for autopsy to determine the cause of death. Six pets have died since being returned to their owners. At 1500 hours on January 19, 2005—thirteen days after the rail wreck, death, and injuries—the chlorine purging process was complete. Air monitoring readings of 0.0 ppm (parts-per-million) were detected at the intake and outtake of the rail tank car. A hole was cut in the tank to allow Norfolk Southern contractors to pressure wash the tank car. The car will be loaded onto a flatcar and transported to Altoona, Pennsylvania, where it was impounded, subject to further investigation by the National Transportation Safety Board.

On February 6, 2005—a month after the Graniteville wreck—the three-man crew accused of failing to switch the railroad track back to the main line before disaster hit were fired by Norfolk Southern Railways. A railroad spokesman stated that the workers were terminated because they "failed to perform their duties properly." Union officials said the three men will appeal, and each man had at least twenty-five years experience. The accident on January 6, 2005, killed nine people and injured hundreds more.

A Senate Resolution

South Carolina General Assembly
116 Session, 2005–2006

TO EXPRESS THE DEEPEST SYMPATHY OF THE MEMBERS OF THE SENATE TO THE FAMILY AND MANY FRIENDS OF WILLIE CHARLES SHEALEY OF GRANTE-VILLE WHO DIED AS A RESULT OF THE TERRIBLE FREIGHT TRAIN ACCIDENT ON JANUARY 6, 2005, IN GRANITEVILLE WHICH SUBJECTED HUNDREDS OF PEOPLE TO DEADLY CHLORINE GAS.

Whereas, the members of the Senate were deeply saddened to learn of the death of Mr. Willie Charles Shealey of Graniteville at the age of forty-three who died as a result of exposure to deadly chlorine gas which was released as a result of a freight train accident in Graniteville earlier this year; and

Whereas, Charles Shealey was an employee of Avondale Mills at the Woodhead plant, was third shift supervisor, and one of the top managers at this facility; and

Whereas, his recent promotion and receipt of corporate awards signified that he was a person who was rapidly moving up in the Avondale ranks; and

Whereas, he was a person who cared deeply about his job, about the quality of the products he helped produce, and about his family and fellow associates; and

Whereas, Charles Shealey was a devoted husband and father of three sons. He kept photographs of his sons at all stages of their lives on his locker at the Woodland plant; and

Whereas, he served the Graniteville community in a number of different ways and also served his country as a member of the South Carolina National Guard 122 Engineering Battalion in Graniteville; and

Whereas, the train accident in Graniteville is one of the worst disasters ever to have occurred in South Carolina which has had many tragic consequences, the most significant of which is the loss of many fine South Carolinians like Charles Shealey; and

Whereas, the members of the Senate, by this resolution, would like to extend their heartfelt condolences to his widow, Mrs. Sherry Randle Shealey and their three sons, Chad, Travis, and Brent, and to other members of his family upon the death of this truly wonderful man. Now, therefore,

Be it resolved by the Senate:

That the members of the Senate express their deepest sympathy to the family and many friends of Willie Charles Shealey of Graniteville who died as a result of a terrible freight train accident on January 6, 2005, in Graniteville which subjected hundreds of people to deadly chlorine gas.

Graniteville was one of the last "company towns" in the United States. In its past history, Graniteville was wholly owned by the "company" including buildings, schools, layout of the land that comprised the town, streets, and houses where employees of the firm lived and paid rent to the company. Everything was provided by the company. As with many company towns, there was no city council, no mayor, and no governing structure whatsoever. As the only landlord in town, Graniteville provided the town's security, sewage treatment, water, power, and town maintenance buildings and structures. Best of all, almost every adult, except retirees, had a job with the company. There was almost no unemployment during most of Graniteville's history.

A Pictorial Timeline of Graniteville, South Carolina (1845–1996) was written by Jean Clark Boyd and produced by the Horse Creek Historical Society: Branch 1 of Vaucluse, Graniteville, Warrenville, and Stiefeltown. According to this source, William Gregg was the founder of Graniteville and the Graniteville Manufacturing Company. He was elected the company's first president on March 7, 1846. Gregg designed his own mill as well as sawmills that created lumber for worker housing and the interior portions of the mill. Granite on the company property was used to shore up the walls of the canal and to construct the exterior of the mill. By November of 1848, Gregg had built forty cottages for his workers, all the same color. About twenty-two of these buildings still remain on Gregg Street.

In June of 1848, two bales of cotton were purchased to christen the new Graniteville factory, and to start the machinery in the new water-driven, turbine-powered mill. By September of 1848, advertisements were taken in the local paper for an additional 300 workers for the Graniteville Manufacturing Company. By November of 1848, the village of Graniteville was a going concern consisting of one hotel, an academy, a post office, five stores, twelve boarding houses, eleven supervisors' houses, and forty cottages for workers. A year later, the village had 325 workers in a population of 900.

On June 3, 1861, the Civil War came to Graniteville, South Carolina. Company F, 7th Regiment of the South Carolina Volunteers was enlisted that date in Graniteville with 121 men and boys. By the close of the war, fifteen had been killed in battle, thirty-one had been wounded, and ten died while in service (46.3 percent killed, wounded, or died in service). In April of 1862, Graniteville got hit with a scarlet fever epidemic; while in October of 1862 with a shortage of workers due to the Civil War, Graniteville Manufacturing Company drew up an advertisement seeking twelve or fifteen "Negro men" to work for wages. By January of 1864, the railroad was coming to Graniteville, and 500 Negroes were working on that stretch of tracks. In June of 1865, William Gregg was pardoned for providing supplies to the Confederate Army by the President of the United States, Andrew Jackson. By mid-1867, the railroad through Graniteville village was completed (currently known as the Norfolk Southern Railway Company).

In the years that followed the Civil War, the Graniteville Manufacturing Company prospered by purchasing the Vaucluse mill, and a new 20,000-spindle mill, known as the Hickman Mill, which meant that additional housing for the mill workers had to be built. One hundred residences were constructed on a hill west of the mill complex that required three new streets to be built. In 1921, the Sibley Mill of Augusta, Georgia, which is located about twelve miles from Graniteville, South Carolina, was added to the Graniteville Manufacturing Company. In December of 1936, the Gregg Dyeing Company merged completely with the parent company. By 1935, Graniteville village consisted of 178 acres of land, 325 company houses, a number of store buildings, a school, community house, ball park with grand stand, ice plant, garages, street lights, fences, saw mill, and farm property. In 1938, the parent company added Enterpriser Mill of Augusta, Georgia to Graniteville Manufacturing Company. In the period of 1943 to 1947, the same company overhauled every worker home in Graniteville, Vaucluse, and Warrenville and provided new slate roofs, brick foundations, indoor plumbing in kitchens, and indoor bathrooms. The Woodhead Division was constructed in 1944 as a coated fabrics mill specializing in awnings. The final construction of houses occurred in 1945 for ten houses. The new, larger Gregg Dye plant was built on Marshall Street in 1950, and the old building was reused by the Corduroy Department and the Research Laboratory.

Additional construction continued with another textile plant in 1963 when the Swint Division was built on Ascauga Lake Road; and in 1966, the Townsend Division was constructed on the same road. In 1969, village houses in Graniteville were sold to Graniteville Manufacturing Company workers who had formerly paid rent for living there. The Graniteville Company in 1974 bought three more mills in South Carolina; in 1976 the original mill, the Graniteville Canal, Gregg Street, and assorted houses on Taylor and Canal Street were placed on the National Register of Historic Places. Graniteville was selected as a National Historic Landmark in 1977, and from 1983 to 1986 the company operated as a wholly-owned subsidiary of GWD, a former cigar company now investing in a wide variety of businesses. In 1984, all public services including the village's thirteen-man police force, trash collections, street maintenance and lighting, and recreational services ended after 161 years.

The Steven Steam Plant began operations in 1943 during the Second World War to provide power to the Granite, Hickman, and Gregg Dye plants. Sixty-two years later, death would visit the Stevens Steam Plant and other mills in what used to be the Graniteville Manufacturing Company with a chlorine weapon, a poison gas that is a common hazardous material throughout the world. This incident killed nine, sickened 554 persons who were decontaminated at the scene, while 75 of those were admitted to various hospitals, and about 5,400 people within a mile radius of the derailment site were evacuated for a number of days. One tank car of chlorine gas leaked into the night air; it was not a pretty way to die or be ravaged with serious respiratory difficulties caused by chlorine gas inhalation. Some victims and others became heroes, some ran like hell, others climbed onto roofs since they recognized that chlorine gas was heavier than air, and a few lucky souls never realized that there was a poisonous gas in Graniteville village. All the dead, and most of the injured, inhaled, and were sickened. Captain Hugh Pollard, in *The Memoirs of a VC (Victoria Cross)* in 1932 remembers in the battle at Ypres where chlorine gas was used. "Dusk was falling when from the German trenches in front of the French line rose that strange green cloud of death. The light north-easterly breeze wafted it towards them, and in moment death had them by the throat. One cannot blame them that they broke and fled. In the gathering dark of that awful night they fought with the terror, running blindly in the gas-cloud, and dropping with breasts heaving in agony and the slow poison of suffocation mantling their dark faces. Hundreds of them fought and died; others lay helpless, froth upon their agonized lips and their racked bodies powerfully sick, with tearing nausea in short intervals. They too would die later—a slow and lingering death of agony unspeakable. The whole air was tainted with the acrid smell of chlorine that caught at the back of men's throats and filled their mouths with its metallic taste."

In The National Archives, Lance Sergeant Elmer Cotton described the effects of chlorine gas in 1915. "It produces a flooding of the lungs—it is an equivalent death to drowning only on dry land. The effects are these—a splitting headache and a terrific thirst (to drink water is instant death), a knife edge of pain in the lungs and the coughing up of a greenish froth off the stomach and the lungs, ending finally in insensibility and death. The color of the skin from white turns a greenish black and yellow, the color protrudes and the eyes assume a glassy stare. It is a fiendish death to die."

Both the survivors and the dead at Graniteville, South Carolina, shortly after 2:39 a.m. on January 6, 2005, learned very quickly how disastrous chlorine gas can be. Chlorine gas is a respiratory irritant with a distinctive odor similar to household bleach. Chlorine gas is detectable at extremely low concentrations, as low as 0.2-0.4 parts per million (ppm). If you inhale chlorine gas at even minor concentrations, you will likely be bothered by tears, a runny nose, and breathing problems as chlorine combines with moisture in the eyes, other body parts, and lungs to form a weak acid. At the first whiff in any concentration, you will know you have to leave the area, if you can. Children, Senior Citizens, and the handicapped can suffer more from such symptoms than active, in-shape adults. It is important to realize that you can die in even small amounts of chlorine gas if you do not vacate the area. Just remember, chlorine gas, also known as bertholite, was first used as a weapon against human beings in World War I on April 22, 1915 at the battle of Ypres.

The Agency for Toxic Substances and Disease Registry (ATSDR) states that persons exposed only to chlorine gas pose little risk of secondary contamination to others (contamination that occurs due to contact with a contaminated person or objects rather than direct contact with agent aerosols; cross contamination). However, clothing or skin soaked with industrial-strength chlorine bleach or similar solutions may be corrosive to rescuers and may release harmful chlorine gas. ATSDR also notes that chlorine is a strong oxidizing agent and can react explosively or form explosive compounds with many common substances.

Chlorine is heavier than air, and tends to concentrate in lower areas on an incident site. It is also highly corrosive as residents, employees, and Avondale Mills—the present owner of the mills in Graniteville—found out to their great disgust (when this writer spent ten days in Aiken County and Graniteville, South Carolina, in mid-June of 2006—roughly seventeen months after the actual incident—Avondale Mills was just completing the work of the contractor that was brought in to do cleanup work on corrosion and other matters). Because of certain reactions, water substantially enhances chlorine's oxidizing and corrosive effects.

Chlorine is often shipped in steel cylinders as a compressed liquid. The standard rail tank car is a 90 ton capacity, Department of Transportation, 105A500W for Chlorine Service (Post 1982). It is a pressure car (those tank cars which are built for transporting liquid commodities with a vapor pressure greater than 40 psig at 105 to 115 degrees F), as differentiated from a non-pressure car (those tank cars which are built for transporting liquid commodities with vapor pressure less than 25 psig at ambient or 40 psig at 105 to 115 degrees F).

According to *Field Guide To Tank Car Identification* put out by the Association of American Railroads/Bureau of Explosives, a guide for firefighters and all other first responders, the DOT 105 tank car, a pressure car, is used to transport liquefied gases and other high hazard or environmentally sensitive materials. Such cars are insulated with foam, fiberglass, ceramic fiber, or cork and have an exterior metal jacket to protect the insulation. They can be distinguished from non-pressure cars since all the loading and unloading fittings are in one location on the top of the car and all are covered by a protective housing. DOT 105 pressure cars are equipped with a spring-loaded safety relief valve or combination device no less than (incorporating a valve and a frangible disc or breaking pin) which is usually set to function at 75 percent of the test pressure of the tank. The liquid and vapor valves may be equipped with excess flow valves to stop product flow if they are sheared off in a derailment. Bottom outlets are not permitted nor are any fitting outside of the protective housing. The DOT 105 cars are typically used to transport such commodities as chlorine, carbon dioxide, sulfur dioxide, and anhydrous ammonia.

"Human error is the largest single factor in train accidents, accounting for 38 percent of all accidents over five years," according to the Honorable Norman Y. Mineta, the Secretary of Transportation, in remarks on May 16, 2005, relative to the deadly Graniteville incident. He noted that few of these types of human errors are actually addressed by Federal Railroad Administration regulations. Some leading causes include improperly lined switches, and leaving train cars on operational track, two human factors that contributed to the Graniteville tragedy.

The National Transportation Safety Board (NTSB) states the railroad brakeman at Graniteville the evening before the tragedy could not remember setting the track switch in a safe position when he quit work about 7 p.m. on January 5, 2005.

Misaligned switches are one of the leading causes of train wrecks. The Federal Railroad Administration (FRA) did report that between January 2001 and December 2003, there were 751 incidents in which switches were not aligned properly and 74 incidents in which the switches were not locked. Hand-operated track switches left in the wrong position caused eight other serious train wrecks since the Graniteville incident. Ten people died and more than 600 were injured in these crashes.

Many reports, photos, videos, personal interviews, publications, and other materials were given to this writer, either with or without comment. Written "After-Action Reports" were provided by the local fire department, the Graniteville-Vaucluse-Warrenville Fire Department, locally known as the GVC Fire Department; the Aiken County Sheriff's Office; and the Aiken County Government. The objectives of each of these reports" is not exactly the same for each agency reporting, since the agencies tend to have slightly different responsibilities.

The following report focuses primarily on the actions and observations of the Aiken County agencies involved in the initial response. Strengths and improvement ideas will be identified for each responding department, so that they will gain ability to recognize, respond to, and control a hazardous materials emergency, as well as to coordinate an integrated response that will protect the health and safety of emergency response personnel, the general public, and the environment. Strengths are those areas in which responders demonstrated exceptional ability or knowledge, or other areas of programmatic solidity. An improvement item by itself does not degrade the response, but demonstrates ways in which the emergency response could be more effective if alternative measures were used. Strengths and improvement items will be identified utilizing objectives that are applicable to the agency's response authority.

Graniteville-Vaucluse-Warrenville Fire Department After Action Report

Objective 1:   Safety

Strength: GVW personnel provided specific directions to responders reporting to the command post (CP). Safety officer appointed at CP per established department policy.

Objective 2:   Protective Actions

Strength: Response personnel were instructed to clear the area by GVW Fire Chief upon realization of imminent danger. Access/Egress zones implemented through quick establishment of roadblocks. Immediate area evacuated (300 yards) shelter-in-place for within 1 mile radius; roadblocks established in timely manner. Savannah River Site provided periodic weather updates for Protection Action consideration.

Improvement Item: Reverse 911 was not activated in timely manner due to access available only by Emergency Management personnel. This weakness has been corrected so that Reverse 911 can now be activated through direction from Dispatch supervisor or authorization of Incident Commander.

Objective 3:   Mitigation

Strength: Responders were thoroughly debriefed when they returned to the CP from operations in the hot zone. Logistical support was timely in processing requests once they were established. Additional maps were available at the CP within the first hour. Railroad consists received at the CP within the first hour. Written preplans were used for searches of mill facilities; GVW Fire Department walks down all Avondale facilities annually.

Improvement Item: GVW Fire Department did not have adequate resources to conduct decontamination activities for mass casualty situation.

Strength: Logistical support was timely in processing requests once EOC (Emergency Operation Center) was established.

Improvement Item: Activation and full operation of the EOC was a slow process due to early hour and lack of a dedicated facility.

Strength: Internal Fire Department communications were successful. Nextel was used as backup communication for privacy of command staff conversations. Primary fire department communications occurred via E-Tower which was restricted to GVW Fire Department use. Dispatcher initiated all-call page for other county fire departments to be on standby. State of South Carolina provided additional communications capabilities through 800 MHz radio. Faxes, phones, etc. available on Haz-Mat units was a key factor in good communications. Twice a day briefings with written objectives were conducted at UCP (Unified CP); status of previously established objectives were updated at each briefing.

Improvement Item: Dispatch should provide more detailed information on location of victims requesting assistance. Dispatch should coordinate received information between positions for distribution to all agencies.

Strength: Recorder position for fire department implemented upon activation of the UCP. Assistant Chief/Chief was available on the scene throughout the event.

Improvement Item: Fire Department should establish recorder position to assist and document IC (Incident Commander) activities; Court recorders were provided but not coordinated with IC. No coordination between Fire Department and EMS during initial incident response. Incident Command System (ICS) process was not followed by all responding agencies.

Strength: South Carolina Firefighter Mobilization plan activated and well staffed. Unified Command provided access to all needed agencies. Federal agencies well-integrated and supportive; EPA continually provided maps once the Unified Command Post (UCP) was established. Mutual aid agreements were in place with SRS (Savannah River Site) and Aiken County. Fort Gordon Haz-Mat resources were briefed to GVW Fire Department approximately six weeks prior to incident through a Fort Gordon community support training activity.

Improvement Item: Formal mutual aid agreement needed with Richmond County. GVW Fire Department personnel need to be briefed on County Emergency Operations plans and procedures. Entry teams from other agencies not coordinated with FDIC (Fire Department Incident Commander) during early hours of the incident. Buses used for transport of evacuees were not coordinated with FDIC. Better integration of law enforcement and EMS personnel into Fire Department ICS (Incident Command System) was needed.

Strength: CP relocated due to wind direction considerations (flag provided visual confirmation of wind direction). Initial responders notified subsequent responders of danger involved.

Improvement Item: Initial Fire Department accountability weak for first thirty minutes due to response from multiple locations; control was regained through radio roll call and telephones. Lack of credentials caused some problems with movement of volunteer responders; county produced generic badges with names, but no photos.

Objective 4: Chemical Monitoring

Strength: Habitability surveys conducted at CP (command post) upon arrival of Haz-Mat team. EPA conducted surveys at CP upon their arrival. SRS (Savannah River Site) and Richmond County Haz-Mat resources arrived on scene within a timely manner and were designated by FDIC (Fire Department Incident Commander) to be responsible for Haz-Mat operations. Haz-Mat personnel assisted in CP location determination. EPA utilized Coast Guard Gulf Coast Strike Team to provide monitoring and on scene response. By comparing consist (a list of all the cars in the train which describes their position in the train, type, contents, destination, etc.) to entry team visual inspection, chemicals involved were accurately identified. Written response plan and safety procedures implemented for Haz-Mat operations. Briefings provided to Haz-Mat responders by Safety Officer on entry considerations; maps were covered for responders unfamiliar with the area.

Strength: SRS (Savannah River Site) and Richmond County Haz-Mat personnel were familiar with Aiken County personnel and integrated seamlessly into Fire Department operations. During UCP (Unified Command Post) meetings, a CTEH scientist explained plume models in such a manner that everyone was comfortable. Ascauga Lake/Bettis Academy Road decontamination unit established and vital signs recorded. Multiple decontamination centers established on perimeter of affected area.

Improvement Item: Decontamination logs were not accurate due to chaotic state at the scene. Gross decontamination performed but quickly overwhelmed; Fire Department did not have adequate resources to conduct decontamination activities for mass casualty situation.

Objective 5: Staff & Activate

Strength: Dispatcher initiated all call pages for other fire departments to be on standby without consulting IC (Incident Commander). Specific directions were provided to responders reporting to CP. Community support to provide facilities (Honda Cars/Johnson Motors, Baptist Church) was very beneficial to command and response operations.

Objective 6:   Public Information

Improvement Item: GVW Fire Department should establish PIO (Public Information Officer) position for adequate representation at joint press conferences. This would allow for better coordination between ACSO (Aiken County Sheriff's Office) and the Fire Department Incident Commander.

Objective 7:   Medical

Strength: Initial evacuees treated and vital signs monitored at decon check points established by Fire Department; additional treatment station established at GVW Fire Department Station 2. Medical communications regarding signs/symptoms were clear and accurate. Haz-Mat/EMT/First Responder training conducted by GVW Fire Department now includes discussion of appropriate actions to this event.

Improvement Item: Development of checklists for mass casualty incidents to record patient information. There was no coordination between fire department and EMS during initial event response; effective coordination between fire department and EMS occurred several hours into incident.

Objective 8:   Recovery

Strength: There was good coordination with Avondale plant officials in developing recovery plans. GVW representative attended daily NTSB (National Transportation Safety Bureau) briefings.

Improvement Item: Development of a recovery checklist may be beneficial for future incidents to address issues such as CIS debriefings, vehicle recovery, and temporary department facilities. Designated individual should be identified to coordinate donations and volunteers.

Aiken County Sheriff's Office After-Action Report

Objective 1:   Safety

Strength: ACSO (Aiken County Sheriff's Office) personnel had Personal Protective Equipment (PPE) in their vehicles and were directed to utilize it. ACSO Sheriff contacted neighboring county Sheriff's directly via cell phone to coordinate safe arrival direction to the staging area. ADPS (Aiken Department of Public Safety) Staging officer directed rescue personnel through specified safety routes. US-OSHA representatives offered support on Day 2 and identified no safety concerns for responders.

Improvement Item: Habitability surveys were not conducted initially at Command Post or Forward Operations. Safety Officer was not initially assigned for the incident, however one was appointed when Command Post (CP) relocated to Kmart parking lot.

Objective 2:   Protective Actions

Strength: Access controlled early through traffic control points established quickly and efficiently due to recent training. Locations determined based on major intersections and information received from 911 distress calls within first fifteen minutes. Roadblock placement reevaluated within first thirty minutes, and determined to be adequate based on wind direction and Haz-Mat input. ACSO/GVW FD/ACEMD agreed to recommend shelter-in-place through utilizing Reverse 911. Air monitoring at checkpoints discussed at 03:00 Command Post meeting.

Improvement Item: Reverse 911 was not activated in a timely fashion due to access available only by Emergency Management personnel. This weakness has been corrected so that Reverse 911 can now be achieved through direction from Dispatch supervisor or authorization of Incident Commander.

Strength: ACSO shift supervisor performed running roll call for those on-duty. Personnel were ordered to go to staging (per their Incident Command System training) and reported to Aiken Department of Public Safety for accountability. ACSO appointed Staging officer to coordinate incoming law enforcement resources. To aid in accountability efforts, an employee roster was developed by plant supervision (from Avondale Mills).

Improvement Item: No formal accountability procedure was utilized; however, handwritten logs were maintained as a result of previous training. Staging checklist would be helpful if individual who normally fills position is unavailable.

Strength: Personnel initially isolated the incident scene and surrounding area through conservative estimation by the ACSO shift supervisor. DOT Emergency Response Guide was used to determine 1.5 mile radius as initial protective isolation distance. Electronic version to be added to CP (Command Post) laptops. Key representatives of fire, law enforcement and emergency services at Command Post actively discussed evacuation versus shelter-in-place.

Objective 3:   Mitigation

Strength: Evacuees were aided by ADPS and ACSO on Aiken/Augusta Highway and other traffic control points. Due to scope of the event, continuity of daily operations was identified as an issue to be addressed in planning. Decontamination stations were set up early at multiple locations outside of hazard area. Federal Homeland Security assets requested to supplement rapidly exhausting resources. GIS relationships previously utilized to aid criminal investigations and fire response resulted in early use of maps. Evacuation for 1 mile based on information obtained early on (grid maps, etc.); populations had already been determined through GIS (Geographic Information System) data. School closures were planned at 03:00 meeting. Plan was developed for safe shutdown at Avondale Mill plant operations. ADPS (Aiken Department Public Safety) personnel utilized at hospitals to conduct triage, treatment and security.

Strength: County-GIS personnel supported operations through continuous production of maps that were distributed to all agencies. EOC (Emergency Operations Center) staff worked to procure buses for initial evacuation and for transport from decontamination sites to hospitals. Salvation Army/Red Cross response implemented through plans developed by Emergency Management staff.

Improvement Item: Credibility of EOC (emergency operation center) hampered by lack of a designated, adequate facility. EOC staff was not available until approximately 09:00–10:00 due to set-up and activation. Many issues the EOC could have helped with were handled at staging.

Strength: 800mz radios were brought in to make sure a common radio frequency was utilized among agencies. Because of the familiarity with State and Homeland Security assets, this request was initiated early on. Aiken County communications center dedicated one channel for fire units operating at the Graniteville incident. IC had constant communications with Haz-Mat, EPA and DHEC personnel.

Improvement Item: Initial incident information was not adequately shared among responding agencies due to incompatible radio frequencies. This issue is being addressed through acquisition of 800 MHz radios for responding agencies command staff. Radios are being obtained through Homeland Security funding.

Strength: The Incident Command System and key positions were implemented early in response. Routine briefings were conducted for participating agencies. When Sheriff left the CP, command was transformed to other ACSO staff.

Improvement Item: Key agency representatives responding to the CP should be clearly identified and remain in the CP throughout the incident to support the IC (Incident Commander).

Strength: Additional emergency response agencies reported to Staging and sort out Sheriff for briefing. Agencies were logged in and a directory of contacts was developed. Private contractors were staged apart from the responder's staging area. Haz-Mat entry team provided video at first light; SLED helicopters utilized for search and rescue and for scene status. Federal/state response agencies integrated into Unified Command and participated in briefings.

Strength: ACSO (Aiken County Sheriff's Office) SWAT (special weapons and tactics) team activated per pre-developed plan, to address possible additional terrorist events at critical infrastructure locations in the county. Due to possible hostile incident indicators, State Homeland Security resources activated by Sheriff upon receipt of initial call, bringing the South Carolina Law Enforcement Division (SLED) on board. FBI responded quickly due to pre-established relationship. Current Mutual Aid Agreements in place for additional response agencies. Curfew implementation was discussed in the early morning and planned. Council Chairman signed a county ordinance to give the Sheriff the authority to impose a curfew prior to Governor declaring a State of Emergency. Sheriff was in contact with Governor's Office and the Attorney General's office to coordinate declaration of State of Emergency. Attorney General arrived at the scene to discuss legal ramifications of declaration.

Strength: Rescue, curfew and evacuation operations were initially planned for a 7-day period. Issues identified and addressed included: food/hydration/shelter/sanitation barricades/shift rotation. ADPS and ACSO provided hurricane stock of bottled water for responders.

Improvement Item: Accountability system (Haz-Mat wristbands) implemented by Fire Department was not communicated to all responding agencies. Coroner had no PPE for entering Haz-Mat zone. This issue is being addressed through acquisition of PPE through Homeland Security funding. Accountability badge system needs to be developed for private venders that respond to incident with no official identification.

Objective 4: Staff & Activate

Strength: Dispatch conducted recall by alpha-numeric pager (all-call) to respond to Staging. Initial Command Post (CP) was at Honda Cars of Aiken for thirty to forty-five minutes before being relocated to Kmart. CP setup was conducted through the on-call Communications Officer from procedure in place. Briefings were conducted at least six times daily with formal two-hour notice; more often if needed. All response agencies were informed of briefing times. Uninterrupted dispatch communications at the CP accomplished by mobile communication vehicle and aided response communications. Hard phone lines were run to the CP by noon on Day 1. Initial access controls were put into place via cones/tape/patrol officers. Day 3, decision makers moved to the antique mall and restricted access through twenty-four-hour security procedures that were implemented.

Objective 5: Public Information

Strength: Initial news release was issued within an hour of the event and contained accurate information. Briefings were scheduled to accommodate newspapers/radio/television deadlines. Community meetings were conducted to provide information on housing, food, and progress on cleanup operations. Mental health agencies were present at these meetings. Issues included pets and re-entry concerns. Rumor control—211 information telephone line was coordinated by the EOC (Emergency Operation Center); rumors were also addressed during news briefings. Spanish interpreter used to provide emergency information to public. EPA and SCDHEC (South Carolina Department of Health and Environmental Control) produced for citizens with information regarding housekeeping and food handling upon return to homes. Public Service Announcements were produced and broadcast regarding housekeeping and food handling upon return to homes. Quarterly media relations meetings conducted by local law enforcement to develop pre-crisis relationships resulted in effective communications.

Strength: ACSO Public Information Officer (PIO) coordinated media through implementation of a media staging area that was clearly identified to media.

Improvement Item: Responding agencies should pre-identify a PIO and participate in Joint Information Center (JIC) briefings.

Objective 6: Recovery

Strength: Law enforcement met with fire, school representatives, EPA and DHEC (South Carolina Department of Health and Environmental Control) on school re-openings. Requested a visible DHEC/SLED/FBI support on re-opening days. Open house conducted day prior to re-opening. Maps were updating in reverse showing reduction in impacted areas. Re-entry was coordinated with DHEC/EPA and companies contracted to perform cleanup. Detailed discussion conducted with cleanup contractor regarding re-opening of roadways, and possible equipment located on Aiken/Augusta Highway. Meetings were held to discuss financial implications of a plant shutdown and other issues of affected businesses and utilities (meeting payroll, phone communications, etc.).

Improvement Item: Recovery plan not formally documented in a written plan. Reimbursement needs should be included in recovery plan (supplies, hours, equipment, etc.). Utilization of business cards or other "quick reference" needed to assist in identifying major players involved in recovery planning. Recovery plan should include animal control/consideration of animal welfare.

Objective 7: Facilities & Equipment

Strength: Command staff made arrangement to utilize nearby vacant building for Unified Command Operations.

Improvement Item: Arrangements need to be made for copy machines/printers/pin boards/grease pencil boards/current maps. Early identification needed to resolve issues with generator smell and noise at CP/UCP. Pre-determined arrangements should be made for potential fuel needs during disasters. This issue has been addressed and agreements have been secured to meet this need. Consideration should be given to developing capability of mobile mapping and GIS capabilities. UCP (Unified Command Post) setup should include rapid setup of Internet capabilities for more effective communication and data sharing between responding agencies.

Aiken County Government After-Action Report

Objective 1: Safety

Strength: ACEMS (Aiken County Emergency Medical Services) personnel experienced no injuries during the response.

Improvement Item: Habitability surveys were not conducted initially at Command Post or Forward Operations. First ACEMS unit responded directly to the scene and had to leave the area due to fumes. Entry should be coordinated with IC (Incident Commander). Safety officer was not designated for EMS operations. Safety Officer responsibilities were defaulted to ACEMS Shift Manager.

Objective 2: Protective Actions

Improvement Item: Reverse 911 was not activated in a timely manner due to access available only by Emergency Management personnel. This weakness has been corrected so that Reverse 911 can now be activated through direction from Dispatch supervisor or authorization of Incident Commander. Capability will also be established at North Augusta Public Safety and Aiken Public Safety dispatch centers. The database used to initiate calls was five years old. This was identified post incident, and updated information is now available for input into the system. Public was unaware that unlisted phone numbers results in not being on 911 call list.

Strength: Key representatives of fire, law enforcement and emergency services at Command Post actively discussed evacuation versus shelter-in-place.

Improvement Item: ACEMD had to contact SC Emergency Management Division (SCEMD) to initiate the Emergency Alerting System (EAS) which only works if radio station is in auto position. ACEMD did not have (EAS) monitoring capability to determine if EAS message had been transmitted to citizens. SCEMD resource issues can impact initiation of EAS. Procedure to confirm dissemination of public protective action notifications should be developed.

Objective 3: Mitigation

Strength: U.S. Environmental Protection Agency (EPA) stated that Geographical Information System (GIS) maps in place when they arrived were very beneficial to planning mitigation activities.

Improvement Item: Plume models and GIS mapping need to use same coordinates.

Strength: ACEMD initiated early request for assistance from SC Emergency Management Division; 75 percent of State Emergency Support Functions were activated. ACEMS equipment was quickly met, once requested through the EOC (Emergency Operations Center). Shift turnovers were pre-planned and worked well. School District representatives notified at approximately 3:30 a.m. and decision was made to close schools prior to EOC activation.

Improvement Item: Credibility of EOC hampered by lack of a dedicated, adequate facility. Lack of coordination between EOC and CP affected logistics, food deliveries, housing, etc. CP was duplicating effort, and information was not being shared effectively. Formal status briefings need to be conducted for EOC staff on a regular basis.

Strength: The Incident Command System and key positions were implemented early in response.

Improvement Item: Better communication between response personnel would have resulted in pertinent information sharing. Unified CP was in place when EPA arrived. Clear lines of authority had not been established, but the right things were occurring, although maybe not as smoothly as they could have.

Strength: SC Department of Health and Environmental Control were well-informed of incident by time of arrival at CP.

Improvement Item: ACEMS supervisor was not present at initial Command Post (CP). Local/National Red Cross point of contact needed at the CP to coordinate food for personnel in outlying areas. National Red Cross may be needed in the EOC (Emergency Operation Center). EPA personnel were initially unaware that the Aiken County EOC was operational. Aiken County GIS resources were not involved in UCP planning meetings. ACEMS observed additional EMS support arrive from outside Aiken County. Additional units were not coordinated with ACEMS. Large numbers of individuals at the CP did not have a reason to be there. Better identification of key command staff would have helped. State Fire Marshals were contacted through SC Firefighter Mobilization Plan without the knowledge of the Damage Assessment Chief in the EOC. A Mutual Aid Agreement is in place with the Building Officials Association of SC, but was not utilized initially. Shelter staffing issues arose when a shelter was opened without EOC coordination and/or knowledge of DSS (Aiken County Department of Social Services)/Red Cross. There is a potential for county liability and financial responsibility if the Red Cross has not been involved with shelter opening.

Strength: EMS Supervisor relayed information to arriving units within ten minutes to stay clear of the incident scene.

Improvement Item: ACEMS access was restricted after first entry due to lack of PPE availability, and to fit incomplete testing on equipment received from the Department of Homeland Security. Accountability system (Haz-Mat wristbands) implemented by Fire Department was not communicated to all responding agencies. Pre-determined accountability system needed for Aiken County emergency response agencies. Agency accountability was being maintained, but not with other agencies.

Objective 4: Chemical Monitoring

Strength: ACHMT (Aiken County Hazardous Materials Team) staged at parking lot near Line Log/Silverbruff Road for safe-area accountability and to determine number of responders available. Staging Haz-Mat at Kroger negated the need to provide specific entry routes to responders unfamiliar with the Graniteville area.

Improvement Item: LEL (lower explosive limit) and standard O2 levels monitored by ACHMT, indicating crash scene impact only. Chlorine could have been indicated with proper monitoring equipment. ACHMT was not effectively integrated into Haz-Mat operations. Decontamination areas were not monitored due to lack of Haz-Mat support at decontamination locations.

Improvement Item: LEL (lower explosive limit) and standard O2 levels monitored by ACHMT (Aiken County Haz-Mat Team), indicting crash scene impact only. Chlorine could have been indicated with proper monitoring equipment. ACHMT was not effectively integrated into Haz-Mat operations. Decontamination areas were not monitored due to lack of Haz-Mat support at decontamination locations.

Objective 5: Public Information

Improvement Item: EOC did not have press releases prior to distribution at CP. Hard copies of press releases were not initially distributed at press conferences. Unmanned radio stations limited ability for local alerts to be made. Initial notification did not go out through NOAA Weather Radio, although it was utilized later in the day. EOC (Emergency Operation Center) PIO (Public Information Officer) could not get response from PIOs at CP (Command Post) to coordinate messages for media at EOC. Citizens in shelters had no official information source.

Strength: Salvation Army provided interpreters for Hispanic population.

Improvement Item: 211 (Aiken County Help Line) received calls immediately but had no information to provide initially. 211 received updated information via television news report. As a result, 211 personnel did not learn key information such as the shelter-in-place message that had been transmitted to residents. 211 was not accessible via cell phone. Additional number needs to be provided. EOC was receiving updated information via television news reports. Media staging area was located too close to CP.

Objective 6: Medical

Strength: ACEMS (Aiken County Emergency Medical Services) utilized PPE from Aiken County COBRA team which allowed EMS personnel to re-enter scene for rapid rescue.

Improvement Item: ACEMS attempted to medical monitor other responders, but they were entering incident area without EMS coordination. Triage tags were not utilized, although they were available. The on-duty EMS supervisor must relinquish control of outside incidents and focus on major incident being responded to.

Strength: ACEMS supported three separate decon sites with medical monitoring. Due to overwhelming number of calls for assistance being received from Graniteville area, decision was made to enter with Level-B suits by Haz-Mat technician-level EMS personnel. Decision to not transport patients prior to decontamination was made by ACEMS Shift Supervisor.

Improvement Item: EMS entry into the hot zone was coordinated through ACSO (Aiken County Sheriff Office) Dispatch who contacted the EMS supervisor at USCA (University of South Carolina at Aiken). There was no coordination with the GVW Fire Department.

Strength: Local hospitals were contacted early on by EMS supervisor informing them of patient potential.

Improvement Item: Mass casualty plan was not implemented initially due to communication difficulties. Communication of patient status at decontamination was not well-coordinated with Red Cross shelter representatives. Persons at shelters were registered, but if they were sent to the hospital or left with friends/family, their status was unknown.

Objective 7: Recovery

Strength: EPA led recovery effort to re-open schools and area businesses. Coordination occurred through UCP (United Command Post). A school representative was onsite for all entries. The County finance office implemented an hour code to assist in tracking costs.

Improvement Item: Not all agencies attended Critical Incident Stress Debriefing (CISD); this needs to be added to the recovery plan checklist. EOC had some difficulty obtaining some resources due to weekend hours. Commercial disaster recovery resource books may be useful in the EOC, as well as emergency contacts for local suppliers. County Damage Assessment official initially left out of planning loop for re-entry. All support agencies (Salvation Army, Red Cross, DSS (Department of Social Services) etc., were not kept informed of recovery status. Although daily status meetings were held at the UCP (United Command Post), the information was not communicated with the EOC.

Objective 8: Facilities and Equipment

Strength: SC Department of Social Services called in individuals to staff shelter at USCA (University of South Carolina at Aiken) campus who had not been previously designated in planning.

Improvement Item: Generator noise made it difficult to communicate at or with the CP. Electric capabilities earlier on would have proven helpful. Ladders were available for phone setup in EOC, but due to the chaos people were standing on chairs to connect the lines. There was difficulty in obtaining contracts for telephone installation; however, once SCEMD (South Carolina Emergency Management Department) became involved, it went smoothly. Procedure is now in place to obtain phones in emergency situations. Field charging capabilities are needed for portable radios and cell phones. Mobile Command Center is obtaining additional radios/batteries. Web EOC communications and tracking system was not utilized due to time consuming effort to set up basic needs in EOC. Lack of copiers at CP significantly hindered information distribution. GIS map plotters being used were 1 mile away at County planning office. Portable plotter capabilities need to be addressed. EOC printing capabilities were limited.

On January 21, 2005, Motley Rice LLC and W. Mullins McLeod, Jr., announced that they had filed a lawsuit seeking relief from persons with property damage resulting from the Graniteville train disaster. Named as defendants is the railroad Norfolk Southern, the Union Tank Car Company which manufactured the tank cars carrying the chlorine in the deadly train crash; the Olin Corporation which manufactured and shipped the deadly chlorine; and the Norfolk Southern employees, who allegedly failed to set the switch after they left their train on an active track. Motley Rice LLC has also been retained to handle a number of the personal injury and death cases as they relate to this catastrophe and those claims are being handled in separate lawsuits. Explained Ron Motley, a founding member of Motley Rice, "We believe this lawsuit will encourage the defendants to accept responsibility, and provide the property clean-up, replacement or payment as they are obligated to do under the law."

According to Mary Schiavo, another member of Motley Rice and a former Inspector General of the U.S. Department of Transportation, "Railroads are entrusted with vast access and right-of-way in our communities and our nation, but they hold such rights only insofar as they comply with the regulations governing safe operations of both rail and hazardous materials transport. Norfolk Southern and others failed to comply with these laws, regulations, and standards, and failed to put in place known and recommended practices which absolutely would have prevented this deadly crash."

Terrence Collins is the Thomas Lord Professor of Chemistry at Carnegie Mellon University who contends that the dangers of chlorine chemistry are not adequately addressed by either academe or industry, and alternatives to chlorine and chlorine processors must be pursued. He notes, "Many serious pollution episodes are attributable to chlorine products and processes. This information also belongs in chemistry courses to help avoid related mistakes. Examples include dioxin-contaminated 2,4,5-T, extensively used as a peacetime herbicide and as a component of the Vietnam War's agent orange; chlorofluorocarbons (CFCs); polychlorinated biphenyls (PCBs; the pesticides aldrin, chlordane, dieldrin, DDT, endrin, heptachlor, hexachlorobenzene, lindane, mirex, and toxaphene; pentachlorophenol for wood preservation; and dioxins-producing wood pulp bleaching with elemental chlorine" (*Chemical & Engineering News*, October 18, 2004, *82*(42), pp. 40-45).

It is very easy to create a daily record of conditions on scene in the tragedy known simply as "Graniteville." People in Graniteville and Aiken County were easy to talk to; and provided much information, reports, photos, commentary, newspaper, and television news features, and sadness. There is an old saying that "Time Heals All Wounds," but people in Graniteville recognize they suffered on Thursday, January 6, 2005, and are still suffering today. However, they do feel that the situation could have been much worse; an untold number could easily have killed and many more injured.

The National Transportation Safety Board (NTSB) has found much evidence to indicate the probable cause of the January 6, 2005, collision and derailment of Norfolk Southern train 192 in Graniteville, South Carolina, was the failure of the crew of Norfolk Southern train P22 to return the main line switch to the normal position after the crew completed work at an industry track. Contributing to the severity of the accident was the puncture of the ninth car in the train, a tank car containing chlorine, which resulted in the release of poison gas.

At 2:39 a.m. on January 6, 2006, one of three tank cars carrying 90 tons of chlorine was ruptured and released 65 percent of its load, which immediately turned from liquid in the tank car to a mass of poison gas because of a 47-to-1 expansion ratio. Some 183 night shift workers in five buildings at Avondale Mills with no warning of a tragedy, before retreating from the mill or waiting for rescue as the poison gas became more severe. Neither the residents of Graniteville, nor the night shift workers in the mill, had any idea what was going to happen. Later, five night shift employees were found dead, and many were injured.

Survivor Number One

A number of Graniteville people called "911" to report that they were trapped and terrified that early morning. Many mill employees were still bothered of the fact that they had lost five mill workers to a train/auto accident just weeks before. Maggie Adams appears to be the first person to dial "911" from the mill. She had worked at the Graniteville mills for more than twenty-two years and loved her job prior to the chlorine wreck. "I was laid off from the mill on May 31, 2006, and I was glad to get out of there because there are a lot of memories of that night of the train wreck. Having to go to the data processing building, where I used to work, is just a reminder of how close to death I came the night of the wreck. I was a computer operator in that building for twenty-one years, and always worked alone at night. That night I heard the screeching of the brakes like thunder when it rolls across the ground. I just held on to my desk and waited for the screeching to stop. I could feel the train like it was in my office, because I work in a large computer room and the wall behind me is only about 30 to 40 feet from the train tracks where the train is now derailing. Rail cars are now rolling around, coming apart, and crashing into each other as they were being shuffled into the spur track which is located in our parking lot.

"After a few minutes of being actually involved a few feet from a train wreck, I left the computer room and went down the hall towards the end door to the parking lot. Opening the door where the wreck occurred, my car was parked right outside the door. Suddenly, a smell almost floored me. This black asphalt, smoke, burning, stinking smell just hit me! I slammed the door, and retraced my steps to the computer room and called 911. I told the operator there had been a train wreck, that I was there all by myself, and I needed someone to come and get me, and they said, 'Okay.' I then told the operator I would be at the lobby door waiting for someone. I got my purse, and for some reason I picked up a little jacket with a hood on it because the computer room is kept cool for the equipment there.

"I got to the front door of the building, was kept waiting and waiting for a long time, and nobody came. All this time, I kept inhaling this burning air, and water started coming out of my nose and my eyes; it was like the air was coming out of my lungs. I kept looking down the hallway and everything was yellow, and the lights were dimming. The air from outside the building was being brought into the building by the air conditioning system, and I was sucking it in. I realized I had to get away from that smell. I called 911 again, and I was told if I would get off the phone they would send help. I knew they would not, and they never did.

"I next called a guy I work with, just because I remembered his telephone number. 'Steve,' I said, 'You have got to come and get me!' He says, 'Give me a moment, and I'll come and get you,' this being about 3:00 a.m. Then I called my supervisor, Doug, woke him up, and he warned me to get out of the building if I could since I thought at that time the building had caught fire. At the same time, the lights went out, and then the generator kicked in, and all this time the poison gas just kept spreading and getting thicker. I then telephoned my sister, plus a lot of other persons.

"About this time, I decided I would have to save myself," remembers Maggie Adams, a good-looking woman in her early fifties. "I took the hood on my jacket and put it over my head, then took my lab coat and wrapped it around the bottom of my face. I had my cell phone in my hand; I could not see since I had to take off my glasses for I am near-sighted, and I was praying hard. Mucus was just pouring out of me as I felt my way down the stairs. My thought was maybe if I stay in the middle of the road, someone will see me. Since I could not see, I was afraid of tripping over a curb, or possibly another body. On my way proceeding very slowly up Main Street where the wreck occurred, I found two abandoned automobiles with their doors hanging open. I walked up and started beating on the doors, and screaming, 'Can somebody help me? I can barely see light.' However, nobody appeared, so I continued my slow walk on Main Street. My boss started calling me, and I sat down at times and rested. It took me about two-and-a-half hours to walk less than a mile.

"Everybody was calling me on my phone. My boss, Doug, called me again, and he asked me if I could make it to a funeral home on Main Street, and I said I would try. He said he knew someone that lived behind the funeral home, and he would try to wake them up by a telephone call, so they could get me out. In any event, I made it to the funeral home and the home behind it. Doug called me again, and told me he could not get anyone in the house to answer the phone. I told Doug I had to get out of this area, because I kept spitting up stuff from my lungs.

"I then made my way out of Graniteville until I reached Route 1. I may have walked a bit more, but I kind of blacked-out for a while. Eventually, I walked underneath the bridge and traveled like I was going to Aiken. At one point, I thought everybody in Graniteville might be dead. I met two guys coming down the road who asked me if I was okay; I told them no, I am not all right. There's been a train wreck and there's a toxic smell in Graniteville. One of the men told me they had heard that news, and said I had better get out of the area. I kept thinking why weren't people evacuated? The rail wreck happened about 2:40 a.m., I had left my office about 3:15 a.m., and had been walking for a long time before eventually reaching Abare's Paint & Body Shop. I remember clearly there was a big mailbox there, and I was just stranding there holding on to the mailbox.

"Finally, a friend called the cops, and they let him into the enclosure area that had been established around Graniteville, to guide the police to where I was at the present time. I saw one car come down the road, and waved to beat the band, but the car went right by me. My sister called me, telling me she was going to run the road block, she was coming after me. I asked my sister to tell my kids and grandkids that I loved them all, and my sister is screaming, 'Please don't leave.' Then Steve, my co-worker, called and said, 'Please tell this officer *exactly* where you are!'

"I made my peace with God, and just lay down on the ground with my head on my lab coat for a short time period. I saw a blue light coming towards me, and the cops were really flying. The police car crossed the median, and two officers got out dressed in camouflage Haz-Mat-type suits. They picked me up, and put me in the back seat of their vehicle, turned around, and sped back to the campus of the University of South Carolina at Aiken, where a decontamination area had been established. At the campus, they sat me on the ground and someone came with a stretcher and an oxygen tank, and asked who I was. I told them, and someone said, 'This lady needs help,' and they put me on the stretcher to remove most of my clothing. They left my socks and panties, and I was scared as well as freezing. I went through the decontamination showers twice, and then was taken to a trauma unit just up the road in Augusta, Georgia.

"At the hospital, I probably went into shock as the doctors looked me over. All I remember is there were tubes going down into my body, and I didn't know anything from Thursday until Saturday. The doctors put me into a coma so they could rest my lungs and breathing. I started coming out a little bit on Sunday, and went home on Tuesday. However, the breathing tubes irritated my esophagus, and I had to go through pulmonary treatments to clear that condition. I have been seen by psychologists or psychiatrists that put me on a very mild antidepressant drug because I was having very bad dreams. The nightmares featured yellow smoke coming through my room at night. My doctor says I am doing fine.

"While in the hospital, the doctors kept me full of medicine to keep my muscles strong with which I had some trouble. I don't remember anything except I was told I was coughing, flirting with the doctors, and having affairs with their wives. Well, I just said I could not help it; I was full of drugs in the hospital, but I never took drugs before the rail wreck. But my best experience in the hospital was in seeing my boss, Doug. He sat by my bedside to see my face, and I talked to him. He cried because he could not get to me when the freight train went off the tracks and crashed near my building. He tried several time to rescue me, but could not succeed in the confusion apparent just after first responders realized they had had a poison gas to deal with [armed roadblocks were installed throughout Graniteville at once].

"We do not know anything about the long-term effects of my experience with chlorine gas. I was out of work for three months working on days, but they were going to put me on night duty again. I can't go back to work at night! I can't even pass by the building where I used to work; I'll dodge it, I just won't go that way. My car got totaled, but I had to force myself to go look at the car months later. The shocks were corroded, green stuff was just flowing out of coins in my console, and I had to break the window because the lock was broken. I signed my car over to Norfolk Southern Railway Company. Basically, the railroad company said, 'How much do you want,' and I told them, and they sent me a check."

Maggie Adams went back to work on April 1, 2005. "The company started me out on half-days every other day, got me working every other day full time, then I went to five days a week. When they tried to put me on nights, I said, 'That's it! I cannot be here by myself working at night.' The company said there will be security guards on duty at night, but I responded, 'I don't care; they can't stop the train.' One day I got stuck in a thunder storm at work, and I almost lost it because it sounded just like the train. I got to the point where I would not go anywhere. Seventeen months later I still do not drive, and I don't like going anyplace after dark. My doctor told this was normal; she said I would eventually get past that. But they also told me I could be doing great, and I would take three steps back. I was a strong person, very independent and progressive before the rail wreck. After the wreck, I did not have any energy for months; I was just dragging and sluggish in whatever I tried to do.

"I used to love my job. For one year I worked days, and then I worked twenty-one and one-half years at night in the computer room by myself. However, after the train wreck, the job was getting to me. It was just so much pressure, and they were laying people off. I once had the perfect job for twenty-something years when I worked three-nights-a-week and then was off work for three days. Suddenly, after the chlorine gas experience, I was working five days a week, and did not have enough time for personnel commitments. I had housework to do, appointments to keep, clean the house every Saturday, cook dinner on Sunday, and then back to work on Monday. There was no time off, and I was unable to get things done. However, on the good side, I do have a wonderful, large family with eight brothers and one sister."

At the end of the interview, Maggie Adams reported how her office looked after the chlorine had done its work on her building.

"They had photographs in the break room when I returned to work. The water fountain was orange, and anything that had water or metal in it was orange. The orange color was running down the walls like blood, and in the bathroom sinks and commodes were completely orange due to corrosion. In my office, the chlorine destroyed the computer printers and servers. They said it still stinks in there. Somebody got the folders off my desk and out of my drawers so I could look at them. My hands became orange and yellow, and I started coughing."

Survivor Number Two

Charles Reyes is an American Indian. His mother is a Cheyenne from the Green River Reservation, and his father is a Comanche, a member of a Shoshonean tribe, formerly ranging from Wyoming to Texas, but now based in Oklahoma. Reyes belongs to two tribal groups and two reservations; and speaks a number of languages; including Apache, Navajo, Cherokee, Comanche, Iroquois, Spanish, English, and German. Charles Reyes, seventeen months after he went through a poison gas attack in his own home town, still shows clear signs of what chlorine gas did to a formerly healthy man. At certain times, his breathing is labored and scratchy, and he has to will himself to speak normally. Sometimes, his speech is reduced to that of a young child. At other times, he scratches himself endlessly, even picking up objects off the ground to get a better scratch. The itching goes on, and on, and on. Charles Reyes, and his wife, Brenda Reyes, have learned much about chlorine, but they sincerely worry about what comes next.

Nowadays, Charles does a lot of volunteer work, although he is still employed fulltime as an electrician for a federal hospital. "I had three tours in Vietnam, the first time with the Fifth Division, the second time with the 82nd Airborne Rangers, and the third time with the 14th Division. For the past twelve years, I have been an officer with the Graniteville-Vaucluse-Warrenville Fire Department. I am also a volunteer interpreter for the Sheriff of Aiken County, as well as the South Carolina Highway Patrol. My wife inspired me because she has been a firefighter as a lady first responder, nurse, emergency medical technician (EMT), and a Public Safety Officer. She has done all of that; and I told myself, 'If she can do it, I can do it.' We were a husband and wife team for years.

"I also travel to Indian powwows where I now play the drums and dance, but since my voice was messed up, I don't sing like I used to. I help the Boy Scouts to become Eagle Scouts, and I am a puppeteer. I also go to schools, churches, Veteran Administration and other hospitals, and colleges where I put on my presentation for the Native American Indian. I just turned sixty years old, but in the weekend of June 10, 2006, for three days I qualified for rope rescue in water rescue operations. I don't really feel sixty years old because the Lord gave my life a purpose.

"Early in the morning of the sixth of January, 2005, I was sleeping soundly. My wife sometime can not sleep at night, and sometimes she gets to clean the house when this happens. However, at 2:39 a.m., I did hear the freight train hit the work train that had been parked on a siding the day before. That crash shook our entire house. I heard the crack and bang of thundering metal, wheels and products slithering everywhere, amid my wife yelling and screaming; I thought I was back in Vietnam for another tour. I picked up my keys, jumped out of my pajamas and into my clothes, and headed for the firehouse just down the street. I always make it the fire station between 38 and 42 seconds, from the house to the firehouse, and my wife always follows me because she is also a first responder. So I took off running, and I only live about a half block away. When I approached the fire station, I had no idea I might be dealing with a poison gas. The only thing I had on my mind was the train derailment, how many people were injured, and what we had to do as first responders. On the way down, I was praying because I always pray.

"When I got to the tree and fence in front of my house, I saw this green cloud coming toward me with a gold and white mist. I thought, 'Oh No. Gas,' so I dropped my gear, and by the time I turned around, the poison gas went through my nose, my eyes, my throat, my lungs, and my heart. The gas felt as though it was burning my body, and I thought what I learned with the Green Berets in Vietnam. I thought, 'chemical agent, water, and dirt.' I was trying to find dirt because it had rained the day before, but the more I moved the more I was burning. My heart and brain seemed ready to explode"

Charles Reyes saw the gaseous cloud, and it poisoned him right away. At this point, he fell down in the road, and may have had an out-of-body experience. "I went on the ground, but I do not know how long I was there, but I saw three angels around my body, like a bubble protecting me. I turned to God, and poured my guts out to him. I saw my body, and I saw the angels. I did not know what was going on, but a voice came out of nowhere and comforted me. It seemed like I stole time, like I robbed time. Those seconds were an hour with my God, and it was the most beautiful thing I ever experienced in my life. When I was talking to him, I felt that he was leaving me, and I didn't really want to come back to my body. I wanted to go home with him. All of a sudden, I felt a vacuum in my body, and an angel picked me up. I ran about a mile-and-one-half, knocking on doors to try to get people out. My head still hurt very much; and I was still gagging, coughing, could not see straight, and was bouncing all over the street and sidewalk. I was just trying to do my job, think about my wife, hopping my wife survived the poison gas because I knew she was right behind me when I ran to the fire department.

"About this time, I was on Main Street in Graniteville and borrowed a telephone to call our dispatcher to dispatch our fire department. He told me that 306 was looking for me, which is my wife's call number with the fire department. I told him to tell her I was fine; but we have gas in the area and we need control, sheriffs, and fire departments. I told the dispatcher I could not breathe, I need help, I can't do this by myself. I added I was trying to get control, but there were people dropping left and right in this area. So after that exchange, I collapsed."

When he came to, Charles Reyes found that first responders had him in a shower bath trying to get rid of any contamination that he might have on his clothes or skin. "They took off my clothes, and I became embarrassed as the guys were scrubbing me. I guess I passed out again, and they put me in an ambulance that took me to University Hospital in Augusta, Georgia. I was released the next afternoon, although I was still having problems. I wanted to get back with the group, but still did not feel very good, so I cannot know why they released me at this time. We were given an apartment to stay in, since our home was probably contaminated. That day or later, I started spitting blood, and my wife had blood coming out here nose and rectum. Subsequently, a man's body was found near our home. The man had run away from the mill where he worked at night to get away from the gas, but he tried to go through the woods, where he died. His body was not found immediately."

Seventeen months after the gas attack in Graniteville, it is easy to tell that Charles Reyes still has symptoms of the poison gas attack he lived through. "My legs hurt, and my heart is enlarged. Recently, I went through a day-and-a-half physical, and told the doctors I can't eat. Every time I eat, it hurts in my belly. It hurts so much that I get upper and lower cramps, so I eat soup and other very light foods. I did go back to firefighting fifteen months ago. I finally got back, was able to pass the physical, because I am the type of guy that does not make excuses; I want to get back in action. I still have 'bubbles,' although I do my exercises and everything my doctors tell me to do."

When asked if he had a suit against the Norfolk Southern Railway Company (NS), Charles Reyes said that he did for himself and his wife. "We are not settled with the railroad on any aspect of the suit. I am not going to sign anything, and not going to release them from anything. Every time I go to the doctor, I get copies of all my reports; the whole thing is a hassle with the amounts of paper work. I have to go to a lawyer, and the lawyer goes to the railroad to get them to pay my medical bills. In the same manner, the railroad has to go through my lawyer to get to me. I end up with a stack of paperwork which I break down month-by-month. At least, I'm not afraid of dying any more because I know what's on the other side.

"My 'day job' is as a civil service electrician for the last thirty-four years at the Veteran's Administration hospital. March 5, 2006, I had thirty-four years with civil service. After duty at the VA hospital, all the rest is a volunteer as a firefighter, rescue worker, or helping the county sheriff and the State Highway Patrol as an interpreter. Basically, I stay pretty busy. The federal government gave us three fire vehicles after the chlorine gas incident because our old equipment was done-in by corrosion caused directly by poison chlorine gas. This gas took the paint and chrome off our fire vehicles so it was unusable for any other purpose; what do you think it did to my lungs and body at the same time?"

Survivor Number Three

Brenda Reyes, the wife of Charles Reyes, could not get to sleep during the early morning hours of January 6, 2005. "I had been up for two days since I have insomnia, so instead of just lying about, I got up and cleaned the house. About 2:30 that morning, I was just finishing up with the house when I thought I could go to sleep, so I was heading for the bedroom when I heard this awful sound of metal hitting metal. It was like thousands and thousands of fingernails coming down the chalk board. I figured that the express freight train had hit another vehicle at the crossing because about five weeks earlier a special train had hit a car with five people from the mill, and they were all killed. I was already dressed, so I told my husband to get up, and I would meet him at our car. The incident had not been called in because it had just happened. I went out the back door, and I stood there looking around. I said to my self, 'This is very odd!' The mist was up to my knees, and it smelled awful, a heavy smell that resembled household bleach. We have been through hazardous materials training with the GVW Fire Department (The Graniteville-Vaucluse-Warrenville Volunteer Fire Department), and I knew that the smell could mean poison. All of a sudden, I knew we should leave the area, and not go to the fire station just down the street; we should go in the opposite direction.

"I don't think the chlorine was meant for our area; it was just on a train going through our area, to the best of my knowledge. However, the mist was so pretty, basically green, a beautiful green with gold specs in it. However, I could smell the bleach at once; like somebody was cleaning his pool and put in too much chlorine into to the pool. I got to thinking that nobody had a pool in our neighborhood. I told my husband to come out of our house; we did not need to go to the fire station because nothing was broadcast yet over our little monitors. We needed to go in the opposite direction to find out what was going on. This is poison, and it's going to kill us. He said, 'No, I am going to the fire station.' I had no other thing to do but follow him. By the time I got out front without the car, the poison gas had already come up to my chest, and I was having breathing problems, hacking and coughing. I went and got in the car, turned it around, and headed for the fire house. I made it only to the convenience store a short way from our house. I couldn't see anything since the mist covered my car; the poison gas covered everything on the street. I could actually see about two feet ahead of my car, and I was having difficulties because of the gas, and I could not see my husband. I called him on my cell phone, but he did not answer.

"I used my cell phone to call all my neighbors and tell them to get out of Graniteville. One was a paraplegic, another an eighty-five-year-old person, and a third, seventy-six years old, and the fourth was a forty-five-year-old person. I told them I did not know the whole story, but we have a toxic gas in town, and it can kill you. I got two other people awake, but I could not get the eighty-five-year-old person across the street because that person has put on to much weight for me to lift. So I called 911 to have somebody come in and help this person. I then drove over to Warrenville to get my breath because I was really coughing, my nose was burning, the same with my throat; everything seemed like it was on fire. Then I drove back to Graniteville, and at every rut I hit in the road I had to get out of the car and look under the car to see if I had hit my husband or another person. I could not see a thing, and was unable to get by the convenience store near our home. The odor and the fog were still on our street, and I had mucus coming out or my nose. I would rub the mucus off, and I had no napkins or tissue paper in the car, so I would rub it off on my shirt and pants. I could not find my husband, so I left Graniteville again to get some fresh air.

"I came back in a second time, and in the fog and mist that made it impossible to see, I hit a big rut in the road just before the funeral home, and I felt sure I had run over my husband. I got out of the car and shined my flashlight under each wheel, but Charles was not there. Visibility was about zero, and my car started acting up like it was out of gas, although I had plenty of gasoline. Again, I made it no further than the convenience store. I did this routine a total of four times, with more mucus, hacking and coughing each time. I looked in the rearview mirror, saw myself, and my image scared me. I seemed to have white blood smeared all over me. I had thought it was just plain mucus from the chemical, but I knew then I had more problems.

"About this time," according to Brenda Reyes, "Warrenville-Langeley Fire Department radioed on the pager, 'We have your husband down here.' I went where they told to go, and both my husband and I got decontamination showers. They took off all our clothes. We had no clothes on whatsoever which can be very embarrassing; just those little, thin sheets. This was something we had to do by fire regulations for some chemicals, including chlorine gas. They also flushed out our eyes because they were red, irritated, and just weeping away; and I washed my own hands and face. We were then taken by ambulance to University Hospital.

"At the hospital, I got all kinds of treatments for my lungs and to keep me breathing, but I told the doctors my husband should be treated first. The doctors then treated him and gave him a shot to open his lungs, then gave me another shot. After we got discharged from the hospital, my husband went to the Station Two fire department on Lake Road, well away from any lingering chlorine gas. I still did not have any clothes on. They wanted us to wear scrub gowns out of the decontamination site, but sometimes scrubs don't work if you do not have underwear on. I called my son who picked me up, and his wife gave me a shirt to wear. It looked like an evening dress hanging down around my ankles. We then learned about a church in Aiken, on Laurens Street, that was helping people in distress due to the poison gas situation. We went there because I didn't want to stay at my son's house since he and his wife needed their privacy, and I needed time to heal. The railroad company provided motels, food vouchers, and necessary medicines to eligible people. I signed up for a motel, actually a Guest House in Aiken, and to get food three-time-a-day. I had to buy nebulizers (an atomizer equipped to produce an extremely fine spray for deep penetration of the lungs) and other medications since all my regular medications were at home. I take a lot of medication since I have a bone disorder, and because of the chlorine incident, doctors had me on 27 different medications. That's a lot of chemicals to mix together.

"I had to go to an eye doctor because the thin membrane covering my eye was burned, plus I coughed so hard and so violently during the gas attack that I developed a deviated septum. My lungs were burned, my esophagus (a tube connecting the mouth or pharynx with the stomach gullet) was burned, and I had to go into the hospital to get cauterized because my stomach was bleeding. In addition, I had to go to a heart doctor to slow my heart down because they said I had post traumatic stress disorder from what I experienced. I had to get more medicine for that to calm me down. Six months later, I told the doctors that I was done with all that medicine. I wanted to get off all the medicine because I felt that it was doing me no good. I asked them to let me go and heal on my own. At this time, I could not eat solid foods, so I had to drink a lot of chicken broth. Milk shakes were good, since they cooled the throat very well, and the doctors gave me a spray for my tongue and mouth because I had blisters inside my mouth even seventeen months later. I still have my nebulizer and my inhaler. I take a pill three times a day to keep me from freaking out when I hear a train coming down the track. We have a train about every two hours, even on Sundays. Thirty years ago, you wouldn't hear tell of a train on Sunday coming through Graniteville; now Sunday does not mean anything, they come and they go!"

Brenda Reyes commented about the increase in hazardous materials the railroads are now hauling. "It's definitely a problem. Seventeen months later, I am still in need of my lungs to be cleaned, but they will *never* be healed. My lungs are scarred, and they are going to be scarred for the rest of my life. My esophagus got burned, and I have to absolutely take medication daily because the poison gas did so much damage to my stomach. I was just lucky it did not get my larynx. Now I have a deviated septum because a little membrane burst in the back of my nose, which gives me apnea (temporary suspension of respiration). When I sleep, I cannot lie on my back, but have to lie on my side. I also have to see a psychologist for traumatic stress disorder."

Brenda Reyes started out with a fifth-grade education, but in 1979 when her kids were raised and out of the house, she decided to go back to school. She started at Aiken Tech, studying to get her GED certificate which she obtained in three months, then entered a class for nurses attempting to qualify as L.P.N. (Licensed Practical Nurse). After she graduated from that class, and was working as a licensed nurse in the local nursing program, she noticed there was an Emergency Medical Technician (EMT) class which she entered. In 1980 the City of Aiken Police and Fire Department had an opening for a Safety Officer/ Medical Officer, and Brenda Reyes was hired. Brenda readily admits she had a little extra pull in getting this job; the wives of two senior officers told their husbands to get Brenda and put her in the fire division, so they could feel better having somebody who's licensed and an EMT to go with to their fires. Brenda was voted into the position a couple of months later as a Lieutenant. The police duty was a bit more dangerous than the firefighters had been. As she notes, "You know, in a burning building, you can run out of the building; but if you have a 38 pistol pointed between your eyes, you cannot do much at all. The guys always kidded me because I was the third woman on the police department. I preferred fire fighting, and for the City of Aiken I worked as an engineer (truck driver/operator). I didn't have to fight many fires, but I had to make sure I had enough water for the guys at the fire. I drove all the fire trucks, the ladder trucks, and whatever they had for emergency vehicles. Later, it got too dangerous for a woman to do police duty, so I went back to nursing at the federal Department of Nursing where I worked in the psychiatry ward. I've had a good life. The Lord was with me the night of January 6, 2005; without him that night, I would be dead."

Chief Phillip A. Napier of the GVW Fire and Rescue Department, who also runs Napier's Hardware Store in downtown Graniteville, South Carolina remembers, "We received a call that the train had possibly hit a building in Graniteville. Upon our response, I told the men to report to the station, and I would try to locate the incident site and find out what was involved. And evidently, by listening to the tapes at a later date, it is evident I was already beginning to be disoriented. I couldn't see. I called for an ambulance, and I couldn't say what I wanted it for. When I pulled up to the man on the railroad track, and rolled the window down, he told me they had had a head on collision with the train, they had a chemical leak, and he couldn't breathe. Then he fell to the ground, and later died.

"He was the engineer of the wrecked train. Immediately, just like something cut off my wind, the chlorine hit me. I vaguely remember making a u-turn at a rail-crossing less than 50 feet from where I was, and headed north. At this time, there was probably a five to eight minute time span when I lost my memory. When I came back to my senses, I was on

Number l Highway at a cross street calling for hazmat teams to assist us. I told them I could not breathe in this area, and that we were going to have a major evacuation in this area.

"So we just went in from there as far as setting up. The first two gentlemen that came in, I put them in charge of hazmat response. One of these men was also Fire Chief with the Augusta Regional Fire Department just down the road in Georgia. Early-on, the rest of the first responders I secured worked under my and their direction. The law enforcement officers started securing the area by blocking off the streets coming into Graniteville. We also had to send some teams in to shut down the mills. People had run out leaving the mill with the steam plant boilers still running, so we had to send teams in to shut them down. It wasn't long before the railroad company gave us information on the chemicals from the train cars that we would have to deal with, basically and possibly chlorine and sodium hydroxide. However, at this time, we did not know what cars had breached. A clean up company hired by the railroad came to the scene, suited-up, went in, and attempted to patch the car to stop the leak. The next morning, we received a call, probably around five in the morning, that there was a victim trapped in an automobile, and we sent teams in to rescue this person. A man was rescued, and to my knowledge, health-wise, he is okay.

"In the days that followed, we had different agencies coming in; including EPA, Federal Railroad Agency, National Transportation Safety Agency, and many others. We started setting up as far as what to do and what not to do. We basically went to a Unified Command situation, which was the best thing that could have happened with this incident. The Sheriff's Department knew they were in charge of law enforcement, the Fire Department knew they were in charge of fire and hazmat, and Emergency Medical Services knew that that they were in charge of triage and medical care. Everybody had their phases of what they needed to do.

"Shortly after daylight on the first day, we were notified that there was a fire in the steam plant. We had to start sending teams in to extinguish the fire in the steam plant. We found the boilers were fed by coal, and fire got into the coal chute, and there was hardly any way to extinguish the fire. We set up deluge guns and automatic monitors to keep a constant flow of water to try to contain it to keep it from getting into the silo. Had we not tried to contain the fire, then the whole town would have been covered with black smoke and soot. It was just one thing after another the entire day. We started receiving calls of natural gas leaks. Just about anything that could happen, happened within that twelve-day span. Worst of all, we started receiving confirmation of fatalities. In the beginning, we started receiving calls that people were trapped in buildings and couldn't get out. Then a long-haul truck driver from Canada sleeping in a mill parking lot waiting for either a pickup or a delivery was found dead from chlorine inhalation. We have put up a fine memorial in town to honor him as well as for eight other workers and residents who were killed by chlorine gas

"I and other members of the GVW Fire and Rescue Department have been to eleven states to provide information about the rail wreck and the deaths of our citizens and workers. We have made a Power Point presentation, and last week I and the GVW Assistance Chief were in Baltimore, Maryland to appear before the National Association of Fire Chiefs Haz-Mat Technical Conference. We have also been to Traverse City, Michigan; Cincinnati, Ohio; the FDIC in Illinois; the State Firemen Convention at Myrtle Beach, South Carolina; Fairfax, Virginia; and other cities. We probably put on our show on 150 times.

"In the early minutes of this incident, we had people begging for help," explains Chief Philip Napier. "We needed good information of what we were actually dealing with, so we sent in a hazmat team made up of firefighters from the Savannah River Site and the Augusta-Richmond County hazardous materials response teams. The railroad company, Norfolk Southern Corporation, was very cooperative through the whole event. Anything we needed, they would assist us. The clean-up companies they brought in were very good to work with, and a medical doctor and a chemist were provided. If you read the National Transportation Safety Board's report of this incident, you can tell the clean-up went well. Overall, probably the best thing that could have happened was when we went to a unified command set-up. That way, it puts every organization in their correct place with their normal duties. By federal law, the fire chief is the incident commander. However, my stance from the get-go is you work together as a team for the good of the people. It is not, 'Who is in Charge;' you need to work together. But the bottom line is the fire chief is in charge, not an elected official. I would always suggest that you try to go into unified command as soon as possible. That way it will put every organization in their place. It probably was two days after the rail wreck when it was decided to go into unified command."

Chief Napier was asked about survivors who say they still have health problems even eighteen months after the train derailment. His answer was, "Nobody really seems to know. We still have people who say they have health problems and are having effects from it. People have basically an on-going damage to their homes in close proximity within Graniteville village. They see the damage to their town, their homes, and loss of good employment which can be a definitely emotional experience. The textile industry says that the train wreckage has caused them to have to close their business. American Textiles was in trouble before the train wreck. It started laying off people and closing different facilities. Now, whether or not the train derailment forced them to close the mill because of down time and the later clean-up processes, is a question that I cannot answer. Once the clean-up company finished their work, the mill company said they were shutting down."

Regarding the people in town that were injured, did they get any long-term help for their suffering and injuries? "I don't know if they or did not," said the Chief. "A lot of them have personal lawsuits with the railroad, and a lot of that is pending. There was a class action suit filed. Also, the railroad company paid everybody within a mile radius of the train wreck $200 a day for every person for every day they were evacuated plus, $2,000 for the household. Some are still seeking legal attention. In addition, the State of South Carolina appropriated $340,000 for Graniteville Disaster Relief, and some of our politicians tried to use it to buy police cars for the county. A lot of the people in the community found out, stood up, and protested the use of that money in such a fashion. Our desire is for the money to come to Graniteville for continued health screening of the people who live or work in the town. At the present time, it looks like we are going to win that battle. If we don't win, we are going to sue. But that's how politics gets involved in everything. They try to use the suffering and pain of poor people to benefit somewhere else. At the present time, as far as state and federal help directed to injured persons, this community have received nothing. During the actual incident, however, the Federal Homeland Security Agency sent transfer trucks to us loaded with all sorts of equipment, including all kinds of Haz-Mat suits and air packs, keeping us in business for a number or days. I believe the transfer trucks came from Columbia in South Carolina, and they appeared in less than one-and-a-half days after the wreck."

When asked about the payment of damages to the GVW Fire and Rescue Department, Chief Napier responded, "Our fire station was within 250 yards of the train derailment. We lost our fire station, two fire engines, a service truck, a First Response vehicle, and all equipment within the station. We had to completely gut the building, rewire it, sheetrock it, and redo the floors. The chlorine gas just penetrated the building; and as a matter of fact, the second door down from the fire station is where there was a fatality in a residence.

"I'm not sure how many people were injured in Graniteville, and no one can tell how long it's going to take for them to get better. It's really an uncertainty for all of us. We don't know what effect this chlorine is going to have on us in the long-term, as far as the environment here, and as far as our health. Personally, I was exposed to chlorine gas which basically took five to ten minutes away from my life. I have no memory of that short time period, no memory at all. Yet, to the best of my knowledge, I seem to have no lasting effects from the chlorine after seventeen months as I talk to you. It was a short, momentary memory loss from the gas, except I know I headed north but I ended up in the south portion of town. It was a lot to put on your plate at one time. I called my home and told my daughter to get her mother and just get in the car and go toward North Augusta. Also, I called a brother and told him to get our mother who lives within a block of the train wreck, probably within a 100 yard dash. Thank God, they're all okay today, with just some long term effects."

When Chief Napier was asked if he had any information where the leaking chlorine gas traveled through Graniteville, he responded, "In the beginning we'd had reports a deputy sheriff saw a green cloud rolling down Main Street toward the fire station. The deputy turned around in the fire station yard, and it seemed like the chlorine gas was on the ground as it passed Dale's Convenience Store. You can tell by the vegetation that was peeled and blistered, and then the gas moved to the creek, and it followed the creek which was on lower ground to the north. When it reached the mill buildings where more deaths and injuries were noted, the deadly gas seemed to follow the creek and the railroad track to the north. You could see across the pond that the blistering effect of the trees was continued. "Everybody says chlorine sticks close to the ground, but it does not in all cases; even the cross on that steeple of the church had to be replaced for it was terribly tarnished. On pine trees 70, 80, 90, 100 feet tall, the needles were completely bleached. There were reports that behind Woodhead mill building, when the U.S. Coast Guard team came in and checked the swamps, they said there were snakes there that were bleached white."

November 10, 2004:   Graniteville, SC. Five workers at a textile mill driving home after working all night were killed at a railroad crossing in Graniteville. A Norfolk Southern train engine pulling two cars hit their car doing 45 M.P.H. in a stretch of track with a maximum speed allowed of 49 M.P.H. Three automobiles tried to beat the train to the crossing, according to witnesses; but two won the race, leaving the third car to get hit with the train engine. All five occupants in the car were killed. Fire Chief Philip A. Napier remembers this incident very well. "I was sitting here at my store one morning and I looked out and I saw the train going by. I thought to myself 'that train is flying,' it's an engine and two passenger cars (this train was testing the rails for safety, yet it killed five people). About a minute or two later, we got a call that the train had hit a vehicle. When I responded, I told the men in the fire department on the radio that it was going to be a bad incident because the train was really moving and there were five fatalities. It was just like it knocked the life out of them. That was in November, and in December I went before Aiken County Council with a presentation. I asked them to try to do something to slow the trains coming through this town before we had a derailment. On January 6, 2005, less than eight weeks later, we had a colossal derailment that killed nine persons from inhalation of chlorine gas and injured many more. I'll always believe that had the chlorine train been going 25 mph, there would have been a derailment and there would have been a collision but I don't think it would have reached the chlorine car. Since then, the train does come here at approximately 25 mph.

"All nine fatalities were from chlorine inhalation," says Napier. "They might have received minor damage from the train derailment, but chlorine inhalation is what killed them. It's amazing how some people were affected differently. Like me, to my knowledge it hasn't bothered me. But we have one gentleman in the fire department that was exposed and was put in the hospital. He had a lung collapse, and he's still out on disability. Within the first few minutes, we were dispatched because a train had possibly hit a building; but within the next few minutes we received another call that sounds like there's been a train wreck, there's a 'bleach' smell, and it is foggy. That information was never relayed to the first responders. The Reverse 911 was mentioned but it was not implemented until hours later. With reverse 911 you can call houses and tell them what you've got, that you need to evacuate, what you need to do, and that wasn't done for hours. In the very first minutes, I gave the order we need to have a major evacuation, and we ended up having a major evacuation, four hundred-plus people. Also, Reverse 911 was told to be implemented. That's two things that weren't done in the close timeframe. But again, it's kind of open for discussion. If the Reverse 911 was not working, if you don't tell the people which way they need to go, they would have come right down into the chlorine gas."

Rodney Cooper was the acting animal shelter manager during the Graniteville train derailment and cleanup operations. "We went in to rescue the animals, and later removed the animals as necessary for their health and wellbeing. As the hot zone started decreasing in area, we were allowed by other groups to go in if someone had an animal in their house, and they could not rescue the animal(s), they came to us. They signed a release for us to go into their house, and left their house keys for us to get into the residence. That is, we went in to the decreasing hot zone at their request, to either pack the animals up, and get them to safe area, or feed and watered their animals and make sure the animals were all right. We did find a few animals in the houses that had died."

Rodney Cooper came up with exact count of how many animals the Aiken County Animal Services provided services for: dogs and cats 338, birds 20, hamsters 2, ferrets 1, rabbits 1; 27 where dead on arrival).

Cooper said they set up their operation in the parking lot of the high school which is located about four miles from Graniteville. "It was a fenced-in area, and we set little cages up where we had some shelter. As we picked up the animals, we brought them back here. Anyone who had animal they wanted to claim, could come to the high school and try to find their animal(s). If they found their animal, the animal was returned to them along with a 'Vet Check' to make sure everything was fine with the animal. We had a lot of volunteer help from the Charlestown Police Department animal control division, and from the Lexington County animal control division. They called us, asked us if we needed help, and we told them we would take any help they could send. Basically, we borrowed their staff. Charleston sent two officers and two vehicles, while Lexington sent three officers and three vehicles. Also, the Animal Control Officer for Aiken City was also here. We had a very good response. If the volunteers were from out of town, we teamed them with a local officer who knew the Graniteville area so they would not be going in blind.

"Once they deemed the outlying area as no longer a hot zone, we were allowed in, which I believe was about forty-eight hours after the actual rail wreck. However, prior to that time, some of our officers had to put on air packs and other protection because they were near the hot zone, but this was only one episode. However, once we were allowed to go into the former hot zone, about the third or fourth day of this incident, we were able to check residences where owners wanted us to go, and check the roads and back lots for dead animals. At this time, we had a couple volunteer veterinarians come to our staging area to check the animals as we brought them back from Graniteville. Once the animal was cleared, they were unloaded and put in carriers for their owners to claim them. If they were not claimed that day, we took them back to our shelter, and the owners were notified to come to claim their animals. Most people picked their dogs up pretty quickly. Each day we probably did not have more than 15 animals that we had to take to the shelter, and most of them were claimed the next day, so we where not overloaded with animals. We had dog food donations from citizens, friends, and a huge donation from the Pedigree Dog Food Company that sent tractor-trailer load that somebody told me consisted of forty pallets of dog food. With a lot of volunteer help who came from all around the area, it was hectic but it was a unified chaos. It went well, compared to what it could have been."

Captain Greg Bailey is a shift supervisor with the Aiken County Emergency Medical Services (EMS) who was on duty the night of the rail wreck in Graniteville. He is also a member of the Aiken County Haz-Mat Team, so he had a problem. "I responded to an EMS page, but on my way to the staging location, I had to make a decision which way I went. I called my supervisor who was on twenty-four hours of duty, and asked him which way I should go. Do you want me to go to EMS or Haz-Mat? He sent me to Haz-Mat, and I went down to join Ed Shuler, the technical resource person for the Aiken County Haz-Mat Team, for an entry into the hot zone in the Graniteville incident. We found one injured survivor, who was in a wrecked automobile near the railroad tracks surrounded by damaged rail cars. I came back, made my report to the central command. I was then reassigned back to emergency medical services (EMS), which was in the middle of a process of doing multiple tasks. One thing we had was a couple of ambulances assigned to go in to an area with Haz-Mat technicians in Level B (liquid splash protection suit for hazardous chemical emergencies), another task was readying all the precautionary stuff we needed; also, we had ambulances removing injured, and our people were active in running two triage areas and two decontamination sites. In the decontamination sites, everybody who appeared there was basically stripped of their clothing, everything but their underwear, then run through showers, and on to a medical setup in case they needed to be transported to a hospital.

"At the site located at the University of South Carolina at Aiken, by the time I got there they already had a hospital tent set up from Fort Gordon in nearby Georgia, as well as medical personnel from our trauma center. Across the street, there was Aiken Regional Medical Center which had set up another decontamination and triage site. Before anybody went to the hospital, they had to go through one of our decontamination stations. My original assignment was working at the decontamination station, washing anybody who needed to be washed. We were dressed in Level B protective clothing and SCBA (self-contained breathing apparatus). A lot of people said, 'I have all ready been home, where I took a shower.' I explained to them, 'If we don't wash you, you are not going to get medical treatment. Drop your clothes right there!' Some people said, 'You are not getting my clothes,' and left. Actually, they were able to take their clothes off in privacy, put their valuables in a bag which they got back after they showered, and then they went through the showers. We did *not* give them back their clothes, which were bagged for later disposal. When they came out of the shower, someone would catch them, put them in a blanket or sheet, than another person through signs and symptoms would determine if that person needed medical attention or not. If they needed medical attention, we sent them to the medical tent where there were multiple doctors treating respiratory conditions. If they did not need medical attention, we sent them to a gymnasium. A few persons just refused to go through the showers for a number of reasons; their excuses ran from, 'I just don't want to do it,' to 'I already had a shower,' to 'I don't want to lose my new clothes.' We made it very clear that we considered their clothes contaminated, but we had to set up a new line for less than 25 persons, men or women, who would not go through the showers; even though we put 350 people through the showers.

"About that time, I was relieved and then placed in an ambulance crew, and later I went into what I believe was called 'the warm area' where we used the Cobra Team bus to evacuate people from these areas. That is, we took a few homebound persons to the top of a hill to avoid the chlorine gas which remained in the low valley. I remember a quadriplegic; we went in and picked him up in what appeared to be a fairly safe area. In an area near Belvidere, we met a private ambulance coming from Augusta, Georgia. Since we were still suited up in protective clothing, we transferred the quadriplegic to the private ambulance, and they took him to the Veteran's Administration hospital because the patient's home area was thought to be contaminated and under quarantine. Over all, it was a very large effort, and EMS people were there through out the whole procedure. I went home that night about 7 because I had to come back for my twenty-four-hour shift the next morning. For a week to ten days, since I had responsibilities for both EMS and the Aiken County Haz-Mat Team, I did not get much sleep. I would go home, get a few hours of sleep, perhaps change clothes, and then take an assignment down at the Command Center. I took only one assignment for the Haz-Mat Team, other than the one mentioned before, but took assignments for EMS a number of times because they were required to have a supervisor on duty. Usually, I went to the Forward Command Post at Honda Cars of Aiken located on Route 1 to make sure we had enough ambulances standing by, and to make sure that anyone who went into the hot zone got a medical examination before they entered, and another examination when the came out. We had private ambulance companies there helping us out, and we went on several days, so my wife did not see me much for a week. About this time, I helped the Aiken County Haz-Mat Team go into the former hot zone to find the body of the last casualty. It was not a pleasant site! Everybody worked in this tragedy; you helped anybody and everybody helped you. If you could help, that is what you did."

When asked about triage, Captain Bailey responded, "We use a standard system of triage. Basically, the worst get treated first. I missed the worst patients (Red Color) since at the start of the incident, I was detailed to the Aiken County Hazardous Materials Team to enter the hot zone in protective clothing and SCBA (self-contained breathing apparatus). Red is for priority patients, patients who could die if they don't get treated first. Yellow is what we call 'Delayed Patients,' they need to be transported to medical care, but they can wait; they are our Priority Two. Green patients could be the walking wounded, or persons with cuts and bruises. Black patients are fatalities, or wounded so much they probably can not survive. Some of these decisions can be the hardest you can possibly make. In a mass casualty incident, you might have to make decisions on who you are going to save, and who you are *not* going to save. You might spend your efforts trying to save someone knowing you cannot save him, and your efforts could be used to save another person."

To change the subject, Captain Bailey mentioned a few people who had no idea they were in the middle of an area where a chemical warfare agent was killing nine persons and sickened hundreds of others. Bailey and Ed Shuler of the Aiken County Haz-Mat Team were dressed in protective clothing and SCBA on their initial tour into the hot zone. "We had walked about a mile into the hot zone with a one-hour air supply around 4:00 in the morning, when a gentleman parked his vehicle beside us. He was cussing us out as he began walking down the street, then he stopped, waved his hands, and said, 'You need to leave.' About that time, a woman pulled her vehicle to a stop, opened her doors, and cried, 'What's going on here?' I told her she needed to get the hell out of here. My assistant told me later he was doing an original search in a mill building along with EMS and Haz-Mat personnel searching for survivors or dead bodies, when he noticed a guy come in carrying his lunch bag. He asked the guy with the lunch bag what he was do in this building, and the worker told him he was coming to work, like he did every day. Since the whole area had been evacuated by law enforcement personnel manning road blocks, the able worker admitted he had driven around a number of roadblocks on his ride to work."

Ed Schuler, CHMM, works fulltime for the Savannah River Site in Aiken County and is also a Lieutenant with the Aiken County Hazardous Materials Team with duties as Technical Resources Officer. He also holds a Bachelors degree from the University of Maryland, and was originally a sailor and instructor in the Navy's nuclear submarine program. "On January 6, 2005 around three in the morning, my pager goes off and the message was something about a possible train derailment," says Ed Shuler. "We first responded to a staging area that was on the other side of town, then we got a message saying we were staging at Honda Cars of Aiken which was just down the highway. We responded in our personal vehicles, and our Haz-Mat truck was already there. We told the party there we had instrumentation on the Haz-Mat truck that could transmit radio signals. We were told to go down close to the hot zone because they really did not know what was going on down there. We drove down Canal Street right after Route 1, got out of vehicle, and soon realized we should not be there. The chlorine gas was very strong at this point. There was a firefighter coming out of the hot zone showing sign of severe respiratory distress, so we called an EMS unit, and they administered to the firefighter. We then situated some of our gas monitors, and then took some back roads to get back to Forward Command.

"At that point, we did not really know what we were dealing with. When we started our entry into the hot zone, I believe we reached ground zero around 5:30 a.m.; that is, nearly two hours between the actual accident and the start of our recon mission. That sounds like a long time to a lay person, but we had to suit up in protective gear, and make sure our PPE [personal protective equipment: equipment provided to shield or isolate a person from chemical, physical, and thermal hazards that may be encountered at a hazardous materials incident; and should include protection for the respiratory system, skin, eyes, face, hands, feet, head, body, and hearing], an electronic Geiger apparatus, four gas monitors, and two radios with one for my partner [Gregory R. Bailey, Shift Manager for EMS] and one for me. It was still dark, warm and foggy, somewhere about 50 degrees F. We had to huff a long way since we did not know exactly where the hot zone began. It was perfect temperature, calm with some slight breeze when we were at the bottom of a hill, the low point in our journey, when we climbed back up to go across a little bridge. There was a slight odor here, but we could not identify exactly what it was. We saw a man who said, 'My son is back at the house,' and I told him he better get the hell out of here."

Ed Schuler and Greg Bailey continued to walk down the pitch dark street in dense fog, crossed the railroad tracks, where they began to hear the crossing light at an intersection going ding, ding, ding, ding, ding, ding, ding in an endless rhythm that haunted that night in Graniteville, South Carolina. "We came down a slight hill where the road crossed over the railroad tracks before going in a straight line again, and the bell going ding-dong seemed to get louder and louder," says Ed Schuler. "I looked over to see this huge oak tree laying down on the highway surface. It took me a few seconds to realize there was an auto-mobile captured and held within the tree branches. I went over to the car but could not get the doors open, and beat on the windows in case there was anybody alive in the car. There was a live man in the car, but suddenly I seemed to be covered with white powder; I thought it was sodium hydroxide powder, but that's not the way they ship it. I told the guy in the car to roll down a window, and I told him that we would get him out of the car. We called a rescue team and got a response; these guys were a block away in a pickup truck, and they were successful in getting a rear window out of the guy's car and extracting the victim.

"We then continued the reconnoitering mission, but when we found the derailment, we could not see any placards, or any other signs of identifying information in the jumbled mess. We did see fire burning with smoke apparent. When we got down to the scene, we has plenty of air; but working with the victim in the car, I ran out of air a bit faster than I antici-pated, slipped in white powder, fell backwards down a short distance, and ended up with crap all over me. My electronic Draeger tubes [Draeger tubes are designed to detect specific compounds in the air. Users will draw a specific volume of air through a tube with a Draeger pump, and then read a color change reaction against a series of quantifying markers on the tube. The glass tubes are filled with one or more substances that undergo chemical reactions in the presence of specific chemicals or types of chemicals. The basis of any direct reading Draeger tube is the chemical reaction of a measured substance with the chemicals of the fill-ing preparation. This reaction will result in a color change which the user can identify and therefore quantify the amount of the measured substance. Each tube has specific interfer-ence that will indicate a positive reading for the compound of interest. The user must have a general knowledge of what the compound is to minimize the number of separate tests] were so coated I could not read it. I also lost my radio, so my partner called for a pickup, since I knew we could not make it out of the hot zone if we had to walk back a mile because of our limited air supply. While waiting for the pickup truck, we crossed to that little park on Canal Street, across from the data processing center that serves the mill company. A lady drove up to where we were standing, and we told her to leave this area. The pickup truck drove us back to Forward Command. It now was about daybreak.

"A little after daybreak, we formed a team with three mill company personnel and brought them down to the hot zone so they could shut down the different mill plants. We dropped off one team, and then went to another building. I saw in the supervisor's office a man sitting in a chair who did not look well at all, and was probably dead. I went into the break room and saw a guy lying on his back, and I went over to check him out. I could not get a pulse. It was now getting light; I say its now about 6:30 a.m. to 7:00 a.m., and I tried to call the other plant, sixteenth of a mile away, where we had left a team. I could not get them on the radio, I could not reach Command, and I could not reach my team. However, I had the victim's cell phone, so I called 911. I reported our status, that we had found victims, and gave a general status report. The operator could hear me, but I could not hear the operator. We got back in the car that took us back to pickup the other team, after which we went back to Forward Command."

Ed Shuler was of the opinion that people from out of town were quick to arrive in Gran-iteville to offer their help in every possible way. "We suddenly had aerial maps, detailed street maps, and every thing was in good order. The FBI, the railroad investigators, and

the environmental monitoring teams from the railroad must have been dispatched immediately, within minutes of the crash. The railroad had some type of alert system, and they
got teams airborne at once. Their monitoring team was also flying in, and they had remote
monitors all over the place, which was pretty amazing. They had capabilities for monitoring that nobody was tapping into; and they were monitoring for possible liability that
might be directed to the railroad, Norfolk Southern Corporation."

On Saturday or Sunday, Ed Shuler was put in charge of a thirteen-man search and rescue team because there was one individual that was not accounted for. "They went through
the accountability system, knew where everybody was except for one mill employee. We
went to the Woodhead Division of the mill, and the fire department said we should enter
the building in Level B protection since the building was very near the incident location.
We split our team into two groups, and each group had one member who knew the building. One group went clockwise around the building, and the other group went counter-
clockwise. My group did a thorough search, and came out. The backup team was chomping
at the bit, and went through the building slowly checking every possible spot for a missing
body. Just on their way out, someone found the body underneath some equipment in a
dark spot within the building."

"After a number of days, there was a lot of back and forth in the newspapers between
the Sheriff and the Fire Chief accusing one or the other of wanting to be the overall commander of this incident. The Sheriff took overall control of this incident, possibly saying,
'This is the way it is going to work.' It was politics in a small place, and in fire departments,
everybody wants to be in command. The county was not too keen in this decision, and
there was a lot of dissent. On a Friday, at one point in time, while railroad crews were trying to re-patch a second chlorine car when the first patch had failed, I was in charge of Haz-
Mat Command for a short period of time. Someone said there was an evacuation order in
effect, so I called Incident Command to ask them what was going on. Basically, Incident
Command did not know about the evacuation order being in effect. I told all my Haz-Mat
crews to return to base. We found out later that the railroad repair crews, because the first
patch had blown off, did not tell anybody working on the site, and did not tell Emergency
Command. That lack of attention and coordination could have killed somebody."

Avondale Mills lost six valuable employees, an untold number of other employees were
injured, clean-up work lasted at least seventeen months, and many trusted and qualified
employees were, as a result of the deadly incident, laid off. On January 6, 2005, Avondale
Mills had 183 employees working the third shift that night in five different buildings, all
located in a depression by a small creek. About 2:40 a.m., that whole area became a hell
hole of suffering humanity. Chlorine gas began issuing from a wrecked tank car with an
amazingly strong chemical odor; survivors described it as strong bleach odor, some said
was a green fog that enveloped them and made them lose their way, and some mentioned
they never knew a toxic gas was present until they felt it. And many people *felt* it. Their
eyes and noses ran with fluid, it stopped their breathing, and each of the breaths they could
get was no blessing because it burned their lungs. It was a world of hacking, coughing and
vomiting like no other world experienced before. It was not a nice way to give up a life as
nine persons learned. It was the worst train wreck in the last thirty years.

Employees in the difference plants were initially able to call supervisors and report
what was going on, and then call 911 for an emergency in progress. However, if you wanted
to escape from the cloud of chlorine gas, you were pretty on your own. No significant
rescue efforts were attempted in the hot zone for hours since law enforcement officers set

up an evacuation zone and installed armed police on roadblocks around the border of the hot zone. Some workers, even one's who had made it out of the hot zone on their own, went back in to the hot zone to rescue other workers. Such conduct is quality behavior of the highest order. Almost no one knew what the gas was; they just knew that it was killing them. Some workers believed it was a chemical used in the mill for normal production. Actually, Avondale Mills and its employees, Graniteville and its citizens, and Aiken County had absolutely nothing to do with responsibility for this tragedy. Yet, they suffered without end.

Avondale Mills had an absolute tragedy to recover from, although the company had no responsibility for such a tragedy. The company's computer system was destroyed by corrosive gas. What used to be done by computer, now had to be done with manpower. This included raw materials, cloth and yarn, payroll preparation, plant scheduling, check writing, cost approval, and ordering had to be done by hand rather than by computer. Both suppliers and customers were easy to deal with in a quick rebirth of Avondale Mills and their mills in Graniteville, but it would take months to repair all the damage done to the various plants. Due to major corrosion, a large number of electrical and metal components had to be replaced. The railroad paid for a contracting company to come to Graniteville to work with mill personnel to do a general cleanup of the property.

However, early in October of 2005, Avondale Mills began laying off up to 350 workers due to apparent problems cleaning up corrosion left from the tragic incident that happened nine months earlier. And the situation worsened. This author visited the mill in January of 2007 and was shocked by what I saw, or rather didn't see. The many large trucks with "Avondale Mills" printed on their sides were all gone. The site appeared to be deserted. The lots were empty, and there was no one walking about on the mill grounds. Whether the plant was purchased or not, I cannot say. After eighteen months of makng so-called repairs, and a year of trying to fix the damage on their own, the company may have decided that the plant was unsatisfactory for their manufacturing purposes.

The Human Costs of Chlorine Gas in Graniteville, South Carolina

Deaths

Rusty Rushton III, employee of Avondale Mills, age forty-one, Warrenville, found dead on loading dock of Stevens Steam Plant where he was a Boiler Operator with twelve years at Avondale. He was a 1982 graduate of Midland Valley High School, a native of Aiken County, loved riding motorcycles, and cared greatly for his three animal buddies. He was a friendly person with an expansive personality, somewhat of a cutup and local wit with a keen sense of humor, and a free spirit. People also said he knew his job very well, and was willing to help anyone with a problem. The father of five children liked Jimmy Buffet songs, and loved beaches. At his funeral, family and friends played a Jimmy Buffet song and later family members placed a small amount of his ashes on a favorite beach.

Willie C. Shealey, employee of Avondale Mills, age forty-three, of Graniteville, was found dead in low, swampy woods near the Woodhead Division Plant. He was the Shift Third Supervisor and had been with Avondale Mills for eighteen years. Mister Shealey was also a member of the National Guard 122 Engineering Battalion in Graniteville, South Carolina. He was a comer in the plant in all respects, and rumor has it he could do any job in the Woodhead Division Plant with ease. Shealey had been one of the first winners of the of the Corporate Zero Defects award at the Woodhead Plant, and he was so good at handling personnel problems that people on the day shift wanted to come to work with Shealey who worked from midnight until eight in the morning. People noticed he did his job very well, wanted production quality, and cared about his fellow workers. He was also a horseman, and belonged to a local riding club. He didn't run from the gas attack on home ground until he had shut down the production machinery, and helped other workers to flee the plant. Shealey finally left the plant with John Laird, Jr. They attempted to help each other outrun the toxic gas cloud, but were unsuccessful. Sometime later, their bodies were found together.

John Laird, Jr., an employee of Avondale Mills, age twenty-four, of North Augusta, South Carolina, was found dead in low, swampy woods near the Woodhead Division Plant. Employed as a Lead Machine Operator, Laird had five years with the company, and had a reputation as a reliable young person with a serious respect for the work he did, who had recently been promoted. Laird was regarded as a good-hearted man who would do any-thing for any person. His new job was a position of responsibility, and he handled it well. He was dedicated to his work, and had the respect of his co-workers of all ages. He was also a good listener, and very mature for his age. If Laird made a mistake, he would not do it again. John Laird, Jr. loved to work with his hands, and was a fan of NASCAR racing.

Steven W. Bagley, an employee of Avondale Mills for two years, age thirty-eight, worked at the Gregg Plant as a fork lift driver who loved to fish and hunt. He lived in Augusta, Georgia about twelve miles from Graniteville, and became a father of a baby son ten days before the Graniteville railroad tragedy. After the incident, his body was found in the break room of the Gregg Plant. Mister Bagley was a hard worker with a cheerful attitude, by all accounts, and his constant smile could light up a whole room.

Allen Frazier, an employee of Avondale Mills for thirty-six years, age fifty-eight, lived in Ridge Spring, South Carolina and worked as a Supervisor in the Gregg Plant. He was a graduate of Midland Valley High School, and a lifelong resident of Aiken County. Mister Frazier's body was later found in an office of the Gregg Division. He was known to be an avid fisherman, and worked at Avondale Mills for his entire adult life. He was well known for his dependability and dedication at work, and for his caring attitude. He was a quiet and soft-spoken man who actually listened to anyone who might have a problem. A private person, Mister Frazier enjoyed spending time with his family, was known to be a good cook, and seemed to be highly regarded by workers and other supervisors. He also had a son who works for Avondale Mills.

Willie Lee Tyler, an employee of Avondale Mills for thirty-four years in the Wood-head Division, age fifty-seven, of Aiken City, was a Chemical Controller grinding pigments and paints in the textile unit. His body was not found until the third day of this tragedy, and his death was the sixth and final fatality of employees working for Avondale Mills. Mister Tyler was a deacon at the Sardis Baptist Church in Salley, South Carolina, and a well-known gospel singer active with the groups Fantastic Melodairs, and Deacon Wil-lie Tyler and The Mighty Gospel Jewels. He also had appeared a number of times on the local Sunday morning television show, "Parade of Quartets." During the year 2000, Willie Lee Tyler was Avondale Mills' nominee for the South Carolina Manufacturing Alliance's "Manufacturing Citizen of the Year." He was an experienced worker, as well as a person who was easy to work with or for. Willie Lee Tyler was always in a good mood, and a joy to be around, no matter what might be going on.

Christopher Seeling, a railroad engineer for Norfolk Southern Railroad, age twenty-eight, was born in Fort Wayne, Indiana and lived in West Columbia, South Carolina due to his job. He had employment that he figured was the best job in the entire world. He loved his job. He graduated from the B.S.N.F. Railroad Academy at Johnson County Community College, and was secretary/treasurer of the local Brotherhood of Locomotive Engineers. On January 6, 2005, Seeling was to die through no fault of his own. The day before, another crew working for Norfolk Southern Railroad made a tremendous, unjustified error by forgetting to change a manual switch for a railroad siding back to the main line track. When Seeling came through Graniteville at 47 to 49 m.p.h. (a legal rate at the time) about 2:40 a.m. on January 6, 2005, with a three-engine train towing forty-two cars loaded with general freight; he probably had about four seconds before he hit an unmanned locomotive and two attached cars on the siding track. The engineer and his conductor were not killed in the crash. They found a mixture of fog and chlorine gas that blinded them, but they knew they were in Graniteville so they blindly followed the railroad ties between the tracks until they found the fire chief, and group of Avondale Mills third shift employees trying to save themselves. Seeling reported to Fire Chief Napier of the GVW Fire and Rescue Department that he had a head-on collision with another train, that there had been a chemical leak, and that he and his conductor could not breathe. The fire chief called an ambulance, and engineer Christopher Seeling died in a hospital some hours later. His conductor was seriously ill for a long time, but survived.

Joseph Stone was a long-haul truck driver, age twenty-two, who lived off-the-road in Sherbrooke, Quebec, Canada. His sleeper-truck was a "Freightliner" which pulled into Avondale Mills too late for unloading on January 5, 2005. Stone, also known as "Rolling Stone," his CB radio name, drove for J.W. Express located in Deauville, Quebec. He called his girlfriend at home in Quebec to tell her he would be held up for the night, and then went to bed in his truck. Runaway chlorine gas killed Joseph Stone by inhalation while he slept on Leitner Street in Graniteville, and he may never have known what killed him.

Tony L. Deloach, age fifty-six, was an invalid who lived in a house on Main Street in Graniteville, He was an Ex-Marine who spent time in Vietnam where he lost much of his vision. He also had emphysema and pain in his knees. His home was near, or in the hot zone, and he died from inhaling chlorine gas. His body was buried in the Graniteville Cemetery with full military honors.

In the nine deaths listed above, it should be noted and understood that all nine persons died from inhaling chlorine gas. They may have had other injuries, but the cause of death was inhalation of chlorine gas.

Injured

In this incident, more than 550 persons sought medical attention, and 75 people were hospitalized in six area hospitals. The South Carolina Department of Health and Environmental Control (DHEC) identified 72 persons who were hospitalized and 525 persons who were treated in hospital emergency rooms or in private physicians' offices after the train crash and chlorine spill in Graniteville on January 6, 2005. In the days immediately following the incident, this agency interviewed 280 people who were known to have received medical care. These interviews were conducted in person at a hospital or by telephone. Of these people, 54 were hospitalized and 226 were treated as outpatients. Their symptoms included the following conditions: Increased or new coughing (81 percent), burning eyes (76 percent), shortness of breathe (73 percent), headaches (62 percent), chest pain (58 percent), nausea (53 percent), nose burning (52 percent), coughing up phlegm (49 percent), choking (46 percent), dizziness (42 percent), and vomiting (34 percent).

In June of 2005, DHEC mailed a follow-up questionnaire to 280 people who were interviewed. Of the 94 respondents: 23 percent had been hospitalized, 83 percent still were experiencing symptoms they felt related to the chlorine spill, 52 percent were taking medication for problems they felt were related to chlorine exposure, 51 percent were under a doctor's care for problems they felt were related to chlorine exposure, and 48 percent screened positive for post-traumatic stress disorder.

Heroes of the Moment

About 5,400 residents and workers in Graniteville were evacuated from normal pursuits in the one mile radius hot zone, which had to be broadened to two zones, Zone A and Zone B. Some people had to stay out of downtown Graniteville for up to twenty-one days. Their homes were examined, sometimes multiple times. They were lucky if they could take the clothes on their back when they were evacuated for as much as three weeks. These people were heroes of the moment—all of them.

In the first minutes of pandemonium, one of the workers in the Hickman Plant, part of the Avondale Mills company, started calling the emergency number 911. When he finally got through to the 911 Operator and reported the situation at Hickman, the 911 Operator told him some unpleasant news. Basically, the police knew all about the disaster, but they were not going to send anyone into the hot zone. The employee did the right thing; he called his department manager, John Albright, at Albright's home. When John's pager went off, he was sound asleep one moment and wide awake the next, as he listened to the caller's report. John Albright also did the right thing; he drove into the chlorine gas in the hot zone to save his employees. However, Albright was thinking he was headed toward a 55-gallon barrel or drum leak, not a very sizable poison gas leak without end. He was coughing, his eyes stung, and he was spitting constantly. He could see where the railroad engines had challenged one another, but he could not see the wrecked railroad tank cars due to the ever-present gas. Unable to drive directly through the gas and fog, John Albright drove around the Hickman Plant in another direction, and found his workers. He loaded all nineteen people into his pickup truck, reported to his supervisor by telephone as to the condition of his employees, and took them all to his home. There, Albright, his wife, and mother did a quick triage maneuver to learn how serious the nineteen workers might be; in the absence of any ambulances, they then used automobiles and vans to take to take nineteen persons to hospitals.

Those nineteen people had some adventures of their own before John Albright arrived. While Melvin Scott called the emergency number of 911, and then called John Albright; Clifford "Bubba" Hastings and Dewey Thompson helped organize the evacuation of the Hickman Plant. Forgetting nobody, and staying together, the nineteen workers left Hickman to seek their fate, good or bad. They soon found the railroad engineer and the train conductor who were both in bad shape (the engineer would die in a hospital within hours, and the conductor would survive after long hospital and home stays). They continued their trek with Bubba and Dewey lifting and supporting the engineer and two other workers responsible for the conductor, as they continued their journey toward clean air, they hoped.

The mill employees were suffering more than before, and just about this time a car came through the gas and fog, and a lady popped out. "I am a nurse; can I be of any help," said Brenda Montgomery. Within a few seconds, she was gone with three injured men, including the conductor and the engineer. By now, the engineer begged for more air, so Brenda drove with all windows wide open to be one of the first vehicles to arrive at Aiken Regional Medical Center after this incident.

At the Gregg Plant, which covered 13 acres and held 120 workers on the third shift who were there that night, many employees were able to get to their cars in the parking lot and drive away. The poison gas outside Gregg was much worse than the gas inside the building, so it took a lot of courage to run through the parking lot. Some groups joined hands and walked or ran to the packing lot, dropping off employees when individuals got to their parking spot; others just got the hell out of there, hoping they knew where they parked their cars on this particular night. John Tillman evacuated his people to the parking lot, and waited while they drove away; then he drove home to his residence. He was worried about other workers, so he went back to the Gregg Plant, running at least one police road-block in his process. Mr. Tillman found some people still in the process lab, who were unaware of the escape route through the Gregg parking lot. In another room, he found a man who could not breathe and could not stand. He exited the building again, picked up his truck, and positioned the truck near a door in the Gregg Plant. John then found a machine in the plant, knowing he could not carry the man who could not walk, and used the machine to transport the injured man to his truck. Two other persons needed help, and John put the three of them in his truck, and drove out of the Gregg plant parking lot as he looked for a safe position where the chlorine gas would be less severe, finding a better situation on the top of the hill near the Townsend and Swint plants. The road-block was manned EMTs who treated the three injured men with John. John Tillman then heard information about another group who had who had gotten out of the Gregg plant but could not fight through the chlorine gas, or could not locates their automobiles. This group had called 911, but where told to stay where they were. However, nobody was coming to save them! Tillman begged for help from authorities, but nobody would be allowed to go into this toxic cloud until authorities knew what gas was present. It was a lock-down situation, and no one was going to argue with authorities, except John Tillman who was not going to let workers die when they could be saved. John Tillman knows his way around Graniteville, and drove down what could be described as a cart path, avoiding road-blocks, and entered the hot zone without being discovered. He merely honked his horn at the Gregg plant and flashed his lights, and he was immediately joined by seventeen survivors. All of them somehow got into John's truck and were driven to a safe area. John Tillman ultimately saved twenty workers.

Michael E. Hunt is the Sheriff of Aiken County. This county has a total land area 1,073 miles, which is the size of the State of Rhode Island, and a population in the county of 140,000 citizens. The total annual county operating budget recently was $108,654,553, and county employees numbered 854. The Aiken Sheriff's Office employs 106 sworn deputies and 38 civilian employees who provide patrol of a wide area within Aiken County. The Aiken County Sheriff's Office also operates a state-of-the-art Drug Identification Laboratory which serves all Law Enforcement agencies in Aiken County, as well as surrounding counties. Career paths open to sworn officers include Uniform Patrol, Criminal Investigations, Narcotics Investigation, Crime Scene Investigations, Civil Process, Juvenile Services, School Resource Officer, K-9 and Bloodhounds, Special Operations, Training, and

Dispatch. The Office of the Sheriff in Aiken County is a well-run organization with an esprit de corps.

Sheriff Hunt had been in office about three years when the Graniteville tragedy occurred. Just like Graniteville Fire Chief Phil Napier, Hunt is a Graniteville resident. Both men have behaved honorably at the time of this incident and the months that followed. There was some anger in town that an early call for "Sheltering-In-Place," and road-blocks set up by police, resulted in very limited rescue being done at first, and possibly increased deaths and serious injuries among the 183 staff who were working the night shift at Avondale Mills. However, a third-party, after-action report conducted by Westinghouse Savannah River Company, and the U.S Department of Energy, claimed the emergency response was well organized in general and done safely.

"In this state, the sheriffs are very involved in local jurisdiction in Homeland Security," according to Sheriff Hunt. "We started very quickly to becoming involved, and trying to build bridges with other agencies; such as, Emergency Management, Emergency Services, FBI, and Law Enforcement. We wanted people to come together and talk, because the sheriff's office had never been involved in planning or dialogue with these other agencies. As a result, this made us successful in this major accident, because we had been people at a table. We made it very clear that we didn't care about turf battles; because we had an obligation to the people we serve. We all should put turf interests aside, and plan together. We became very involved in planning, and we received a lot of Homeland Security money in this state.

"Our State Law Enforcement Division disperses the money, but each county needs an assessment. The purpose of this is that the Sheriff, Police Chief, Fire Chief, Emergency Medical Services, and Emergency Management, and a few other agencies have to decide the needs of each agency, Homeland Security plan and response, as well as the actual needs of each agency, and the manner of each agency about how they are going to spend their money that was given. All money given out includes a packet of information. First of all, for the first couple of rounds, we purchased equipment that we thought would be beneficial to everybody in a county, and we pre-staged massive amounts. In Aiken County, we really went out of the box, worked very closely together. As this Graniteville train wreck occurred, all that pre-staged equipment responded to the staging area and assisted in this fourteen-day operation. As an example, every police officer in South Carolina has an emergency kit including masks, Tyvek suits, and a respirator.

"The early morning of the alarm, the Sheriff's office received Command through our pager that said there was a train accident in downtown Graniteville, with mass casualties possible," says the Sheriff. "I live in Graniteville and called the dispatcher asking, 'Is this crash near the train track-switch in town.' The dispatcher told me it was at the switch, and that two trains had collided. We immediately activated a number of actions, including quickly our Homeland Security response enforcement division, notified the FBI, and had them all responding to our staging area because we took this act as a possible terrorist action or a Homeland Security attack since it involved a railroad switch at 2:34 in the morning. We did a full recall of Sheriff Personnel that included dispatchers, and sealed the area very quickly, in about twenty minutes.

"When I came out of my house, the report was we were getting a green cloud and a smell like bleach which was irritating to the lungs of bystanders. We controlled the sick people first, getting them to decontamination centers. At decontamination, all of our supplies and equipment were funded by the Homeland Security project. We set up a Unified Command where the Fire Chief, Emergency Management, Emergency Services, and everybody involved in this action, so that Unified Command became larger because people from all over were dispatched to the scene.

"We also received during the first day of the incident our Homeland Security PODS (truck boxes without wheels filled with badly needed incident supplies and equipment that stay with response teams for days at a time). The PODS were shipped in from Columbia, South Carolina, and were very impressive in the equipment and supplies they contained. During the first day, we held an orderly evacuation (The Sheriff's office directed the mandatory evacuation of those who has been sheltering-in-place within a mile radius of the accident scene, which amounted to as many of 5,400 persons. Some people had to wait eight days to return home, while others had as long as fourteen days).

"A curfew was used about seven days strait," adds Sheriff Hunt. "The curfew (an order establishing a specific period of time, usually at night, during which certain restrictions apply, especially that no unauthorized persons may be outdoors or that places of public assembly must be closed) was from dusk to dawn. We also had a No-Fly-Zone so helicopters would not get too close to the incident scene.

"When I took office as Sheriff, in South Carolina law enforcement agencies never participated much in incident command. As a result of 911 and the presidential directive, Incident Command and Unified Command training was a mandate. As an example, we spent a lot of money on our SWAT team (special weapons and tactics) paid for by Homeland Security, and the morning of the event we activated our tactical team. We sent half of that team to our staging area, and the other half of the team was sent to the other end of the county. Also, in this state, there is a regional COBRA team that is responsible for dealing with biological and chemical incidents, who were very helpful in the early stages of the Graniteville tragedy." (The large Wackenhut Savannah River Site near Graniteville could be a site of terrorist interest.)

The mission of Aiken County Hazardous Materials Team includes protection of public safety, protection of the environment, protection of property and assets, and safety of all emergency responders. Besides dealing with typical emergency response challengers, the team intervenes in chemical, biological, and radioactive incidents. The team is trained for weapons of mass destruction (WMD) incidents. Strict guide lines of the Occupation Safety and Health Administration's (OSHA) regulation 29 Code of Federal Regulations 1910.120 are followed, which include an Emergency Response Plan, Chain of Command, Training, Medical Surveillance, Personal Protective Equipment, and Post-Emergency Response (PPE).

The Operations Captain for the Aiken County Hazardous Materials Team is Fred Wilber, who works a fulltime job at the Savannah River Site. "What the title gives me is day-to-day operations of the team. My job in an emergency is in Operations, within the Incident Command structure. I approve the Decontamination Officer, based on training records with all the training up to this year, that they've have passed their medicals, wear protective clothing, and other responsibilities; and pretty much run our Hazardous Materials Team. On the Graniteville incident, we were extremely lucky to have the Fort Gordon U.S. Army Decontamination Team from over in nearby Georgia to assist with the mass decon we had to face. Lots of federal money paid for this. This team was already there with their tractor trailers and shower systems plus personnel to run the system; this was the first time they had used this procedure, except in training. They were very organized, and knew what they were doing.

"I was working on my regular job when I got alerted for this incident by my mom, "said Fred Wilber. "We are normally alerted by pager, and it's my understanding that the pagers did go off, but I work in a very large building made of concrete at Riverside, so radio signals do not come in on the telephone. Her message was, 'There's been a Haz-Mat, and they will probably alert you.' I called my boss at Riverside and told what is going on, and he told me I had permission to leave. Out in the open air, my pager stated going off at federal site three. The pager told us we needed information on what was in the rail cars, and to deploy a team in there.

"We found the train wreck, and one of the guys walked into a couplet in the dark. The team could not see anything except a cloud near the earth with vapors coming down. When they could see more, they found a man still alive in a car trapped under a tree. On that first try, the scout team saw a car coming, stopped it, and radioed Command with a report of, 'Suspected Chlorine.' This team went in two more times, and eventually was able to take good photographs of general damage and the leaking tank car.

"I went to Honda Cars of Aiken, an out-of business car dealership, to establish Forward Command for the Haz-Mat team, and get things organized. We radioed information back to Command, which was placed in another location on Route 1, and very busy trying to set up an Incident Command system. This type of incident had never been faced before in this county. There were multiple response agencies, and we had to learn a lot quickly. We had been using Incident Command among ourselves, having been trained in this procedure, but it was a horse of a different color dumped on us really quickly. There were a bunch of kinks our first time, as well as a number of people's first time, which took us a little time to get acclimated to. Experts were doing it too, and the railroad had there own experts. We were put on shifts to monitor housing during the evacuation next morning making sure everybody was out, doing searches for unknown victims, and we had our own protective equipment which set to Red for respiratory protection needed, and Yellow for meter readings but you could still dress up if desired.

"The bodies were found the first day," remembers Captain Wilber. "Once there was daylight, bodies were fairly easy to find. The last body was found only after three entry teams had no luck. Veterinarians trained in hazardous materials and others, who were not local, were brought in to deal with pets left behind during the evacuation. They had certain drugs they needed, bags of cat and dog food, checked on the health and well being, fed the animals, and stockpiled food; they did that every twelve hours, I was told. I think they had one house where in a closet or small room they found nine Pomeranian dogs that were wild after being in there for three days."

"At the very beginning of this incident, there was a weather inversion (a reversal in the normal temperature lapse rate, the temperature rising with increased elevation instead of falling) that helped a lot because that kept the toxic gas packed down to the ground, of just above it. This tragedy would have been worse if it had occurred later in the morning with traffic moving on the streets, a lot of shift changers at the mill and elsewhere, schools opening, and everybody going this way and that. If the cloud moved that way, we would have to evacuate that way."

The Massacre at Columbine High School

2

Two hate-filled, dysfunctional youths, one seventeen years old and the other eighteen, who were masters of logistics and hated mankind, attacked their classmates at Columbine High School in Littleton, Colorado, killing thirteen and seriously wounding twenty-one. Some of the wounded will be in wheelchairs for the rest of their lives. They prepared for their unholy terror for at least twelve months, unbeknownst to all. They brought confusion, carnage, and chaos to the campus on April 20, 1999, and caused our nation to take a much closer look at our children. They dreamed of death and destruction and fulfilled their fantasies in less than seventeen minutes of killing and forty minutes total time before committing suicide by gunshot wound.

Eric Harris and Dylan Klebold staged a terrorist-style attack that was incredibly well thought out. They had brain power, and maintained top-notch security for as much as twelve months. Similar to other terrorist procedures, they planted diversionary bombs some distance away from Columbine High School to draw police away from the scene of their assault so they would have longer to kill and maim fellow students and faculty members before law enforcement personnel arrived at their killing field. Harris and Klebold made at least ninety-nine secondary explosive devices to endanger emergency responders and to add to the death toll of classmates. All told, there were thirty exploded devices and forty-six unexploded bombs at Columbine High, thirty explosive devices in the boys' two automobiles in the parking lot, and eight explosive and incendiary devices at the Harris and Klebold residences. How could two "normal" boys secretly build ninety-nine bombs without their parents or their friends ever noticing?

Of the explosive devices that actually exploded at Columbine High School, thirteen were placed outside of the school building, five were situated in the library, six exploded in hallways and classrooms, and six went off in the cafeteria. Of the explosive devices that failed to function at the school, two were placed outside the school, twenty-six were situated in the library, fourteen in classrooms or hallways, and four were positioned in the cafeteria.

When searching the Harris residence, sheriff's deputies found three significant homemade videotapes. A tape produced during 1999 shows Klebold and Harris gloating over the explosive devices they had built just before the massacre. That videotape provided evidence on twenty-one pipe bombs, and twenty-nine CO_2 cartridge bombs referred to as "crickets" or "grenades." In a tape made in April of 1999 just before the actual assault, the soon-to-be-murderers talked about making bombs and mentioned working with propane and home-made napalm. In a spiral-bound notebook also found at the Harris home, with an entry date of April 3, 1999 — seventeen days before the actual attack — states, "We have six time clocks ready, thirty-nine crickets, twenty-four pipe bombs, and the napalm is under construction." On another undated page there appears to be a list of bombs now available: it lists fifty-three crickets and super crickets plus twenty-four pipe bombs of various sizes. General information was located in Eric Harris's journal, a Day Timer notebook, and a calendar found at the Harris home after the massacre.

Actual explosives devices recovered at Columbine High School were CO_2 cylinders, 1×3 inch pipe bombs, 2×12 inch pipe bombs, and 2×12 inch pipe bombs with a pipe bomb, and two car bombs. Ignition devices included large kitchen matches doused with model rocket igniter, and clocks and batteries. Incendiary devices included two twenty-gallon propane tanks, white gas, kerosene, "napalm," one pound propane tanks, and four gallon, two-and-one-half gallons, and one gallon gasoline tanks filled with fuel. Klebold and Harris carried four knives in their gear but did not use them in the killings. However, they used their firearms in gay abandon. During the actual assault, Dylan Klebold used an Intratec TEC-DC-9, nine mm semi-automatic handgun attached to a strap over his shoulder; and a Stevens 12-guage, sawed-off double-barrel shotgun. Eric Harris used a Hi-Point nine mm carbine rifle on a strap hidden under his black trench coat or "duster," in addition to a Savage-Springfield 12-guage, sawed-off pump shotgun. In the actual assault, the perpetrators fired a total of 188 shots from these four weapons. Harris fired a total of 121 shots while Klebold fired 67 rounds.

	Harris	Klebold	Total
Shotgun rounds	25	12	37
Library	21	6	
Inside	4		
Outside	0	2	
9mm rounds	96	55	151
Library	13	21	
Inside	36	31	
Outside	47	3	
Total fired	121	67	188

Taken from "Columbine High School Shootings — April 20, 1999" of the Jefferson County Sheriff's Office Report, Golden, CO.

The killers were able to access their weapons-of-choice very easily. A friend, Robyn Anderson, was of legal age to buy guns and joined Kebold and Harris at a gun show. The eighteen-year-old girl purchased two shotguns and one rifle which were later used to kill Columbine students and staff. Anderson could not be charged with any crime since no state of federal law prohibits the purchase of a long gun from a private individual (that is, no federally licensed dealing was involved). Mark Manes sold his Intratec, model TEC-9, 9 mm pistol to Dylan Klebold for $500. In addition, he also purchased two boxes of 9 mm ammunition for Eric Harris on April 19, 1999, the day before the massacre. Mark Manes was eventually sentenced for one count of providing or permitting a juvenile to possess a handgun, and one count of possession of a dangerous or illegal weapon because he had gone shooting with the young killers and had fired one of their sawed-off shotguns.

Dylan Bennet Klebold

Dylan Klebold was the second child of Thomas and Susan Klebold who was born September 11, 1981, in Denver, Colorado. He had a brother, Byron, who was three years older. His parents told police investigators Dylan was a gentle kid who failed to give any evidence of a violent streak. In elementary school, he was assigned to the "Challenging High Intellectual Potential Students" class for gifted and talented children. Dylan was also viewed as quiet and shy, but during his early school years he was active in T-ball, baseball and soccer. At Columbine High School he continued to be shy and never had a girlfriend but he did socialize with different groups of friends. Along with some friends, he was interested in video games, midnight bowling and fantasy football games. Like Eric Harris, he was also involved in video productions and the Columbine Rebel News Network. He became a computer assistant at school and helped maintain the Columbine computer server. He did enjoy spending time with groups of other teenagers, and was deeply involved with computers. By all accounts, *nothing* in his behavior or manner indicated what he and Eric Harris were planning.

His father told police that Dylan never seemed all that interested in guns. As a high school senior, Dylan was interested in attending the University of Arizona, and had already been accepted by this university as a computer science major. In fact, on March 25, 1999, less than a month before the Columbine massacre, his parents left on a four-day trip with Dylan to pick out his university dormitory room; they detected nothing unusual about his behavior or demeanor. Earlier, on January 30, 1998, Klebold and Harris were arrested for breaking into a vehicle in Jefferson County. In April of 1998, they were placed in a juvenile diversion program. They had to pay fines, attend anger management classes and counseling, and complete the diversion program. All charges were then dropped, and the two of them were released from the program on February 9, 1999. Friends and teachers thought Klebold was a "nice," normal teenager according to police investigators. In various formats, Dylan Klebold stated some of his problems, hates, and loneliness. *"Fact: People are so unaware ...well, ignorance is bliss I guess ...that would explain my depression."* Klebold began a journal on March 31, 1997, in which he noted, *"was a weird time, weird life, weird existence."* He described himself as not fitting in, being depressed and generally hating his existence and hating his life. He talked about suicide. In a further entry, he wrote, *" swear — like I'm an outcast, and everyone is conspiring against me."* Again in 1997, he wanted to die, and mentioned having someone buy him a gun so he could kill himself. In November, 1997, Klebold elucidated getting a gun and going on a killing spree. In their junior year at Columbine High, a full year before their vengeful assault, Klebold made four entries in Eric Harris's yearbook. One entry read, *"the holy April morning of NBK (natural born killers)."* A further entry says in part, *"Killing enemies, blowing up stuff, killing cops!! My wrath for January's incident will be godlike. Not to mention our revenge in the commons.* Police theorize Klebold considered his "January incident" meant their arrest for breaking into a vehicle. The Columbine cafeteria is also called the commons.

Klebold had a notebook with his math homework inside. He jotted down some comments, apparently the day before the April 20, 1999, massacre. *"About 26.5 hours from now the judgement will begin. Difficult but not impossible, necessary, nerve-wracking & fun. What fun is life without a little death? It's interesting, when i,m (sic) in my human form, knowing I'm going to die. Everything has a touch of triviality to it."* In addition to more writings and drawings in his notebook, Dylan's Klebold final entry was a schedule: *"Walk in, set bombs at 11:09 for 11:17. Leave, drive to Clement Park. Gear up. Get back by 11:15. Park cars, set car bombs for 11:18, get out, go to hillside area, wait. When first bombs go off, attack. Have fun!"*

Eric David Harris

Eric Harris was born April 9, 1981, in Wichita, Kansas, to Wayne and Kathy Harris. His father was a Major in the U.S. Air Force, and moved his family a number of times before retirement in July of 1993 when he moved them to Littleton, Colorado. Like Dylan Klebold, Eric Harris had a brother who was three years older. Eric first met Dylan at Ken Caryl Middle School in Littleton. As a youngster, Eric played soccer and baseball, had an interest in computers, baseball cards, computer games, and videos. In high school, he was involved in video productions and the Columbine Rebel News Network, and computer labs. He used the nickname "Reb," whereas Dylan Klebold used the nickname "VoDKa" with his initials capitalized. On March 18, 1998, the Jefferson County Sheriff's Office wrote a suspicious incident report coming from Randy Brown, father of Brooks Brown who was an acquaintance of Eric Harris. His son had received death threats via Eric Harris's Web pages. Since Brown wanted to remain anonymous to protect his son, a very limited amount of investigation was done on this report before it was forgotten. Brooks Brown and Eric Harris were off-again/on-again friends and/or enemies. Brooks Brown and Eric Harris must have been friends on April 20, 1999, because when Harris passed Brown in the schoolyard before he started killing on the day of the massacre, he told Brown to leave the Columbine campus, "because he liked him."

Harris began writing in a journal during April of 1998, a year before the killings at Columbine High School. Among his writing in the journal, he noted how much he hated the world, and that he and Dylan Klebold were different because they had self-awareness. *"I will sooner die than betray my own thoughts, but before I leave this worthless place, I will kill whoever I deem unfit ..."* Harris also wrote that he sought revenge against anyone who had ever wronged him. Sometime later, he wrote, *"I want to burn the world, I want to kill everyone except about 5 people .... if we get busted anytime, we start killing then and there ...I ain't going without a fight."* He mentioned "choices" and that he chose to kill. *"It's my fault! Not my parents, not my brothers, not my friends, not my favorite band, not computer games, not the media, it's mine."* He also stated at another point in time, *"I'm full of hate and I love it."* There was one last comment in his journal during 1999. He assessed the preparations including weapons and bombs for the proposed massacre, and finished his assessment with the quote, *"I hate you people for leaving me out of so many fun things."*

In late 1998, Eric Harris wrote that he would have been a good Marine. *"It would have given me a reason to be good."* In actual practice, the U.S. Marine Corps did not consider Eric Harris as among "A Few Good Men." During 1999, the "natural born killer" as Harris had named himself applied for enlistment in the Marine Corps. The recruiting sergeant, following standard procedures, asked Harris if he was taking any drugs. Harris said no. A couple of weeks later, in a follow-up meeting with Harris and his parents, the father and mother admitted Eric was taking "Luvox," and anti-depressant medication commonly used to treat patients with obsessive-compulsive disorder. Eric was then disqualified for entry into the Marine Corps five days before the massacre at Columbine High School. Luvox is one of a class of drugs called selective serotonin reuptake inhibitors (SSRI). This drug works by interacting with a chemical in the brain called serotonin, which affects mood. Prozac, Paxil, and Zoloft are additional anti-depressant medications. Prozac is the most commonly prescribed anti-depressant in the United States. Luvox is generally prescribed to patients whose obsessions or compulsions cause to them to distress, consume time, or interfere with their daily activities. Perhaps in response to the Columbine killings and other school killings that have plagued the United States in recent years, the Colorado State Board of Education sought a higher standard of care when prescribing mind-altering drugs for school children, and in October/December of 1999 passed the following resolution by a vote of 6 to 1.

"WHEREAS: only medical personnel can recommend the use of prescription medication: and WHEREAS: the Colorado State Board of Education recognizes there is much concern regarding the use of appropriate and thorough diagnosis and medications and their impact on student achievement; and WHEREAS: there are documented incidents of highly negative consequences in which psychiatric prescription drugs have been utilized for which are essentially problems of discipline which may be related to lack of academic success......."

A year before the massacre, Eric Harris wrote in Dylan Klebold's Columbine yearbook, *"God I can't wait till they die. I can taste the blood now — NBK (Natural born killers). You know what I hate?......MANKIND!!!! ...*

Kill everything ......Kill everything......" In Eric Harris's 1998/1999 academic planner on a page for Mother's Day, he wrote, *"Good wombs have born (sic) bad sons."* Also recovered by the Sheriff's Office was an undated diagram of the Columbine High School cafeteria with two "Xs" near the pillars, and a timeline that Harris had drawn during actual observation of how many students were in the cafeteria prior to and during the first lunch period, apparently as a guide as to when to set off the two propane bombs destined for the cafeteria on the day of the massacre. Like Klebold, Harris also had a written itinerary for April 20, 1999, as follows:

5:00	up	10:30	set up four things
6:00	meet at KS	11:00	go to school
7:00	go to Reb's house	11:10	set up duffle bags
7:15	he leaves to fill propane	11:12	wait near cars, gear up
	I leave to fill gas	11:16	HA HA HA
8:30	Meet back at his house		
9:30	practice gear ups		
Chill			

Harris and Klebold also had three videotapes that were found after the "natural born killers" had committed suicide on April 20, 1999: a two-hour tape shot on three different occasions in March of 1999; a tape twenty-two-minutes long taped on April 11 and April 12, 1999; and a third tape forty-minutes long that was taped in eight separate occasions in early April, 1999 to the morning of the massacre. Harris and Klebold, who apparently made the three videotapes, shot a tour of Eric Harris's bedroom which showed their weapons and bombs, recorded each other conducting a dress rehearsal of their planned attack, and filmed the drive in Harris's car to purchase supplies needed for the massacre. They commented on how easy it was to make other people believe what they wanted them to believe. They spoke of how "evolved" they were, and stated they considered themselves "above human." They predicted that they would be successful because they were going to die. They stressed that they had been planning the Columbine killings for more than eight months. They spoke of anger and rage that they had built up over the years, and stated they would destroy the world if they could. Eric Harris stated, "*There is noting anyone could have done to prevent this. No one is to blame except me and VoDKa* (a nickname for Dylan Klebold).... *A two man war against everyone else.*" Did Klebold and Harris initially plan the murders to occur on April 19 rather than April 20? They stated "Monday" on one occasion, and on another occasion they mentioned, "Today is the 11th, eight more days." (April 19th was the anniversary of Waco, Texas and the Oklahoma City bombing of the federal building.)

The Killing Begins: April 20, 1999

When the actual killings began, some victims did not have any idea of what was actually happening. Some thought the initial gun reports were firecrackers, perhaps related to Senior Prank Day. At about 11:22 a.m., the Littleton Fire department, the West Metro Fire Department, and a patrol deputy from the Jefferson County Sheriff's Office were dispatched to the intersection of Chatfield and Wadsworth about three miles southwest of Columbine High School. A road crew in this area had noticed two bags and tossed them aside. One of the bags exploded, starting a small brushfire. The bombs had been placed there that morning as a diversionary device by Dylan Klebold and Eric Harris to draw away law enforcement officers from the killers' assault on Columbine High School. In the meantime, the teenage gunmen had arrived at Columbine, parked their separate automobiles which were both loaded with car bombs, and carried two duffle bags into the school cafeteria. Then they returned to the parking lot, donned their trench coats, or "dusters," and checked their weapons. Immediately, they began shooting at their classmates sitting by a fence and outside the cafeteria near the west entry doors. There was no hesitation by either Harris or Klebold as they began the actual assault. During their initial attack, they killed two students and injured five, and threw at least one explosive device onto the roof of the school. Then, they entered the school building. Even at this time, a major proportion of students, faculty and staff had no idea their lives were in danger, or what was happening to their school.

The attackers threw pipe bombs, fired inside the school doors, and then entered the school through the west wing entry doors. Shots were fired down hallways at running students, Harris and Klebold continuously reached into canvas bags they carried to secure explosive devices which they threw as they walked. Next, the killing kids entered the library and remained there for a few minutes, killing ten classmates and wounding eleven more. Then they went to the cafeteria and partially detonated one of the initially deployed duffle-bag bombs. They then proceeded to the science area, shooting through doorway glass panels, placing additional pipe bombs, and starting a fire. The killers then returned to the library and aimed and fired their weapons to outside the school where paramedics were rescuing victims they had shot previously. Now, because of the haphazard route taken by Klebold and Harris, law enforcement personnel were under the impression that as many as eight attackers may have taken over Columbine. An incident reconstruction seems to indicate Harris and Klebold committed suicide shortly after shooting at the paramedics at approximately 12:05 p.m. Prior to that, while in the cafeteria, the killers' actions were recorded by a surveillance camera that would provide additional information of the gunmen's dress, their armament, their handling of pipe bombs, and their shooting at a pre-deployed bomb made from a twenty-pound propane cylinder with an attached two, one-gallon containers of flammable liquid. The explosive they were shooting at did partially ignite, activating the school's sprinkler system at 11:52 a.m. The detonator was made of a pipe bomb with a combined mechanical and electrical activator. The propane cylinder was not breached, and the container did not vent in the ensuing fire. Only the flammable liquid in the one-gallon metal can ignited.

The Assault in Slow Motion

Sheriff's Deputy, Neil Gardner, a fifteen-year member of the Jefferson County Sheriff's Office, was a community resource officer at Columbine High School. Although Officer Gardner normally ate lunch with the students during first lunch period in the cafeteria, on the warm and sunny morning of April 20, 1999, he sat in his cruiser with Andy Marton, an unarmed security officer employed by the school district. Shortly after he finished his lunch, Gardner received a call on his radio from a school custodian. "Neil, I need you in the back lot," said the caller. Gardner put his car into gear and headed for the south student parking area close to the cafeteria. Then Gardner heard another report being dispatched on his sheriff's radio, "Female down in the south lot of Columbine High School." At that precise moment, 11:23 a.m., he put on his lights and siren. As he drove, he saw kids running in all directions. Smoke was coming from the west end of the parking lot, and Neil Gardner heard several loud explosions, plus gunfire coming from inside the building. As he alighted from his car, he got another call on the school radio, "Neil, there's a shooter in the school!" There was suddenly so much traffic on the sheriff's office radio that Gardner could not tell the dispatcher that he was on the scene.

At that moment, Gardner came under fire from either Eric Harris or Dylan Klebold. Gardner leaned over the top of his police vehicle about sixty yards from Harris or Klebold. The gunman seemingly had his rifle jam and was apparently trying to clear the weapon. Gardner fired four shots, and saw the gunman spin hard to the right, making Gardner feel he might have scored a hit. However, the gunman soon began firing again. The Deputy could distinguish other shots being fired inside the high school but he had no idea who else might be shooting, how many shooters might be in action, or where they might be situated. Neil Gardner commented later the gunman who fired about ten shots from a rifle could possibly have been Dylan Klebold. However, the Jefferson County Sheriff's Office report issued during the year 2000 stated that Eric Harris was the gunman who shot at Gardner. Gardner said the shooter was tall, about six feet, skinny, with collar-length hair, and was wearing a blue flack jacket with a baseball cap worn backwards. Klebold was wearing a dark T-shirt and a cap worn backwards when caught in the videotape by the surveillance camera in the cafeteria. He was much taller than Harris and had longer hair.

A few minutes later, the teenage killer initially identified as Eric Harris, but later possibly identified as Dylan Klebold, fired at Deputy Sheriff Neil Gardner from inside the double doors of the high school. Deputy Sheriff Paul Smoker, a motorcycle patrolmen seeking more cover behind a vehicle, saw Gardner with his weapon drawn from another vantage point and heard him yell that the shooter was leaning out the broken window of the west side doors, using the doorframe as a cover. Garner returned fire. Smoker could not immediately see the shooter for a half-fence and a dumpster partially blocked his view. Smoker left his cover behind the vehicle and went out into the open. He saw the shooter, and shot three rounds before the shooter disappeared back into the building. At 11:30 a.m., within minutes of Gardner's report of shots being fired in the building and the need for additional units, six Jefferson County Deputies were on the scene. Gardner again called dispatch and reported that all units around him were under fire. An explosion by the south double-doors blew out two windows. A fireball then exploded in the cafeteria. "You could feel the explosion," according to Deputy Sheriff Walker, who was also a member of the bomb squad. "You could see the windows flexing out and then being sucked back again. As the cafeteria was bombed, six students ran out of the south cafeteria door towards Walker. He motioned for them to take cover behind nearby, parked vehicles. One hysterical young woman kept asking Walker, "Are we going to die?" He reassured her that she was not going to die.

Walker radioed to dispatch that he had six students with him, but did not have any safe path to evacuate them from the parking lot. To the east of the cafeteria area, on the second floor, Walker saw a young woman pounding on the glass with a sign in her hands. The sign was too far away and the glass too reflective for Walker to read the message displayed. He momentarily saw a shooter through the window on the upper level in the southeast corner. He called dispatch and described the shooter as wearing a white T-shirt with some kind of holster set. Two Littleton Fire Rescue Units responding to Deputy Neil Gardner's earlier call for medical assistance, entered the south parking lot and drove toward Walker's position. Being unaware that the position was still under fire, the firefighters — who do not normally enter an active crime scene — drove close to the downed students lying outside the cafeteria and jumped out of their vehicles. They saw the law enforcement personnel creeping in to provide cover, and one of the firefighters mouthed, "Is it Safe?" One of the police officers responded, "NO!!! It's not safe. Get 'em and go!!! Get 'em and go!!!" Gunfire then came from the second floor library windows. Walker and Deputy Gardner returned fire as Denver Police Officers joined in providing suppression fire, allowing firefighters to remove the students (Sean Graves, Lance Kirklin, and Anne Marie Hochhalter. The fourth student, Daniel Rohrbough, was determined to be deceased). The firefighters rushed the living students to medical attention, and police officers could still hear shots inside the school building, but the gunfire from the library had ceased.

An ad hoc, hastily formed SWAT (Special Weapons and Tactics) team made up of officers from the Jefferson County Sheriff's Office, and Denver and Littleton Police Departments now was ready to enter the Columbine High School. This SWAT team was split into two groups; one went in to the east side, and one went in the building from the west side. Just after the SWAT team(s) entered the building, Deputies on the outside of the building and another Deputy in a helicopter flying overhead observed an injured male figure who seemed definitely intent on coming out from a second floor library window even though there was nothing but a concrete sidewalk to break his fall. They waved and yelled at the Lakewood SWAT team and armored truck in the south parking lot trying to get the attention of Lakewood Police. Lakewood SWAT responded to their entreaty with the armored truck, learned the game plan, and planned an operation to get the job done. Two SWAT members mounted on the roof of the armored truck caught wounded student Patrick Ireland as he fell through the window. That image remained in the minds of all persons who watched television that day; it became one of the most significant images of the Columbine massacre.

The deadly journey through Columbine High School in Littleton, Colorado, by Dylan Klebold and Eric Harris can be identified by ninety-one critical witnesses who actually saw or heard the killers doing their massacre assignments as any other students might do their algebra homework. Witness information was broken down into three chronological time segments including "immediately before" the incident, "during," and "after" the incident. According to the well-done investigation report of the Jefferson County Sheriff's Office, "Columbine High School Shooting April 20, 1999," and various investigation teams assigned to different segments of the case; and relying on witness statements, evidence, ballistic reports, pathology findings, and the totality of the case overall, Harris and Klebold were said to be the only individuals apparently involved in the shootings at Columbine.

On the outside of the school, two students, Rachel Scott and Daniel Rohrbough, were killed and eight were wounded including teacher Patricia Nielson. August 20, 1999, was a sunny spring day that lured a number of students from inside the building to go outside and sit in the warm sun while eating their lunches. The first shots hit Rachel Scott, who died, and critically injured Richard Castaldo with multiple gunshot wounds. Eric Harris took off his trench coat, or "Duster," and placed it on the ground near the top of the hill shortly after he began shooting. He was warming to his job of killing, whereas Klebold wore his black trench coat while outside the school. They started firing down the slope outside the cafeteria; Harris steadied his gun on the chain link fence at the top of the steps while firing at students at the bottom of the steps: Daniel Rohrbough, Sean Graves, and Lance Kirklin. The students saw Klebold and Harris with weapons, but thought they were joking. The three walked out of the northwest side entrance of the cafeteria and headed up the dirt slope between the steps and the cafeteria wall toward Harris and Klebold who started firing. Daniel Rohrbough eventually died, and the other two students were wounded.

Five students were sitting behind a few small pine trees when Harris and Klebold shot at them. Michael Johnson and Mark Taylor went down, but Johnson was able to reach an outdoor athletic shed for cover; Taylor fell to the ground and could not flee with the others. Dylan Klebold went down the stairs and shot and killed Daniel Rohrbough, and then shot Lance Kirklin at close range. Sean Graves was able to drag himself back down to a cafeteria doorway and attempted to get back inside the school. He managed to get the door propped half-open with his body, and then played dead while Dylan Klebold, who had just shot Rohrbough and Kirklin, stepped over him and went briefly into the cafeteria , holding his weapon in a "ready-to-fire" sweeping motion. Klebold, now inside the cafeteria, looked around but then walked back outside the building to the top of the chain of steps.

While the Mutt and Jeff killers were shooting from the top of the stairs, William "Dave" Sanders, a faculty member, along with two custodians, Jon Curtis and Jay Gallatine, can be seen in the cafeteria surveillance camera tape directing the students to get down, to hide under the cafeteria tables, or get out of the cafeteria. Realizing the imminent danger the students decided for themselves to hastily exit the cafeteria and run up the stairs to the second level. At the time Dylan Klebold had stepped over Sean Graves in the doorway and entered the cafeteria for a few moments, most of the students in the cafeteria had evacuated this area. Anne Marie Hochhalter was wounded as she stood to run to the cafeteria. A friend dragged her away from the gunfire to the cafeteria wall. He then hid behind vehicles in the senior parking lot, where he heard someone yell, *"This is what we always wanted to do. This is awesome!"* Outside the buildings, Harris and Klebold were observed lighting and throwing explosive devices into the parking lot and toward the grassy hillside. One gunman took four bombs from inside his coat and threw them. Another critical witness described a gun attached to a gun strap that hung to the gunman's side, while other witnesses saw bombs taken from bags at the gunmen's feet. Additional shots were fired toward the ball field where other students were fleeing. At this time, only four minutes had passed since the first shot fired outside the building. It was 11:25 a.m.

Most students outside the school had scattered so there were no students readily visible to the killers, so they walked toward an entrance and fired their weapons into the school as they walked. Faculty member, Patricia Nielson, was on her way outside to tell the shooters to "Knock It Off" because she believed some student prank was going on, or students were making a video. When she got near a set of double-doors at the west side of the building, the glass windows shattered when hit by bullets shot by Klebold and Harris who were outside the building. Nielson was hit by flying glass in her shoulder, forearm, and knee. Student Brian Anderson and a friend were in the school hallway. When they went through the first door into the air-lock and reached the second door to the outside, Anderson saw Eric Harris shoot out the window from the outside which rained shattered glass on him. Both Nielson and Anderson were able to flee because Harris and Klebold got distracted by the arrival of Deputy Neil Gardner, the school community resource officer, in the senior parking lot with the lights on his cruiser flashing and the siren blaring. Nielson and Anderson fled to the school library on the second floor. Patricia Nielson , hiding under the front counter, called 911, while Anderson hid in the magazine room.

There were 488 students and staff inside the cafeteria when the shooting began within this area. Fifty-one were identified as critical witnesses, people who saw a gunman, saw a victim get injured, or saw duffle bags inside the cafeteria which may have contained explosive devices. Critical witnesses were mainly positioned along the west side, adjacent to the windows of the event happening outside. No one other than Klebold or Harris was identified as being involved in the cafeteria shooting. There was no one injured as a direct result of Harris and Klebold's actions.

There were twenty-six witnesses to significant events in the hallways of Columbine High School on April 20, 1999. The twenty-six also included two teachers, one secretary and a campus supervisor. Witnesses Stephanie Munson and teacher Dave Sanders were also victims among the twenty-five living witnesses. Eight persons positively identified Dylan Klebold as the person they saw in the school hallway, while nine additional witnesses reported seeing a suspect who matched his description. Of the twenty-five living witnesses, five positively identified Eric Harris as the person they saw in the school's hallways, and seven additional witnesses described seeing a suspect who matched the description of Eric Harris. Class was in session for those students who did not have first period lunch, so there were a lack of students out and about in the school at this time. Up to this time, the entire sequence of deadly events had taken place outside and no one within the school, other than those in the cafeteria, yet realized there was a massacre in progress. A student saw Brian Anderson hit by flying glass, and observed teacher Patricia Nielson run into the main entrance to the library before she herself decided to run. She met teacher Dave Sanders at the top of the library stairs who told her to, "get down stairs." Students who entered the north main hallway from adjoining classrooms saw Klebold and Harris standing just inside the school's west main level entry doors. Klebold fired a semi-automatic weapon toward the students to the east and south down the library hallway. Student Stephanie Munson was shot as she ran through the main lobby. Another student in the gym hallway saw Klebold and Harris walk east down the main hallway firing their weapons and laughing. A student in the counseling hallway saw students in the north hallway running past the lobby with Dylan Klebold running behind them. Klebold stopped near some telephones at the entrance to the main lobby area. The student could see the sleeve of a black trench coat shooting a TEC-9 toward the main entrance of the school. She dropped the telephone and hid in a nearby bathroom until she could no longer hear any sounds of commotion in the hall-way, then went back to the bank of telephones and called her mother. "Mom," she said into the phone, "there's a shooter in the school. Come pick me up." She managed to escape outside through the east door. Her mother's cell phone indicated the call had been made sometimes between 11:23 a.m. and 11:26 a.m. and lasted 3.8 minutes.

Klebold was last seen running west down the north main hall in the direction of the library. Teacher Dave Sanders had come up the stairway from the cafeteria to the south main hall to direct as many students as possible to safety. A student ran up the steps with Sanders and turned to the right (north) in the hallway in front of the library. The student said she saw Dylan Klebold standing by the west doors holding his weapon with both hands; Klebold began firing a "rapid firing black gun" in their direction, and they both turned around and ran back toward the stairway. Teacher Dave Sanders was mortally wounded, but crawled to the corner of the science hallway where another teacher helped him into one of the science classrooms. A group of students, including two Eagle Scouts, Aaron Hancy and Kevin Starkey, tended Sander's injuries, tried to stop the blood flow, and provided first aid. At this time, there was no longer anyone in the hallway since they had fled to classrooms to hide, or had run out of the east exit of the school. Others hid in the auditorium. When police later investigated a 911 call made from the Columbine library, they confirmed that the killers spent almost three minutes in the library hallway firing weapons and lighting and throwing pipe bombs, and threw pipe bombs over the stairway railing down into the cafeteria.

There were a total of fifty-six people present in the library including four faculty and staff females. All of the witnesses in the library were determined to be critical witnesses as all of them observed movement and actions of Klebold and Harris. In the library, the teenage killers wounded twelve and brought death to ten of their classmates.

Dead and Injured within the Columbine Library, April 20, 1999

Killed	Wounded
Cassie Bernall	Jennifer Doyle
Steven Curnow	Steven Eubanks
Kelly Fleming	Makai Hall
Corey DePooter	Patrick Ireland
Mathew Kechter	Mark Kintgen

Daniel Mauser   Lisa Kreutz

Isaiah Shoels   Nicole Nowlen

John Tomlin   Kacey Reusegger

Lauren Townsend   Jeanna Park

Kyle Velasquez   Valeen Schnurr

                         Danny Steepleton

                         Evan Todd

Normally, the library would be much more crowded than it was that day due to good weather outside the building on that particular day. At about 11:20 a.m., persons began to hear noises coming from the northwest area outside the building. Teacher Patricia Nielson who was already injured ran to the library and dialed 911. She told the dispatcher what was happening, and to the students in the library yelled there are students with guns in the school, and everyone should hide under the tables. She knew they had nowhere else to go since the last she had seen of the teenage killers they were near the emergency exit from the library and were heading inside the hall leading to the library. When the gunman had been outside at the very beginning of the massacre, teacher Peggy Dodd had looked out a window in the library and recognized Dylan Klebold standing on a hill shooting. He had been a student in one of her computer classes the previous year. She remembered him as a troublemaker who hacked into computers and wore tall Nazi-style boots and an overcoat.

Teacher Patricia Nielson hid under the front counter of the library. Harris and Klebold were then directly outside the library shooting weapons and throwing pipe bombs. Smoke began pouring into the library starting fire alarms. At 11:29 a.m., the "natural born killers" entered the library, shouting "Get up! All jocks stand up. We'll get the guys in the white hats." Then one witness heard one of the killers say, "Fine, I'll start shooting." Eric Harris fired shotgun shells down the length of the front counter. Kyle Velasquez was the single student not hidden under a desk or table who was sitting at a computer table. Klebold fired as he walked by Velasquez, killing him. Klebold dropped a carry-on bag filled with ammunition and Molotov cocktails on one of the computer tables while Eric Harris was down on one knee shooting out the west library windows towards law enforcement officers and fleeing students. Klebold then took off his trench coat about the same time teacher Patricia Nielson who was using a telephone under the front counter of the library to talk with the 911 dispatcher reported the massacre as it was going on, dropped the phone and cut-off conversation with the dispatcher. Klebold then shot Daniel Steepleton, Makie Hall, and Patrick Ireland. As Ireland was attempting to give first aid to another student, he was shot a second time. At about the same time, Eric Harris killed Steven Curnow who was hiding under the south computer table, and wounded Kacey Reugsegger who was crouched next to Curnow. Harris then walked over to table nineteen, bent down when he saw a frightened young woman. He twice slapped the table, said "Peek-a-boo," and the killed Cassie Bernall. Harris them made a comment about hitting himself in the face. Police believe he broke his nose from the "kick" of his weapon when he fired under the table. Bree Pasquale was sitting on the floor because there was not enough room under a nearby table. Harris asked her if she wanted to die. She saw Harris was bleeding from his nose, and seemed to be disoriented for a few seconds. When Klebold yelled at his at table sixteen position that one of the two boys hiding there was an African American, Pasquale heard Harris laugh and say, "Everyone's gonna die," and "We are going to blow up the whole school anyway." While Harris and Klebold stood on opposite sides of table sixteen, Klebold was heard to make a racial comment and grabbed Isaiah Shoels to pull him out from under the table. Harris shot under the table, killing Shoels; and Klebold fired killing Mathew Kechter as well. Harris then heaved a small CO2 cylinder which rolled under table fifteen. Makie Hall retrieved the CO2 cylinder and threw it away from the students where it exploded. Harris relocated south to the west bookshelves in the library. He jumped up on the shelves and shook them back and forth while swearing. Klebold and Harris then walked to the toward the library entrance where Klebold shot a display cabinet near the front door, and then fired at table one injuring Mark Kintgen. Klebold also fired under table two injuring Valeen Schnurr and Lisa Kreutz. He then shot and killed Lauren Townsend. Eric Harris fastened his attention on table three where two young women had fled. He looked under the table, said "Pathetic," and then walked away. Harris next stopped at table six, fired under the table injuring both Nicole Nowlen and John Tomlin. Tomlin fought his way out from under the table, and Harris shot him again, killing him with the second shot. Next, he walked around table six killing Kelly Fleming who was near table two where three students had already been shot. Harris again fired under table two killing Lauren Townsend and wounding Lisa Kreutz and Jeanna Park for the second time.

Harris and Klebold reloaded their weapons as they entered the center portion of the library. An acquaintance of Klebold's hiding under table eleven asked Klebold what he was doing. "Oh, just killing people," responded Klebold. The student wanted to know if the killer was going to kill him. Klebold told him to get out of the library, which he did. Harris shot under table nine murdering Daniel Mauser. Both the killers shot under table fourteen, killing Cory Depooter and wounding Jennifer Doyle and Austin Eubanks. With regard to Cory DePooter, he was the last person murdered in the Columbine massacre. The initial shooting outside the school had begun at 11:21 a.m., and it was now 11:35 a.m. The child killers' had killed twelve classmates, one teacher, and wounded or injured twenty-six others in approximately fourteen minutes. Soon they would commit suicide, but no one left living knows the exact timeframe when Harris and Klebold killed themselves in the Columbine library.

Klebold and Harris now went behind the library counter where they were confronted by Evan Todd who had already been wounded. The two taunted Todd and discussed whether they should kill him. However the gunmen walked away although Klebold fired one shot back into the kitchen area of the library and struck a television set. The murdering teenagers talked about going back to the commons (cafeteria) area, and Klebold threw a chair on top of a computer terminal on top of the main counter under which teacher Patricia Nielson was still hiding. They exited out the library's main entrance at about 11:36 a.m. There had been fifty-six people in the library. Harris and Klebold had enough ammunition to kill them all, but they failed to do so. Now, it was very quiet in the library and the room was filled with smoke and the fire alarms were blasting. The dead lay among the living and wounded, and nobody looked at anyone else. They dreaded that Klebold and Harris would come back and finish the work they had started. One by one, or in small groups, they got up and sought a safe area, or just a place to hide. Teacher Patricia Nielson crawled out from behind the library counter and into the library kitchen area where she hid in a closet. Library Technician Carol Weld and assistant Lois Kean hid in the television studio, and teacher Peggy Dodd hid in the periodical room. They all remained hidden for long periods of time until found and evacuated by SWAT personnel. Patrick Ireland lost consciousness several times but forced his way to a library window and attracted an outside SWAT , and fell into the arms of two SWAT team members who had climbed to the roof of the team armored truck positioned under the second floor window. The last of the wounded, Lisa Kreutz, was rescued by inside SWAT team personnel at 3:22 p.m. and transferred for care.

Exiting from the library, Klebold and Harris had done exactly what they had planned to do over the previous twelve months; now, they seemed to have lost direction as they wandered aimlessly through the Columbine hallways. In the science area, their behavior seemed strange and out-of-place for the "natural born killers" they thought themselves to be. Perhaps, they were looking for more victims, or maybe just tired of killing since the excitement and thrills had worn off. Perhaps the thought of their own death which would soon occur set them back a bit. They were aimlessly walking down the science hallway, firing into empty classrooms, checking locked rooms, making eye contact with some students in locked rooms, and taping a Molotov cocktail to a storage room door.

The Columbine murders demanded much of law enforcement personnel as well as emergency medical services and fire department members. Responders, as well as students, were in danger of death. As an example, if the propane bombs previously placed in the cafeteria before first period lunch had gone off when they were supposed to explode at 11:20 a.m., the actual count of potential victims in the cafeteria at that time amounted to 660 persons. The explosion of the two propane bombs would turn just about anything in the cafeteria into shrapnel, and a giant fireball would suck up oxygen making death welcome to many victims, and ensuring that survival was all but impossible. It could have amounted to the worst domestic terrorist killing in the United States, dwarfing the death rate at the Oklahoma City Federal Building bombing, if Klebold and Harris had a better understanding of explosive reactions, electrical wiring, and current. Overall, the Columbine incident was short-lived but drew an amazing response from many different agencies which had to deal with an ultra-complex, totally infamous situation in a mass casualty attack. Terrorist-style incidents are different than response personnel prior experiences. They demand new, nontraditional methods of response. In its conclusions in the Federal Emergency Management Agency/United States Fire Administration report, "Wanton Violence At Columbine High School, Special Report," the authors state: "Unfortunately, such violent incidents perpetrated against young, innocent victims have become all too frequent. 'It can't happen here' *is* happening here, and the emergency service providers who may next be called must be prepared for that possibility."

The Aftermath

Two years after the incident, some Columbine survivor families are suing the sheriff's department charging it mishandled the response to the attack. They feel the department could have prevented the running amuck incident if law enforcement officers had truly investigated earlier complaints against Eric Harris. In addition, in April of 2001, Jefferson County District Judge Brooke Jackson ordered Sheriff John Stone to release drafts of previous affidavits deputies prepared a year before the actual attack in hopes of getting a search warrant to use against Harris. The affidavits were drafted after allegations that Eric Harris was threatening people over the Internet and making pipe bombs, but they were never submitted to a judge or the district attorney. Another issue in lawsuits filed by victims and their families is whether deputies could have prevented the massacre if they had investigated a violent essay Dylan Klebold wrote for his English class at Columbine High School.

In another court case brought by three dozen families of Columbine victims against the gunmen's parents and the persons who provided guns for the teenage killers to use in the Columbine massacre, the families on April 19, 2001, (on the anniversary of the Oklahoma City Federal Building explosion as well as the anniversary of the Waco, Texas fire that destroyed the Branch Davidian religious cult, and one-day before the second anniversary of the Columbine massacre) agreed to a $2.5 million settlement. The parents of Brian Rohrbough who was killed in the incident, did not settle. Robyn Anderson, who is accused of buying some weapons used by Eric Harris and Dylan Klebold during the attack, is still discussing settlement with the families. Other lawsuits are still pending at this writing against the sheriff's office, the school district, and the operator of the gun show where Anderson allegedly got the weapons four months before the actual attack.

The criminal case regarding the Columbine Massacre will remain open indefinitely because there is no statute of limitations for the crime of murder. However, questions about protecting our children at school will remain forever. No one expected that our sense of security would be threatened — by our own children. It has been threatened time-and-time-again, and there seems to be no solution or remedy at hand.

Guide to Chemical/Biological Agent Response

3

The purpose of this book is to provide a well-researched guide for emergency response to chemical and biological agents and weapons for a professional and technical audience. The audience includes emergency medical personnel from physicians to EMS technicians, firefighters, law enforcement officers, emergency management and civil defense officials, hospital staff, engineers of all types, laboratory personnel, city managers and safety officers, military officers and personnel, industrial managers and plant fire chiefs, commercial response contractors, business firms who deal with manufacture/transport/regulation/ response or reporting of chemical and/or biological agents or weapons, health regulatory persons, government agencies who engage in matters related to chemical/biological matters and concerns, industry hazardous materials response teams/first responders/and fire brigades, government and private consultants, private security forces, maritime personnel, power and industry staff, transportation and distribution personnel, counterterrorism professionals, environmental specialists, and possible victims.

Contamination/Exposure Limits: The various occupational exposure limits found on Material Safety Data Sheets (MSDS) are based primarily on time-weighted average limits, ceiling values, or ceiling concentration limits to which a worker can be exposed without adverse effects. However, the exposure limits that follow were designed to provide worker safety in industrial and/or occupational situations. It is admittedly difficult to determine how such values should be applied in chemical or biological agents or weapons alarms for emergency first responders. Such measures that follow were designed for a healthy adult working population; thus, their utility is limited when dealing with incidents that involve biological and chemical weapons, mass casualty alarms, and other training scenarios that were designed subsequent to September 11, 2001. The following values that are explained below should be used as benchmarks for determining relative toxicity, and possibly to aid in selecting appropriate level of chemical protective clothing and equipment.

Minimum Effective Dose: The minimum effective dose for anthrax, for example, has been stated in the past to be 8,000 to 50,000 spores in aerosol. The new, "super-anthrax" found in Florida and Washington, D.C., beginning about October 15, 2001, may have a minimum effective dose in spores that is much smaller.

Threshold Limit Value (TLV): TLV is a value developed by the American Conference of Governmental Industrial Hygienists (ACGIH) that refers to airborne concentrations of substances and represents conditions under which it is believed that nearly all workers may be repeatedly exposed day after day without adverse effect.

Threshold Limit Value: Time-Weighted Average (TLV-TWA) is an ACGIH value that is a time-weighted average concentration for a normal eight-hour work day and a forty-hour work week, to which nearly all workers may be repeatedly exposed, day after day, without adverse effect.

Threshold Limit Value: Short-Term Exposure Limit (TLV-STEL) was also developed by ACGIH and represents the concentration to which workers can be exposed continuously for a short period of time without suffering from (1) irritation, (2) chronic or irreversible tissue damage, or (3) narcosis of sufficient degree to increase the likelihood of accidental injury, to impair self-rescue, or to materially reduce work efficiency; and provided that the daily TLV-TWA is not exceeded.

Threshold Limit Value: Ceiling (TLV-C), also an ACGIH term, means that the concentration that should not be exceeded during any part of the working exposure.

Permissible Exposure Limit (PEL) developed by the Occupational Safety and Health Administration (OSHA) means the same as the TLV-TWA developed by ACGIH.

Immediately Dangerous to Life and Health (IDLH), also developed by OSHA, means a maximum airborne concentration from which one could escape within thirty minutes without any escape-impairing symptoms or any irreversible health effects.

Recommended Exposure Limit (REL) was developed by the National Institute for Occupational Safety and Health (NIOSH) and defines the highest allowable airborne concentration that is not expected to injure a worker; expressed as a ceiling limit of time-weighted average for an eight- to ten-hour work day.

Emergency Response to Biological Agents or Weapons

Attack Indicators/Biological and Chemical

Biological: The U.S. Army Medical Research Institute of Infectious Diseases (USAMRIID) at Fort Detrick in Frederick, Maryland, writing in their newly updated *USAMRIID's Medical Management of Biological Casualties Handbook* (4th ed., February 2001) provides information of characteristics of biological weapons and warfare: "Potential for massive numbers of casualties; ability to produce lengthy illnesses requiring prolonged and extensive care; ability of certain agents to spread via contagion; paucity of adequate detection systems; diminished role for self-aid and buddy aid, thereby increasing sense of helplessness; Presence of an incubation period, enabling victims to disperse widely; ability to produce non-specific symptoms, complicating diagnosis; ability to mimic endemic infectious diseases, further complicating diagnosis."

Further characteristic include mysterious illnesses; large numbers of insects or unusual insects (vectors) for the area, a number of dead and dying victims; numbers of dead or dying domestic or wild animals; many victims with similar symptoms or unexplained symptoms; previously healthy persons who now show the same symptoms and signs of illness not normally compatible with this area; victims who do not respond to usual standard medical treatment; a single victim or two with an unusual disease; artillery shells and bombs with less powerful explosions, or that go "pop" rather than explode; immunocompromised persons (like those with heart or kidney transplants) who show first susceptibility and rapid progression of illness; anything strange or unusual with victims' conditions such as questionable seasonal or geographic distribution; mist or fog sprayed by aircraft or vehicles; multiple, simultaneous outbreaks of some "strange" illness; any epidemic seeming caused by a multi-drug resistant pathogen; or the same "disease" being reported from different areas.

USAMRIID also cites epidemiologic clues of a biological warfare of terrorism attack as follows:

- The presence of a large epidemic with a similar disease of syndrome, especially in discrete population
- Many cases of unexplained diseases or deaths
- More severe disease than is usually expected for a specific pathogen or failure to respond to standard therapy
- Unusual routes of exposure for a pathogen, such as the inhalational route for diseases that normally occur through other exposures
- A disease that is unusual for a given geographic area or transmission season
- Disease normally transmitted by a vector that is not present in the local area
- Multiple simultaneous or serial epidemics of different diseases in the same population
- A single case of disease by an uncommon agent (smallpox, some viral hemorrhagic fevers)
- A disease that is unusual for an age group
- Unusual strains or variants of organisms or antimicrobial resistance patterns different from those circulating
- Similar genetic type among agents isolated from distinct sources at different times or locations
- Higher attack rates in those exposed in certain areas, such as inside a building if released indoors, or lower rates in those inside a sealed building if released outside
- Disease outbreaks of the same illness occurring in noncontiguous areas
- A disease outbreak with zoonotic (a disease communicable from animals to humans) impact
- Intelligence of a particular attack, claims by a terrorist or aggressor of a release, and discovery of munitions or tampering.

The impact of a biological agent or weapon attack will include a pronounced potential to overwhelm the medical systems' capacity; incidents will be of long duration requiring continuous rotation of staff and equipment; resources may become hard to obtain (medicines, vaccines, adequate detection capabilities); standard practices, isolation and quarantine may become long-term; medical services may have to be prioritized on a need-to-have basis; psychological and physical impacts will grow over time; and most incidents involving biological agents or weapons will be a crime scene requiring sample and collection of evidence.

Chemical: Chemical agents or weapons leave potential and slightly different clues to their use:

- Unusual numbers of dying animals may be present
- There may be no insect sounds and birds may be all dead
- Victims may have unexplained watery blisters, wheals, or rashes
- A number of persons have serious medical problems such as nausea, disorientation, breathing problems, convulsions, and possible death.
- There is a definite pattern of casualties, and victims are distributed in a manner that mimics agent dissemination methods

- Unusual liquid droplets may be present; there is no recent rain, but surfaces and bodies of water seem to be covered with an oily film

Life on Scene: The priorities in chemical or biological incidents will be to protect life, environment, and property; alleviate damage, loss, hardship and suffering caused by such incidents; restore essential government services; and provide relief to affected government agencies, services, businesses, and individuals. Local fire departments and hazardous materials response teams (HMRTs) are trained and equipped to isolate and secure a scene, deny entry, and establish control zones. They establish command, evaluate scene safety and security, stage incoming units, gather information regarding incidents, and assign Incident Command System positions to deal with an emergency that endangers life and property. Local agencies initiate notifications to hospitals, law enforcement personnel, commercial response contractors, and federal agencies that may be required at an incident scene. Staff know how to request additional resources from private sources, are trained to use appropriate personal protective equipment and maintain a minimum number of personnel exposed, plus initiate additional public safety measures. Locals will control, isolate, and provide emergency medical services to patients, control activities in law enforcement with police officers, begin or assist with triage, and provide mass decontamination to survivors. In certain circumstances, local Haz-Mat responders will establish and maintain a chain or custody system for evidence protection.

Federal response teams cannot be on site in time to save lives or treat victims. Local response personnel need, but cannot afford, the same assets and training as federal employees are now receiving. Haz-Mat mitigation is a local responsibility. Someone must be on scene at once to isolate the area of a chemical or biological agent attack, evacuate, and care for the injured and the dead, and deny further entry. Any and every chemical or biological attack will be a crime scene that has to be isolated and guarded.

Execute Command Authority: The major goal of first responder agencies doing emergency response to chemical and biological agents or weapons is to quickly identify, evacuate, decontaminate, and treat victims. Evacuation of victims and removal of clothing is the most effective means of decontamination. After liquid or solid hazardous materials contamination, removal of clothing will remove 80 percent or more of the contaminant.

Recommendations

Chemical agents can cause immediate effects; biological agents cause delayed effects.
Position uphill and upwind; stay away from building exhaust/ventilation systems.
Stage at a safe distance away from the site; consider using multiple staging areas.
Avoid any wet or moist areas.
Locate and identify command post.
Check weather reports often for potentially dangerous conditions (rain, changes in temperature, etc.)
Establish Incident Command System; appoint personnel to handle specific ICS positions
Evacuate survivors from the scene.
Segregate symptomatic injured in one area and asymptomatic in another area; use triage if necessitated by numbers and conditions of the victims.

Move ambulatory persons away from the source of the contamination. You have to separate the walking wounded from the prone (e.g., "All those who can walk, come to me!"). Particularly with chemical agents, victims may have eye problems that blind them at least temporarily; tell every victim that appears disoriented to put his right hand on the back of the shoulder of the victim in front of him and "hang on" to that person. In this way, a whole line of survivors can be led out of a contaminated area.

Confine all contaminated and exposed victims to a restricted/isolated area at the outer edge of the Hot Zone.

Secure and isolate the entire incident area and deny entry to all but accredited responders. Assign control areas: Hot/Warm/Cold zones.

Victims exposed to chemical agents require removal of clothing, gross decontamination, and medical care.

Proceed with intensive monitoring of the site.

Complete a hazard and risk assessment to determine if it is acceptable to commit responders to the site.

Handle complete size-up and operational planning.

Enforce rules for entry into the total area involved, including Hot/Warm/Cold zones.

Identify exposure problems, potential difficulties and hazards, equipment needed and available, make appropriate notifications to obtain outside help, study pre-plans if they exist, develop action plan (priorities, tactics, assignments to the ICS (Incident Command System) and Unified Command.

Attempt to identify agent or weapon, type of dissemination (aerosol/liquid/vapor/solid), and characteristics of the agent and threats posed. Search the area for secondary explosive devices or booby-traps designed to kill or maim first responders. Limit the number of personnel and exposure time on scene.

Hold a briefing for entry personnel and support crew. Ensure that all staff on scene understand emergency signals, preplanned escape routes, locations of Hot/Warm/Cold zones, operational hazards at this site, the evacuation system, the buddy system, the need for appropriate personal protective equipment (PPE); the potential need to decontaminate rescuers and their equipment, and contaminated persons; location of rehabilitation area, possible air space closure over incident site, etc.

Establish a public information/media area.

Use soap/water decontamination beginning with the most symptomatic victims.

Establish isolation distance and restrain entry; law enforcement officers should establish an outer perimeter, and completely secure the scene.

Once good information is available on the offending agent(s), this information shall be relayed to emergency medical services personnel and hospitals.

Eventually first responder personnel on scene will have to answer the basic question: Do you have an attack or not? Unlike exposure to chemical agents, exposure to biological agents does not require immediate removal of victims' clothing and gross decontamination. With biological agents, inhalation is the most common route of entry to a victim's contamination.

A major issue is whether SCBA (self-contained breathing apparatus) and structural fire-fighting clothing, known as "bunker gear," provide adequate protection for first responders responding to alarms where chemical or biological agents or weapons may be present. Unless you are talking about **QUICK** rescue actions for known live persons, the answer is **NO!** For an unknown method of dissemination, quantity, or identification of the agent or agents; entry personnel should begin with Level A and then gear down to lower levels of protection as the entry personnel learn more about this particular incident. **First responders equipped with bunker gear and SCBA should refrain from entering areas of high concentration, unventilated areas, or low areas (except for quick rescue efforts of known live persons).**

There is a shortage of biological agent detection equipment that can be used in the field, but the market has undergone vast changes since September 11, 2001, that should lead to additional units.

Gather important information: the agent used; was it aerosol, liquid, gas, powder or vapor; location; method of delivery; do you have the necessary personal protective equipment (PPE) to deal with the hazard, or have you called for assistance by a specialized team; are you sure that anyone who enters a contaminated area has the proper PPE and is trained in its use; be sure to establish control — keep all victims, non-victims and bystanders at the crime scene (if there is any suspicion of an attack) until it is determined who among them may be a terrorist or a witness; perform decontamination, triage if necessary, isolation, quarantine, search and locate evidence, maintain chain of control, and collect samples.

Hospitals should be notified immediately that contaminated victims of an attack may arrive or present themselves at a hospital whether or not they have been decontaminated, or not.

If available, Haz-Mat team personnel in proper protective equipment should be used for rescue, recon, and/or agent identification.

Consider some differences between a chemical or biological (C/B) agent attack and a hazardous materials alarm with industrial chemicals: C/B is designed to inflict mass casualties, has high lethality, and may have an extremely toxic environment. Chemical items have a relative ease and manner of production; and biological weapons have an incubation time which allows terrorists to release them and then be gone in plenty of time before victims' symptoms appear. There is a serious potential for attackers to "mix" industrial/explosive/radioactive/chemical/biological weapons, two or a number, together at a single incident, and fortify C/B weapons so they won't respond effectively to medicines such as antidotes. There is a very narrow response time to render lifesaving antidotes for chemical agents and antibiotics for biological substances.

Chemical Protective Clothing and Equipment

Levels of Protection: The U.S. Environmental Protection Agency (EPA) has designated four levels of protection in order to help in determining which combinations or respiratory protection and protective clothing should be employed.

    Level A protection should be worn when the highest level of respiratory, skin, eye, and mucous membrane protection is needed. This level consists of a fully-encapsulated, vapor-tight, chemical-resistant suit, chemical-resistant boots with steel toe and shank, chemical-resistant inner/outer gloves, coveralls, hard hat, and self-contained (positive pressure) and SCBA.

    Level B protection should be chosen when the highest respiratory protection is necessary, but a lesser degree of skin and eye protection is required. It differs from **Level A**

only in that it provides splash protection through use of chemical-resistant clothing (overalls and long-sleeved jacket, two-piece chemical splash suit, disposable chemical-resistant coveralls, or fully-encapsulated, non-vapor-tight suit, and SCBA).

Level C protection should be selected when the type of airborne substance(s) is known, concentration is measured, criteria for using air-purifying respirators are met, and skin and eye exposure are unlikely. It involves a full face piece, air-purifying, canister-equipped respirator and chemical-resistant clothing. This level of protection provides the same degree of skin protection as **Level B,** but a lower level of respiratory protection.

Level D is basically a work uniform. It provides no respiratory protection and minimal skin protection, and it should NOT be worn on any site where respiratory or skin hazards exist.

OSHA mandates specific training requirements that require eight hours of initial training or sufficient experience to ensure competency for personnel in the First Responder Operations Level. Also, each agency must develop a health and safety program and provide for emergency response. Such standards are also intended to provide particular protection for firefighters, police officers, and emergency medical service personnel. OSHA's final rule put out on March 6, 1998, as 29 CFR 1910-120, states training shall be based on the duties and functions to be performed by each responder or an emergency response organization. It is true that no single combination of protective equipment and clothing is capable of protecting against all hazards. PPE should be used in conjunction with other protective methods since the use of PPE can itself create significant worker hazards: psychological stress, heat stress, physical stress, as well as impaired mobility, vision, and communication. Response personnel should not be expected to use PPE without adequate training. A comprehensive training plan for personal protective equipment should include, at a minimum: hazard identification; medical monitoring; environmental surveillance; selection, use, maintenance, and decontamination of PPE; and training. There are a number of complications with the use of PPE: reduced dexterity, limited visibility, restricted movement, claustrophobia, suit breach, dehydration, insufficient air supply, and the effects of both heat and cold. Only first responders who are physically fit and have OSHA, NIOSH, and National Fire Protection Association (NFPA) training requirements should be allowed to wear PPE during an incident. Medical surveillance should be conducted on all entry personnel both before and immediately thereafter their use of PPE. Proper donning and doffing procedures must be followed while being closely followed by the on-site Safety Officer. Upon completion of technical decontamination, all personnel should be examined to find marked changes in health, and medical care should be provided when illness or injury is found. Both the EPA and NIOSH recommend that initial entry into unknown environments or into a confined space that has not been chemically characterized be conducted wearing at least Level B, if not Level A, protection. When considering the proper level of protection, it will include the potential routes of entry for the chemical(s), the degree of contact, and the specific task assigned to the user.

Decontamination: Decontamination is the process of removing or neutralizing harmful materials that have contaminated people and/or equipment during response to an incident. Decontamination is vastly crucial to the overall response to a chemical agent response; due to the incubation period problem, there is some doubt that emergency first responders will be called to a biological agent alarm except to do cleanup. Speaking mostly about chemical agents, responders protect all hospital personnel by sharply limiting the transfer of chemical agents from a contaminated zone into clean zones. They protect communities by preventing transportation of persons contaminated with chemical agents to other sites in communities through secondary contamination. They also protect workers by reducing contamination and resultant permeation of, or degradation to, their protective clothing and equipment. Responders also seek to protect non-contaminated persons already receiving care at hospitals in the area. Avoiding contact is easy; that is, do not allow contaminants to get on responders or bystanders by following a preconceived standard operational procedure. If contamination cannot be helped, then appropriate decontamination and/or disposal of a person's, or responders,' outerwear will have to be done. Segregation and proper placement of clothing or items of protective clothing in a polyethylene bag or steel drum will be needed until thorough decontamination has been completed. With extremely hazardous materials or chemical agents, it may become necessary to actually dispose of some contaminated items.

Defensive Measures: The great majority of inhalation anthrax cases where treatment was begun after victims had gone through an incubation period and become symptomatic have died, regardless of the treatment offered. Most anthrax strains are sensitive in vitro to penicillin, but penicillin-resistant strains exist naturally, and it's not too difficult for a government or terrorists to induce resistance to penicillin (or to tetracyclines and erythromycin which are usually recommended to patients who are penicillin-sensitive). However, all naturally occurring strains tested to date have been sensitive to erythromycin, ciprofloxacin, chloramphenicol, and gentamicin; with a lack of information regarding antibiotic sensitivity, treatment should begin at once at the earliest signs of disease with oral ciprofloxacin with a dosage of 1000mg initially, followed by 750 mg orally twice-a-day) or intravenous doxycycline (200 mg initially, followed by 100 mg every twelve hours). Supportive therapy for shock, fluid volume deficit, and adequacy of airway may all be needed.

Response On Scene: The U.S. Army Soldier and Biological Chemical Command (SBC-COM) at Aberdeen Proving Ground located in Maryland put together a study group in 1999 aided by firefighters from Montgomery County Fire Department, and the Baltimore City Fire Department. The focus of the study and testing that went on examined how well turnout gear (commonly worn by firefighters) with SCBA protects the firefighters against vapor adsorption at the skin. Firefighters were exposed, while wearing turnout gear with SCBA, to a chemical agent simulant (methyl salicylate) to measure what was called the Physiological Protective Dosage Factor (PPDF). An assessment was conducted using the measured PPDFs to determine quick rescue times for known living victims. It was leaned that standard turnout gear with SCBA provides sufficient protection from nerve agent vapor hazards inside interior or downwind areas of the hot zone to allow thirty minutes rescue time for known live victims. The study came up with "General Guidelines" as follows:

"Standard turnout gear with SCBA provides a first responder with sufficient protection from nerve agent vapor hazards inside interior or downwind areas of the hot zone to allow *thirty minutes rescue_time for known live victims.*

"Self-taped turnout gear with SCBA provides sufficient protection in an unknown nerve agent environment for a *three-minute reconnaissance to search for living victims,_*(or a two-minute reconnaissance if HD (mustard, blister agent) is suspected."

Treatment on Scene

Aerosolized Atropine (MANAA)   is effective for nerve agent poisoning; including tabun (GA), sarin (GB), soman (GD), GF and VX. Atropine solution in a pressurized container with an inhaler, MANAA contains about 240 puffs for inhalation each holding 0.43 mg of atropine sulfate, equivalent to 0.36 mg of atropine. MANAA is designed to be used by ambulatory nerve agent casualties with respiratory symptoms as a supplemental treatment after adequate injection of atropine has been given. It is to be used under medical supervision and not for self-buddy aid. Limited side effects are usually deemed insignificant in a nerve agent casualty.

Nerve Agent Pretreatment Pyridostigmine (NAPP)   provides a countermeasure to soman (GD) and/or tabun (GA). NAPP consists of twenty-one 30-mg pyrdostigmine bromide tablets in a blister pack within a sealed pouch. (In the military, pyrdostigmine bromide tablets can only be used under what is called an "activated contingency protocol" — if it is determined that soman or tabun are an actual threat, permission is given that pretreatment be given.) It has been deemed that NAPP treatment substantially increases the chemical components of the Mark I kit against soman and tabun. Tablets are taken every eight hours, and personnel are issued a total of forty-two tablets each. Although a number of side effects could be apparent, they are still considered insignificant versus the tremendous improvement of the Mark I kit's effectiveness against the nerve agent soman (GD) offered by NAPP.

Convulsant Antidote for Nerve Agent (CANA)   is used as a countermeasure to soman (GD). CANA is a single auto-injector holding 10 mg of diazepam to be used for the control of convulsions and to prevent brain and cardiac damage. CANA is to be used in conjunction with NAPP (Nerve Agent Pretreatment Pyridostigmine and the Mark I kit Nerve Agent Antidote Kit). Military personnel can be issued one CANA for self/buddy aid of nerve agent casualties. CANA may cause drowsiness, considered insignificant in a nerve agent casualty. CANA is manufactured by Survival Technology, Inc., Rockville, Maryland.

Dimercaprol (British Anti-Lewisite or BAL)   is a colorless, viscous oily compound with an offensive odor used in treating arsenic, mercury, and gold poisoning. It displaces the arsenic bound to enzymes. The enzymes are reactivated and can resume their normal biological activity. When given by injection, BAL can lead to alarming reactions that seem to pass in a few hours.

Nerve Agent Antidote Kit (NAAK or MARK I)   consists of an atropine auto-injector (2 mg), a pralidoxime chloride auto-injector (2-Pam-Cl, 600 mg), the plastic clip joining the two injectors, and a foam case. The kit serve as a countermeasure to nerve agents, including tabun (GA), sarin (GB), soman (GD), GF, and VX. Military personnel can receive three MARK I for self/buddy aid. Possible side effects of atropine and/or 2-PAM-Cl are deemed insignificant in a nerve agent casualty. Intravenous atropine and 2-PAM-Cl can also be made available. The MARK I kit is manufactured by Survival Technology, Inc., Rockville, Maryland.

Antidote Auto-Injectors: Meridian Medical Technologies, Inc. (10240 Old Columbia Road, Columbia, MD 21046. Tel. 410-309-1477, or 800-638-8093, FAX: 410-309-1475, online at   http://www.meridian.com), a leader in the development of auto-injection drug delivery systems, currently responds to the domestic preparedness market. This company introduced the concept of auto-injector self-administration to the military medical community. Presently, their products include a range of auto-injector delivery systems and drug formulations for the domestic preparedness (civilian) market pertaining to nerve agent antidotes to be sold to first responders, firefighters, police officers, civil defense agencies, and cities and states developing anti-terrorist programs. In the event of a terrorist attack or criminal activity involving nerve agents such as VX, sarin, or soman, first responders must administer antidotes quickly and safely. Meridian Medical Technologies' MARK I kit (NAAK) allows first responders to conveniently and effectively to treat not only contaminated victims but also themselves. A syringe can be slow and cumbersome, but auto-injectors are compact, self-contained antidote delivery systems. The company is the only FDA-approved supplier of nerve agent antidotes to the U.S. Department of Defense and U.S. allies. Domestic preparedness nerve agent antidotes auto-injectors come in the following applications.

AstoPen: NSN 6505-00-926-9083. Contains 2 mg atropine sulfate equivalent in 0.7 ml.

ComboPen: (2-PAM-Cl) NSN 6505-01-125-3248. Contains 600 mg pralidoxime chloride in 2 ml.

Nerve Agent Antidote Kit: (NAAK)/ MARK I. NSN6505-01-174-9919. A kit containing the AstroPen and ComboPen.

Diazepam: (CANA) NSN 6505-01-274-0951. A kit containing 10 mg diazepam in 2 ml.

Morphine: NSN 6505-01-302-5530. Contains 10 mg/0.7 ml morphine sulfate. Each auto-injector is a disposable, spring-loaded device pre-filled with a prescribed drug. The concealed needle and speed of injection render it quick, easy, and convenient applications for self and buddy use. To activate, remove the safety cap found on the end of the unit, place the front end of the unit against the victim's outer thigh and push firmly. The pressure releases a spring which drives a concealed needle into the muscular tissue of the outer thigh. The auto-injector can be administered on bare skin, but it can also penetrate through heavy clothing. When using the MARK I kit, the AstroPen would be administered first followed by the ComboPen. In addition to the classic symptoms, nerve agents also induce convulsive seizures in some victims; the diazepam auto-injector would treat this indication. The morphine auto-injector is available for pain management needs.

Skin Decontamination Kit (M291):   This kit available to the military and through contractors to the civilian market is designed to work on liquid nerve agents including tabun (GA), sarin (GB), soman (GD), GF and VX, and vesicant agent sulfur mustard (HD) The kit contains six packets, each containing a pad filled with a mixture of activated resins that both absorb and neutralize liquid agents from a victim's skin. One pad will decontaminate both hands and the face, or an equivalent area of skin. Contact with open wounds, eyes, and mouth should be avoided, but the reactive and absorptive resins in the M291 Kit are nonirritating and nontoxic, even after a prolonged contact with the skin. Manufactured by Rohm and Haas Co., Philadelphia, the skin decontamination kit is FDA approved.

  Decontamination Kit, Skin (M258A1) to avoid confusion, carries six packets within a kit, three Decon 1 carrying a mixture of hydroxyethane and phenol which adsorbs and neutralizes the G-series nerve agents, and three Decon 2 with a mixture of chloramine B and hydroxyethane which are meant to absorb and neutralize the nerve agent VX and liquid mustard. For military troops, the 251A1 kit is currently the standard issue for personal decontamination of liquid agents on the skin. A soldier would use a Decon 1 packet, fold on the solid line, tear open quickly at the notch, remove pad from envelope, unfold quickly, and wipe hands, neck and ears for one minute. Next, the soldier would take a Decon 2 pad, crush ampoules, fold packet at the solid line, tear open quickly, remove pad letting screen fall away, and wipe exposed area for two to three minutes. If soldiers are certain they have an agent on their face, they would use a Decon 1 pad to wipe their hands. Then they would hold their breath, lift off the protective mask from their chin, decontaminate the lower half of the face, then decontaminate interior sections of the mask that contacts the skin. They would then re-mask using the same Decon 1 pad to wipe neck and ears. They would then repeat the same procedure using a Decon 2 pad on hands, face, mask, neck and ears. **WARNING:** The ingredients of the Decon 1 and Decon 2 packets of the M258A1 kit are poisonous and caustic and can permanently damage the eyes. The wipes must be kept out of the eyes, mouth, and open wounds. (The M259A1 kit will eventually be replaced with the M291 kit.)

Toxicity: Toxicity is influenced by four factors — chemical, exposure, the exposed person, and the environment. Factors related to the chemical area include composition; physical characteristics (size, liquid, solid), physical properties (volatility, solubility); presence of impurities; breakdown products, and carriers. In the exposure area are included dose, concentration, route of exposure (inhalation, ingestion, dermal), and duration. Factors related to the exposed person include heredity, immunology, nutrition, age, sex, health status, pre-existing diseases, and pregnancy. Finally, factors related to the environment tend to be media (air, water, soil, additional chemicals that may be present, temperature and humidity, air pressure, and fire. In most instances as the dosage increases, the severity of the toxic response increases; but the severity of the toxic effect also depends on the duration of the exposure. Toxicity information available to emergency first responder personnel is provided primarily by test animals. As an example, LD 50 (lethal dose 50 percent) is the dose that is lethal to 50 percent of the test animals from exposure by a specific route (not including inhalation) when provided all in a single dose. Another usual term when discussing toxicology is the LC 50 (lethal concentration 50 percent); this term means the concentration of material in air that is expected to be lethal to 50 percent of the test animals when concentrated in a single exposure, generally over a time period of one hour. For instance, for sodium cyanide the acute lethal dose used orally on rats is given as LD 50 6.4–10 expressed as mg/kg (milligrams of the compound administered per kilogram body weight of the test animal); while the LD 50 on the same type of rat using the same terminology as before would be 2,600–7,000 mg/kg for the industrial chemical toluene. As with different types of test animals, toxic chemicals will treat different human beings in varying effects. People can cause different toxic effects because of physiological variability that is present in all human beings — if you smoke, drink, have a bad diet, are extremely young or a senior citizen, have a reduced ability to metabolize a specific toxic compound, have a pre-existing medical condition (like a low immune system as transplant survivors often have due to the medicines that have to take for life, renal disease, diabetes and other situations), previous exposure to toxic chemicals, and certain medications — effects will vary. **NOTE: Not all chemicals have a threshold level. Some carcinogens (cancer-causing chemicals) may produce a response in the form of a tumor at any dose level, and exposure to such compounds may be associated with some risk of developing cancer.**

Transportation: Regulations on the transportation of biological agents are an attempt to ensure the public and workers in the transportation media are protected from any exposure to such agents. Protection is built around rigorous packaging, labeling packages with the biohazard symbol, and documentation of the hazardous contents of all biological agents shipped in commerce, and training of workers who handle biohazard shipments. Various laws in the United States control the importation of etiologic agents of human disease; etiologic agents of livestock, poultry and other animal diseases; plant pests; transfer of select biological agents of human disease; and export of etiologic agents of humans, animals, plants and related materials.

Field Decontamination: The question, when do you decontaminate, and when do you not decontaminate is often asked. First, decontamination can be accomplished by neutralizing or removing a contaminant. Simple weathering could also remove a contaminant, but time is of the essence when dealing with chemical and biological agents or weapons. Responders and victims exiting the hot zone at a chemical or biological incident should immediately undergo water (high volume–low pressure) decontamination. For example, if a responder is exposed to nerve or blister agents, he or she should begin decontaminating action within one minute of contamination. For some agents, you will want to decontaminate contaminated persons, equipment, and materials as quickly as possible. For other agents, such as arsine, cyanogen chloride, and hydrogen cyanide, there is little need to decontaminate a victim exposed only to vapor; if liquid agent is present, remove clothing and wash the liquid off the victim's skin. For diphosgene and phosgene, no decontamination is likely to be needed in the field; however, provide aeration in closed spaces. The blister agent ethydichloroarsine (ED) does not usually require decontamination in the field; however, for enclosed areas, use HTH, STB, household bleach, caustic soda, or DS2 to decontaminate. Decontamination of most biologically contaminated victims and equipment can also be accomplished with soap and water. Ideally, skin decontamination should start within one minute of contamination. Within two minutes, some chemical agents can cause serious injury to a victim. Phosgene oxide, a blister agent, can damage skin within seconds. It is necessary to either neutralize or remove the contaminant as rapidly as possible following good practice and procedures. In many instances, chemical agents will undergo natural decontamination following the passage of sufficient time, but this type of withering will not normally be acceptable at an incident scene, and certainly not when people or first responders have been contaminated. When chemical agents are used or abused, the basic good response practice can be stated in four words: first responders have to *detect, identify, control,* and *decontaminate* chemical agent(s). Controlling the scene is of the utmost importance. Anyone in a contaminated area should be detained, checked for contamination, and decontaminated if necessary. Contamination should be contained within the smallest possible area, and no one should be allowed to carry contamination into a wider area. Responders will have to deal with contamination that is solid (dust, powder, dirt), liquid (rain, mist, vapor), or gas (toxic gas clouds or residue). It is very difficult to stop the spread of contamination. You can walk through it, sit on it, wipe your nose, inhale it, or eat it to spread your contamination from one surface to another. The more you spread it, the more you will have to decontaminate people including yourself, equipment, and materials in formerly "clean" areas. Most chemical agents evaporate or disperse within a short period of time, but liquid agents on surfaces, equipment, and materials may release toxic gases for days.

The "enemy" may use two or more agents, chemical and/or biological, at the same time to confuse efforts to detect and identify exactly what the victims should be responding to. Response personnel should decontaminate the chemical agent(s) first, since these agents often comprise the most lethal, fastest-acting types of contamination. At the present time, no one in North America really knows what will be the reaction of average citizens under an enemy attack or terrorist action involving chemical or biological agents or weapons. We all have opinions, but practical knowledge is lacking for all except the "super-anthrax" attacks that began in Florida and Washington, D.C., about October 15, 2001. We may have people running amok, losing self-control, and creating panic; or we may have trained and prepared citizens who have the necessary equipment, supplies, and knowledge needed to make a bad situation better than it might have been. History seems to illustrate that military units trained, prepared, and equipped to respond to chemical agents have better defense and survival rates. If the civilian responders are also trained, prepared, and equipped to respond to chemical agents (many public safety agencies, commercial response contractors, and industry response teams already meet these important requirements for survival), we will be much better off than we were on September 11, 2001. However, if the funding and resolve necessary to provide special tactics training, preparation. and equipment to the many responders who have not yet benefited from such preparations; the necessary resulting task will be very expensive, time consuming, and very possibly a failure.

Decontamination Solutions

The military services use a number of standard decontamination solutions, and some not so standard. Even the standard ones have problem with which emergency first responders should be aware and prepared to handle. Civilian fire responders often have to decontaminate victims, but with chemical and biological agents and weapons they will have to decontaminate more equipment, materials, surfaces, buildings and grounds than they might have done in the past. **Here, we are NOT talking about decontamination of human beings in the rest of this paragraph; we are talking about decontamination of objects rather than people (Do not use such items on people, <u>except</u> for hot soapy water in the listings below).** For the nerve agents sarin (GB) and tabun (GA), the following decontamination solutions are listed in order of preference: (1) caustic soda solution (sodium hydroxide), (2) DS2 (decontaminating solution 2), (3) washing soda solution (sodium carbonate), (4) STB (supertropical bleach) slurry, and (5) hot soapy water. With the following blister/vesicant agents, sulfur mustard (H), mustard/lewisite mix (HL), and lewisite (L), the decontamination solutions to be used in preferential order would be (1) HTH-HTB calcium hypochlorite, (2) DS2, (3) STB slurry (supertropical bleach), (4) commercial of household bleach solution (sodium hypochlorite). For another blister agent, phosgene oxime, the solution of choice is DS2 used by the military. The nerve agent VX should be handled by the following decontamination solutions in order of preference: (1) HTH-HTB solution (calcium hypochorite), (2) DS2, (3) STB (supertropical bleach) slurry, (4) commercial or household bleach solution (sodium hypochlorite). For the choking agent phosgene (CG) and the blood agents, cyanogen chloride (CK) and hydrogen cyanide (AC), the first choice in the military would be DS2 solution and the second choice caustic soda solution (sodium hydroxide).

In preparing smaller amounts of either STB (supertropical bleach) or HTH (calcium hypochlorite) solution, the following mixture would apply. For a 5 percent solution, mix

0.6 pounds of STB or HTH with one gallon of water; or mix 3.6 pounds of STB or HTH with five gallons of water. For a ten percent solution, mix 0.75 pounds of STM or HTH with one gallon of water; or mix 4.5 pounds of STB or HTH with five gallons of water. A five percent solution (in water) of either sodium hypochlorite (household bleach) or calcium hypochlorite (HTH) will provide an effective decontamination of most chemical and biological agents or weapons **(such mixtures can be used to decontamination human skin if a soap and water bath follows the application of the decontamination solution, and the decontamination solution is kept away from the victim's eyes, mouth, and face).**

Decontaminating Solution No. 2 (DS2) can be used against all known toxic chemical agents and biological materials (except for bacterial spores) when allowed to remain in contact with contaminated surfaces for approximately thirty minutes and rinsed with water. DS2 is most effective when scrubbing action is used. However, a number of cautions need to be followed. Decontamination workers should wear protective masks and rubber gloves and vapor should not be inhaled. The solution is combustible and extremely irritating to the eyes and skin. When DS2 contacts the skin, it is necessary to wash the area with water. The solution ignites spontaneously with STB and HTH, and will cause a black color change with M8 detector paper and a false/positive result with M9 paper.

Supertropical Bleach (STB) is appropriate for both chemical and biological agents, being most effective against lewisite, V and G nerve agents, and biological agents. STB must have contact with the contaminated surface for at least thirty minutes, then be washed off with water. Problems areas with STB includes spontaneous ignition with liquid blister agent or DS2, emission of toxic vapors on contact with G nerve agents, and corrosion of most metal and damage to most fabrics so that after use, it is necessary to oil surfaces and rinse fabrics. Both dry and slurry mixtures of STB do not decontaminate mustard agents in a satisfactory manner if such agents have solidified at low temperatures. STB should not be inhaled or come in contact with the skin, and a protective mask or other respiratory protection should be donned when preparing slurry. STB can be used as a slurry paste (approximately equal parts by weight of STB and water) dry mix (two shovels of STB to three shovels of earth or inert materials), and slurry mix (for chemical decontamination, use forty parts of STB to sixty parts of water; for biological decontamination use seven parts of STB to ninety-three parts of water by weight).

Soaps and Detergents handle both chemical and biological decontamination when contaminated surfaces are scrubbed or wiped with hot, soapy water solution or when items are soaked. Soaps and detergents physically remove contamination, but the runoff water must be considered and handled as contaminated, as it may cause casualties. For smaller amounts of soap solution, use about one pound of powdered soap per gallon of water; for larger amounts, mix 75 pounds of powdered soap in 350 gallons of water. Laundry soap could also be used; cut 75 pounds of laundry soap into one-inch pieces and dissolve it in 350 gallons of hot water. Using detergent, the mixture would be about one pint of detergent to 225 gallons of water.

Sodium Hypochlorite Solution (household bleach) can be used against both chemical and biological agents. It is effective and fast-acting against blister and V nerve agents, but requires a contact time of ten to fifteen minutes for all biological materials. It is available in food stores as a 5 percent solution under various brand names, and as a 14 to 10 percent solution at commercial laundries. Undiluted, household bleach is harmful to skin and clothing, and corrosive to metals unless the metal is rinsed, dried and lubricated following decontamination. For decontamination spray, dilute half and half with water. No mixing is required for chemical decontamination; for biological decontamination, add two parts bleach to ten parts of water.

Calcium Hypochlorite (HTH, HTB or high test hypochlorite) can decontaminate lewisite, V nerve agents, and all biological agents, including bacterial spores. It reacts within five minutes with mustard agents and lewisite, but must have fifteen minutes contact time for biological agents. HTH is more corrosive that STB. Undiluted, it will burn on contact with VX, HD, or DS2. It can be used as a slurry or dry mix, has a toxic vapor and will burn the skin. Therefore, a mask and rubber gloves are the minimum protective equipment required when dealing with this substance. For chemical decontamination, mix five pounds HTH to six gallons of water for a 10 percent solution; for biological decontamination, mix one pound HTH to six gallons to six gallons of water for a 2 percent solution. A slurry with three parts HTH and ninety-seven parts water can be used on flat surfaces by spreading one gallon per eight square yards.

Sodium Hydroxide (caustic soda or lye) decontaminates G nerve agents, neutralizing them on contact. It also decontaminates lewisite and all biological materials including bacterial spores. Contact time with the contaminated surfaces should be about fifteen minutes. Sodium hydroxide can damage skin, eyes, and clothing on contact, can damage lungs or the respiratory system via inhalation of dust or concentrated mist, and will cause a red color change with M8 detector paper. In dealing with this decontamination solution, full rubber protective clothing, gloves, boots, and mask are required. Flush with large volumes of water and rinse with acetic acid or vinegar if contact with skin occurs. The runoff is highly corrosive and toxic. Mixtures for 10 percent solution are ten pounds of lye to twelve gallons of water. Never use zinc, aluminum, or tin containers for mixing; use strictly iron or steel mixing containers, and add the lye to water since excessive heat will be formed. Do not handle the mixing utensil with bare hands. Lye should NOT be used as decontamination solution if less caustic solutions are available.

Sodium Carbonate (washing soda, soda ash, or laundry soda) normally reacts within five minutes against G nerve agents or weapons. Sodium carbonate is unable to detoxify VX nerve agent, and when mixed with it creates very toxic by-products. Sodium carbonate solution is also ineffective against distilled mustard (HD) blister agent as it does not dissolve this agent. For a 10percent solution, mix ten pounds of sodium carbonate with twelve gallons of water. **This is a chemical decontamination solution and *should not be used on biological agents.***

Potassium Hydroxide (caustic potash) will work on certain chemicals and biological agents. Remarks under sodium hydroxide apply across the board to potassium hydroxide. **Ammonia or Ammonium Hydroxide** (household ammonia) can be used to decontaminate G nerve agents but takes longer than sodium hydroxide or potassium hydroxide. SCBA or a special purpose mask is required when working with this product. Ammonium hydroxide needs no further mixing; it is a water solution of ammonia.

Perchloroethylene (tetrachloroethylene) is a nonflammable solvent of low toxicity that dissolves and removes H blister and V nerve agents but does not neutralize them. NIOSH has recommended that this substance be treated as a potential human carcinogen. It does not work with G nerve agents.

Decontamination Solutions Used by the U.S. Environmental Protection Agency

For both known products and unknown products, please refer to "Guideline for Selecting Degradation Chemicals for Specific Types of Hazards" below:

Solution A: Five percent sodium carbonate and five percent trisodium phosphate. Mix four pounds of commercial-grade trisodium phosphate with each gallons of water.

Solution B: Solution containing 10 percent calcium hypochlorite. Mix eight pounds with ten gallons of water.

Rinse Solution C: A general purpose rinse to be used for both solutions A and B. Five percent solution of trisodium phosphate with each ten gallons of water.

Solution D: A diluted solution of hydrochloric acid (HCL). Mix one pint of concentrated HCL into ten gallons of water (acid to water only). Stir with wood or plastic stirrer.

Guideline

Solution A: Inorganic acids, metal processing wastes.

Solution B: Heavy metals; mercury, lead, cadmium, etc.

Solution B: Pesticides, chlorinated phenols, dioxins, PCPs.

Solution B: Cyanides, ammonia, and other non-acidic inorganic wastes.

Section C or A: Solvents and organic compounds such as trichloroethylene chloroform, and toluene.

Solution C or A: PBBs and PCBs.

Solution C: Oily, greasy, unspecified waste not suspected to be contaminated with pesticides.

Solution D: Inorganic bases, alkali, and caustic waters.

Casualties and Field Decontamination

Precise diagnosis of biological agent casualties is likely to be "by guess or by-golly" because casualties and response personnel, including on-scene paramedics and EMTs, may be in protective clothing, biological attack casualties might be mixed in with conventional, Nuclear/Biological/Chemical (NBC) casualties, appropriate laboratory services may not be readily available, and the scene may be a mass casualty incident. In any event, the treatment required for biological agents or weapons casualties will not differ much in the basic management from that provided to patients suffering from the same disease obtained by natural means. "Field Decontamination" means **immediate** decontamination with a mass of water at low pressure. It does **not** need to include scrubbing or scouring human bodies with brushes of other implements (which may be fine way to decontaminate response equipment, gear, or vehicles). The major premise of field decontamination is to get water on the victims and do it as quickly as humanly possible. In the absence of agent-specific guidance, casualties should be disrobed if at all possible (**this cannot be overstressed**), showered with fire hoses or other equipment for at least ten minutes if possible, be provided with a temporary change of clothing such as hospital scrubs or anything else to maintain dignity, have their normal clothes bagged and stored until the actual agent is known (and a decision can be made to decontaminate such clothing or discard it in a lawful manner), and have their normal personal gear (keys, wallets, rings, jewelry, photos, money, etc.) bagged and stored until a decision can be made if it can be returned, decontaminated, or discarded). Once these actions have been completed, exposed areas or response equipment should be cleaned with the use of diluted sodium hypochlorite solution at 0.5 percent.

Isolation Procedures: Once a victim of chemical or biological agents or weapons gets to the hospital or another healthcare facility, that person may have to abide by patient isolation procedures. These include Standard Precautions, Airborne Precautions, Droplet Precautions, and Contact Precautions. These precautions are spelled out in the third edition of *Medical Management Of Biological Casualties Handbook* published by the U.S. Army Medical Research Institute of Infectious Disease located at Fort Detrick in Frederick, Maryland.

Standard Precautions: Hand washing after patient contact. Use gloves when touching blood, body fluids, secretions, excretions, and contaminated items. Use of mask, eye protection, and gown during procedures likely to generate splashes or sprays of blood, body fluids, secretions, or excretions. Handle contaminated patient-care equipment and linen in a manner that prevents the transfer of microorganisms to people or equipment. Practice care when handling sharps and use a mouthpiece of other ventilation device as an alternative to mouth-to-mouth resuscitation when practical. Place the patient is a private room when feasible if they may contaminate the environment.

Airborne Precautions: Standard Precautions plus: Place the patient in a private room that has negative air pressure, at least six air changes/hour, and appropriate filtration of air before it is discharged from the room. Use respiratory protection when entering the room. Limit movement and transport of the patient, and use a mask on the patient if they need to be moved.

Droplet Precautions: Standard Precautions plus: Place the patient in a private room or with someone with the same infection if possible. Use gloves when entering the room. If not feasible, maintain at least three feet of space between patients. Use a mask when working within three feet of the patient. Limit movement and transport of the patient, and use a mask on the patient if they need to be moved.

Contact Precautions: Standard Precautions plus: Place the patient in a private room or with someone with the infection if possible. Use gloves when entering the room, and change gloves after contact with infective material. Use a gown when entering the room if contact with patient is anticipated or if the person has diarrhea, a colostomy or wound drainage not covered by a dressing. Limit the movement or transport of the patient from the room. Ensure patient-care items, bedside equipment, and frequently touched surfaces receive daily cleaning. Dedicate use of non-critical patient-care equipment to a single patient, or patients with the same pathogen. If not feasible, adequate disinfection between patients is necessary.

Other Standard Precautions: After an invasive procedure or autopsy, all instruments used and locations need to be disinfected with a sporicidal agent. Recommended for use is iodine, but it must be used at disinfectant strengths since antiseptic-strength iodophors (a complex of iodine and a surface-active agent that releases iodine gradually and serves as a disinfectant) are not usually sporicidal. Chlorine (sodium or calcium hypochlorite) can also be used but with due caution since the action of hypochlorites is greatly reduced when used with organic material.

Psychological Impacts: Responders, as well as victims, at chemical and biological incidents or warfare may share symptoms or reactions that are similar to battle fatigue or combat stress, according to the U.S. Army Center for Health Promotion and Preventive Medicine at Aberdeen Proving Grounds in Maryland. Common physical signs that signal stress reaction include tension; jumpiness; cold sweats; heart pounding and light-headedness; breathing, stomach, bowel, and bladder problems; loss of energy; and a distant, haunted "1,000-yard" stare. There are also common mental and emotional signs that include anxiety (keyed up, worrying, expecting the worst) irritability, attention span, thinking (being unclear, having trouble communicating), sleep problems, feelings of grief, guilt and anger; and lack of confidence. First responders and citizens who get involved with chemical and biological attacks, spills, and/or terrorism can control stress, at least to some extent. The U.S Army has some tips. Try to look calm and in control, follow standard operating procedures, focus on success, realize that stress reaction is normal, breathe deeply and relax, and maintain open communication with your coworkers. Gather facts, avoid rumors, avoid alcohol but drink plenty of fluids, eat well-balanced meals, maintain personal hygiene, and practice quick relaxation techniques. First responders and others should debrief after unusually stressful events, share grief with a friend, keep active, and stay physically fit. Some serious stress reactions include hyper-alertness, fear, anxiety, trembling, irritability, anger, rage, grief, self-doubt, guilt, complaints, inattention, carelessness, loss of confidence, loss of hope and faith, and depression. Also included are symptoms such as insomnia, impaired job performance, erratic actions, outbursts, immobility, "freezing" under stress, terror, panic, total exhaustion, loss of skills and memory, impaired speech or muteness, impaired vision/touch, weakness and paralysis, plus hallucinations and delusion.

Detectors for Chemical and Biological Agents

The two-volume report, *Guide for the Selection of Chemical Agent and Toxic Industrial Material Detection Equipment for Emergency First Responders*, done by the National Institute of Justice, is available on the Internet (http://www.ojp.usdoj.gov/nij/pubs-sum/184449. htm). This guide for emergency first responders provides information about detecting chemical agents and toxic industrial materials and selecting equipment for different applications. Because of the large numbers of items identified in this guide, Volume I features the guide, and Volume II contains the detection equipment data sheets. The full text of the report is available in PDF format and an be downloaded without cost. The *Guide* contains the following contents: Foreword, Executive Summary, Introduction, Introduction to Chemical Agents & Toxic Industrial Materials, Overview of Chemical Agents and Toxic Industrial Materials (point detection technologies, standoff detectors, analytical instruments), Selection Factors (chemical agents detected, toxic industrial materials detected, sensitivity, resistance to interferants, response time, start-up time, detection states, alarm capability, portability, power capabilities, battery needs, operational environment, durability, procurement costs, operator skill level, training requirements) Equipment Evaluation (equipment usage categories, evaluation results), Recommended Questions on Detectors, and References. There are also a number of tables and figures. This book is a good source for persons needing to know basic information of the current market of chemical agent and toxic industrial materials

Detection Equipment for Emergency First Responders

SensIR Technologies

The TravellR, a portable Fourier transform infrared spectrometer (FTIR), can be a key component for a hazardous materials response team, and can serve as a complement to existing gas meter technology. The FTIR spectography offers a simple, rapid solution for any sample type. This chemical identifier handles solids, liquids, pastes, and powders in seconds. According to Director of Sales and Marketing Sensir Technologies Jim Fitzpatrick (based out in Danbury, Connecticut), "Sensir has since September 11th been called upon to help with the remediation of hazardous material incidents. For about twenty years, it has been a manufacturer of spectrometers that are used mainly in the pharmaceutical, drug enforcement, petroleum and petrochemical industries. We are actually considered experts in our field. Haz-Mat response was somewhat new to us. I can describe a little bit about how that came about.

"The basis behind FTIR (Fourier transform infrared) spectroscopy is that infrared energies pass through a sample and the infrared light interacts with the sample, and it's this interaction that helps us identify the compound. Every sample has a different interaction unless it's an identical sample. For example, cocaine versus procaine, two very similar chemicals, but in our system two very different spectra. We would have no problem at all differentiating one from the other. What we look at with our system is when an infrared light passes through the sample, some light is absorbed and doesn't come out the other side. We are looking at the difference between what we started with and what we ended up with, and the absorption of light is what we register; the absorption of light is what identifies the compound. With our system, we can run hundreds of thousands of compounds, and store their interaction with light on the hard drive of a computer. Then when we want to run something that no one has seen before, we compare against the interactions of all these other compounds and we get an identification based on that. This is a physical constant of the compound, so it doesn't change. If cocaine gives us a spectrum today, that's what we call this interaction, we can run cocaine twenty years from now and get the same spectrum. It doesn't change for that compound. It's highly specific for that compound, so once we get an identification of a material, put in the library, we could instantly identify that. It takes about eight seconds so the system is very simple and easy to use. We have a sensor on the front, and the Haz-Mat technician could procure some samples, whether it's a liquid or a solid, walk to the system, put it onto the diamond; the diamond is about a millimeter in size, so it's not a whole lot bigger than the head of a pencil. You just lay the sample on the diamond. If it's a solid, you compress it with a little pressure device that's on the system. If it's a liquid, you just put a drop on, hit the button, and within 8-l0 seconds it will have gone through, registered the absorption of light, compared it against the library, and come back with an answer.

"The reason it's been successful in sales is that it requires no calibrations in the field. It requires no consumables. The firefighter can put it away for six months, take it out again when he needs it, and be operable within minutes. The reason we got involved at all was the Haz-Mat teams were telling us there was no detector for solids or liquids. There *are* systems that can look up solids and liquids that are giving off gases, but they are only identifying the gas. Firefighters needed something that would identify the actual chemical in its solid and liquid form, and there was nothing out there that would do that. That's where we came into the fold. We got significant usage during the white powder incidents because we were able to identify baking soda, baby formula, and talcum powder and sugar; all of the things that were being used in a hoax situation, and we were doing them in eight to ten seconds. The other option was for the Haz-Mat responder to send them away to a laboratory, and then wait twenty-four hours for the information. So, literally within seconds, they were doing that, then off to the next call. We had back-to-back calls since we could remediate the problem very quickly. It was only if the Haz-Mat responders got hit by some biological substance that they would then have to go to another level of identification, and use alternate means to identify the biological material. Once again, the FTIR spectrometer is a **chemical identifier;** it recognizes biological material, but it doesn't identify biological material. As soon as you get a hit in our system that says 'biological,' you would then reach for your other tool kit and identify the biological. The big deal is the speed of analysis.

"Yesterday, we had forty different groups come through. I'd show them how to do it, they would walk up and put a sample on the diamond. With the diamond sensor, one of its advantages is that it is *diamond*, so you could put any liquid whether it's corrosive and/or abrasive; you don't have to worry about, 'is this going to damage my system?' The front end of the FTIR is stainless steel with diamond; so it is very rugged, a very do-able system. Even in a Level A suit, which is quite a cumbersome suit, the system is quite easy to operate. Use a spatula to scoop up a very small sample, and just place it on the diamond. That was an incident of a powder. There are also a number of incidents involving liquid. Any liquid, whether it's an acid or just a shampoo, there's quite a bit of interest from the airport authorities. They want to be able to look at your cosmetics in your bag. Is it what you say it is or are you carrying something onto the plane in a container that says 'After Shave' but it might be flammable liquid, or much worse? You may be carrying a water bottle, but it could contain almost anything. There might be a pool of liquid in a corner of the airport building; the FTIR could quickly identify exactly and specifically the chemical agent involved.

"In terms of solids, anything that's found in the clandestine drug lab can be run in our system, including any of the precursors, any of the drug tablets, the Sudafed's that are used to manufacture methanphedamine, chemicals that are used in the cooking process; any of those items can be used, and all can be identified in eight seconds since we have libraries of all of those compounds. There are any number of applications; if it's a chemical, everything we're wearing, everything we eat, the medicine we take are all chemicals. We can identify them, we can put it in our system, and identify it.

"The smaller the sample, the less clean up. We can identify a single grain of sugar. For liquid, we need one tenth of a drop. We can greatly decrease the amount of sample needed, which means the cleanup is easier, the transport of the sample is safer. The least amount you use, the less concern it causes. However, you have to present that sample to the system, so at some point you take a toothpick, dip it in the liquid and then transfer that drop to the diamond. You're talking about such a tiny amount of liquid that cleaning it up, you're restricting it to a tiny area for analysis. Clean the tiny area with bleach, hydrogen peroxide, anything you want to clean it with; the system can take it. If you learn the sample is biological, then you would want to bleach that area or use hydrogen peroxide. There's a lot of toxic stuff, but you would have to wash off the system using whatever was appropriate for that.

"If you're dealing with micro drops, everyone's safety is enhanced. You might take a large sample for evidentiary purposes, but from that sample you would take a single half-drop, or a single flake of powder. There are libraries with hundreds of thousands of compounds available; but speaking frankly, they were not very useful for the Haz-Mat responder because there were plastics, pharmaceuticals - everything that the firefighters are not necessarily concerned with. Who cares if it's a Rubber A or Rubber B. The guys in the Haz-Mat teams are concerned about much more dangerous things. As part of our commitment to the Haz-Mat market place, we have actually run these libraries of interest to first responders. We went and procured the list of the NIOSH hazardous materials where there's about 600 compounds. We went to local universities and ran every chemical that they had in their stockroom. We worked with Major Ireland at Fort Gordon with the Georgia National Guard Civil Support Team. He was able to get us access to the nerve agents, so we now have a library of GA, GB, GD, GF, VX - all of those nasties. We have about 500 drugs and drug precursors in our library. When I say drug precursors, I mean the type of things that a Haz-Mat team would run into if they were trying to assist the police. Our FTIR spectroscopy system will identify every chemical, and then they'll know how to take care of them. We have about forty explosives in our library; we have about fifty what we call common bipowder to assist with their remediation.

"The way a library search works is highly specific. Response personnel are not chemists; their job is not to sit there and say, 'I see the band at 1720.' We don't want them to have to worry about getting involved in the interpretation. There are people that have been doing interpretation of spectrums for forty years, and they're very good at it, but that's all they do. We can't expect the firefighters to be an expert in spectroscopy, and we don't. We take care of that problem through software. Our libraries will take peak pattern; the pattern of peaks are highly specific of that compound, and identify the location of the peak. Every peak location has a number, and every peak has a certain height. You might look at different drugs and think they're identical, and a doctor might even think they're identical; but we might see that because if it's been sitting on the shelf for too long, some of the ingredients have actually started to break down, so some of the bands will be shorter. Even though it's chemically the same, some of the bands have gotten smaller. Our software looks at every peak location, every single one of them, four thousand potentials, and it takes all of that into account and says, 'I found the match.' If it's a match if it's 90 or more percentile, if you look at them you'll see that every single peak lines up, the same height and it's in the same location. Let's say you get a 70 percent when you overlay, you'll see exactly that some of those peaks are actually missing. What you have to then is say is we're in the right family but we don't have the right to say it's a 'match.' We tell people if you're **not** in the 90 percentile, then you haven't hit it. We really want to remove the decisions from the responder because he's on-scene, potentially in a Level A or Level B suit. He's got a lot of stuff happening around him. We want to help him make it as clear as possible that this is the right answer. I can set the system so that if it doesn't come back with greater than 90% percentile response, it is not good enough for the emergency responder to hazard what would have to be a guess; that's what we recommend to the emergency responders.

"If I were caught tomorrow with drugs anywhere in this country, and they confiscated the drug for analysis, somebody would be using an FTIR, Fourier transform infrared, to run it no matter which state I was caught in. The FTIR is used in every crime lab in this country. It's computerized infrared which used to be done by hand until a gentleman named Fourier came up with the method, so we now use his Fourier transform to come up with this pattern. It's just a mathematical function we call an FTIR. For all intents and purposes, it's an analyzer that is court accepted identification. In my example, my attorney can't get me off by saying the court didn't use a proper technique. Every court in the county accepts this analysis as proper identification of the compound because it's so highly specific.

"Now, with portable FTIR, we can bring that strength to the emergency responders. We've now made it smaller, more portable and fireman proof; we can now take it out to the field for them to do the same types of analysis off the back of a fire truck. When a Haz-Mat team goes to an incident scene and members see something they have never seen before, they are able to learn from that situation, and the next time they see the same situation they have knowledge to bring to the table. Our system does the very same thing; if we run a compound against the library, and we don't get the right answer, or we get no answer because we set the system up to only return very confident answers. So, we'd say the system did not identify the compound. We have to bring in a lab, send a sample away, and then wait for it to come back. Say the sample proves to be dried extinguisher powder from a fire extinguisher. Now we know what it was, we can now go back to the library and say I know you were dry powder extinguisher and I'm listing you in my library as dry powder. That's done by pushing two buttons; you say add to library, you pick the library you want, and it's in there. If you ever run into that anywhere again, you'll get a perfect identification because it's now in your library. That's the strength of the system, it evolves as the threat evolves.

"We have a Web site access for library updates. What we call 'Administrators,' a key person in the fire department or others agencies that have the FTIR system, handles the security of the system. He has password for protected access to our web site so not everybody can post there, only him. If he runs into a blue liquid that he's never seen before, he identifies it; he says, 'this is Listerine.' He or she puts it in the library, and he says he wants to share this, 'what I saw today, I don't want anyone else in the country having to learn the way I learned.' The Administrator puts it on our Web site; when that happens, every other Haz-Mat responder who is an Administrator who has password and protected access to our Web site can come on the Web and could constantly share spectrum information. They're not having to reinvent the wheel every time. The Web site can really assist in the development of libraries and the dissemination of information. We're facilitating that because we can't run all compounds since there's twenty million of them. We want to be able to run the ones that are causing the most scare at the moment, and the Haz-Mat guys know what's causing the scare so we wanted them to be able to share the information.

"A couple of costs are involved in having the FTIR system, number one is the initial price; the initial price of the system is about $55,000. That includes a tone key system, the lap top computer, all of the software needed to operate the system, and all of the libraries that I described: weapons of mass destruction, toxic industrial chemicals, common chemicals, white powders, drugs, drug precursors, and explosives. It comes with a carrying case that protects the system in transport. It comes with a battery pack. There is also a one-year warranty which includes access to the Web site, access to the 247 Reach Back, and then a twenty-four-hour response if your system goes down. We'll have a system to you within twenty-four hours for you to use while we're fixing yours. To maintain your Reach Back to your normal program, and access to the Web site, you would buy warranties ongoing to extend the warranty for $4,000 per year. You get the things that we described, all parts, fix the system, a full warranty on the system, full access to the chemist, full access to the Web site, and all library updates within the normal program. If your system breaks today and you call us by 3 o'clock, we'll have one in Federal Express to you the next morning by 10 o'clock; we'll take yours back while you use ours. A Haz-Mat team can't be without a working system, so we want them to have instant access to our system.

"Every three to four years, there's a laser and a source in the system, and either one or both of these can go," continues Jim Fitzpatrick. "The cost of a laser is about $700–800, including the power supply, and the cost of the infrared source, which is literally the light bulb of the system, runs about $250. Both of those items can be replaced by the user in the field. The customer would just put in a new source and install a new laser; there's no alignment necessary, no difficult mechanics that need to be done. Just open the top, take out the old one, break the connection, make the connection of the new one, and put it into the same place. Into the third year, you might want to budget $1,200–1,500 or replacement parts, but you would only do that every three to five years. It costs you nothing for the sample, because literally you could use a cotton ball or a toothpick or something else to pick the sample, put it on the diamond, and get your spectrum. There's no need to have expensive sampling tools. If you want to have other tools, you certainly can get yourself some tools, but I use a dental pick as my transfer tool. Admittedly, a toothpick does the same job for either a solid or a liquid. I often use coffee stirrers that are disposable; a coffee stirrer is ideal because it picks up liquid or powder, deposits it on the diamond, and then I can dispose of it.

"There's a pressure device on the system because the sample must make contact with the diamond. If you squirt a liquid on the diamond, it makes perfect contact; if you sprinkle a powder on the diamond, powders don't make perfect contact. To make perfect contact with a powder, take two to three turns of the pressure device. You just take a little piece off the intended sample. You could actually break a pill, and the dust that falls off is enough for the analysis. You must make contact and the pressure device squashes the sample down to make contact. If you sprinkle a little piece of powder on the diamond, you wouldn't get spectra. Squash that powder, and you can actually see the powder being squashed when you use the little video camera. You can immediately see the spectrum show up. You have all the pressure you need: do not get worried about crushing the sample as you are making it touch the diamond.

"You might ask how our system compares to other detection systems. We don't do anything with airborne sources, whereas they do, and so we kind of fill a hole that's been left on the solids and liquids. In speaking to some Haz-Mat responders, the next scare for them is toxic industrial chemicals (TIC). There are hand held tubes that you can use but the problem is you need to have ten or twelve of those tubes, and you need to put liquid in all of the tubes, and then you need to identify the compound based on that. But if it's not one of the twelve things that you're analyzing, you're out of luck. We can identify one of thousands and thousands of compounds because our library can grow. You can't make yourself up a new tube tomorrow because the threat has changed, but we can. If the threat tomorrow is blue liquids, we can make a blue liquid library and off we go. So, there really isn't competition as such, but there are other complementary technologies. Gas chromatography is a great complementary technology because we do solids, they do gases; to have both is the perfect world. Any gas detector with our system is a nice combination. But to my knowledge, other than a Draeger tube or a pH-paper test or a simple chemistry test, there isn't another instrument analyzer. There are HazCat kits which are an excellent tool for trying to identify an unknown, but that's where you actually do a lot of chemical tests. The reason you want to keep your HazCat kit is that we don't have all the answers. Spend the time to get the right answer; our tool doesn't replace any tool, it just adds to the battery of tools that are already available.

"One of the things that we have done is to make a commitment to the Haz-Mat community. Every person that a caller could have contact with, or every person that is involved in the design of a new piece of equipment or design of a software packet, employees literally from the front desk help to the people who make the boxes, every person that would have any thing to do with customer interaction or developing the new products will be trained to the Haz-Mat Specialist level which calls for twenty-four-hours of Haz-Mat training. The reason that's important is that as we design new instruments, we will now understand what it means to be in a Level A suit. If an engineer is going to make a little rinky- dinky screen, or make buttons to push, he will now understand that if you're in a Level A suit with three pairs of gloves on, you can't push that button. It's something we believe we have to walk in the shoes of the responder before we can design instrumentation, before we can design software, before we can answer questions. We need to know how cumbersome it is to operate in personal protective equipment, how difficult it is to read a screen when you have two visors to look through, how hard it is to contact the people who work with you about the instrument because you may not have a two way communication system. All of that effort means that we can design systems that are easier, better and more suitable for Haz-Mat responders."

Contact: Jim Fitzpatrick, SensIR Technologies, 15 Great Pasture Road, Danbury, CT 06810-8153. Tel. 203-207-9727; fax: 203-207-9780; e-mail: jfitzpatrick@sensir.com.

Smiths Detection and Protection Systems–Barringer

APD 2000: A portable, hand-held chemical detection and monitoring device that detects chemical warfare agents, recognizes pepper spray and Mace, and identifies hazardous compounds.

SABRE 2000:   The portable SABRE 2000, using a scanning system based on IMS (Ion Mobility Spectrometry) can detect drugs, explosives, and chemical warfare agents. More than forty substances can be simultaneously detected and identified in seconds.

Lightweight Chemical Detector (LCD-S):   The LCD-S acts as a local warning alarm system for individuals and small groups of persons within the domestic or military market. It simultaneously detects, identifies, and differentiates between type of chemical warfare agent at below attack concentration, and warns users when to don personal protective equipment.

Ionscan CENTURIAN Detection System: This system provides fixed site continuous ambient air monitoring for chemical warfare agents and toxic industrial chemicals.

The author met with Scott Goetz, Director of Marketing for Barringer instruments, which is a part of Smiths Industries. "Within Smiths industries, we basically have three companies who do nuclear/biological/chemical protection, instruments, environmental technologies, and dynamics. We've been doing this for thirty to forty years. That's what we do for a living, primarily for the military. However, for the last five years there have been more and more civilian applications as well. With the first responder market, most of our products have evolved from our military products.

"What we discovered early on was that the civilian battlefield is much different from the military battlefield, at least in terms of what users want and what their requirements are, even simple things like batteries. The military detectors run off lithium batteries that are expensive; you can't recharge them, can't dispose of them because they're hazardous waste. False alarms are a big problem since they are more tolerant of false alarms in the military battlefield than in the environment. The consequences of evacuating a stadium, mall or building are more severe than having a false alarm on a battlefield. Using the same technology we've used for the military, we've basically designed products specifically for first responders. We went out and did a user survey and found out what people wanted, what they did not want; that result turned out to be something called the APD 2000. This detector detects all the chemical warfare agents, all the blister agents, all the nerve agents, does radiation detection as well in the same machine. People have told us over and over that it would be nice to be able to detect pepper spray and Mace that are irritants. However, a lot of Haz-Mat calls involve irritants. Pepper sprays are used in schools to get the rest of the day off. We programmed pepper spray and Mace into the APD 2000, the ability to detect irritants, since we realized that the odds of nerve agents blowing through downtown Baltimore were not that great. The APD 2000 can be used daily by police and firefighters for irritants. This has evolved into pretty much the detector of choice amongst all the major metropolitan areas, in pretty much every major city as in Chicago, New York, DC, LA, Baltimore. There's about a thousand of these things fielded over the last several years. It detects pepper spray and Mace, but it also detects all the chemical warfare agents as well, it detects nerve agents, it detects mustard and radiation. This chemical agent detector is in use by first responders throughout the country. The FBI uses them in every field office. It's used by the EPA and other major federal agencies.

"Another product that we have is a product from Barringer called SABRE 2000. This is a device that's extremely versatile, detects chemical warfare agents, nerve agents, blister agents, it also detects explosives; it can also be programmed to detect drugs. Here's a nice tool that law enforcement for Coast Guard agencies, people who have multi-jurisdictional, multi-tasking response duties. The Coast Guard certainly is involved in drugs; it's involved in explosives as well, and they more and more have a chemical warfare role as well. So in one machine we're able to detect those three classes of compounds: explosives, drugs and chemical warfare in a hand-held portable, battery powered SABRE 2000 detector. The SABRE 2000 is an off shoot of the device called the Ionscan 400B, which is the explosive detector in about 80% of the airports worldwide. When you go through the airport terminal and they take your briefcase or baggage, put into that explosive detector machine, that's our machine. The SABRE 2000 is a portable version of the IONSCAN 400B.

"For twenty-four hour continuous monitoring, we have a product called the ionscan CENTURION system. The idea behind this device was we could integrate this into the HVAC system [heating, ventilation, and air conditioning] of the building, or around the parameter of the building, so that if a terrorist decided to introduce chemical agents into fresh air into the building, we would detect that in a matter of seconds. Management could turn off the HVAC system so they are not drawing more fresh air into the building, notify appropriate personnel that agent has been detected, and respond appropriately so the building supervisor or security, or Haz-Mat team or police can provide any sort of sensor outfits that people require to take the appropriate action. This unit not only detects chemical warfare agents, it detects about twelve toxic industrial chemicals [TIC] as well. TIC's are commonly available, things that are carried on railways and highways in every major city. Chlorine and cyanide gases and other chemical agents are readily available to terrorists. We also detect those things as well, about 12 toxic industrial chemicals as well as all the chemical warfare agents. This runs twenty-four hours a day seven days a week, and provides computerized display for the building supervisor to look at, identifies by compound what's been detected as well as the concentration.

"For biological agents we're developing a product ("Bio-Seeq PCR Detection System") that is a hand-held detector about the size of a notebook that detects biological agents, anthrax, smallpox things like that, based on something called PCR technology which is a chain-reaction technology. What we're measuring is DNA (any of the class of nucleic acids that contains deoxyribose, found chiefly in the nucleus of cells; responsible for transmitting hereditary characteristics and for the building of proteins) so we're measuring a finger print of biological agents. In theory, this technology is capable if you introduce one single anthrax spore into this detector. Again, this is theory, not so much practice. We first open that spore of anthrax extract from the DNA, replicate that DNA millions of times in a period of about fifteen minutes, and then with a DNA-probe measure that DNA. It's very specific because we're looking at the genetic structure of that material so there's very low false alarm; it's very sensitive for that same reason. In practice, we're probably capable of detecting less than a one hundred spores of anthrax, which is probably an order of magnitude more sensitive than anything that's out there right now. This is going to be introduced into the market in late summer or early fall. Nothing out there of this size, has these capabilities for first responders.

"Our parent company is called Smiths Aerospace, and Smiths has three companies in our division. Our division is called the Detection and Protection Systems Division; in that division are three companies, one is Environmental Technology and that's based in Baltimore. The second is Barringer and they're based in Toronto and New Jersey, and the third is Grazby Dynamics based in the United Kingdom just outside of London. The products I just talked about are all designed for first responders. There's other products that are designed specifically for the military application as well, battlefields, urban as well as military battlefields. This particular product LCD is really for both. It's a small detector it's about half the size of a tape recorder that detects nerve agents and blister agents. In the LCD, the technology is the same for both the military and the first responder unit; it's just a different implementation. Originally, the LCD was designed for the UK military. It's been redesigned somewhat for the first responder market. This is a product that's just being introduced as well; this will be introduction in earnest in the USA in the next month or two."

Scott Goetz was asked whether the 911 attack change the market for chemical agent detectors. "We've been in this market for thirty years; this is what we do for a living, we were making detectors before 911, making more detectors after 911. We had sold the APD 2000, we'd sold probably 500 or 600 detectors before 911, so none of these are new instruments or new markets for us. It's basically what we do for a living, but we certainly have seen an increase in sales since 911.

Contact: Scott Goetz, Director of advertising Smiths Detection & Protection Systems – Barringer Instruments, 2202 Lakeside Blvd., Edgewood, MD 21040; tel. 1-410-510-9370; fax: 1-410-510-9491;. e-mail: sgoetz@barringer.com; Web site: www.barringer.com.

Environics USA. Inc.

Environics produces chemical sensors, detectors, and detection systems for protection of people, the environment, and for space research. Their sales network covers more than thirty countries all over the world. Currently, in the United States, they are marketing the ChemPro 100 for both military and civil defense, as well as other chemical detection gear.

President of Environics USA, Michael Phillips, explains, "We're a Finnish-based company that deals with developing and manufacturing of chemical and biological detection systems for military and civil defense applications. We currently have a product out that we're selling to the first responder and military community called ChemPro 100 which is a hand-held mobility spectrometer-based device which is capable of detecting traditional chemical weapons agents, nerve agents and blister agents at levels that are significantly lower than the current threshold. The ChemPro 100 is a less than a two pound device that offers a lot of flexibility to the military as well as first responders in that it has the ability to have multiple libraries of chemical agents programmed into the device's chemical warfare libraries. The ChemPro 100 detector also can employ toxic industrial compound (TIC) libraries that can be tailored to whatever the user wants; the particular application these libraries are operator-selectable. The operator can actually monitor several different compounds with one device without having to go back and do any switching of libraries. In addition, their maintenance can be done on the fly. The ChemPro 100 also offers the ability to do some reconnaissance monitoring where you go into a building, not knowing if anything is there, and monitor the ion activity of our detector and be able to tell if there's something in there, and by using the manual library switch. Operate the switch through the various libraries. If it matches one of the libraries as you are switching through there, it will pull up and identify the agent. This is all done in real time, and response times are typically two to ten seconds. Clear down times are very quick as well. The device can respond to a very large dynamic range from .004 up until ten to twenty milligrams per cubic meter at the high concentration. Our ChemPro 100 detector does not have the traditional clean up problems as other devices. We don't use a membrane as traditional IMS [ion mobility spectrocity] devices do, so we are able to quickly respond and quickly clean up.

"There essentially two main technologies being used in chemical detection today. One is IMS and the other is surface acoustic; we are dealing in the IMS side. The ChemPro itself is actually a series of four good sensors that are integrated inside the device. We take data from four different sensors and use that as part of our process, so it's essentially using data fusion to take these various inputs, and take the spectrum that's generated. With these other sensors, we are able to make agent 'alarm/no alarm' decisions. The other advantage of our device over some of the traditional ones is costs are very low. We do not have cartridges or suit packs like some of the other IMS devices do which require routine changing after so many hours. Our expendables in our devices filter the dust; the filter that we have in the device provides positive indication to the user when there's a problem because we monitor the filter with sensors. Should the filter become clogged due to contaminated areas, it will give an alarm to the operator. One of the other advantages we have is that we can also update the libraries very easily. Our system is set up with an interface if new libraries are required, or other compounds want to be added. It's as simple as hooking your Palm Pilot to a computer with some software that we sell, and be able to upload libraries that I could e-mail to you. One of the other advantages we have, on-board the device there is a data log capability to keep track of data alarms on board the detector. During any event, there could be a 'history file' stored on-board the device that can be downloaded to the computer after an event for report purposes or close analysis after the fact.

"The memory on our device is set up so that we could store and display the information on our detector," says Michael Phillips. "The is also a sixteen channel IMS spectrum which has enough memory to store four hours worth of one second data. It can store data for a continuous four hour time period that could be downloaded after a Haz-Mat incident to track exactly what time the detector encountered various compounds, or the firefighter(s) or first responder(s) were contaminated and what time they were decontaminated. Another advantage of the ChemPro 100, it's always running and sampling air from the outside, and the sensitivity levels are such that you can catch very small agent doses as long as you're down wind from the site. Another factor in marketing this device is the ChemPro 100 gives the operator flexibility's. We can store up to fifty libraries onboard our detector. Each library has enough memory for one hundred compounds. The reality is you'll never do that because it's not a spectrometer having five thousand compounds, but it does give a lot of versatility to the user who can selectively pick based on what is desired.

"If you're a firefighter and you have a refinery or chemical company in your area program using certain chemicals on-hand that you know are going to be potential hazardous materials the Chempro 100 operator can set up and select libraries on the fly as he's responding, and put in the library set that best suits the area that they're responding to. There's also an ability onboard our device that is different from our previous ones is that the Chempro detector has been going through ongoing development of over fifteen years of this technology with Environics. Our predecessor, the M90 Chemical Warfare Agent Detector, which was used by U.S. Air Force, U.S. Army Technical Support Unit, and other agencies and countries all over the world. It has been a very reliable chemical agent detector; it was based on the same technology, the only difference was the M90 was a six channel IMS cell device. The Chempro is sixteen channels IMS device that gives us better selectivity, an ability to help identify more agents, and a chance to identify the various chemical structures of specific chemical agents. The U.S. Army Technological Support Unit uses M9 as their main detector, the one(s) they use on a daily basis. They're probably one of the few people that actually buy hardware they actually use against live agents. They're very happy with the performance and sensitivity of our devices.

"We are continually improving the performance of our system, adding additional sensors, putting in other types of sensors to help sensor upgrades, become more introduced to the production we're offering our customers," adds Michael Phillips. "We have the ability to get our detectors updated within the first year free if they buy now, so they're not buying obsolete equipment. As this new sensor integration gets incorporated, we will upgrade their units for that with no additional cost to them."

Michael Phillips was asked to define "false positive" and "false negative" dealing with chemical detector operation. "The false positive is defined as an interferent if the unit goes into alarm. False positives are kind of tough because traditional false positives in the military are chemicals that is not a chemical warfare agent; if it is ammonia, that's a false positive. In a Haz-Mat response by first responders, however, ammonia might be a compound of interest. The latest new rules say the false positive is something that may be a Haz-Mat responder's ideal compound. Traditionally, in the military, a false positive is anything that requires soldier to put a mask on when he doesn't have to.

"The false negative is when you're in this background interference ability, you don't have a false positive, and you still have to be able to detect agent. If you're in background interfering and agent challenge happens to come along, you're supposed to be able to still identify the agent; if you don't, then that's considered a false negative. It's kind of an unusual thing in Haz-Mat first responder practice because there are so many compounds of interest. All the traditional ones are false positives for military, but are compounds of interest to civilian response personnel. That's where the Chempro detector offers some advantages and in that you can select the libraries you want to detect. If you want to look for chemical warfare agents, you set it for chemical warfare agents; if you want to look for hazardous materials, you set the device for toxic industrial compounds (TIC). Toxic industrial materials have different lists for different groups; firefighters have a list, the Army has a list, everybody has a list, and no two lists are exactly the same. Where the Chempro gives an advantage is it can program in those specific compounds for various customers.

Contact: Michael R. Phillips, President, Environics USA, Inc. P.O. Box 290699, Port Orange, FL 32129-0699; tel. 386-304-5252; fax: 386-304-5251; e-mail: mrphillip@aol.com

Other Manuals Pertaining To Chemical and Biological Agents

The *North American Emergency Response Guidebook* (*NAERG*) is a 354-page soft-cover guidebook that deals with most industrial chemicals and some infectious substances. It is updated every few years. First responders must be trained in the use of this guidebook, and the manual has been delivered to every fire apparatus in the United States. The *NAERG* is structured to apply standard practices for first responders during the initial phase of a hazardous materials/dangerous goods incident in the United States, Canada, or Mexico. It is meant to aid first responders in quickly identifying the specific or generic hazards of the material(s) involved in the incident, and protecting themselves and the general public during the initial response phase"(that period following arrival at the scene of an incident during which the presence and/or identification of dangerous goods is confirmed, protective actions and area securement are initiated, and assistance of qualified personnel is requested). It is primarily designed for use at a dangerous goods incidents occurring on a highway or railroad. **It is not intended to provide information on the physical or chemical properties of the dangerous goods.** There are sixty-one emergency response guides numbered from Guide 111 to Guide 172. Neither the order in which the guides are presented nor the guide number itself is of any significance. Since many materials represent similar types of hazards that call for similar initial emergency response actions, only a limited number of guides are required. **The guides are not applicable when materials of different classes (Explosives, Gases, Flammable Liquids, Flammable Solids, Oxidizers and Organic Peroxides, Toxic Materials and Infectious Substances, Radioactive Materials, Corrosive Materials, and Miscellaneous Dangerous Goods) and/or divisions are involved in an incident and are intermingled.**

The specific guides that applies to chemical and biological agents and weapons in the *NAERG* are listed below; first is the guide number, second is the title, and third is the type of chemical or biological agent represented.

Guide 117:   *Gases-Toxic-Flammable (Extreme Hazard)*   Blood Agents
Guide 119:   *Gases-Toxic-Flammable*   Blood Agents
Guide 124:   *Gases-Toxic and/or Corrosive-Oxidizing*   Choking Agents

Guide 125:   *Gases-Corrosives*   Blood Agents, Choking Agents
Guide 153   *Substance-Toxic and/or Corrosive (Combustible)*   Nerve Agents, Blister Agents
Guide 158   *Infectious Substances*   Biological Agents

Since every fire department in the United States, Canada, and Mexico, and many businesses and companies, have the 2004 *NAERG,* responders should know the following:

- The Criminal/Terrorist use of Chemical/Biological/Radiological agents includes differences between a chemical, biological, and radioactive agent (page 354).
- Provides indicators of a possible chemical incident (page 354).
- Deals with indicators of a possible biological incident (page 355).
- Deals with tips about personal safety considerations (pages 356 and 357).
- Provides a glossary of definitions that is agent-specific (Pages 358 to 365).

There are a number of **Military Field Manuals** (and NATO Field Manuals) and other military manuals, guides, and handbooks dealing with both chemical and biological agents and weapons that were researched in the writing of this book. Since the tragedy of September 11, 2001, the availability of such has increased. They are very well written, contain a great deal of useful information, and are not difficult to obtain by a normal citizen (please refer to the Bibliography in this book).

The *Emergency Response to Terrorism Job Aid* is designed, produced, and distributed through a joint partnership of the Federal Emergency Management Agency, the United States Fire Administration, the National Fire Academy, and the United States Department of Justice/Office of Justice Programs. It contains a wealth of good and factual information regarding "B-NICE" (Biological, Nuclear, Radiological, Incendiary, Chemical, Explosives), and it is free. This document is not a training manual, but is designed to assist the first responder from the fire, EMS, Haz-Mat, and law enforcement disciplines. This *Job Aid* includes both tactical and strategic issues that range from line personnel to unit officers and the initial incident commander. It is expected that first responders have appropriate training and experience to address the identified tactics.

To secure this *Job Aid*, telephone 866-512-1800 (toll free), or fax 202-512-2250. This publication is available free of charge from the United States Fire Academy Publications Center to fire departments wanting five or fewer copies. Other emergency response agencies may order one free copy. Other interested persons can buy copies for $11.00, including shipping charges, from the Government Printing Office, Washington, D.C.

As an example of what specific fire departments and firefighters are doing to combat weapons of mass destruction and hazardous materials with emergency response, the Orlando, Florida, Fire Department has produced *Hazardous Materials Standard Operational Procedure-Tactical* guidelines to explain procedures to emergency medical services technicians and paramedics as well as to other fire department members. The mission of the department is "to protect the life and property of the citizens and visitors of Orlando by providing the highest possible levels of service through fire prevention, public information, fire suppression, emergency medical services, and mitigation of the effects of natural and man-made disasters or emergencies consistent with the resources provided." The area in and around Orlando certainly gets its share of visitors with the Walt Disney World resort complex, Epcot Center, Universal Studios, the Orlando Convention Center, the Magic Kingdom, Disney/MGM Studios, and many other attractions. Established in 1885, the Orlando Fire Department has been providing services for almost 115 years. The department covers about 95 square miles, utilizes 12 fire stations, and approximately 450 civil service and civilian personnel. Response equipment includes six aerial tower apparatus, sixteen fire engines, five command vehicles, nine rescue units, one rescue boat, an ARFF truck, one water rescue vehicle, a hazardous materials response vehicle, three rural fire fighting units, and eighteen support vehicles. The normal work schedule for fire suppression and emergency medical services personnel is twenty-four on duty and forty-eight hours off duty (civilians work forty hours a week while civil service personnel work forty-eight hours per week). During a recent year, the Orlando Fire Department hazardous materials response team (HMRT) responded to thirty to thirty-five hazardous materials alarms per month out of 25,846 total fire/rescue, medical, and Haz-Mat alarms per year. The thirty-six-member HMRT responds from Station 1 on Engine 101, Towers 2, 8, and 9, Rescue 1, and Haz-Mat 1. All medics were trained by the EMS Division to comply with National Fire Protection Association standard NFPA 473, "Competencies for EMS Personnel Responding to Hazardous Materials Incidents."

District Chief Armando S. Bevelacqua was a key figure in producing the department's medical standard guidelines. He has been involved with hazardous materials response since 1980 when he started out as a firefighter in Tallahassee, Florida. He came to the Orlando Fire Department twenty years ago and is now officer-in-charge of Special Operations which includes the hazardous materials response team and other units. Bevelacqua is a very professional and active fire officer, an experienced paramedic, a noted trainer, and a successful writer with a number of books with Delmar Learning. His new book with Richard Stilp is entitled *The Citizen's Guide To Terrorism Preparedness* (Delmar Learning; www.firescience.com/citizens-guide). The District Chief is a member of the Mayor's Terrorism Task Force for the City of Orlando, the Federal Terrorism Preparedness State and Local Advisory Committee, as well as the Health and Medical Committee Members of the Regional Terrorism Task Force. He is adjunct instructor and program developer for both the National Fire Academy and the International Association of Firefighters. He has just finished a new effort, *Hazardous Materials Medical Protocol*, that covers medical protocols for Bronchospasms, Tachydysrhythmias, Chemical Burns, Carbon Monoxide, Nitrogen Compounds, Cyanide and Hydrogen Sulfide, Organophosphate and Carbamate (Nerve Agents), Hydrofluoric Acid, Phenol, Chloramine and Chlorine Products, Lacrimators, and Phosgene. Chief Bevelacqua has given this writer permission to publish the Organophosphospates and Carbamates (Including Nerve Agents: Tabun (GA), Sarin (GB), Soman (GD), and VX) section of the book (Chief Bevelacqua can be contacted at HZMT-Medic1@AOL.com).

Organophosphospates and Carbamates

Nerve Agents: Tabun GA, Sarin GB, Soman GD, VX

Neurotoxins can be inhaled, ingested or absorbed. Once in the body it binds with critical neurotransmitters (acetylcholinesterase) initially causing excitation of the nervous system, conduction, then paralysis. Common signs are Diarrhea, Urination, Miosis, Bronchospasm, Emesis, Lacrimation, and Salivation.

General Symptoms/Organophosphospates

- **Vapor** — Can cause symptoms within seconds to a few minutes.
- Mild to Moderate — Immediate miosis (eye pain, headache behind the eyes, dim, blurred or tunnel vision), rhinorrhea, mild to severe mucous secretions and dyspnea.
- Severe — Seizures, dyspnea to apnea, loss of consciousness, flaccid paralysis.
- **Liquid** — Can cause symptoms within minutes up to several hours.
- Mild to Moderate — Fasciculation's and sweating, at the site of contact. Nausea, vomiting, diarrhea, and weakness.
- Severe — Seizures, dyspnea to apnea, loss of consciousness, and flaccid paralysis within minutes.

General Symptoms/Carbamates

Dyspnea, rhinorrhea, mild to severe mucous secretions, local fasciculations, loss of consciousness, and weakness.

There are many pesticides, and pesticide chemical derivatives. The specific chemical must be referenced before an antidote treatment is administered.

Basic Life Support: Patient Management

- Ensure rescuers do not make physical contact with the patient or doffed clothes until the appropriate level of PPE is donned.
- Remove clothing and isolate.
- Decontaminate the patient immediately.
- Decontaminate the skin using soap and water. These products are toxic materials; therefore, decontamination is extremely important. The removal of clothing containing liquid or vapor laden clothing is a part of the decontamination process.
- Repeat the decontamination process as often as necessary.
- Ensure administration of high flow 100 % oxygen
- Document the time of exposure vs. when the oxygen therapy was started.
- *Do not* induce vomiting.
- When hypotension occurs place the patient in trendelenburg position.
- Maintain the patient in a warm environment.

Advanced Life Support: Patient Management

- Follow BLS procedures.
- If the patient is conscious with vitals compromised, consider Rapid Sequence Intubation.
- Apply Positive Pressure Ventilation using PEEP at 4 cm/water or CPAP mask.
- If patient is unconscious, consider intubation.
- Apply Positive Pressure Ventilation using PEEP at 4 cm/water or CPAP mask.
- Start Intravenous Fluids.
- Administer Atropine 2–4 mg. IVP at minute intervals until atropinization occurs. There is not a maximum dose.

- *Use extreme caution in a hypoxic patient. Giving Atropine to a hypoxic heart can cause ventricular fibrillation.*
- Administer Pralidoxime (2 – PAM) IVP 1 gm. over 2 minutes. ***DO NOT GIVE IN CARBAMATE POSIONING.***
- Treat seizures with Valium 5–10 mg.

It should be noted that space permits only one sample of District Chief Armando Bevelacqua work. However, his production also includes medical protocols for Bronchospasms, Tachydysrhythmias, Chemical Burns, Carbon Monoxide, Nitrogen Compounds, Cyanide and Hydrogen Sulfide, Organophosphate and Carbamate (Nerve Agents), Hydrofluoric Acid, Phenol, Chloramine and Chlorine Products, Lacrimators, and Phosgene.

Introduction to Biological Agents and Toxins

4

The Chairman of the National Intelligence Council released a report in Washington, D.C., entitled, "The Global Infectious Disease Threat and Its Implications for the United States." This report deals with warnings of growing possibilities for American citizens to come down with infections that run rampant is other parts of the world, since the United States is a sizable hub for world travelers, immigration, and commerce. Also, we have a high percentage of American military service personnel serving in all sections of the world. The Asian continent has seen steady increases in infectious diseases such as the spread of HIV (human immunodeficiency virus) and AIDS. In addition, an estimated thirty diseases, unknown in the past, have appeared globally since the early 1970s. These diseases include Hepatitis C, Nipah virus that is encephalitis-related, and Ebola hemorrhagic fever, and other diseases that so far remain incurable. In like manner, infectious diseases such as malaria, cholera and tuberculosis have rejoined our nation since the 1970s (also, terrorist use of biological agents has increased with the 140 anthrax hoaxes perpetrated in the United States in the late 1990s, and the actual anthrax attacks the country seemed to be so unprepared for in late 2001. The report also gives a warning that most infectious diseases originate in other countries but are brought into the United States by travelers, immigrants, imported animal and foodstuffs, and military troops who have served in far corners of the earth.

September 11, 2001, was the date on which nineteen Middle Eastern hijackers with brilliant planning and a forceful presence took over four different airliners from three separate airports and crashed two of the planes into the World Trade Center complex in New York City, and one into U.S. military headquarters at the Pentagon outside of Washington, D.C. Add that attack on United States soil to a series of "super" anthrax attacks that began five weeks later on October 15, 2001, in Florida and Washington, D.C., which changed the entire concept of emergency response to hazardous materials and chemical/biological agents. We thought we knew how active domestic terrorism would change our entire society, but we really had no idea of how many freedoms we would lose in the process of adapting to a subsequent world order, or lack of world order. We thought we were learning about dealing with biological warfare weapons and terrorist attacks, but until September 11, 2001, our learning problems seemed minor. Now they are major.

The response to biological warfare weapons, terrorism attacks, and even influenza mini-epidemics can require an immense number of federal/state/county/city/private workers and medical personnel ranging from physicians to registered nurses to emergency medical technicians and paramedics. As an example, this writer has actually seen an incident where thirty-one agencies from all levels of government responded with an untold number of persons who were assigned at the site. In the recent past, the incidents at the federal building at Oklahoma City, Oklahoma, and the terrorist attacks of September 11, 2001, as well as railroad wrecks like the one in Graniteville, South Carolina, on January 6, 2005, told new tales in emergency response in North America. A series of anthrax hoaxes, as well as a number of severe anthrax attacks using a new level of danger for this bacteria, can bring and require innumerable specialists in emergency hazardous materials response to an incident scene with the threat of terrorism we have today in North America. Terrorism may have arrived late for us in North America, but now it is with us in full force. We all have a lot to learn.

Any terrorist incident that uses chemical or biological agents is basically a hazardous materials incident. Hazardous Materials Response Teams (HMRT) and first responders to such incidents use the following principles and characteristics. They deal with abatement, action checklists, after-action reports, antibiotics, antidotes, biological agents, blister agents, blood agents, breakthrough time, as well as briefings and critiques. Such responders know Computer Aided Management of Emergency Operations (CAMEO), case histories, chemical agents, chemistry of Haz-Mat, command post operation, compatibility, computers, and containment. Containers, contingency planning, databases, decision making, decontamination, detectors, dispersants, disposal, emergency response plans, and evacuation are "must know" information for HMRT. Size-up & Evaluation, hazard analysis, funding/cost recovery, incident command system, incident vigilance & discipline, industrial agents, leak/fire/spill control, and "lessons learned" are daily facts of life. They consider levels of incidents, levels of protection, manuals, material safety data sheets (MSDS), medical surveillance, monitoring, and national and local contingency plans among their S.O.P.s. HMRT members know that nerve agents, neutralizers, no-fight situations, offloading/transfer, patching/plugging, perimeter control can help them or hurt them. They study personnel safety, pH, physics, post-fire residue, protection, reactivity, recon, and resource materials to be better prepared. They also understand responder liability, scene management, secondary emergencies, S.O.P.(s) and protocols, sorbent materials, staging areas, standards, storage, tactics, team concept, testing methods, toxicology, training, triage, vaccine, vectors, vehicles, weapons of mass destruction (WMD), and zonal delineation.

Hazardous materials response personnel of all types — firefighters, law enforcement, medical staff of all varieties, military personnel, federal/state/county/local governments, commercial response contractors, and industrial response brigades are all learning new factors added to biological agents response since September 11, 2001, and weeks following. The ability to inflict mass destruction has been modernized as advances in technology have made Nuclear/Biological/Chemical (NBC) devices more available to the common man. Local response personnel will be the first to arrive at the scene of a chemical and/or biological incident, make the important life or death decisions, and be the last to leave the scene. Anyone can be a victim. Innocents are prime targets of terrorist acts; not recognizing innocents means an infinite number of targets. Medical triage and treatment of victims is the weakest link in a terrorist incident involving chemical or biological agents; it is a crisis, but there is no way we can deal with everything, treat everybody, and save them all. The medical system will have visits from a lot of people who are not ill or contaminated; the system will be severely overtaxed. Uncontaminated persons will be walking into hospitals and clinics demanding medical care; control (triage) must be established. Vaccines and antibiotics can be defeated by genetic engineering. The absolute basic tactics from an incident involving chemical or biological agents are as follows:

Protection:	TIME — DISTANCE — SHIELDING.
Initial Actions:	ISOLATE — EVACUATE — DENY ENTRY.
Medical A/B/Cs:	AIRWAY — BREATHING — CIRCULATION

Many emergency medical service providers do not have respiratory protection as good as those available to firefighters. By the time biological agent attack victims begin showing symptoms, it is often too late to save their lives. Plague vaccine has no effect on inhaled aerosol. Many biological agents are not contagious person-to-person, but there are exceptions; plague and smallpox are very contagious person-to-person. Any vaccine can be overwhelmed by the dose of the biological weapon used. An intact skin is a very effective barrier against biological toxins except for micotoxins; T-2 micotoxins, such as in mold or fungus, can come through intact skin. Any chemical or biological agent attack will be a crime scene. There is a need to balance response to a crime scene with control of a chemical or biological agent. A two to six micron (a unit of measurement equal to one-millionth of a meter) range is perfect for aerosols to get toxins through your nose to infect you. Much of the federal government's critical operation functions, including those of the military and the medical services, rely on services provided by the private business sector. No United States citizen had any experience with a massive cleanup of anthrax spread by aerosol until after September 11, 2001.

Problems and Opportunities

Some problems, or opportunities, depending on the manner in which individual readers look at problems, about emergency responders to chemical and biological agents will be presented below. We are speaking here of "emergency responders" of all possible types and talents including firefighters, police officers, emergency medical technicians and paramedics, ambulance crews, decontamination crews both in the field and at hospitals and clinics, emergency management officials, military troops, health regulatory personnel, commercial response contractors, engineers and construction employees, industry HMRTs, county/local/state/federal government responders, consultants and specialists, specialty teams, and mortuary workers.

We will deal with *General* (meaning situations that apply to both chemical and biological agents and weapons), *Chemical* (meaning situations that apply mainly to chemical agents and weapons), and *Biological* (meaning situations that apply mainly to biological agents and weapons). Since we do not have any vast history in the United States relative to chemical and biological agents or weapons, our information had to be gleaned from factual information from wars and incursions, hoaxes featuring threats of contamination within the United States, treatment of infective diseases within the United States, military experience and field manuals, and news reports. This exercise is meant to improve the safety and information available to emergency responders.

In this writer's experience stretching over twenty-nine years, in some incidents with chemical and biological agents or weapons, there has been a hesitancy to provide rapid, early decontamination; lack of consistent attention to control of the ABCs (airway/breathing/circulation); and adequate pain management for ill persons. In some mass casualty situations, there has been a lack of immediate triage, and failure of hospitals to verify that decontamination was done on the incident before victims arrived at hospitals. There also have been failures to report a chemical or biological incident to proper authorities when it appears there may be a terrorist or criminal act involved, and a tendency to recognize that very young persons as well as senior citizens who are contaminated with chemical or biological agents are more in danger than healthy, in-shape people of other ages. There have also been situations where failure to provide protective clothing and equipment fully in-line with the requirements demanded by the involved chemical or biological agents, and a number of incidents where vaccines and anti-viral medicines were in short supply when and where they were needed.

There have been failures to intubate an unconscious patient, failure to bring a "container" to the hospital personnel, if appropriate, so they can identify the chemical or biological agent or pesticide; or discharging a chemical or biological patient without adequate observation. Medical service staff has also failed to obtain an adequate record about the chemical or biological agent affecting a patient, not being alert to problems with angina or myocardial infarction, and not holding a patient where respiratory support capability is immediately available. There has been a failure to re-triage victims at an incident site at certain points in time to see if their condition has changed for better or worse so their medicine or care can be changed as well, inability to move quickly to decontaminate victims except in very limited situations, and failure to incinerate or autoclave contaminated material. Other problems, or possibilities, if you believe that identifying problems makes them easier to correct, include failure to use strict controls or protocols to ensure quality care for victims even in a mass casualty situation; failure to heed a patient who states that he or she undergoing severe pain, failure to recognize or acknowledge high-risk chief complaints, and failure to do complete and total vital signs on victims.

In the rush and excitement of a serious chemical or biological agent incident, response personnel may forget that doffing protective clothing in Levels A, B, C, and D in a less than careful manner may allow contamination or recontamination of medical and first responder personnel. Emergency response personnel choose personal protective clothing based on the identified properties, threats and damage potential of the present hazard. If a first responder does *not* know the hazards involved, he or she should always use the highest possible level of protection, Level A (vapor protective suit for hazardous chemical emergencies). Terrorists seemingly are using more technically adaptive methods and techniques in their weapons-of-interest than in the past as evidenced by the super-anthrax that became readily apparent in the United States beginning on October 15, 2001. The anthrax attacks were real this time, totally unlike anthrax incident hoaxes that have kept firefighters and police officers running for the preceding three years. In some cases, failure to use all possible means and methods to identify the agent and thus allow the best protocols to overcome the threat in a timely manner is a cross that all emergency response personnel have to fear. They also must pay attention to the possibility of secondary explosions, mixed agents, binary agents, time-of-day, weather conditions, temperatures, flora and fauna, which can all affect the damage done by chemical and biological agents. It seems curious that, for unknown reasons, the citizens within the United States seem to be most afraid of radioactivity rather than chemical or biological agents. This probably goes back to our years of the Cold War with Russia, yet radioactivity is easy to detect, seldom lethal, and cannot cause epidemics in the manner of bacteria or viruses. Preparation is the key to being ready for chemical or biological warfare or terrorism, but recent history has shown that the United States is woefully unprepared to respond to a serious mass casualty incident involving either or both chemical and biological agents.

Supply and logistics have changed drastically in the United States over past years. For whatever reasons, hospital, clinics, and drugstores have moved to an order-only-when-you-need-it, do-not-store -much-of-anything way of doing business for all types of drugs including antidotes, vaccines, painkillers, antibiotics, and corticosteroids. Such drugs are available only to very limited degree, and replacement takes weeks or months. First responders may now be a generic term that is out-of-date. In a major terrorist attack or in warfare using chemical and biological agents or weapons of mass destruction (WMD), Medical Responders will assume a new, very crucial role including emergency medical technicians, paramedics, physicians, and nurses as well as medical support staff, and hospital and clinic personnel. That is, in incidents to come, first responders will be dealing with *not* just pre-hospital care but with a full range of medicines, isolation of patients, immunization, exposures, decontamination, air monitoring, and the "anti's" (antibiotics, anticonvulsants, antidotes, antiserum, antitoxins), detection, degradation and penetration, personal protection, compatibility charts, contamination, control, degree of hazard, detoxification rate, double gloving, emergency response plans, exposures, hot zones, IDLH, inoculation, micron, monitoring, off-gassing, oximes, plumes, pyridostigmine bromide, quarantine, respirators, specific gravity, spores, symptoms, synapse, toxicity, triage, vaccines, variola, and vectors.

In addition, there has been failure to do a complete and total statement of the medical and triage procedure in the medical record, and mistakes made to first secure the incident scene and identify the exact nature of the existing threat.

Moving on to situations that apply to mainly to chemical agents and weapons, it is noted that nerve agent medical treatment, in general terms, is similar to the treatment of persons contaminated by organophosphate insecticides such as malathion, which can be fairly well known to medical first responders in rural areas. When engaging in decontamination with chemical agent victims, first responders should refrain from hard scrubbing and use of hot water. Instead, they should **immediately** decontaminate nerve agent contaminated persons with copious use of water, then rinse obviously contaminated victims with 0.5 percent hypochloric solution made from one portion of five-percent household bleach and nine-parts of water. **First responders must immediately decontaminate victims contaminated by chemical nerve agents.**

Nerve agent **vapors** are heavier than air and tend to seek out low areas or hug the ground, while all nerve agents rapidly penetrate clothing and the skin underneath. Medical personnel dealing with chemical agent contaminated persons should be forewarned that surgical masks and air-purifying respirators can be inadequate versus nerve agent **vapors;** dual donning of latex gloves for self-protection is virtually useless action against nerve and blister agents. Regarding the same theme as addressed above, chemical agent **gas** as in phosgene, chlorine, or cyanide; no protective clothing *may* be required as off gassing should be insignificant. However, chemical agent **vapor** from volatile liquid such as vesicant or nerve agent vapor *would require protective clothing as off gassing may present some level of exposure to medical and first responders.* When handling victims with contamination from nerve and vesicant agents, medical and first responders should be **required** to wear *both* respiratory and skin protection.

Chemical agent incidents may result for both warfare and terrorist actions in addition to agent stockpile storage at eight chemical weapons stored around the United States, disposal of toxic chemical weapons at these facilities, or from industrial accidents. The chemical agents used today are normally liquids or solid and only exceptionally gases; the media often refer to nerve "gases" which is a misnomer; tabun, sarin, soman, GF, VX, and V-Sub X are liquids delivered in an atomized form as an aerosol. Tear gases (CS, CR, CN, CH and pepper spray) and psychological agents (LSD, mescalin, benzilate) are aerosolized mists, not gases, and cannot be detected using Draeger tubes. The most toxic of nerve agents, **VX**, is most difficult to decontaminate effectively due to its low volatility. Mustard agents present **both** a liquid and a vapor threat for which there may be no symptoms of damage even though such agents cause serious damage within minutes of contamination. Therefore, it is crucial to decontaminate at once, within the first one or two minutes of contamination if possible. If not carried out within 60 or 120 seconds, decontamination on mustard agents does not prevent later blistering. No specific treatments or antidotes can cure the cellular effects of a mustard agent.

Regarding the biological agents and weapons, it may be fairly common to diagnose an illness as influenza when the victim has just returned from an endemic area where Venezuelan equine encephalitis (VEE) is a serious problem. That is one reason why if you show up **at a hospital and complain of an "unknown" disease, a nurse, or student doctor in a teaching hospital, will take a history from you. One of the questions on your history will include your recent travel history. Another question might be relative to exactly what you do for work. If you skin rabbits or are a postal worker, the nurse** is suddenly interested. Other errors noted when responding to biological agents or weapons, include failure to remove casualties from further exposure to either biological or chemical contamination, and failure to provide oxygen to victims when necessary. However, the major problem with biological agents is there's a lack of field detectors for first responders. There are some field detectors for biological agents, but the current supply is small and there are troubles with false negatives and false positives. Prior to September 11, 2001, the U.S. military services used to be the only customers for biological detectors; after September 11, 2001, the entire market opened up and everybody everywhere became a potential customer. Industrial companies are very busy trying to develop a field detector for biological agents, but the race has yet to be won for a reliable, easy-to-use, minimum-of-training-required detector that can deal with false negatives and false positives.

Right now, we have to deal with incubation time and wait for victims to develop symptoms. The incubation time for anthrax is one to six days. In this example, anthrax victims would have one to six days between exposure and the onset of symptoms. Anthrax is *not* transmissible from person to person. Compare this to the incubation time with that of the virus, smallpox, which is ten to seventeen days. Smallpox is highly transmissible from person to person. After exposure to smallpox, a person could travel by air around the world a number of times and contaminate many people before developing any symptoms. However, naturally occurring smallpox has been eradicated worldwide since 1977. Terrorism could rapidly change that eradication to an attack since samples of the smallpox virus have been stored in both the United States and Russia.

In emergency response to biological agents at the federal level, particularly in the anthrax attacks, there has been a failure to always dispose of contaminated clothing and equipment in an approved fashion, and a failure to utilize strict precautions to protect worker health and life. Medical first responders who handle persons who are contaminated with biological agents require respiratory protection, but skin protection is generally not needed *except* for trichothecene mycotoxins. Such responders to biological agent **aerosol** contamination may not be required to don protective clothing because secondary aerosolization of remaining biological agent from victims should be insignificant. However, when and if victims are contaminated with biological agent liquid or powder, universal precautions should required which include donning at least Level C protection with a powered air-purifying respirator(PAPR) and a high-efficiency particulate air (HEPA) filter if contamination may include mycotoxins.

At the time of this writing, it is virtually impossible to judge the number of biological agents, or count the number of biological warfare agents that can be developed from such agents. With the variola (smallpox) virus, containment is crucial and strict blood/body fluid/droplet protection are required for personnel that treat or transport with known or suspected smallpox victims because of its easy transmissibility from person to person. In addition, personnel even exposed to a contaminated smallpox victim require vaccination and quarantine. A single case of smallpox almost guarantees that a larger number of people have been infected by this disease. Both a vaccination and antisera are available for smallpox, but antimicrobial therapy is not effective. Vaccinia virus is a live poxvirus vaccine that induces strong cross-protection against smallpox for at least five years and partial protection for at least ten years or more, and is administered by dermal scarification injection. Patients with smallpox should be treated by vaccinated persons using universal precautions (methods for healthcare workers to avoid infection from blood-borne diseases first developed by the Centers of Disease Control and Prevention (CDC) in 1987; their guidelines include use of protective gloves, masks, and eyewear when in contact with blood or body fluids). Objects in contact with a victim, including bed linens, clothing, ambulance gear and equipment and other such items; require disinfection by fire, steam, or sodium hypochlorite solution.

As is the case with many diseases, particularly with infectious diseases of biological origin, a number of cases may tend to be misdiagnosed and/or not correctly reported. It is very difficult to properly diagnose disease cause by a biological agent. A patient may be best served by either a physician who has a long history of dealing with infective diseases, or a medical student or newly licensed doctors because such medical students or new doctors are still in the micro area of medicine and may better recognize specific symptoms. Also, a veritable host of biological agents produce symptoms that duplicate, or almost copy, naturally occurring diseases. Since there are so few *field* biological detectors (and not as yet any field "generic" detectors that can identify causative agent such as a specific bacterium or virus), correct diagnoses of a biological-disease-related-incident illness can be a serious and lengthy problem. It is a fact of life that many biological agents produce diseases that are foreign to many U.S. physicians in their normal rounds. Too many biologically-related diseases are actually diagnosed at the time of an autopsy. Anthrax can often provide a ready clue by a chest X-ray of a victim with evidence of a widened mediastinum (a part of the space in the middle of the chest between the sacs containing the two lungs; it extends from the breastbone to the spine, and contains all the chest organs except for the lungs). But a very obvious clue to the cause provides likely proof that this victim suffers from anthrax. If you have a serious illness that leads to an immunocompromised condition, you may have a strike against you in a biological agent situation. This example includes any person in an immunocompromised state who takes necessary medicine every day for his or her remaining life to keep their bodies from rejecting a new heart or other transplanted body part, a person with HIV obtained by transfusion, a drug addict with used and dirty needles, or other persons. **Such persons cannot take the vaccine currently available for smallpox.** Another person who is immuncompromised should avoid contact with standing water and soil in parts of the world where melioidosis is endemic. Biological agents could be the world's next master weapon. Such weapons have a high psychological fear factor; actually, they scare the Hell out of people. They have a low visibility and cannot be easily traced, have a significant time delay between infection and the development of symptoms, are highly potent, readily obtainable, and most are able to be made into a stable aerosol with ease. They are also invisible, odorless, tasteless, and have silent dispersal. You will never know what hit you until it is too late.

Background to Biological Incidents

Both biological agents and weapons go back to before the fourteenth century when the Black Death (bubonic plague) killed over 25,000,000 people in Europe. Bubonic plague featured symptoms that included painful swollen lymph nodes called buboes in the armpit, groin, or neck, fever as high as 106 degrees F, low blood pressure, exhaustion, confusion, and bleeding into the skin from surface blood vessels which produced a rose-colored ring.

This guide deals with biological agents that could be used by terrorists or others as weapons of mass destruction. There are all kinds of bacteria, viruses, rickettsiae, chlamydia, fungi, and toxins available in the world; but only a limited number which may be used as weapons. Some *bacteria* are round (cocci), rod-shaped (bacilli), spiral (spirochetes), or comma-shaped (vibrios), and are capable of reproducing outside living cells. Some examples include anthrax, brucellosis, cholera, plague (pneumonic), shigella, tularemia, and typhoid fever. The nature, severity, and outcome of any infection caused by bacterium depend on the particular species, but diseases caused by bacteria often respond positively to the use of antibiotics. *Viruses* are tiny organisms that can only grow in the cells of another animal; more than two hundred viruses are known to cause disease in humans. Antibiotics are not much of a help for virus-produced diseases, although viruses may be at least partially responsive to a few antiviral compounds that are available. Examples would include Crimean-Congo hemorrhagic fever, dengue fever, ebola, eastern equine encephalitis, influenza, HIV, and Rift Valley fever. *Rickettsiae* are small, round, or rod-shaped special bacteria that live inside the cells of fleas, ticks, lice, and mites and are transmitted to humans by bites from such pests. They are similar in one respect to viruses in that they grow only within living cells, but dissimilar in that treatment of disease caused by rickettsiae often includes the use of broad-spectrum antibiotics. Some of the world's worst epidemics such as scrub typhus, Q fever, and Rocky Mountain spotted fever have been rickettsial in nature. *Chlamydia* are microorganisms that live as parasites within living cells. Two species cause disease in humans: chlamydia trachomatis, and chlamydia psittaci (also known as parrot fever). *Fungi* are simple parasitic plants that lack chlorophyll and reproduce by making spores. Of the 100,000 known species of fungi, approximately ten cause disease in humans. Fungal infections tend to be mild but hard to cure. *Toxins* are non-living poisons that come from living animals, plants, or microorganisms although some toxins can be produced or altered by chemical means. Examples include botulinum toxins, mycotoxins (trichothecene), ricin, and staphylococcal enterotoxins.

If you are a first responder, HMRT member, firefighter, police officer, or emergency medical service person called to an "incident," *any* incident, remember the following information about biological incidents. Except in unusual circumstances, you can't see a biological agent; they are odorless, colorless, and tasteless. There is a delay in incubation; even the much feared ebola fever which has a moderate transmissibility from person to person takes seven to nine days from exposure to the time when symptoms actually appear. How many persons could you expose during those seven to nine days? No other weapon, including chemical agents, nuclear warfare, military ordnance, or bows and arrows can compare to the sticker price of biological agents: anthrax could cost you slightly under one U.S. dollar per casualty. Such weapons can be easy to prepare; in some cases, you could even grow or produce your own. Since many biological agents will be disseminated as aerosol, the enemy can be long gone before any reaction occurs; they can be almost untraceable.

A biological incident is very difficult to defend against because of ease of concealment, lack of knowing who the "enemy" is, highly lethal potency, ready accessibility, and relatively simple means of dissemination. A small quantity can do you in; a lethal aerosol anthrax dose could be as little as a millionth of a gram. As of today, biological response is beyond the capability of local government unless they have a fully trained and equipped Level A hazardous materials response team, possibly beyond the ability of state government unless it has a top notch medical services program; a mass casualty biological incident may be even beyond the experience, training and equipment levels of the national government. Now that we have experienced the "super-anthrax" attacks in Florida and Washington, D.C., that began about October 15, 2001, we have learned another hard-earned lesson.

There is little or no **field** detection equipment for biological agents at the present time available to local first responders or local hazardous materials teams; the possibilities are being studied by defense consultants, but delivery may take a long time. There is a very apparent lack of vaccines, antibiotics, assisted ventilation devices for victims, lack of experienced and trained medical personnel, detectors or meters for detection and identification of biological agents, top-heavy spending on federal programs with trickle-down economics to vastly important local programs. Federal programs will "assist" in biological response while local programs will be called upon to do the actual work of clearing, evacuation, control, triage, medical care, and urban search and rescue.

One of the central problems in the copycat anthrax threats, other than the very high cost involved in responding to such hoaxes, is that biological detection *in the field* is just not available to many local response personnel. The military may or may not have a couple of vehicle-mounted biodetectors that may or may not be currently reliable and past the "developmental" stage. "Human beings are a sensitive, and in some cases the only, biodetector," (Department of the Army Field Manual FM 89, *NATO Handbook on the Medical Aspects of NBC Defensive Operations*, AMedP-6(B), Part II - Biological). There is little or no field detection equipment for biological agents at the present time. The military does have Biological Integrated Detection System (BIDS), a vehicle-mounted system that can identify a limited number of biological agents through antibody-antigen combinations by exposing samples of air to antibodies The process takes about thirty minutes and can currently detect the presence of botulism, anthrax, bubonic plague, and staphylococcus enterotoxin B.

Biological and chemical agent detection systems for field use are a key factor in the country's domestic preparedness programs in the event of terrorist attacks which may use weapons of mass destruction in light of the approximately 110 "anthrax" hoaxes that have occurred in the United States during six months. If local response forces do not have reliable field detection equipment, they could be forced to use mitigation techniques ranging from decontamination to therapeutic drugs to vaccines that may not be necessary. One of the problems with so-called field detectors for biological agents is a possible "false positive" and/or "false negative" readings on present detection equipment.

Biological agents may be alive, they can spread through infection, may be able to duplicate themselves, some may be persistent, others may be transmissible from person to person. The United States closed down its offensive biological weapons program in 1969; there has since been a loss of knowledge, experience, and data on new developments in offensive biological weapons as knowledgeable persons sought other employment, retired or died over the last forty years. In an attack with chemical agents, there are immediate casualties along with knowledge that there has been an attack. When you have knowledge of an attack, you are much more aware of the dangers presented by an emergency response. Training for a chemical attack is irrelevant in regards to a biological attack; they are two streets with different zip codes.

In a biological attack, there is a delay due to the incubation period required by biological agents — a possible lack of knowledge even that an attack has occurred — until hours or days later when people arrive at hospitals with flu-like symptoms. The United States does not presently have reliable disease surveillance programs that would be necessary to identify the biological agent(s) and provide the correct treatment. Doctors, nurses, hospital personnel, and public health-care workers would be the first line of defense against biological warfare. Many such people work for private firms rather than government agencies. Is the civilian medical community in all areas of the nation really ready to deal with a biological mass casualty incident? It will be the civilian health care system, plus local firefighters, police officers, and emergency medical technicians and paramedics that will manage and do the work required by a biological attack in the United States — at least in the first thirty-six hours when most of the life and death decisions will be made.

Biological agents are most likely to be disseminated by aerosol, a fine aerial suspension of liquid, fog or mist; or solid; dust, fume or smoke with particles small enough in size to be stable. The perfect size for human exposure is between 0.5 to 5 microns (or micrometers) which are a unit of length equal to one millionth of a meter. Larger measurements might be naturally filtered out by the inhalation process, while smaller sizes might be inhaled but not retained in an efficient manner. Aerosol exposure can also contaminate food, water, and skin. Although healthy, intact skin can resist the entry of many but not all biological agents, skin with wounds, cuts or abrasions provide an opening for infection. Protective clothing must be worn along with respiratory tract protection provided through use of a mask with biological filters, or the use of self-contained breathing apparatus (SCBA) with positive pressure when responders do not know exactly what the threat(s) are (a single bio agent, both a chemical agent and a bio agent — decon for the first before you decon for the second, two bio agents with different incubation times, etc.).

The best time for spraying aerosol is late at night or just before the first rays of dawn. The attackers want both security and a chance to get away with a dastardly deed; but they also need weather and atmospheric conditions as their unpaid assistants. They need a time when conditions offer minimum interference from ultra-violet radiation, and maximum assistance from an atmospheric inversion which can assist a toxic cloud to move along the surface of the land. As an example, the early morning hours tend to be a time of slowest wind speeds. The slower the wind speed, the higher the dosage, the smaller the area of coverage, and the higher the toxic effects. Dosage is a very important factor in relation to biological agents. Chemical agents have an "effective dose," the amount of a substance that may expected to have a specific effect. Biological agents have a comparable term, "infective dose," which refers to the number of microorganisms or spores necessary to cause an infection (spores are a form taken by some bacteria making them resistant to heat, drying, and chemicals; in some circumstances, the spore may change back into the active form of the bacterium; anthrax and botulism present examples of diseases caused by spore-forming bacteria). For means of comparison, the average lethal chemical agents in storage today are thousands of times **less** lethal, by weight, than equivalent amounts of biological warfare agents. Because of very high toxicity, the lethal biological agent dose can be far smaller than that required from chemical agents.

However, biological agents can be used against plants, animals, or materials rather than just against humans. Local responders will probably have no early warning of a biological attack; there are many more detection devices for chemical agents than there are for biological agents. It is entirely possible that local first responders (firefighters, police officers, and emergency medical personnel) will not even be called to the scene. Sooner or later, due to the incubation time delay, "everyone will be coming down with the flu." Always use the highest level of personal protective equipment available as protection for the respiratory tract by using a full-face mask with SCBA with positive pressure, at least until you know the specific threats.

Interim Recommendations for the Selection and Use of Protective Clothing and Respirators Against Biological Agents (provided by the Centers for Disease Control and Prevention (CDC))

The approach to any potentially hazardous atmosphere, including biological hazards, must be made with a plan that includes an assessment of hazard and exposure potential, respiratory protection needs, entry conditions, exit routes, and decontamination strategies. Any plan involving a biological hazard should be based on relevant infectious disease or biological safety recommendations by the Centers for Disease Control and Prevention (CDC) and other expert bodies including emergency first responders, law enforcement, and pubic health officials. The need for decontamination and for treatment of all first responders with antibiotics or other medications should be decided in consultation with local public health authorities.

This interim statement is based on current understanding of the potential threats and existing recommendations issued for biological aerosols. CDC makes this judgment because:

Biological weapons may expose people to bacteria, viruses, or toxins as fine airborne particles. Biological agents are infectious through one or more of the following mechanisms of exposure, depending on the particular type of agent; inhalation, with infection through respiratory mucosa or lung tissues; ingestion; contact with the mucous membranes of the eyes, or nasal tissues; or penetration of the skin through open cuts (even very small or abrasions of which employees might be unaware). Organic airborne particles share the same physical characteristics in air or on surfaces as inorganic particles from hazardous dusts. This has been demonstrated in military research on biological weapons and in civilian research to control the spread of infections in hospitals.

Because biological weapons are particles, they will not penetrate the materials of properly assembled and fitted respirators or protective clothing. Existing recommendations for protecting workers from biological hazards require the use of a half-mask or full facepiece air-purifying respirators with particulate filter efficiencies ranging from N95 (for hazards such as pulmonary tuberculosis) to P100 (for hazards such as hantavirus) as a minimum level of protection. Some devices used for intentional biological terrorism may have the capacity to disseminate large quantities of biological materials as aerosols. Emergency first responders typically use self-contained breathing apparatus (SCBA) respirators with a full facepiece operated in the most protective, positive pressure (pressure demand) mode during emergency responses. This type of SCBA provides the highest level National Institute for Occupational Safety and Health (NIOSH) respirator policies state that, under these conditions, SCBA reduces the user's exposure to the hazard by a factor of at least 10,000. This reduction is true whether the hazard is from airborne particles, a chemical vapor, or a gas. SCBA respirators are used when hazards and airborne concentrations are either unknown or expected to be high. Respirators providing lower levels of protection are generally allowed once conditions are understood and exposures are determined to be at lower levels.

When using respiratory protection, the type of respirator is selected on the basis of the hazard and its airborne concentration. For a biological agent, the air concentration of infectious particles will depend upon the method used to release the agent. Current data suggest that the self-contained breathing apparatus (SCBA) which first responders currently use for entry into potentially hazardous atmospheres will provide responders with respiratory protection against biological exposures associated with a suspected act of biological terrorism.

Protective clothing, including gloves and booties, also may be required for a response to a suspected act of biological terrorism. Protective clothing may be needed to prevent skin exposures and/or contamination of other clothing. The type of protective clothing needed will depend upon the type of agent, concentration, and route of exposure.

The interim recommendations for personal protective clothing, including respiratory protection and protective clothing, are based upon the anticipated levels of exposure risk associated with different response situations, as follows:

Responders should use NIOSH-approved, pressure-demand SCBA in conjunction with Level A protective suit in responding to a suspected biological incident where any of the following information is unknown or the event is uncontrolled:

- the type(s) of airborne agents(s);
- the dissemination method;
- if dissemination via an aerosol-generating device is still occurring or it has stopped but there is no information on the duration of the dissemination, or what the exposure concentration might be.

Responders may use Level B protective suit with an exposed or enclosed NIOSH-approved pressure-demand SCBA if the situation can be defined in which:

- the suspected biological aerosol is no longer being generated;
- other conditions may present a splash hazard.

Responders may use full facepiece respirator with a P100 filter or powered air-purifying respirator (PAPR) with high efficiency particulate air (HEPA) filters when it can be determined that:

- an aerosol-generating device was not used to create high airborne concentration,
- dissemination was by letter or package that can be easily bagged.

These type of respirators reduce the user's exposure by a factor of 50 if the user has been properly fit tested.

Care should be taken when bagging letters and packages to minimize creating a puff of air that could spread pathogens. It is best to avoid large bags and to work very slowly and carefully when placing objects in bags. Disposable hooded coveralls, gloves, and foot coverings also should be used. NIOSH recommends against wearing standard firefighter turnout gear into potentially contaminated areas when responding to reports involving biological agents.

Decontamination of protective equipment and clothing is an important precaution to make sure that any particles that might have settled on the outside of protective equipment are removed before taking off the gear. Decontamination sequences currently used for hazardous materials emergencies should be used as appropriate for the level of protection employed. Equipment can be decontaminated using soap and water, and 0.5% hypochlorite solution (one part household bleach to 10 parts water) can be used as appropriate or if gear had any visible contamination. Note that bleach could damage some types of firefighter turnout gear (one reason why it should not be used for biological agent response actions). After taking off gear, response workers should shower using copious quantities of soap and water.

All information presented in these pages and all items available for download are for public use. Provided by the Center for Disease Control and Prevention, 1600 Clifton Road, Atlanta, GA 30333 (Tel. 404-639-3311. Public Inquiry Hotlines: 888-246-2675 in English and 888-246-2857 in Spanish).

Biological Containment Labs Go Downtown

"Humans face grave risk from animal diseases," according to the Associated Press in a statement on February 20, 2006. "Each year for the last 25 years, one or two new pathogens (biological agents that are disease- producing microorganisms, such as bacteria, mycoplasma, rickettsia, fungi, or viruses) and multiple variations of existing threats have infected humans for the first time." The author, Andrew Briges, states there are 1,407 pathogens that can infect humans, 58 percent come from animals, while 177 of the pathogens are emerging or re-emerging. The fear is that some will cause pandemics (denoting a disease affecting or attacking the population of an extensive region, country or continent), but most will not. However, bird flu could be an exception. New strategy seems to be directed to containing the disease in the animal world, rather than the human world. Bird flu has killed at least ninety-one persons, and it seems to kill half the people it infects. Questions remain about why so many of the animal diseases seem to be infecting humans at the present time. One possibility involves changes in how people interact in a more densely populated world that is becoming warmer where many people travel extensively.

Another problem is that the federal government is withholding more information than ever from the public, as well as expanding ways of shrouding data, according to a report by Open The Government.org. In 2004 the government spent $7.2 billion with 15.6 million documents marked "top secret," "secret," or "confidential," thereby almost doubling the 8.6 million new documents classified as recently as 2001. In 2004, the number of pages declassified declined for the fourth straight year to 28.4 million. In 2001, 100 million pages were declassified. These statistics cover forty-one federal agencies (the Central Intelligence Agency classifications are secret, and are not represented).

Unheard of amounts of federal funds is going into the construction of biocontainment laboratories, or "hot labs," where the deadliest biological agents and potential bioweapons can be studied, researched and analyzed. These new hot labs would be mainly Biosafety Level 4 and Biosafety Level 3 laboratories dealing with infectious agents and toxins, the worst kind as far as danger and safety is concerned.

Biosafety Level 4 (BSL-4): Important procedures for BSL-4 and BSL-3 laboratories (as well as BSL-2 and BSL-1) are contained in a valuable 250-page manual, *Biosafety in*

Microbiological and Biomedical Laboratories, jointly produced by the U.S. Department of Health and Human Services/Public Health Service, the Center for Disease Control and Prevention, and the National Institutes of Health. This procedure is finely written, but sometimes not followed in practice. The 4th ed., May, 1999, quoted here is available by contacting the Government Printing Office, Washington, D.C (priced at $12.00 per copy in 1999).

"**Biosafety Level 4** is required for work with dangerous and exotic agents that pose a high individual risk of aerosol-transmitted laboratory infections and life-threatening disease. Agents with a close or identical antigenic relationship to Biosafety Level 4 agents are handled at this level until sufficient data are obtained either to confirm continued work at this level, or to work with them at a lower level. Members of the laboratory staff have specific and thorough training in handling extremely hazardous infectious diseases and they understand the primary and secondary containment functions of the standard and special practices, the containment equipment, and the laboratory design characteristics. They are supervised by competent scientists who are trained and experienced in working with these agents. Access to the laboratory is strictly controlled by the laboratory director. The facility is either in a separate building or in a controlled area within the building, which is completely isolated from all other areas of the building. A specific facility operations manual is prepared or adopted. Also adopted, in a number of pages, are the following practices: Standard Microbiological Practices, Special Practices, Safety Equipment (Primary Barriers), and Laboratory Facility (Secondary Barriers).

"**Biosafety Level 3** is applicable to clinical, diagnostic, teaching, research, or production facilities in which work are done with indigenous or exotic agents that may cause serious or potentially lethal disease as a result of exposure by the inhalation route. Laboratory personnel have specific training in handling pathogenic and potentially lethal agents, and are supervised by competent scientists who are experienced in working with these agents. All procedures involving the manipulation of infectious materials are conducted within biological safety cabinets or other physical containment devices, or by personnel wearing appropriate person protective clothing and equipment. The laboratory has special engineering and design features." Standard Microbiological Practices, Special Practices, Safety Equipment (Primary Barriers), and Laboratory Facility (Secondary Barriers) are often different for **Biosafety Level 3** than they are **Biosafety Level 4.**

Biological agents may be divided into two broad groups, infectious diseases and toxins. The infectious agents, such as bacteria and viruses, may be disease-causing elements that can grow and reproduce inside the infected individual's body. The infectious agent can also be transmitted to other persons by an infected human. The ability to spread, similar to other contagious diseases, makes infectious biological agents a frightening item to deal with for both research scientists and citizens.

Biological toxins, on the other hand, are simply poisonous substances produced by living things that act quickly (toxins tend to be similar to chemical weapons due to their speed of action). Biological toxins, such as outline toxins, staphylococcal entertoxin B (SEB), racing, and the T-2 mycotoxins are the most likely to be used as biological weapons. They are not capable of reproducing, but can be very potent, and are deliverable as aerosols. Biological toxins could be effective biological weapons because they tend to be more toxic by weight than many chemical weapons. However, they are not likely to be a source of secondary transmission.

Past History

"August 1942: George Merck, president of Merck & Co. pharmaceutical company, accepts the position as head of the newly-created War Research Service (WRS); the coordinating agency joins government and private institution resources to carry out the US biological warfare program. Headed by a small cadre of well-connected individuals, the WRS begins to conduct research at dozens of American universities. Simultaneously, the WRS encourages the Chemical Warfare Service (part of the US Army) to expand its examination of biological weapons potential and construct research facilities. The initial allocation in 1942 for the WRS totals $200,000. Meanwhile, the Chemical Warfare Service receives millions of dollars to construct research facilities." (Credit: The Henry L. Stimson Center, "History of the US Offensive Biological Warfare Program (1940–1973)"; http://www.stimson.org/?sn=cb2001121275)

From 1940 to 1969, the United States researched, developed, built, and stored **offensive** biological weapons. When President Nixon abolished offensive biological weapons in 1969, media personnel were invited to watch the United States get rid of "any existing stocks of biological weapons." Included were Category A agents such as anthrax, botulism, plague, smallpox, tularemia, and viral hemorrhagic fevers; and some Category B agents like Q Fever, brucellosis, glanders, Venezuelan/eastern/western encephalomyelitis, ricin toxin, and staphylococcus enterotoxin B. These germ warfare weapons were mixed with caustic soda, or heated to extremely high temperatures, to destroy their killer instincts, and make them docile once again. Also, one of the manufacturing plants located at Pine Bluff, Arkansas, had their equipment burned down to scrap metal, and then took guided tours to view the destruction in what used to be a "germ factory."

Chemical agents did not get the same treatment, as of March 1, 2006, we have them, to a large extent, still with us after six decades. Problems of what to do with the old and "touchy" chemical weapons, plus the poisonous atmosphere in the plants where they were manufactured, earned our government no good will whatsoever. First of all, almost all Allied governments that had chemical agents on their hands, either those of an enemy or their own, at the end of World War II, dumped them in oceans around the world. Everyone did it. The U.S. is responsible for about sixty sea dumping incidents equaling roughly one hundred tons of chemical weapons along the American coast and elsewhere. In addition, in the 1960s some 50,000 nerve gas (sarin) M-55 rockets were dumped off the coast of New Jersey and Florida. Recently, the United States had the number two collection of chemical weapons in the entire world, while Russia held down the number one position.

Johnston Atoll is an unincorporated U.S. territory located in Oceania where the destruction of our many chemical weapons was engineered, planned, and actually conducted for the first time. What was learned at Johnston Atoll (geographic coordinates: 16.45 N, 169.31) was meant to be used at the eight locations of the U.S. Unitary Chemical Stockpile (Please see below for the location of the nine depots and the variety of chemical weapons stored at each facility.) In July of 2001, the U.S. Army Chemical Activity Pacific left the atoll after thirty years of storing and handling chemical weapons stockpiles, and the destruction of such weapons since 1990. During the destruction of chemical weapons, U.S. Army experts at Johnston Atoll disposed of the very dangerous sarin and VX nerve agents while assisting the Johnston Atoll Chemical Disposal System (a contract civilian team). The two teams destroyed more that 400,000 rockets, bombs, projectiles, mortars, and mines; as well as 2,000 tons of nerve and blister agents (about 6.6 percent of the total U.S. stockpile).

The original chemical stockpile at Johnston Atoll consisted of the following weapons:

Agent	Item	Quantity
HD Sulfur Mustard-Blister	155mm Projectiles	5,670
HD Sulfur Mustard-Blister	105mm Projectiles	46
HD Sulfur Mustard-Blister	M60 Projectiles	45,108
HD Sulfur Mustard-Blister	4.2 Mortars	43,600
HD Sulfur Mustard-Blister	Ton Containers	68
GB Sarin-Nerve	M55 Rockets	58,353
GB Sarin-Nerve	155 Projectiles	107,197
GB Sarin-Nerve	105 Projectiles	49, 360
GB Sarin-Nerve	8" Projectiles	13,020
GB Sarin-Nerve	MC-1 Bombs	3,047
GB Sarin-Nerve	MK-94 Bombs	2,570
GB Sarin-Nerve	Ton Containers	66
VX-Nerve	M55 Rockets	13,889
VX-Nerve	155mm Projectiles	42,682
VX-Nerve	8" Projectiles	14,519
VX-Nerve	Land Mines	13,302
VX-Nerve	Ton Containers	66

Credit: GlobalSecurity.org at http://www.globalsecurity.org/wmd/facility/johnston_atoll.htm

As a side note, beginning in 1964, a huge open-air biological weapons test was conducted downwind from Johnston Atoll in which a number of ships were positioned around the atoll, upwind from a number of barges stocked with test subjects (rhesus monkeys) who were exposed by agents dispensed from aircraft.

There are seven sites in the United States where U.S. Unitary Chemical Stockpile exists. Below they are listed by chemical weapon as follows: GB, sarin; H, mustard; HD, mustard; HT, mustard; VX, persistent nerve agent; GA, tabun; L, lewisite; TGA, thickened tabun; TGB, thickened sarin.

Site	Agents	Agent Tons
Anniston Army Depot, AL	GB, HD, HT, VX	2,253.63
Aberdeen Proving Ground, MD	HD	1,624.87
Blue Grass Army Depot, KY	GB, HD, VX	523.41
Newport Chemical Activity, IN	VX	1,269.33
Pine Bluff Arsenal, AR	GB, HD, HT, VX	3,849.71
Pueblo Depot, CO	HD, HT	2,611.05
Tooele Army Depot, UT	H, HD, HT, GA, GB, L, TGA, TGB, VX	13,616.00

Credit: "Waging Peace," Chemical and Biological Weapons: Use in Warfare, Impact on Society and Environment, by Gert G. Harigel; http://www.wagingpeace.org/articles/2001/11/00_harigel_cbw.htm

Recent History

In October of 2001, American society learned all they needed to know about biological weapons on home ground when they were attacked with military-grade anthrax. A funny thing happened when we realized the anthrax that killed five United States citizens and sickened others was our own. The biological agent was transferred to the victims by contaminated letters, and the anthrax was reported to contain one trillion spores per gram. That is, U.S. Senate Majority Leader Tom Daschle received a tainted letter that was found to contain two grams of anthrax, roughly enough to kill 200 million persons if certain means of dispersal had been used. Such high concentration levels of the anthrax actually used probably means that the biological weapon was originally taken from a national weapons program, probably from our country's own biological stockpile. Now, many years later, the strongest country in the world has been tragically unable to determine who was responsible for this attack.

After the anthrax attacks, and the anthrax hoaxes that followed, the U.S. Congress earmarked $6 billion for bio-safety programs. On June 3, 2004, Anthony S. Fauci, M.D., Director of the National Institute of Allergy and Infectious Diseases, made a presentation before the House Committee on Appropriations, Subcommittee on Labor, HHS, and Education. The title of his speech was, "The National Institutes of Health Biomedical Research Response to the Threat of Bioterrorism." He opened with these comments, "The destruction of the World Trade Center, the attacks on the Pentagon and an airliner over Pennsylvania, and the anthrax attacks in the fall of 2001 clearly exposed the vulnerability of the United States to acts of terrorism. In particular, the anthrax attacks made it very clear that the possibility of the use by terrorists of deadly pathogens or biological toxins such as those that cause anthrax, smallpox or botulism represents a serious threat to our Nation and the world. The Administration and Congress responded aggressively to this threat by significantly increasing funding for bio-defense preparedness and research."

In a single paragraph, Fauci hyped terrorism, the 911 attacks, and the almost perfect anthrax score that bit into the American experience with biological weapons on home ground. Is this really responsible for the role of the National Institutes of Health (NIH) in the execution of a national bio-defense research strategy, or just waving a red flag? The NIH/NIAID is right now planning to build, with noted universities, two National Bio-containment Laboratories (NBLs) that will be Bio-safety Level 4, plus nine Regional Bio-containment Laboratories (RBLs) with Bio-safety Level facilities. At the time of this writing, however, nobody in America seems to know how many Level 4 and Level 3 labs are in the fifty states. We seem to be asked to consider money and prestige as enough reason for building eleven hot labs (at least that we know about). The Arab terrorists did not use biological weapons; they used jet fuel and fire, and a great deal of planning. We could ask the FBI how many planners and support members of the 911 airliner terrorist attack they have sentenced so far, or even brought to trial in United States. However, the FBI has had a number of problems since October of 2001.

The FBI has a "Timeline" of events and activities of interest from May 27, 1908, until October 26, 2001. It seems to end on October 26, 2001. Did the FBI lose interest? Admittedly, they had some problems.

February 18, 2001: FBI Agent Robert Philip Hanssen was arrested for conspiracy to commit espionage. The affidavit in support of an arrest warrant for Hanssen charged that he had engaged in a lengthy relationship with the KGB (Russia) and its agencies.

September 11, 2001: Following massive terrorist attacks against New York and Washington, the FBI dedicated seven thousand of its eleven thousand Special Agents and thousands of FBI support personnel to the PENTTBOM (short for Pentagon, Twin Towers Bombing).

October 18, 2001: in conjunction with the U.S. Post Office, the FBI offered a reward of $1,000,000 for information leading to the arrest of the person who mailed letters contaminated with anthrax to media organizations and congressional officers.

The Sunshine Project (http://www:hammond@sunshine-project.org; P.O. Box 41987, Austin, Texas 78704) keeps track of biological containment labs in the United States. This group published a listing of High Containment Labs and Other Facilities of the US Biodefense Program in November of 2004 (biological containment labs 3 and 4 not known to be heavily dedicated to bio-defense are not indicated on the listing).

Operational BSL-4 Facilities:

USAMRIID Fort Detrick, Frederick, MD
Centers for Disease Control, Atlanta, GA
University of Texas Medical Branch, Galveston, TX
Southwest Foundation for Biomedical Research, San Antonio, TX

Operational BSL-3 Facilities:

Harvard University, Cambridge, MA
Cornell University, Ithaca, NY
DHSW/USDA Plum Island, NY
CALSPAN-UB, Buffalo, NY
SUNY, Stony Brook, NY
PHRI, Newark, NJ
Wadsworth Center, Albany, NY
University of Pennsylvania, Philadelphia, PA
Thomas Jefferson University, Philadelphia, PA
Armed Forces Institute of Pathology, Washington, DC
Naval Medical Research Center, Silver Spring, MD
US Army SBCCOM (2), Aberdeen, MD
University of Maryland, Baltimore, MD
Southern Research Institute, Frederick, MD
Versar, Gaithersburg, MD
The Pentagon, VA
American Type Culture Collection, Manassas, VA
George Mason University, Manassas, VA
Naval Surface Weapons Center, Dahlgren, VA
Commonwealth Biotechnologies, Richmond, VA
Virginia Commonwealth University, Richmond, VA
University of Kentucky, Lexington, KY
Oak Ridge National Laboratory, TN
Wake Forest University, Winston-Salem, NC
Emery University, Atlanta, GA
USDA FSIS/MOSPL, Athens, GA
Midwest Research Institute, Palm Bay, FL
University of Miami, FL
US EPA, Cincinnati, OH

Battelle Memorial Institute, West Jefferson, OH
IITRI, Chicago, IL
University of Wisconsin (x2), Madison, WI
Midwest Research Institute, Kansas City, MO
Southern Research Institute, Birmingham, AL
Louisiana State University, Baton Rouge, LA
University of Texas Health Sciences Center, Houston, TX
University of Texas Southwestern, Dallas, TX
University of Texas Health Science Center, San Antonio, TX
Lackland Air Force Base, San Antonio, TX
Texas Technological University, Lubbock, TX
Texas A&M University, College Station, TX
Oklahoma State University, Stillwater, OK
Centers for Disease Control, Fort Collins, CO
Los Alamos National Lab, Los Alamos, NM
Lovelace Institute, Albuquerque, NM
University of New Mexico, Albuquerque, NM
US Army Dugway Proving Ground, UT
San Diego State University, CA
Scripps Research Institute, La Jolla, CA
University of Washington, Seattle, WA

Planned BSL-4 Facilities:
Boston University, Boston, MA
NIH, Fort Detrick, Frederick, MD
DHS NBACC, Frederick, MD
Centers for Disease Control, Atlanta, GA
University of Texas Medical Branch, Galveston, TX
Rocky Mountain Labs, Hamilton, MT

(Major) Planned BSL-3 Facilities:
UMD of New Jersey, Newark, NJ
University of Pittsburgh, PA
University of Maryland, Baltimore, MD
Duke University, Durham, NC
Medical University of South Carolina, Charleston, SC
University of Alabama at Birmingham, AL
University of Tennessee at Memphis, TN
Tulane Primate Center, Covington, LA
University of Missouri, Columbia, MO
University of Iowa, Iowa City, IA (RCE planning)
USDA / Iowa State University, Ames, IA
Argonne National Lab, Argonne, IL
Agricultural Biosecurity Center, Manhattan, KS
University of Minnesota, Minneapolis, MN (RCE planning)
University of Texas as El Paso, TX
U.S. Army Dugway Proving Ground, UT
Centers for Disease Control, Fort Collins, CO

Colorado State University, Fort Collins, CO
Lawrence Livermore Lab, Livermore, CA

Biodefense Aerosol Facilities:
CALSPAN-UB, Buffalo, NY
US Army Aberdeen Proving Ground, MD
George Mason University, Manassas, VA
Midwest Research Institute, Kansas City, MO
Lovelace Institute, Albuquerque, NM
US Army, Dugway Proving Ground, UT

Open-Air Testing Facilities:
US Army Dugway Proving Ground, UT
Nevada Test Site, NV (Proposed)
White Sands Missile Range, NM (Probable)

Classified/Secret Research:
US Army, Aberdeen Proving Ground, MD
USAMRIID, Fort Detrick, Frederick, MD
Versar, Gaithersburg, MD
Commonwealth Biotechnologies, Richmond, VA
Southern Research Institute, Birmingham, AL
Battelle Institute, Columbus/West Jefferson, OH
Southwest Federation for Biomedical Research, San Antonio, TX
Texas Technological University, Lubbock, TX
DTRA et. al., Kirtland/Albuquerque, NM
US Army Dugway Proving Ground, UT
DOE, Nevada Test Site, NV

(This listing is derived from public sources and provides the most up-to-date information available as of November 4, 2004; it is a rolling list subject to frequent revision.) Credit: The Sunshine Project, at http://www.Sunshine-project.org/biodefense.

What Is a Select Agent?

Select agent is a term given to certain viruses, bacteria, fungi, and toxins that have been identified by the Centers for Disease Control and Prevention and the Department of Agriculture having the potential to pose a severe threat to public health and safety, agriculture and environment. The list is reviewed routinely and can change over time. The list contains wide varieties of agents such as smallpox virus, the bacterium that causes anthrax, and the butulinum neurotoxin. Depending on the agent, certain bio-safety precautions must be used, and all research done with all of the agents comes under specific Federal requirements for work with these materials, including federal background checks of the researchers and other personnel involved with the research.

The National Institute of Allergy and Infectious Diseases (NIAID) is spending great amounts of taxpayer's money into bringing about two National Biocontainment Laboratories (NBLs) and nine other Regional Biocontainment Laboratories (RBLs). The two NBLs will be at the University of Texas Medical Branch in Galveston, Texas, and Boston University in downtown Boston, Massachusetts. The nine new RBLs will be placed at Colorado State University, Fort Collins, Colorado, Duke University, Durham, North Carolina, Tulane University, New Orleans, Louisiana, University of Alabama at Birmingham, Alabama, University of Chicago, University of Medicine and Dentistry of New Jersey in Newark, University of Missouri, Columbia, Missouri, University of Pittsburgh, Pennsylvania, and the University of Tennessee at Memphis.

One-time grants of approximately $120 million each will fund construction of the NBLs, while the nine RBLs will receive one-time grants of between $7 and $21 million each in construction funds. However, each of the universities will be required to provide matching funds. The NBL/RBL construction program was meant to provide funding to design and construct a series of up-to-date laboratories designed and built, it was promised, using the strictest federal standards incorporating special engineering and design features to present microorganisms from escaping from the labs. Safety and decontamination features will attempt to supply multiple layers of protection for scientists as well as nearby populations and environments. At least in theory, the NBLs and the RBLs will be available and prepared to assist national, state and local officials in the event of a bioterrorism or infectious disease emergency.

A variety of scientists, politicians, and interested observers are in favor of the eleven NBLs and RBLs, but others are completely against the whole idea for reasons of safety, price, "Not In My Backyard," absolute fear, wonder of where we will get the trained and experienced scientists to staff these new labs, worry about getting into a biological weapons war, and a number of other reasons.

Scientific research, either good or bad, is extremely competitive, striving for recognition at all costs, as well as for additional funding. Basically, it's a question of money and careers. Laboratories at Level-4 and Level-3 could be a high profile terrorist target, particularly in a densely populated area. Training and experience for new scientists to deal with the "worst-of-the-worst" diseases could provide a killer atmosphere in a hot lab. If the U.S. government tells universities to work on bio-warfare or bioterrorism, many people with long memories believe they will work on such programs as they did during World War II and into the Cold War years. There are still people who believe that "Government-Funded Research" is a dirty word. Remember, we are on the way to "Disease by Design" at the present time.

The feds have just a rough idea of how much bio-containment lab-space is necessary, how much it will cost to run these eleven labs each year, or how to handle and train new scientists in a safe fashion. There is just a lack of competent research staff at the present time to deal with Category A and Category B biological agents.

Category A agents: There are slight differences between the CDC and NIAID categories, although both agencies are very close in the overall rankings. There are six categories as follows; Anthrax, Botulism, Plague, Smallpox, Tularemia, and Viral Hemorrhagic Fevers. According to CDC, viral hemorrhagic fevers include Ebola, Marburg, Lassa, Rift Valley fever, Hanta viruses and New World Arenaviruses. According to NIAID, viral hemorrhagic fever includes ebola, marburg, lassa, Rift Valley fever, hanta viruses, Dengue, New World arenaviruses and lymphocytic choriomeningitis virus.

Category B agents: There are eleven Category B agents as follows: brucellosis, epsilon toxin (clostridium perfringens), glanders, melioidosis, psittacosis, Q fever, ricin toxin, staphylococcus enterotoxin B, typhus fever, viral encephalitis, and water safety threats.

We are talking about **training and experience of new scientists** who will fill these employment slots in eleven hot labs as well as private labs, which seem to be going through another growth spurt. The new labs will probably have to train five-times the number of reliable scientists and workers that are now on hand. Where are they going to come from, and who will train them and give them some experience before they start working in a hot lab? Much building of biological labs is considered to be overkill, almost certain to kill non-biological research in the years ahead.

How bad can it be? Recently, these stories appeared in newspapers and trade papers.

In 1998, a research assistant at the Yerkes Primate Center in Atlanta, Georgia, part of the National Institute of Health's Primate Research Program, died six weeks after being exposed to simian herpesvirus in the laboratory.

In March of 2000, a microbiologist working with infectious diseases in a Biosafety Level 3 facility at USAMRIID contacted glanders due to accidental exposure. Between 1987 and 1990, two other workers acquired infectious diseases at the same facility.

April of 2002, a researcher at U.S. Army Medical Research Institute of Infectious Diseases (USAMRIID) tested positive for exposure to anthrax spores, which were also released in small quantities into an adjacent hallway and office.

In December of 2002, a three-hour total power failure undermined the containment systems at an infectious disease laboratory at Plum Island, New York. Workers had to resort to sealing the doors with duct tape, as the air compressors failed.

Also in 2002, government scientists revealed that over two dozen dangerous biological agents including anthrax, and Ebola went unaccounted for in the early 1990s at the US Army Medical Research Unit (USAMRIID) in Fort Detrick, Maryland. The location of these agents, which were subject to removal without authorization, remains a mystery.

On March 20, 2003, a package containing the West Nile virus exploded in a Federal Express building in Columbus, Ohio. exposing workers to the possible infection and causing offices to be evacuated.

In June of 2003, the U.S. Army unearthed 113 bacteria-containing vials, including live strains of brucellosis and non-virulent anthrax, during excavation of its Fort Detrick site to eliminate toxic chemicals and hazardous waste. (Credit: The Council for Responsible Genetics)

In Oakland, California, several lab workers were suspected of exposure to anthrax when a Maryland lab supplier mistakenly sent live anthrax spores rather than dead microbe strains.

In another case in February of 2004, a researcher was exposed to Ebola virus at the United States Army laboratory at Fort Detrick, MD.

Going some years back in time, Kanatjan Alibekov (or Ken Alibek, the name he held during his time in the United States), the mastermind for two decades behind Russian efforts with germ warfare, in his 1999 book, *Biohazard,* reports that one of his first jobs of quality in the Russian program was when he was assigned to develop a production process to weaponize tularemia. One night, he got a call from an assistant telling him, "We have a problem." The air pressure in one of the tularemia rooms had begun to drop precipitously. He mistakenly entered the room, turned on the light in the darkened room, and realized he was standing in a puddle of tularemia. "The puddle at my feet was only a few centimeters deep, but there was enough tularemia on the floor to infect the entire population of the Soviet Union," he remembers. Ken Alibek defected to the United States in 1992.

On April 23, 2004, in Fremont, California a twenty-nine-year-old instrument research engineer at Ciphergen Biosystems wrote a suicide note at his place of work about 6:00 am, stepped into the yard and inhaled potassium cyanide, thus taking one of the speediest methods to end his life. Firefighters called to the scene could not find the chemical agent vial the suicide victim had used, so they evacuated the premises while searching for the missing vial of the highly toxic chemical. The lost vial was found about noon.

In Newark, New Jersey, on September 15, 2005, three mice infected with the bacteria responsible for bubonic plague disappeared from a laboratory for roughly two weeks ago. The incident came to light only because federal authorities were investigating possible corruption at the University of Medicine and Dentistry of New Jersey. The missing mice were unaccounted for at the Public Health Research Institute that conducts bioterrorism research for the U.S. government and is on the campus of the university. "If the mice got outside the lab, they would have already died from the disease," according to state Health Commissioner, Fred Jacobs. Theft is a primary possibility, and the institute interrogated two dozen of its employees, as well as conducting lie detector tests. However, the mice may have been eaten by other lab animals or just misplaced. Both the FBI and the Centers for Disease Control and Prevention are investigating, but the mice will probably never be found.

On October 6, 2005, a Federal Express cargo plane carrying small amounts of flu virus crashed on railway tracks near Winnipeg's (Canada) city center, killing the female pilot. Among cargo in the plane were research samples of frozen influenza and herpes viruses. The crash landing started a fire which consumed the plane and cargo.

On November 18, 2005, in Houston, the Texas Department of Health Services asked the public to help locate some radioactive vials that disappeared earlier that month from a shipment out of New Mexico. The vials contain antimony-124, a radioactive material for use in the oil and gas industry, with the label "RADIOACTIVE." It is believed by authorities that the vials were removed within Texas when the carrier's tractor trailer stopped in Abilene, Austin, Dallas, and Tyler. Anyone finding the missing vials should not touch the box or the actual vials, and should stay at least ten feet away from them. Anyone with factual information should call 512-458-7460.

Taking Hot Labs Down Town

Boston University Medical Center BSL-4 Containment Laboratory had already contaminated three scientists before they even got funded. This incident brought up again the debate on the Bio Lab 4 need and placement in American communities, as well as a lack of public trust shown by Boston University Medical Center for not informing the public, or anyone else it seems, the contamination of three workers with tularemia in September of 2004 until the press started asking questions in January of 2005. Trust by people can be very short-lived; once it is lost, it is very difficult to regain. One group has sued the Bio Lab to be located downtown in the southern part of Boston. They charge various groups, including the Boston Redevelopment Authority and Boston University, violated state law by misjudging the effect a contamination by tularemia of three scientists would have on the Boston community. Mayor Thomas M. Menino failed to go public with the contamination at Boston University Medical Center when he was informed. After a doctor was relieved of his job as Chief of Infectious Diseases in November of 2004 for reported failure to display leadership, it was alluded to that lab workers did not follow procedures and rules citing tularemia was to be handled in an enclosed chamber (glove-box) that filters infectious agents in the air. The tularemia exposure is being investigated by state and federal public health components, city agencies, Boston University, and the FBI. The university did report the infection in November of 2004, but failed to inform state and local officials who were in the midst of reviewing environmental documentation. Dr. Thomas Moore, provost of Boston University Medical Center admitted in January of 2005 that lab workers failed to observe proper safety precautions when working with tularemia samples. O.S.H.A. started investigation of the contamination of three lab workers only after the agency read about in the newspaper. Boston University Medical Center did actually receive funding for a $128 million bio-defense laboratory dealing with Biological Level-4 agents, the worst-of-the-worst in biological agents. (Such agents can easily be developed as biological agents, can be easily disseminated or transmitted from person-to-person, are stable in aerosol form and can be delivered in this form, have the potential for large-scale dissemination, have the potential to cause high mortality and morbidity, would cause widespread public panic, and pose the greatest threat to national safety and security.)

Tulane National Primate Center BSL-3 Containment Laboratory is expanding thanks to a $13.6 million grant secured from the National Institute of Allergy and Infectious Diseases that will go toward the building of a new Level-3 bio-containment lab. Tulane University will invest $5 million in the project. The center already has a ten-year old Level-3 lab where scientists have been involved with deadly infectious diseases; such as AIDS, SARS, botulism, plague, Lyme disease, smallpox, anthrax and tularemia. The new funding will also build a Level-3 lab that will allow Tulane's National Primate Center to expand its focus into bio-defense-related work. Except, they did not consider Hurricane Katrina would savage the campus. The college administrators were evacuated to Houston, Texas, four hundred Tulane students were evacuated to Jackson State University in Mississippi, and the college's fall semester was cancelled since many students went elsewhere to find a college that would accept them. The uptown campus is covered with debris from fallen trees and shrubs, making it almost impossible to drive or even walk on campus. There is no power in any of the buildings other than the few where the college controls the power. There is devastation of the city and its infrastructure that is deteriorating further due to the flooding from the hurricane. Of the eight national research centers, Tulane's is the largest.

The second National Bio-containment Lab is at the University of Texas Medical Branch on Galveston Island. This is a barrier island in the Gulf of Mexico, actually outside the Intra-coastal Waterway that can experience serious weather during hurricane season. In Hurricane Rita during September of 2005, Alison McCook writing in *The Scientist* (http://www.the-scientist.com/news/20050923/01) stated, "Gulf coast researchers working in the path of Hurricane Rita — featuring winds clocking in at well over 100 mile per hour—began preparing for her arrival far in advance, destroying and freezing samples, ending experiments, and locking away dangerous and unique materials. Still, weeks or more of experiments could be lost, they say." The writer went on to say, "(they) shut down the institution's bio-safety level 3 and 4 facilities by terminating experiments in progress, destroying active cultures, and placing stocks of select agents, such as anthrax and hemorrhagic fever, along with other viruses and bacteria in 'very secure, locked freezers' on site." Why are they building a $167 million six-story bio-safety Level-4 lab in a hurricane alley? It really does not make any sense unless you are attempting to construct a "Bio-Boondoggle" in the coming years. In February of 2006 at the Texas A&M Bio-containment Laboratory, a female researcher was contaminated by brucella (in 1954, brucella suis was the first biological agent weaponized by the United States in efforts to undertake the manufacture offensive biological weapons, even though brucella melitensis may produce more severe human disease) when an aerosol chamber finished a run on exposed mice, and had to be disinfected. The victim may have been contaminated through her eyes. The University then failed to report this incident to the Centers for Disease Control as required by federal law.

On December 7, 2005, Andrew Roberts of the Associated Press wrote how a new federal agency is going to handle an attempt to encourage vaccine development. "By creating a federal agency shielded from public scrutiny, some lawmakers think they can speed the development and testing of new drugs and vaccines needed to respond to a bioterrorist attack or super-flu pandemic. The proposed Biomedical Advanced Research and Development Agency (BARDA) would be exempt from long-standing open records and meetings laws that apply to most government departments, according to motions approved October 18, 2005, by the Senate health committee. The legislation also proposes giving manufactures immunity from liability in exchange for their participation in the public-private effort. The agency would provide the money for development of treatments and vaccines to protect the United States from natural pandemics as well as chemical, biological and radioactive agents. But it is the secrecy and immunity provisions of the legislation that has alarmed patient rights and open government advocates. The agency would be exempt from the Freedom of Information and Federal Advisory Committee acts, both considered crucial for monitoring government accountability."

Guides for Emergency Response: Biological Agent or Weapon: Anthrax

5

The Centers of Disease Control and Prevention (CDC) and the National Institutes of Allergy and Infectious Diseases (NIAID) grouped potential biological agents and toxins during July of 2003 into three categories which were divided by the levels of concern that such agents could be used as biological weapons against the United States. Three categories were derived at levels A, B, and C.

Category A Agents:

Have already been developed as biological weapons.

Can be easily disseminated or transmitted from person-to-person.

Are stable in aerosol form and can be delivered in this form.

Have the potential for large-scale dissemination.

Have the potential to cause high mortality and morbidity; would cause widespread, public panic.

Pose the greatest threat to national safety and security.

There are six categories A biological agents: anthrax, botulism, plague, smallpox, tularemia, and viral hemorrhagic fevers. At the end of World War II, the United States along with allies, Britain and Canada, wanted a biological weapon that could kill people, possessed no natural immunity among victims, was singularly infectious, and could be guaranteed to have a high kill-ratio. The trouble with anthrax was it was too good a killer and biological agent. American and British scientists knew from tests conducted on Gruinard Island off the coast of Scotland in 1942 that anthrax would absolutely ruin any area or region when applied for generations.

History: In a different type of "Lend Lease," in March of 1944, Britain placed an order with the United States for a half-million anthrax bombs to use on the Germans in case they ever used chemical and/or biological weapons in their V-rockets which at the time were wreaking havoc on London. Camp Detrick in Frederick, Maryland, supplied a batch of trial anthrax bombs developed in the laboratory, and sent them to the manufacturing plant. The anthrax bombs ordered by Britain were planned to be manufactured at Vigo, Indiana, slightly southwest of downtown Terre Haute. The war in Europe ended in May of 1945, and the production plant in Vigo, built for a large amount of money, never delivered the British order.

The anthrax actual attacks in the United States in mid-October of 2001 (and the multiple anthrax hoaxes that followed) rocked our nation. To date, there has been no solution to the actual attacks. However, the anthrax actually used in the attacks was determined to be military-grade containing one trillion spores-per-gram. That is, two grams of that anthrax was figured to be sufficient to kill 200 million persons, it was reported. The anthrax used in the attacks was also proposed to be taken or stolen from the United States national weapons program. Basically, our own bacterial anthrax was used to kill five Americans in their own country and injure a number of others. Will we ever be free from the anthrax threat that our federal government researched, manufactured and stored years ago?

Agent: Anthrax. This biological agent (bacteria) has been called a perfect germ for bio-terrorism and is thought of an excellent weapon for mass destruction. Anthrax is quite deadly if not detected and treated early. However, if treated at once, anthrax can respond well to antibiotics. Its spores can be produced in large quantities and can be stored for years without losing potency. Anthrax spores can be disinfected with chlorine or iodine, and all medical personnel should practice standard universal precautions with any anthrax patients. Exposure is difficult to detect; it has no color or cloud, no smell, no taste, and there is no indication of attack when dispersal is done by aerosol spray. Antibiotics will suppress infection only if administered early after exposure, and anthrax is more than 90 percent lethal to unprotected persons. The disease begins with an incubation period of from one to six days presumably dependent upon the dose of inhaled organisms. Anthrax is an ill-ness with acute onset characterized by several distinct clinical forms such as cutaneous, inhalation, intestinal and oropharyngeal. Chest x-rays may reveal a dramatically widened mediastinum, and shock and death usually follow within twenty-four to thirty-six hours of respiratory distress onset. Planned release of anthrax would probably be done by aerosol since the spore form of the bacillus is quite stable. Anthrax is viewed as the single greatest threat for use in biological warfare; it is quite contagious with a high mortality rate (but is not contagious from person-to-person, except in rare circumstances). **However, caution should be used with drainage or secretions from a patient to present dermal (skin) infection of medical services staff; soiled equipment or other items that have come into contact with anthrax victims should be disinfected or burned, and inhalation anthrax patients should be isolated.**

Anthrax can easily be produced in large quantities, is relatively easy to weaponize, can readily be spread over a wide area, may be stored safely, and remains lethal for a long period of time. About 95 percent of natural anthrax infections are cutaneous; that is, they affect the skin. Additional routes of entry may be by inhalation or ingestion. It can also occur naturally; zebras are very much affected by anthrax. Anthrax spores can settle in the soil; some herbivores may become infected in this manner but humans are unlikely to be affected. Bleach will kill anthrax spores. When using aerosol dissemination in a terrorist or warlike act, inhalation-type anthrax will be the result, a much more dangerous disease than the natural form. It must be treated before symptoms appear with high dose antibi-otic treatment. With anthrax, treatment must come quickly, within twenty-four-hours, or most victims will die. Untreated, the mortality rate of inhalation and intestinal cases is about 95 percent, while untreated cutaneous (skin) anthrax can be up to 25 percent. A unique feature of anthrax is a treatment "eclipse" when patients start feeling better just before they die. At the present time, 2 million military personnel in the United States have been or are being vaccinated.

Classification: Bacterial

Duration of Illness: Three days (usually fatal).

Probable Form of Dissemination: Spores in aerosol.

Detection in the Field: Limited.

Infective Dose (Aerosol):   8,000 to 50,000 spores.

Signs and Symptoms: Fever, fatigue, cough, and mild chest discomfort is followed by severe respiratory distress with difficult or labored respiration, diaphoresis, a harsh vibrating sound heard during respiration in cases of obstruction of the air passages, and a bluish/purplish discoloration due to oxygenation of the blood. Shock and death occurs within twenty-four to thirty-six hours of severe symptoms.

Incubation Time: One to six days (anthrax which is not transmissible from person-to-person). Compare this incubation time with that of the virus, smallpox, which is ten to seventeen days. Smallpox is highly transmissible from person-to-person. After exposure to smallpox, a person could travel by air around the world a number of times and contaminate many people before developing any symptoms. However, naturally occurring smallpox has been eradicated worldwide since 1977.

Diagnosis: Physical findings are non-specific. There is possible widening of the space in the chest between the pleural sacs of the lungs that contain all the viscera of the chest except the lungs and the pleurae which can be detected by a Gram stain of the blood and by a blood culture late in the course of illness.

Differential Diagnosis: An epidemic of inhalation anthrax in its early stage with non-specific symptoms could be confused with a number of viral, bacteria, and fungal infections. Progression over two to three days with sudden development of severe respiratory distress followed by shock and death within twenty-four to thirty-six hours in essentially all untreated cases eliminates diagnosis other than inhalation anthrax. Other diagnosis to consider would include aerosol exposure to staphylococcal enterotoxin B (SEB), plague, or tularemia pneumonia.

Vaccine Efficacy (for aerosol exposure)/antitoxin: Currently no human data.

Persistency: Spores are highly stable.

Personal Protection: Protective clothing must be used as well as protection for the respiratory tract by using a mask with biological filters or self-contained breathing apparatus (SCBA) with positive pressure, at least until you know the specific threat(s). Also, Time/Distance/Shielding.

Routes of Entry to the Body: Inhalation, skin, and mouth. A biological warfare attack or a terrorist incident utilizing anthrax spores disseminated by aerosol would cause inhalation anthrax, a very rare form of this naturally occurring disease. Normal infection could occur through scratches or sores, wounds, eating insufficiently cooked infected meat, or by flies. All human populations are susceptible. The good news is that if you survive anthrax disease the first time, you would be immune to anthrax disease in the future.

Transmissable For Person-To-Person: No (except cutaneous).

Duration of Illness: Two to five days (often proves fatal).

Potential Ability to Kill: High.

Symptoms & Effects: Flu-like, upper-respiratory distress; fever and shock in third day to fifth day, followed by death.

Defensive Measures: Immunizations, good personal hygiene, physical conditioning, wearing a protective mask, and practicing good sanitation.

Vaccines: Yes. BioPort, Lansing, Michigan. This vaccine is a cell-free filtrate vaccine (it contains no dead or live bacteria). Vaccine schedule is zero, two, and four weeks for the initial series, followed by booster shots at six, twelve, and eighteen months and then a yearly booster shot. Five years after anthrax attacks using the U.S. Mail killed five victims and injured seventeen others, in October of 2001, it seems at least one new vaccine for anthrax may be showing promise. In a small test study using 111 healthy adults, a possible anthrax vaccine made by Avecia has impressed some people. BioPort has a contract with the government to provide "the old vaccine," that was provided to U.S. troops who served in possible harm's way during recent wars in Iraq and other countries. The BioPort vaccine was based on knowledge of fifty years ago, long before we had to deal with biological weapons that could be delivered by aerosol at nearly any time and place. The BioPort vaccine requires time and doses to deliver its punch, requiring six shots over eighteen months to provide protection, and had some side effects. The government gave a California company, VaxGen, a $877 million federal contract with a new approach developed by the Army's research laboratory at Fort Detrick located in Frederick, Maryland. However, VaxGen, in an effort to produce their vaccine using three doses in six months has failed to meet deadlines, and said delivery is now scheduled for a minimum of 2008 or later. Not one to hedge its bets, the federal government also gave Avecia a $71 million dollar contract to develop a different vaccine using the same approach. Since the October of 2001 successful attacks made with anthrax five years ago have never be solved at this writing, this tends to worry a lot of people in the United States. Word is the experimental vaccine made by Avecia in its first test on human beings, which required two doses in a single month, no safety problems were noticed. Some immune response was found, but it may be lower than the BioPort vaccine which is in use at the present time. Time will tell the tale if we have an enemy attack against the United States with anthrax delivered by aerosol.

Drugs Available: Yes. (Ciprofloxacin, Doxycycline, and Penicillin), although such drugs are usually not effective after symptoms are present. Supportive therapy may be necessary. Oral ciprofloxacin can be used for known or imminent exposure.

Decontamination: Soap and water, or diluted sodium hypochlorite solution (0.5 percent). Drainage and secretion precautions are necessary. After invasive procedures or autopsy, decontaminate instruments and surfaces with 0.5 percent sodium hypochlorite or with a sporici   Anthrax, after symptoms have became apparent, can be very deadly. Although the death rate for dermal anthrax is roughly about five to 20 percent, the fatality rate for inhalation anthrax after symptoms progress is almost always fatal, regardless of treatment.

Characteristics: In general, characteristics for biological agents that key their possible use as weapons of mass destruction(WMD) depend upon nine different areas of interest: incubation period, infectivity, lethality, pathogenicity, stability, toxicity, transmissibility, virulence, and "other" characteristics. These characteristics are somewhat unique in that, unlike chemical agents, a number of biological agents have the power to multiply in a victim's body over time and enable the biological agents, in some cases, to increase their overall effect.

Duties of First Responders on the Scene

Fire Department   Isolate/Secure the Scene, Deny Entry, Establish Control Zones. Establish Command, Evaluate Scene Safety/Security. Stage the arriving units. Gather Information regarding the Incident, Number of Patients, and Other Basic Information. Assign Incident Command System (ICS) as needed, Initiate notifications (i.e., state/federal/assisting agencies/specialized agencies, law enforcement, and hospitals, etc.). Request Additional Resources, and Use Necessary Defensive Measures (Time/Distance/Shielding, Proper Protective Equipment, Minimize Number of Responders Exposed to Danger, etc.). Initiate Public Safety Measures (Rescue, Evacuate, Protect in Place, etc.), Establish Water Supply, Control and Isolate Patients in a Safe Area, and Coordinate Activities with Law Enforcement. Begin and/or Assist with Triage, administering antidotes and treatment. Begin gross mass decontamination operations. As the incident progresses, prepare to initiate the Unified Command System. Establish Unified Command Post, including the following organizations: emergency medical devices, law enforcement, emergency management, public works, and hospitals/public health. Establish and maintain chain of custody for evidence protection.

Classification: **Anthrax** is classified as a bacterial agent (other possible classifications in addition to bacterial are as follows: viral agents, rickettsia, and biological toxins). Bacteria are small, single-celled organisms, many of which can grow on either solid or liquid culture media. Some, but not all bacterial agents, can change into spores more resistant to chemicals, drying, heat, and radiation than the bacterial agent itself; that is, their spores can become more adaptive to become a more dangerous biological (bacterial) agent. A limited number of bacterial agents can cause disease in humans, but most do not. If they can cause disease in humans, they either invade the tissues of humans or produce toxins that poison the human body. Anthrax could be a near "perfect" biological weapon because it has qualities that endear it to the twenty-one nations around the world that are stockpiling this particular biological weapon. It has a high lethality, its spores are highly stable, it has a long and dangerous shelf-life, and its aerosol has a low rate of biological decay.

Response on Scene by First Responders

Caution: **Although there is a vaccine for anthrax exposure, there is currently no human data for *aerosol exposure.***

Aerosol exposure to spores causes inhalational anthrax, a form of disease of great concern to the military services because anthrax has a unique usefulness as a biological agent. Inhalational anthrax begins with nonspecific symptoms followed in two to three days by sudden onset of respiratory disease with dyspnea, cyanosis, and strider. **It is rapidly fatal!** A chest X-ray often reveals a characteristic mediastinitis with hemorrhagic meningitus frequently coexisting. Standard treatment is to provide massive doses of antibiotics along with supportive care. Field tests may suggest a suspicious substance is bacteria but laboratory tests really are needed to find out if that substance is anthrax. Anyone potentially exposed should be given antibiotics.

Field First Aid: Remove victim(s) to an area of safety (away from the Hot Zone). Remember: patients may contaminate you and/or other emergency responders if you fail to don proper personal protective equipment. Provide victims with emergency medical care as soon as possible. Unless otherwise recommended, remove victim(s) clothing, shoes, and personnel belongings for later return. If the victim was obviously in contact with infectious substance(s), flush skin and eyes for fifteen to twenty minutes. Route victim(s) to hospital for a physician's professional opinion. Ensure that hospital staff is fully aware of the medical situation and the poison or infectious substance that may be involved. An enzyme-linked immunosorbent assay test (ELISA) is now approved for anthrax use in hospital laboratories.

Vaccine is available for cutaneous, possibly inhalation, anthrax. Cutaneous anthrax responds to antibiotics (penicillin, terramycin, chloromycetin), sulfadiazine, and immune serum. Pulmonary (inhaled) anthrax responds to immune serum in initial stages but is of little use after disease is well established. Intestinal, same as for pulmonary.

Drugs: Ciprofloxacin hydrochloride is an antibacterial drug and is bactericidal; that is, it interferes with DNA replication in susceptible gram-negative bacteria preventing cell reproduction. It has been successful in treatment of infections caused by susceptible gram-negative bacteria, including prevention of anthrax following exposure to anthrax bacilla in areas thought to have engaged in germ warfare. Contraindication for victims include allergy to ciprofloxacin, norfloxacin, pregnancy, or lactation. Caution should be used on patients with renal dysfunction or seizures. In case of anthrax exposure of adult persons, the usually recommended dose is 500 mg orally, by mouth, four times a day, for three to six months. It is NOT recommended for pediatric patients because studies have shown it produced lesions of joint cartilage in immature experimental animals. Patients should take oral drug on an empty stomach one hour before or two hours after meals; drink plenty of fluid while on this drug; and report rash, visual changes, severe GI problems, weakness or tremors.

Tetracycline hydrochloride is a member of both the drug classes antibiotic and tetracycline; it is bacteriostatic which means it inhibits protein synthesis of susceptible bacteria, preventing cell replication. It is used for infections cause by rickettsiae, *Mycoplasma pneumoniae*; agents of psittacosis, ornithosis, Brucella, *Staphylococcus aureus,* and a number of other diseases. When the use of penicillin is contraindicated for some diseases including *Bacillus anthracis,* the drug **tetracycline hydrochloride** is normally used as an alternative. Contraindications include pregnancy (as the drug is toxic to the fetus), lactation (the drug causes damage to the teeth of an infant), and cautious use with renal or hepatic dysfunction. The normal adult dosage is one to two grams orally by mouth in two to four equal doses (up to 500 mg orally by mouth four times a day). The dose for **Brucellosis** is 500 mg orally by mouth four times a day for three weeks with one gram of streptomycin twice a day intramuscular the first week and every day the second week. Medical personnel should culture infection before beginning drug therapy; should not use out-dated drugs (degraded drug is highly nephrotoxic and should **NOT** be used); should not give oral drug with meals, antacids, or food; should arrange for regular renal function tests with long-term therapy; and use topical preparations of this drug only when clearly indicated (sensitization from the topical use may preclude its later use in serious infections).

Antibiotics: Ciprofloxacin at 500 milligrams orally by mouth twice a day or doxycycline at 100 milligrams orally by mouth twice a day is the recommended drug and dose when dealing with anthrax. If personnel are not vaccinated, they should be given a single 0.5 millimeter dose of vaccine which should be given subcutaneously. If the attack or terrorist incident is actually confirmed, antibiotics care needs to be continued for at least four weeks for all the exposed individuals until they all have received three doses of vaccine. In the event that vaccine is just not available, antibiotics should be continued beyond four weeks and withdrawn under medical observation

Medical Management: Inhalational anthrax disease situations where treatment was started after victims were significantly symptomatic have almost always resulted in death of the patient, regardless of the actual treatment provided. It needs to be recognized, that the incubation period for inhalational anthrax depends on the dose received. A great majority of naturally occurring anthrax strains are sensitive *in vitro* to penicillin. However, penicillin-resistant strains exist naturally. It essentially might not be too difficult for an enemy or a terrorist group to induce resistance to penicillin, erythromycin, tetracyclines, and other antibiotics in a laboratory. All naturally occurring strains tested so far have been sensitive to ciprofloxacin, erythromycin, chloramphenicol and gentamicin. Intravenous ciprofloxacin (dose: 400 milligrams every eight to twelve hours) or intravenous doxycycline (dose:200 milligrams initially, followed by 100 milligrams every twelve hours) should be offered immediately. Supportive therapy for shock, fluid volume deficit, and adequacy of airway may all be needed.

Fire: To control small fires use dry chemical, sand, soda ash or lime; or fight fire around the biological agent or weapon with normal fire extinguishing materials, but do not spread or scatter infectious substance. If without risk, move any containers containing anthrax or any other infectious substance from the area of the fire.

Personal Protective Equipment: At all times wear protective **Level A until you know what you are dealing with;** upon learning the exact type of infectious material(s) you are dealing with, consult proper toxicological levels and downgrade as necessary for safety and comfort of operation.

Spill/Leak Disposal: Outright cleanup or disposal of Infectious Substances (biological agents or weapons) should **ONLY** be handled by trained and experienced personnel. To contact such help, call any or all of the following numbers.

1. SBCCOM Operations — The U.S. Army Soldier and Biological Chemical Command, Edgewood Chemical Biological Center (ECBC), ATTN: AMSSB-RCB-RS, Aberdeen Proving Ground, Maryland 21010-5424. Tel. 410-436-2148 (twenty-four-hour emergency number).
2. U.S. Army Medical Research Institute of Infectious Diseases, Fort Detrick, Frederick, Maryland 21702-5011. Tel. 1-888-872-7443.
3. Center for Disease Control and Prevention Emergency Response Line. 1-770-488-7100.
4. ATSDR — Agency for Toxic Substances and Disease Registry, Division of Toxicology, 1600 Clifton Road NE, Mailstop E-29, Atlanta, Georgia 30333. Tel. 1-888-422-8737. Fax: 404-498-0057.
5. USACHPPM — U.S. Army Center for Health Promotion and Preventive Medicine, Aberdeen Proving Grounds, Maryland 21010-5422. Tel. 410-671-2208.

6. National Response Center (for chem/bio hazards b& terrorist events); Tel. 1-800-424-8802 OR 1-202-267-2675.
7. U.S. Public Health Service: 800-USA-NDMS.
8. National Domestic Preparedness Office (for civilian use). Tel. 202-324-9025.

Symptoms: Anthrax can fool many victims as symptoms are quite non-specific; namely fever, malaise, fatigue, cough, and mild chest discomfort followed by severe respiratory distress and labored breathing, profuse perspiration, harsh vibrating sounds during respiration, and bluish discoloration from lack of oxygenation in the blood. A patient can go into shock, and death can occur within twenty-four to thirty-six hours after onset of severe symptoms. However, in some instances, there can be a short period of seeming before the victim dies.

Vaccines: A licensed alum-precipitated preparation of purified *B. anthracis* protective antigen has been effective in preventing or significantly reducing the incidence of inhalation anthrax. Somewhat limited human data suggest that after the first three doses of the recommended six-dose series of shots spaced at zero, two, four weeks, then six, twelve, and eighteen months; protection against both skin and inhalation anthrax has been provided (the recommended dose for this series of six shots over a total of eighteen months is 0.5 milliliter per dose subcutaneously). Regarding *all* vaccines, the degree of protection depends greatly on the magnitude of the actual dose received. A high spore challenge can overwhelm a vaccine-induced protection factor. This vaccine needs to be stored at refrigerator temperature, but *not* frozen.

Anthrax by Mail

We will deal with the "anthrax by mail" attacks that have confounded our federal leaders. Seventeen persons have contracted anthrax to the date of this writing, either from inhalation or as a skin disease, five have died, and thirty thousand persons have had to take antibiotics. The situation has changed, and again we were unprepared for change. Federal agency employees have been accused of lack of coordination as well as issuing conflicting statements of the potency of anthrax in the letter received by then Senator Daschle. They have not yet been accused of being proactive in their preparations for responding to a severe, weaponized anthrax attack in the United States.

An anthrax letter addressed to Senate Majority Leader Tom Daschle was postmarked on October 9, 2001, and opened by a government employee on October 15, 2001, in Daschle's office. F.B.I. and Environmental Protection Agency employees renovated a building outside the national's capital to process and investigate 286 drums of letters sealed in plastic bags confiscated after the Daschle letter was found to contain finely ground anthrax that posed a dangerous threat to humans. They built and sealed a large room with a filtered air flow; workers at the site were offered antibiotics while workers in the decontamination hot zone were required to take antibiotics. For "suspect letters," workers cut the plastic bags and swabbed to locate anthrax spores. Reportedly, they found one contaminated letter addressed to Senator Patrick J. Leahy of Vermont on Friday, November 16, 2001.

According to one spokesperson, Senator Leahy's letter, in potency and appearance, was a twin to Senator Daschle's letter. It contained twenty-three thousand anthrax spores, a number of spores roughly equal to two lethal doses of inhalation anthrax (i.e., the infective dose for aerosol dissemination varies widely from eight thousand to fifty thousand spores). This test was taken *outside* the Senator Leahy envelope. Within twenty-four hours, another test was taken of the still unopened envelope, and the results of this test floored officials and media representatives. The Leahy letter contained billions of anthrax spores rather than the twenty-three thousand spores reported broadly just the day before. It is now thought the Leahy letter was so toxic that it contaminated more than 50 of the 630 bags of letters quarantined for a month since anthrax was known to contaminate mailrooms in the nation's capital. The Leahy letter was taped on all four sides, and spores from that letter were so fine they could have penetrated through tiny openings in the envelope. Although investigation had been going on relative to actual toxic anthrax assaults within media and U.S. government offices for at least five weeks, the F.B.I. had been seeking the deadly strain of anthrax (the Ames strain) involved in such assaults for a month, the taxpayers didn't get the word until the Sunday, November 25, 2001, copy of the *Washington Post*. The newspaper stated that the U.S. Army Medical Research Institute of Infectious Diseases at Fort Detrick, Maryland, partnering with the F.B.I. in investigations of the anthrax scare, was the strain's major distributor to research laboratories throughout the United States.

Overall, the effects of the Daschle and Leahy anthrax letters will be with us for a long time. Until November 19, 2001, the Federal Bureau of Investigation had not revealed the Leahy letter even contained anthrax or that its cargo was as potent as that contained in the Daschle letter opened thirty-two days before. The Hart Senate Office building was opened in January of 2002, while the Dirksen and Russell Senate Office buildings reopened on November 19, 2001, after being monitored for presence of anthrax. At the Pentagon military headquarters, there were additional precautions put in place against contaminated mail: all mail was to be opened, visually inspected, X-rayed, and tested for both chemical and biological materials. Once inspected, all mail is held for three days for arrival of test results before such mail is actually delivered to Pentagon offices.

Some local emergency response personnel and hazardous materials response teams have had to deal with numbers of copycat anthrax hoaxes since October 15 media attention to the Senator Daschle letter. What follows represents some basic information on anthrax to local hazardous materials response crews.

Manchester, New Hampshire, Fire Department Deals with Anthrax Scare

On November 5, 2001, Captain Nick Campasano who serves with the hazardous materials response team of the very active Manchester Fire Department within the largest city in New Hampshire, population 99,600, addressed approximately 350 business leaders from throughout the State of New Hampshire. He was one of three very knowledgeable persons who provided their personal views based on their own experience at a free, three-hour training session called "Terrorism and Disaster Planning Practical Knowledge for Managers," an educational forum put on by Occupational Health + Rehabilitation Inc.. The two other speakers were Dr. James Katz of Occupational Health + Rehabilitation Inc, and Dr. Jesse Greenblatt, Chief of Epidemiology for the State of New Hampshire.

Campasano is an experienced responder to hazardous materials incidents who tells it like it is. The subject of his address was "Safe Mail Handling, Hazardous Material Removal."

Dr. Katz had already spoken on "Clinical Bio Terrorism," and Dr.Greenblatt had handled "Investigation and Identification of Anthrax Exposure."

"Thank you for allowing the Manchester Fire Department to address you this morning," Captain Campasano said in his opening remarks. "I'd also like to thank Dr. Greenblatt and Dr. Katz because what they are doing this morning, and this overall event they assisted in organizing, helps me. It makes my job a lot easier. There has been very steep learning curve when this all began, and when I say this all began I'm referring to the date of October 9th, 2001, when here in Manchester we received our first suspicious mail call. Back then, around the nineth or tenth, people were reacting more from the level of fear and anxiety than from knowledge. Hopefully, it's through programs like this that we can reverse that reaction. Now, almost four weeks later, people base their actions more on knowledge than fear and anxiety. But don't get me wrong, fear and anxiety have a place. If we all lost our fear, emergency room departments would be very busy. So, fear and anxiety do have a place. They help us keep up our alertness and our level of caution, but people should not hide their reactions. I have a unique position in that I can see the civilian side, their response, and as a trainer for the fire department I can also look at the perspective of our emergency responders. In the beginning, they also were acting out of fear and anxiety. When this all began we would respond to a suspicious mail call, arrive on the scene, call in an engine company and our hazardous materials team. On scene, they would typically find a package, or parcel of mail, double-bagged and out on the front step of a residence. Our responders would then don protective equipment, respiratory protection, protective suit, gloves, gingerly take that package and hand it to a police officer standing in his uniform or her uniform who would then reluctantly take it, place it in the back of the cruiser and transport it up to Concord, the capital city of New Hampshire, for analysis.

"In the three or four days following October 9, 2001, we responded to approximately 12 calls; of those 12 calls, about nine of them were transported up to the state lab. That's just from Manchester; magnify that by all of the communities in New Hampshire. There was a pile of packages waiting to be opened in Dr. Greenblatt's office that grew each day. We had someone vacuum up a suspicious powder, so we sent the vacuum up. But, again, we were reacting more as a learning curve. In fact the organizations that we were relying on for information were going through that learning curve. The Center for Disease Control initially came out with a directive that when you're handling mail, if you were to find something suspicious you should take that parcel of mail, place it into a plastic bag, place it down, and if anything spilled out onto the counter you should clean it up, and then call 911 after washing your hands. From the fire department perspective, we were a little concerned because we felt that people would not clean the area appropriately and could in fact spread the contamination. We did a little experiment at the station. We took an envelope, filled it with a fluorescent powder that would shine brightly under ultraviolet light, put it in some mail and we had one of our office staff open the mail. When she got to the powder envelope, we had her stop after she opened it up following the CDC recommendations. We were very surprised by the results. When she took the envelope and placed it in the plastic bag, then placed the bag down, she inadvertently blew powder onto her clothing. It wasn't apparent until we put the ultraviolet light on her that she could see the amount of contamination that she didn't realize she had spread. The CDC, within about three days, revised their recommendations to not putting something into a plastic bag. They then asked everyone to cover the item. Again the process of covering something could in fact cause air currents to blow the material around. So, that procedure again was changed. The learning curve has been experienced, not only by the civilian population but also the emergency responders, as well as those organizations that are providing us with the information.

"So, what are the procedures? What should you do within your work place or at home in dealing with mail? What we are recommending now is that you first look around your facility, or your home, and decide upon a location to open the mail. Keep in mind that if you should open something that is "suspicious," or contains a powder, that package will be sent off to the lab for analysis. You would do without that space for the length of time it takes to get results back from the lab. So, for instance, if you're opening up your mail in your administrative offices, can you do without that space until results come back? The second consideration is if you want to open your mail in an area that is free from air drafts and traffic, can you easily segregate your mail? You, the audience, the first step you could do today; when you return to work, locate and isolate that one area in your facility where you will open your company mail. The next step is to bring your mail to that location, separate it into two piles, one pile to be the mail that you expect, the invoices, the bills, and the 'normal' mail.

"In fact, the Manchester Fire Department had a call from an individual who received a 'suspicious' letter. It was a $4,000 telephone bill, and I guess he thought if that mail went to the state lab he would have some excuse as to why he didn't pay it on time. However, your problems will be easier to deal with; when you divide your mail, put the mail that you expect into one pile and then the suspicious mail in the other pile. What do we mean by 'suspicious mail'? We mean mail that you are *not* expecting, mail from a foreign country, mail that has been addressed incorrectly, mail from someone you do not know, mail addressed to someone no longer at your address, mail that is handwritten with no return address, mail that is lopsided or lumpy in appearance, mail that is sealed with excessive amounts of tape, mail that has excessive postage, or mail with restrictive endorsements, such as 'Personal,' or 'Confidential.' I think your list in your handout talks about a 'ticking sound,' that's probably a good tip; it means there's something definitely suspicious in there. That pile of mail should be handled last. Open up your normal mail, and then go to that second pile.

"When you open the mail in the 'suspicious' pile, do so gently. Obviously if you see powder coming out, if you see a stain, a liquid material, don't open it. Unless there's something as obvious as that, go through your mail, open it gently, use a commercial type blade letter opener, avoid mechanized letter openers because that could be the energy source if there was powder in an envelope to get it up and aerosolize it. Very gently open the top of the envelope at a distance, peer into it. If you see anything, any foreign material, basically take that envelope, place it down gently, step away, keep anyone else from the area, wash your hands and call 911. If there's one thing I can emphasize, it is hand washing. The is a habit that we should all be doing at the present time, whether we're opening suspicious mail, or mail from relatives, to wash our hands every time we handle a piece of mail. As we've seen in some cases, anthrax material can 'tag along'; it can be put onto another piece of mail and get to an individual who was not intended as the target. Good hand washing technique is something that we want to develop into a habit.

"People commonly ask, 'What they would expect when they call 911?'" notes Captain Nick Campasano. "At least here in Manchester, I'm familiar with what will happen. We will dispatch on the fire department side the closest engine company and our hazardous materials team. They will not be responding with lights and sirens. They will be driving with traffic, they will park in a discrete area, they will enter your facility with all of their equipment packaged into a small drum, a little fiber drum; they won't be walking in with the big Level A suits. A police officer will respond also, and then we will go through a series of what we call credibility lists. We will see and try to determine how credible that threat is. It's much more difficult when the letter is unopened. Someone feels that it is a suspicious letter because it is from Singapore and they don't know anyone in that country. The letter has not been opened. Remember, our difficulty in the beginning, looking at that learning curve again, was we didn't want to open the envelope because if there is something in there we did not want to release it to the environment. We would be left with the option of packaging it up in a plastic bag and sending it up to the state health laboratory. Since that first weekend of the anthrax scare, the fire department has gone out and we have purchased what we call isolation bags, and I'll just pull one out for you. This is a fairly heavy plastic bag with glove hole-arms in it.

"What you could expect today when we arrive at your facility' we would take that unopened mail, place it inside this bag, seal it and then through the use of the glove hole-arms to actually open up the mail within the heavy plastic bag. In most cases, if not all that we've handled here in Manchester, the letter is a solicitation for money, it's a bill, or it's a letter from someone looking for some information. To date, with the exception of the Family Planning Clinic we had a year and a half ago, we have not had anything of a threatening nature. We have not had any suspicious powder within the envelope. Since October 9, 2001, the Manchester Fire Department has received four to five calls per-day regarding suspicious mail. Our goal is to respond to your facility, dispose of the threat, or determine that it's not a threat, and then leave within a ten minute timeframe. If we look at the intention of the terrorists, it is to disrupt our commerce. So, from a response side, if we were to come to your facility, evacuate your building, put on Level A suits, set up a big shower decontamination outside, close your facility down for a day, who's winning? We're handling four to five of these calls per-day, so that would be a dramatic impact on the economic base of this area. So through these processes, we are attempting to minimize the impact on your business and allow you back to your normal flow within 10 minutes. There is no need to evacuate a building. There will be no showers set up outside. We will not be asking you to come out and disrobe in the parking lot, and walk through ice cold water in December.

"If you feel that you have gotten some of the material on your clothing, or if you just have material on your clothing, we will ask you to go into another room, remove your clothing, we will give you a very fashionable pair of hospital scrubs that come in assorted colors, one size fits all though, and we will ask you to put those on, go home and take a personal shower. Your clothing would be placed in a plastic bag, sealed and just held until the test results come back, and that's probably the one safeguard that we have. Regardless of how we handle the mail, regardless of what happens if it is something suspicious, it will go to state laboratory in Concord to be tested. In the event that something comes back positive, it's 100 percent treatable.

"The last scenario that you might be exposed to is one in which you feel that someone has introduced something into the air handling system of your facility. The biggest safeguard against that is preplanning who has access to those areas, where is the air intake in your building, is it secure, can anyone wearing coveralls with Mike's Heating and Air Conditioning on the back of their shirt walk in with a toolbox and gain access to your mechanical room. Those are the things you want to address now, as well as how do I shut the air handling system off. One of the problems we are addressing with the school system is many times it's the custodian, and he or she is the only one that knows how to do it. If the problem happens when they're on vacation, the necessary process can't be accomplished. But in the event you should determine, or feel that someone has introduced something into your air handling system, we would then recommend that the system be shut down, and everyone should leave the building, and await the arrival of the emergency responders.

"I mentioned earlier that the police department would also be responding. They are doing this from the crime scene aspect. If you receive a letter that is threatening that contains a foreign material, even if it is talcum powder, that's a crime. The police role is to preserve the chain of custody when that item leaves your hands, and then obviously goes up to the state lab for analysis. They're also are working to assist the health department here in the City of Manchester. My contact in the health department and I have, in the beginning, responded together simply because of that learning curve. His office no longer responds to the calls. His office would be relying on the police officer's report. That police officer will be asking you who handled the mail, who else may have been in contact with the mail, who was in the area? Should something come back from the state lab as a positive, the health department would be working in conjunction with the police department going through that list to then start all the appropriate treatments or testing.

"There are some recommendations for the use of gloves for handling mail," continues Captain Campasano. "At this point, we're leaving that up as a personal preference. If you handle a large volume of mail and you feel comfortable, or uncomfortable not wearing gloves, by all means do so. But in most cases it's not necessary. Good hand washing habits will be enough. I just want to mention some people have asked questions about cream. Should we put on an antibacterial cream? Is that sufficient? We're actually recommending hand washing. We want to remove anything that might be on your hands. If you use gloves, I will caution you for two things. The first, most gloves contain a powder inside so that the glove material doesn't stick to itself. So it's very possible after handling mail, and after wearing your gloves and removing them, if you were wearing dark clothes and you touched yourself, you put your hand on your clothing, you could have a streak of powder which in many cases will again raise that anxiety level in most people. The second is, if you're using latex gloves there are many people that are not aware of an allergy that they may have to latex. We noticed that in the fire service when we made the move to giving everyone gloves and it might not become apparent right off where you're wearing them for a few minutes a day; it may take a period of time before that allergy develops. You could have reddening, chafing, and soreness. To the uninformed, that factor might trigger a response, 'I have cutaneous anthrax.' So, just be aware of the way latex gloves wear, and you'll sidestep that problem completely.

"Keep aware of all of the changes. If you go to Internet sites like the government run Centers for Disease Control and Prevention located in Atlanta, Georgia (CDC) to download mail handling procedures, don't do it once and put it up on the wall and forget about it; visit that site on a frequent basis because that information changes frequently. Stay current, and if in doubt, don't be ashamed to call 911. If you think you have a suspicious piece of mail, and you put it in a bag and you throw it away, you may have ended the hazard to other people; but had that piece of mail contained something dangerous, you've broken the chain in that safety net. Unless it goes to a lab, unless it's analyzed, no one will ever know, and anyone who may have come in contact with that material again will not be tracked, will not be treated, and will not be tested.

"Referring back to notification. 'Should I be calling my local police department? Should I be calling 911?' The State of New Hampshire is now covered entirely through the

911 system (Emergency Telephone System). So, unless your local jurisdiction is putting out some different information, and they may be depending on jurisdiction, your first action should be to call 911. They will then get you in touch with the proper agency whether it's police first, or fire first, but all of those agencies will ultimately be notified. There was a question regarding standard responses. 'Will the response in northern New Hampshire be the same as the response in southern New Hampshire, or out in Keene versus on the seacoast?' And my answer to that, unfortunately, is 'probably not.' The fire departments are all individuals. We have some regional Haz-Mat teams. Manchester is fortunate in that we have a dedicated Haz-Mat team twenty-four hours a day standing by. We have over a 100 hazardous materials technicians who may be on duty at any one time so we have been able to go through that learning curve. To date we've had one hundred calls. You may live in a community that to date has had two calls, or no calls, so you are still at the lower end of that learning curve. So, unfortunately, I cannot say that the response will be the same in all areas.

"Your best protections factor is the person you hire to handle your mail. If you follow those procedures that we laid out; if you're alert to the signs and symptoms, that is your best safeguard. Nationally, if we look back at those individuals that have died from anthrax, in most cases it was not recognized because we didn't expect it. But one of the postal workers that went to an emergency room and was seen by a physician never said he worked in a post office. Had he relayed that information, a red flag would have gone up. So, if you have someone in your mail handling area who is vigilant, who is competent, and alert to those signs and symptoms, the first signs of something developing, is probably your best and most reliable safeguard.

"The final question that I have heard, which is on a completely different note, is what's being done about facilities that might have potentially hazardous chemicals in the community? That's a very good question. Because terrorist groups do not stay with the same tactics, as we have also seen in the news. With the attempts made by some of these organizations to obtain hazardous materials drivers licenses and permits, the thought is definitely out there. The EPA several years ago had the thought of requiring all SARA Tier II reporting facilities who had extremely hazardous substances on their property had to file and post on the EPA's Web site their worst case scenarios; which was kind of like a menu, if you will, for terrorists. Now all they needed to do was go to the EPA Web site, pick the town and location of choice, and they would have the worst case scenario laid out for them: what weather conditions were needed, which tank at that facility they needed to target. Our government looked at that policy again and thankfully rescinded it, but that was a concern. What we can do in our communities now is to preplan. Fire departments should be looking over their Tier II lists. There are federal regulations out there that require facilities to report to fire departments, to the state emergency response commission and to the Local Emergency Planning Committee LEPC what chemicals they have within that jurisdiction. They do have to prepare worst case scenarios. Fire departments are required to come up with emergency plans, what to do in the event a release should occur. As we tell fire departments, they should never be surprised by a chemical incident at a chemical facility within your jurisdiction. They're there, we know they're there, and somebody should be preplanning for an incident there. This is a serious concern, and hopefully it is being addressed by most fire departments and emergency responders throughout the state."

Guides for Emergency Response: Biological Agent or Weapon: Botulism

6

AGENT: Botulism is caused by intoxication with any of seven distinct neurotoxins produced by bacillus Clostridium botulinum. The toxins are proteins with molecular weights of about 150,000, which bind to the presynaptic membrans of neurons at peripheral cholineric synapses to prevent release of acetylcholine and block neurotransmission. A biological warfare attack with botulinum toxin delivered by aerosol could be expected to cause symptoms similar to many aspects to those observed with food-borne botulism. A group of seven related neurotoxins (types A-G), botulinum toxins are typically found in canned foods and produce toxin when eaten. Such toxins block acetylcholine release in a similar manner to chemical nerve agents. Botulism can cause paralysis that can lead to respiratory failure requiring assisted ventilation until the paralysis passes. This toxin is not volatile and not dermally active. Botulism is usually NOT transmissible from one person to another. A supply of antitoxin against botulism is maintained by the Center for Disease Control and Prevention (CDC) located in Atlanta, Georgia. Such antitoxin is effective against the severity of symptoms if administered early in the course of disease, and most patients eventually recover after weeks to months of supportive care.

History: The secret was released some time after World War II when it became apparent that British forces provided a grenade specially concocted with botulinal toxins to kill Reinhard Heydrich, the much feared head of the Nazi Security Service, better known as the SD. After the successful killing in the Prague, Czechoslovakia, the stunned Germans shot and burned the town of Lidice (where free-Czechs from Britain had been parachuted in kill Heydrich), and arrested an estimated 10,000 Czechs.

In August of 1942, George Merck, president of the Merck & Co. drug company, began his rein as supervisor of the new War Research Service (WRS), the agency that brought together private and government agencies and institutions to create and fulfill the United States biological warfare program. A myriad of U. S. universities were funded to do research on biological warfare agents. George Merck will be remembered as the man who created the major drug company that bares his name, which in 2005 found it defending itself against thousands of lawsuits against their handling of the popular painkilling drug, Vioxx.

During World War II, the WR S, under George Merck's guidance, in 1943 investigated strengths and weaknesses of the offensive warfare of two leading biological warfare agents, botulism and anthrax. The Chemical Warfare Service (CWS, part of the U.S. Army) in early 1944 received $2.5 million to build 250,000 or more botulium bombs, or one million anthrax bombs, per month. Work would be done at Horn Island near Pascagoula, Mississippi; Vigo, near Terre Haute, Indiana; and Dugway Proving Ground in Utah. Late in 1944, George Merck notified the CWS of research of four other deadly biological weapons including brucellosis, psittacosis, tularemia, and glanders. In 1946, the CWS had its name changed to the U.S. Army Chemical Corps, and retains authority over biological warfare agents.

Botulism is thought to be one of the most deadly known toxins in the modern world; it is able to kill an average adult with an estimated tiny dose. Botulism can also be turned into a biological weapon through creating an aerosol that could cause death through in inhalation. However, during the years 1990–1995, the full-funded Japanese cult Aum Shinrikyo attempted to use botulinum toxin as an aerosol without satisfactory results.

With botulinum toxins, a biological warfare attack would be expected to be delivered by aerosol, and symptoms of inhalation botulism could begin in twenty-four to thirty-six hours to as late of several days. Death, without treatment, comes by respiratory failure or paralysis of respiratory muscles. However, with ventilation assistance and tracheostomy of modern medicine, dead patients should amount to less than 5 percent of total patients, although nursing care can last for days and even months. Botulinum toxin is not transmitted from person-to-person, and duration of illness can be twenty-four to seventy-two hours, and might hang on for months if lethal. It has a high mortality rate of 60 to 100 percent if not treated. Symptoms include weakness, dizziness, dry mouth, nausea, vomiting, difficulty talking and swallowing, vision problems, eyelids that droop, progressive paralysis, and asphyxia (a lack of oxygen or excess of carbon dioxide that is usually caused by interruption of breathing and that causes unconsciousness). With botulinum toxins, diagnosis with routine laboratory findings does not mean a thing, because the cerebrospinal fluid will be normal.

Medical Classification: Toxin.

Probable Form of Dissemination: Sabotage of food/water supply; or aerosol.

Detection in the Field: None.

Infective Dose (Aerosol):   0.001 micrograms/kilograms

Sign and Symptoms: A sagging or prolapse of an organ or part, such as dropping of the upper eyelid, generalized weakness, dizziness, dry mouth or throat, blurred vision and a disorder if vision in which two images of a single object are seen because of unequal action of the eye muscles (also called, "double vision"), difficulty in articulating words due to disease of the central nervous system, defective use of the voice, and loss of or deficiency in the power to use and understand language as a result of injury to or disease of the brain followed by symmetrical descending flaccid paralysis and development of respiratory failure. Symptoms begin as early as twenty-four to thirty-six hours, but may take several days after inhalation of toxin.

Incubation Time: Variable (hours to days).

Diagnosis: Clinical diagnosis. No routine laboratory findings. Biowarfare attack should be suspected if numerous collocated casualties have progressive descending bulbar (involving the medulla oblongata), muscular, or respiratory weakness. As for treatment, provide intubation and ventilatory assistance for respiratory failure. Tracheostomy may be required. Administration of botulinum antitoxin (IND product) may prevent or decrease progression to respiratory failure and hasten recovery.

Differential Diagnosis: With single cases, rather than clearly epidemic cases, the illness could be confused with Guillain-Barre' syndrome, myasthenia gravis, or tick paralysis. Other possible considerations may include enteroviral infections, as well as nerve agent and atropine poisoning.

Persistency: Stable.

Personal Protection: Protective clothing must be used as well as protection for the respiratory tract by using a mask with biological filters or self-contained breathing apparatus (SCBA) with positive pressure, at least until you the know the specific threats.

Routes of Entry to the Body: Inhalation, mouth, wound.

Person-To-Person Transmissable: No

Duration of Illness: Twenty-four to seventy-two hours (months if lethal). Therapy consists mainly of supportive care; such as, intubation and assisted ventilation for respiratory failure.

Potential Ability to Kill: High

Defensive Measures: Immunizations, good personal hygiene, physical conditioning, use of repellents for arthropods, wearing protective mask, and practicing good sanitation. Spores can be killed by pressure-cooking food to be canned.

Vaccines: Yes. Investigational New Drug (IND) Pentavalent Toxoid A-E.

Drugs Available: Yes. IND Heptavalent Anti-toxin A-F (equine despeciated); also, Trivalent Equine anti-toxin A, B, and E.

Decontamination: Soap and water, or diluted sodium hypochlorite solution (0.5 percent). If contamination of foodstuffs is suspected, boil for ten minutes to kill toxin. Botulism is not dermally active and secondary aerosols do not endanger medical personnel.

Ability To Kill: Low. Over recent years, botulism's fatality rate has fallen to about 5 to 8 percent.

Characteristics: Botulism is *not* an infectious disease. However, it is considered a potential terrorist weapon. It is a toxin with a lethal dosage of 0.02 milligrams per minute per cubic meter which can cause poisoning following absorption through any mucous membrane. If disseminated as an aerosol, this toxin could be absorbed through the skin and lungs. Its action within the human body would resemble that of a chemical nerve agent in that it would inhibit the release of acetylchlorine.

Botulism is a potent neurotoxin produced from Clostridium botulinum that is an anaerobic, spore-forming bacterium. There are three different types of botulism: Foodborne botulism occurs when a person ingests a pre-formed toxin that leads to illness within a few hours or days. Foodborne botulism is a public health emergency because the contaminated food may still be available. Infant botulism occurs in a small number of susceptible infants each year who harbor C. botulinum in their intestinal tract. Wound botulism occurs when wounds are infected with C. botulinum that secretes the toxin. Approximately 100 cases of the three types of botulism are reported within the United States each year; about 5 percent are wound botulism, 25 percent are foodborne botulism, and a full 70 percent are infant botulism. Death can result from respiratory failure, but those who survive may have fatigue and shortness of breath for years.

Botulism is usually *not transmissable* from one person to another. A supply of anti-toxin against botulism is maintained by the Center for Disease Control and Prevention located in Atlanta, Georgia. Such antitoxin is effective against the severity of symptoms if administered early in the course of disease, and most patients eventually recover after weeks to months of supportive care.

Emergency Medical Services: Isolate/Secure the Scene, Establish Control Zones, Establish Command, Evaluate Scene Safety/Security, and Stage Incoming Units. If Command has been Established: Report to and/or communicate with the Command Post. Gather Information Regarding Type of Event, Number of Patients, Severity of Injuries, and Signs and Symptoms of Medical Concerns. Assign Medical Incident Command Positions as Needed, and Notify Hospitals. Request Additional Resources as Appropriate and Necessary, Including as Needed: Basic Life Support (BLS) and Advanced Life Support (ALS), Medivac Helicopter (trauma/burn only), Medical Equipment and Supply Caches, Metropolitan Medical Response System (MMRS), National Medical Response Team (NMRT), Disaster Medical Assistance Team (DMAT), and the Disaster Mortuary Response Team (DMORT). Use appropriate self-protective measures (proper personal protective equipment, Time/Distance/Shielding, minimize number of people exposed to danger). Initiate Mass Casualty Procedure, Evaluate the Need for Casualty Collection Point/In-Patient Staging Area, Control and Isolate Patients (away from the hazard, at the edge of the hot/warm zone), Ensure Patients are Decontaminated Prior to Being Forwarded to the Cold Zone. Triage, Administer Antidotes, Treat and Transport Victims. Engage in Evidence Preservation/Collection (recognize potential evidence, report findings to appropriate authorities, consider embedded objects as possible evidence, secure evidence found in ambulance or at hospital). Establish and Maintain Chain of Custody for evidence preservation, and Ensure Participation in Unified Command System when implemented.

Response on Scene by First Responders

Caution: Aerosol dissemination of botulism is a new and potential danger in the terrorist world of today. If aerosol dissemination is used, first responders should wear SCBA and personal protective equipment (PPE) without fail.

Field First Aid: "Natural" (foodborne, infant, and wound) botulism is a natural pathogen that strikes approximately one hundred people a year in the United States, killing about 5 to 8 percent. It is fairly controllable; it cannot usually be transmitted from person to person so patients do not have to be isolated, and there is an antitoxin available. Also, "regular" botulism is not spread through the air or by contact. Botulism toxins spread by aerosol, however, are an entirely different matter. Botulinum toxin is believed to be the most poisonous substance known to man; the lethal dose/50 percent (LD50) is estimated to be one or two nanograms/kg. An attack of botulism disseminated by aerosol on the home front is suspected of being somewhat like the effects caused by food-borne botulism, but with some critical differences. As a potential terrorism agent, botulism would *not* be used for its natural pathogens; instead, the actual tosin produced from the bacterium in a laboratory would do the dirty deed in any such attack. After aerosol exposure, an antitoxin stored at the Centers for Disease Control and Prevention that has been around for thirty years can be an effective treatment, even sometimes after symptoms of intoxication have become apparent. Standard treatment primarily provides supportive therapy and antibiotics for care of secondary bacterial infections. In sharp contrast to botulinum toxin's ability to invade mucosal surfaces, intact skin is impermeable to botulinum toxin. Isolation of patients with botulinum toxic exposure is *not* required, although standard precautions should be followed because the toxin and/or the organism can be found in the patient's feces. Decontamination of victims who have been exposed to botulism is *not* required.

Drugs: There is an antitoxin stored at the CDC. To arrange to use this antitoxin, call your state health department (or CDC at 404-639-2206 or 404-639-3753 workdays, or call weekends or evenings at 404-639-2888). This chemotherapy (antitoxin) available from CDC is a licensed trivalent equine antitoxin for serotypes A, B, and E. There is no reversal of botulism disease with this drug, but the antitoxin does usually prevent further nerve damage. The U.S. Department of Defense (DOD) has a heptavalent equine despeciated antitoxin for serotypes A – G (IND). DOD also has pentavalent toxoid (vaccine) for serotypes A – E (IND). The currently recommended schedule is for use at zero, two, and twelve weeks with a one year booster. This vaccine is supposed to induce solidly protective antitoxin levels in greater that 90 percent of those vaccinated after one year. Contact: USAMRIID, (U.S. Army Medical Research Institute of Infectious Diseases), Fort Detrick, Maryland. Tel. 301-619-2833.

Medical Management: When reaching a diagnosis, physicians or other medical procedures staff in hospitals may confuse symptoms of contact with botulinum toxin(s) with certain neuromuscular problems such as Guillain-Barre syndrome or myasthenia gravis. Also, laboratory findings tend to be of little or no value in reaching a diagnosis. Despite the severity of botulinum toxins, a classic poison, the recognized treatment is fairly straightforward: provide an antitoxin available from the CDC *and* use assisted ventilation for neuromuscular paralysis.

Fire: For an infectious disease or weapon, use available methods and equipment on surrounding fires. Appropriate extinguishing agents: dry chemical, soda ash, sand, or lime. Do not use high pressure water streams

Personal Protective Equipment: At all times wear protective Level A until you know what you are dealing with; upon learning the exact type of infectious material(s) you are dealing with, consult proper toxicological levels and downgrade as necessary for safety and comfort of operation.

Symptoms: The classic symptom of botulism include double vision, blurred vision, drooping eyelids, slurred speech, difficulty swallowing, dry mouth, and muscle weakness. Infants with botulism appear lethargic, feed poorly, are constipated, and have a weak cry and poor muscle tone. These are all the symptoms of the muscle paralysis caused by the bacterial toxin. If untreated, these symptoms may progress to cause paralysis of the arms, legs, trunk and respiratory muscles. In food-borne botulism, symptoms generally begin eighteen to thirty hours after eating contaminated food, but they can occur as early as six hours or as late as ten days. The number of food-borne and infant botulism cases has changed little in recent years, but wound botulism has increased because of the use of black-tar heroin, especially in California.

Vaccines: DOD pentavalent toxoid for serotypes A – E (IND).

Guides for Emergency Response: Biological Agent or Weapon: Brucellosis

7

Brucellosis is a systemic zoonotic disease caused by one of four species of bacteria: Brucella melitensis, Brucella abortus, Brucella suis, and Brucella canis with virulence for human beings decreasing somewhat in the order used here. These bacteria are small gram-negative, aerobic, non-motile (not exhibiting or capable of movement) coccobacilli that grow within monocytes (a large leukocyte with finely granulated chromatin dispersed throughout the nucleus that is formed in the bone marrow, enters the blood, and migrates into the connective tissue where it differentiates into a macrophage) and macrophages (a phagocytic tissue cell of the reticuloendothelial system that may be fixed or freely motive, is derived from a monocyte, and functions in the protection of the body against infection and noxious substances). They reside quiescently in tissue and bone-marrow, and are extremely difficult to eradicate even with antibiotic therapy. Their natural reservoir is domestic animals, such as sheep, goats, and camels for B. melitensis; cattle for B. abortus; and pigs for B. suis. Brucella canis is primarily a pathogen of dogs, and only occasionally causes disease in humans.

They have abraded skin or conjunctival surfaces that come in contact with the bacteria. Laboratory infections are quite common. Brucella species have long been considered potential candidates for biological warfare. The organisms are readily lyophilized (freeze-dried), perhaps enhancing their infectivity. Under selected environmental conditions such as darkness, cool temperatures, high carbon dioxide; persistence for up to two years has been recorded. When used as a biological warfare agent delivered with aerosol, Brucellae could be expected to mimic natural disease. Natural infection by humans is acquired through ingestion of unpasteurized milk or cheese, aerosol present in farms and slaughterhouses, or by inoculation of skin lesions in people in close contact with animals. Intentional exposure would be likely by aerosol, or possibly by contamination of food.

Medical Classification: Bacterial

Probable Form of Dissemination: Aerosol; sabotage of the food supply. Brucellosis is a 'hindrance" bacteria; symptoms can take months to appear, and deaths are few and far between even without medical care.

Detection in the Field: None.

Infective Dose: (Aerosol):   10–100 organisms.

Signs and Symptoms: Incubation period is highly variable. Acute and subacute brucellosis are non-specific; irregular fever, headache, profound weakness and fatigue, chills, sweating, arthralgias (severe pain in a joint, especially not inflammatory in character), myalgias (muscular pain). Depression and mental status changes. Osteoarticular findings (i.e., sacroiliitis, vertebral osteomyelitis). Fatalities are uncommon.

Diagnosis: Blood cultures require a prolonged period of incubation in the acute phase. Bone marrow cultures produce a higher yield. Confirmation requires phage-typing, oxidative metabolism, or genotyping procedure. Enzyme-Linked Immunosorbent Assays

(ELISA) followed by Western blotting are used. As for treatment, doxicycline and rifampin for a minimum of six weeks. Ofloxacin plus rifampin is also effective. Therapy with rifampin, a tetracycline, and an aminoglycoside is indicated for infections with complications such as endocarditis or meningoencephalitis.

Persistency: Long persistence in wet soil and food.

Personal Protection: Protective clothing must be used as well as protection for the respiratory tract by using a mask with biological filters or self-contained breathing apparatus (SCBA) with positive pressure, at least until you know the specific threat(s).

Routes of Entry to the Body: Inhalation, mouth, skin, and eyes.

Person-To-Person Transmissible: No (except where open skin lesions are evident).

Duration of Illness: Varies greatly.

Potential Ability to Kill: Very low.

Defensive Measures: Immunizations (which are questionable in efficiency), good personal hygiene, physical conditioning, wearing protective mask, and practicing good sanitation. Avoid unpasteurized milk products or raw meat.

Vaccines: No approved human vaccine is available in the United States. Killed and live attenuated human vaccines have been available in many countries, but are considered of unproven efficacy within the United States.

Drugs Available: Doxycycline, and Rifampin (Trimethoprim-sulfamethoxazole may be substituted for rifampin, *but* relapse may reach 30 percent).

Decontamination: Soap and water, or diluted sodium hypochlorite solution (0.5 percent) for environmental contamination. Drainage and secretion procedures are necessary. Standard precautions for healthcare workers should be followed. Person-to-person transmission via tissue transplantation and sexual contact has been reported but are insignificant.

Characteristics: Brucella is an encapsulated, non-motile bacteria containing short, rod-shaped to coccoid, gram-negative cells. Such organisms are parasitic, invading all animal tissues and causing infection of the genital organs, the mammary gland, and the respiratory and intestinal tracts, and are pathogenic for man and various domestic animals. If used as a biological warfare agent, it would probably be disseminated as an aerosol and the resulting infection would be expected to resemble natural disease. It was originally thought *not* to be transmissible from person-to-person, but its infectivity is high and there have been recent reports of person-to-person transmission via tissue transplantation and sexual contact. Its incubation time (refers to the period between exposure and the appearance of symptoms and signs of poisoning) can stretch on for days to months, and its duration can last for weeks to years. It has long persistence in wet soil and food (which possibly could make it a sabotage weapon used in the food supply). Humans, cattle, sheep, goats, pigs, the family dog, and a few other creatures can all catch this infectious disease. Brucellosis is a "hindrance" bacteria; symptoms can take months to appear, and deaths are few and far between, even without medical care.

Duties of First Responders on Scene

Law Enforcement: Isolate/Secure the Scene, Establish Control Zones, Establish Command, Stage Incoming Units. If command has been established: Report to Command Post. Evaluate Scene Safety/Security(ongoing criminal activity, consider victims to be possible terrorists, scan for secondary devices, and consider additional threats). Gather Witness Statements and Document Same. Initiate Law Enforcement Notifications (Federal Bureau of Investigations, Bureau of Alcohol, Tobacco, and Firearms; Explosive Ordnance Disposal/Bomb Squad, Private Security Forces). Request Additional Resources, Secure Outer Perimeter, Handle Traffic Control Considerations (staging areas, entry/egress), and Use Appropriate self-protective measures (Time/Distance/Shielding, minimize number of people exposed to danger, proper personal protective equipment if provided). Initiate Public Safety Measures (evacuate, protect in place), Assist with Control/Isolation of Patients, Coordinate Activities with Other Response Agencies, and Engage in Evidence Preservation (diagram the area, photograph the area, prepare a narrative description, maintain an evidence log). Participate in a Unified Command with: Fire/Rescue Services, Emergency Medical Services, Hospitals/Public Health, Emergency Management, and Public Works.

Response on Scene by First Responders

Caution: Avoid drinking unpasteurized milk products and practice good veterinary vaccination practices to avoid natural forms of brucellosis. In a terrorism attack with aerosol, livestock could possibly become contaminated. If this occurs, animal products should be pasteurized, boiled, or thoroughly cooked prior to eating. Water would have to be treated by boiling or iodination after any intentional contamination with brucella aerosols.

Field First Aid: Brucella is typically an acute, non-specific feverish illness with chills, sweats, headache, fatigue, myalgias, artthralgias, and anorexia (loss of appetite). Cough occurs in 15 to 25 percent of cases but a chest X-ray is usually normal. Complications may include arthritis, sacroiliitis, and vertebral osteomyelitis. Untreated disease may persist for month to years, often with relapses and remissions. Disability may be pronounced, and lethality may approach six percent. Brucellosis may be indistinguishable clinically from the typhoidal form of tularemia (see Guide For Emergency Response for Tularemia) or from typhoid fever itself.

The recommended treatment is doxycycline (200 mg/day) plus rifampin (600 mg/day) for six weeks. An alternative effective treatment is six weeks of doxycycline (200 mg/day) plus streptomycin (1 gm/day) for three weeks. Trimethoprim-sulfamethoxazole given four to six weeks is less effective. In 5 to 10 percent of cases, there may be a relapse or treatment failure. Regarding prophylaxis, killed and live attenuated human vaccines are available in many countries but are considered of unproven efficacy. There tends to be no information on the use of antibiotics for prophylaxis against human brucellosis.

Drugs: Doxycycline is both an antibiotic and a tetracycline antibiotic medicine used against brucella, rickettsiae, E.coli, shigella and a host of other infectious diseases. Patients should take this drug throughout the day for best results; if GI upset occurs, take drug with food. The following side effects may occur: sensitivity to sunlight and diarrhea. Patient should report any rash, itching, breathing difficulty, dark urine or light colored stools, or pain at injection site.

Rifampin is an antibiotic and antituberculous drug that inhibits DNA-dependent RNA polymerase activity in susceptible bacterial cells. Patient should take drug in a single daily dose on an empty stomach one hour before meals or two hours after meals. The following side effects may occur: reddish-orange coloring of body fluids (tears, sweat, saliva, urine, feces, sputum; stain will wash out of clothing, but soft contact lenses may be permanently stained; do *not* wear these contact lenses); nausea, vomiting, epigastric distress; skin rashes or lesions; numbness; tingling, drowsiness, or fatigue. Patient should have periodic medical checkups, including eye exams and blood tests, to evaluate this drug's effects. Report fever, chills, muscle, and bone pain, excessive tiredness or weakness, loss of appetite, nausea, vomiting, yellowing of skin or eyes, unusual bleeding or bruising, or skin rash or itching.

Medical Management: Standard precautions are adequate in managing brucellosis patients since this infective disease is not generally or readily transmissible from person-to-person. BLS-3 practices (Indigenous or exotic agents with potential for aerosol transmission; disease may have serious or lethal consequences.) should be used when handling suspected brucella cultures in the laboratory because of the danger of inhalation in such a setting! Oral antibody therapy alone is sufficient in most cases of brucellosis. Hospitals should adhere to the following infection control procedures for Isolation, Patient Placement, Patient Transport, Cleaning/Disinfecting Equipment, Discharge Management, and Post-Mortem Care for patients who have brucellosis or are suspected of having this disease:

> **Isolation:** Standard precautions for all aspects of care, contact precautions, and strict hand washing with antimicrobial soap.
> **Patient Placement:** Private room.
> **Patient Transport:** Limit movement, essential purposes only.
> **Cleaning/Disinfecting Equipment:** Terminal cleaning required with phenolic, disinfect surfaces with 1:9 bleach/water solution (10 percent), linen double bagged, disinfect equipment before taking it from room, change air filter before room terminally cleaned.
> **Discharge Management:** Teach caregivers standard precautions.
> **Post-mortem Care:** Follow standard precautions.
> **Fire:** For an infectious disease or weapon, use available methods and equipment on surrounding fires. Appropriate extinguishing agents: dry chemical, soda ash, sand, or lime. Do not use high pressure water streams

Personal Protective Equipment: At all times wear protective Level A until you know what you are dealing with; upon learning the exact type of infectious material(s) you are dealing with, consult proper toxicological levels and downgrade as necessary for safety and comfort of operation.

Spill/Leak Disposal: Outright cleanup or disposal of Infectious Substances (biological agents or weapons) should *only* handled by trained and experienced personnel. To contact such help, call any or all of the following numbers:

1. SBCCOM Operations — The U.S. Army Soldier and Biological Chemical Command, Edgewood Chemical Biological Center (ECBC), ATTN: AMSSB-RCB-RS, Aberdeen Proving Ground, MD 21010-5424. Tel. 410-436-2148 (twenty-four-hour emergency number).

2. U.S. Army Medical Research Institute of Infectious Diseases, Fort Detrick, Frederick, MD 21702-5011. Tel. 1-888-872-7443.

3. Center for Disease Control and Prevention Emergency Response Line. 1-770-488-7100.

4. ATSDR — Agency for Toxic Substances and Disease Registry, Division of Toxicology, 1600 Clifton Road NE, Mailstop E-29, Atlanta, GA 30333. Tel. 1-888-422-8737. Fax: 404-498-0057.

5. USACHPPM — U.S. Army Center for Health Promotion and Preventive Medicine, Aberdeen Proving Grounds, Maryland 21010-5422. Tel. 410-671-2208. Email: kwilliam@aehal.apgea.army.mil

6. National Response Center (for chem/bio hazards & terrorist events); Tel. 1-800-424-8802 or 1-202-267-2675.

7. U.S. Public Health Service: 800-USA-NDMS.

8. National Domestic Preparedness Office (for civilian use). Tel. 202-324-9025.

Symptoms: Fever, headache, myalgias, arthralgias, back pain, sweats, chills, and generalized malaise. Additional signs or symptoms may include depression, mental status changes, sacroiliitis, and vertebral osteomyelitis; however, fatalities tend to be uncommon.

Vaccines: There is no human vaccine available for brucellosis, although there are a number animal vaccines since this disease is one of the most important veterinary diseases in the entire world.

Guides for Emergency Response: Biological Agent or Weapon: Glanders (includes Melioidosis)

8

AGENT: Glanders: This illness is a contagious and destructive disease, especially of horses, caused by a bacterium of such animals. It is characterized by nodular lesions, especially of the respiratory mucosae and lungs. Glanders may occur in an acute localized form, as a septicemic rapidly fatal illness, or as an acute pulmonary infection. Combinations of these syndromes commonly occur in humans. A chronic cutaneous form with lymphangitis (inflammation of the lymphatic vessels) and regional adenopathay (any disease or enlargement involving glandular tissue, especially involving lymph nodes) is also frequent. **Aerosol infection from a biological warfare weapon containing B. mallei (glanders is a gram-negative bacillus, called B. mallei, with horses, mules, and donkeys serving as reservoirs) could produce any of these syndromes.** The incubation period is from ten to fourteen days, depending on the inhaled dose and agent virulence. The septicemic form begins suddenly with fever, rigors, sweats, myalgia (muscular pain), pleuritic chest pain, photophobia, lacrimation, and diarrhea. Physical examination may reveal fever, tachycardia (rapid heart action), cervical adenopathy (any disease or enlargement involving glandular issue, especially one involving lymph nodes) and mild spenomegaly. Blood cultures are usually negative until the patient is moribund. Mild leukocytosis (an increase in the number of leukocytes in the circulating blood that occurs normally, or abnormally as in some infections) with a shift to the left or leucopenia (a condition in which the number of leukocytes circulating in the blood is abnormally low, mostly due to decreased production of new cells in conjunction with various infective diseases) may occur. The pulmonary form may follow inhalation or arise by hematogenous (arriving in the blood, or spread by the blood) spread. Systemic symptoms may occur. Chest radiographs may show miliary (resembling or suggesting a small seed or many small seeds) nodules (.5 to 1cm) and/or a bilateral bronchopneumonia, segmental, or lobal pneumonia and necrotizing (causing, associated with, or undergoing necrosis) nodular lesions.

Medical Classification: Bacterial.

Probable Form of Dissemination: Aerosol (humans can also be infected by transmission from infected horses, but the transmission rate seems to be extremely low).

Detection in the Field: None.

Infective Dose: Assumed low.

Signs and Symptoms: During a biological warfare attack or terrorist action using glanders, the chronic form is unlikely to be present within fourteen days after an aerosol attack. It is characterized by cutaneous and intramuscular abscesses on the legs and arms. These lesions are associated with enlargement and induration (an increase in the fibrous elements in tissue commonly associated with inflammation and marked by loss of elasticity and pliability) of the regional lymph channels and nodes. Rare cases develop osteomyelitis, brain abscess, and meningitis. Recovery from chronic glanders may occur, or the disease may erupt into acute septicemic illness. Nasal discharge and ulceration are present in 50 percent of chronic cases. The Japanese deliberately infected horses, civilians, and prisoners of war with glanders in World War II, and, during 1943 and 1944, the United States studied glanders as a possible biological warfare weapon but never actually did weaponize it.

Incubation Time: Ten to fourteen days (via aerosol).

Diagnosis: Cultures of autopsy nodules in septicemic cases will usually establish the presence of B. mallei. **Occurrence in the absence of animal contact and/or in a human epidemic form is presumptive evidence of a biological warfare attack.** Mortality will be high regardless of antibiotic use. Remember, size of dose in biological agents' infections can be crucial. In regards to treatment, Standard Precautions ("See Standard Precautions" in this manual) should be used to prevent person-to-person transmission in proven or suspected cases. Sulfadiazine 100 mg/kg per day in divided doses for three weeks has been found to be effective in experimental animals and in humans. Other antibiotics that have been effective in experimental infection in hamsters include doxycycline, rifampin, trimethoprin-sulfamethoxazole, and ciprofloxacin. *The limited number of infections in humans* has precluded therapeutic evaluation of most of the antibiotic agents; therefore, most antibiotic sensitivities are based on animal in vitro studies. Various isolates have markedly different antibiotic sensitivities, so that each isolate should be tested for its own individual resistance pattern.

Persistency: Very stable.

Personal Protection: Personal protection must be used as well as protection for the respiratory tract by using a mask with biological filters or self-contained breathing apparatus (SCBA) with positive pressure at least until you know the specific threats.

Routes of Entry to the Body: Inhalation of aerosol, transmissible from horses to humans although the transmissibility rate is low.

Person-To-Person Transmissible: Low transmission from man-to-man although "Standard Precautions" should be followed when working with patients; also transmissible from horses-to-humans although the transmissibility rate is low.

Duration of Illness: Death in seven to fourteen days via aerosol.

Potential Ability to Kill: Greater than 50 percent.

Defensive Measures: Wearing protective mask, practicing "Standard Precautions" with infected patients, and good personal hygiene.

Vaccine: There is no human or veterinary vaccine.

Drugs Available: Post-exposure prophylaxis may be tried with TMP-SMZ (trimethoprim/sulfamethoxazole).

Decontamination: Standard Precautions for healthcare workers. Person-to-person airborne transmission is *not* likely, although secondary cases may occur through improper handling of infected secretions. Environmental decontamination can be treated with 0.5 percent hypochlorite solution.

Characteristics: Glanders is a gram-negative bacillus, called B. mallei, with horses, mules, and donkeys serving as reservoirs. Transmission is by the organism invading the nasal, oral, and conjunctival mucous membranes, by inhalation into the lungs, and by invading abraded or lacerated skin. This disease is not widespread in humans at the present time, usually inflicting only veterinarians, persons who handle horses and donkeys, slaughter-house workers, and laboratory personnel. But since September 11, 2001, it is rumored to be a potential biological agent that could be disseminated by aerosol. At the present time, there is no pre-exposure or post-exposure prophylaxis, and no vaccine, available for humans. Glanders and melioidosis are quite similar in their effects, and both these infectious diseases are almost always fatal without treatment. Therapy will vary with the type and severity of the individual case as presented. Victims with localized disease may be managed with oral antibiotics for duration of 60 to 150 days, while more severe illness will require longer treatment. With regard to diagnosis, a methylene blue stain of exudates may reveal scant small bacilli; a chest x-ray may show miliary lesion, small multiple lung abscesses, or broncopneumonia. Standard cultures can be used to identify both B. mallei for glanders and B. pseudomallei for melioidosis. Leukocyte counts can be normal or elevated.

Duties of First Responders on Scene

Hazardous Materials Response Team(s): Establish the HazMat Group, and Provide Technical information/Assistance to: Command, EMS Providers, Hospitals, and Law Enforcement. Detect/Monitor to Identify the Agent, Determine Concentrations and Ensure Proper Control Zones. Continually Reassess Control Zones, Enter the Hot Zone (with chemical personal protective clothing) to Perform Rescue, Product Information, and Reconnaissance. Product Control/Mitigation may be implemented in Conjunction with Expert Technical Guidance. Improve Hazardous Environments: Ventilation, Control HVAC, Control Utilities. Implement a Technical Decontamination Corridor for Hazardous Materials Response Team (HMRT) Personnel. Coordinate and Assist with Mass Decontamination. Provide Specialized Equipment as Necessary. Assist Law Enforcement Personnel with Evidence Preservation/Collection, Decontamination.

Response on Scene by First Responders

Caution: Both glanders and melioidosis may occur in an acute localized form, as an acute pulmonary infection, or as an acute fulminant, rapidly fatal, sepsis. Combinations of these syndromes may occur in human cases. In addition, melioidosis may remain asymptomatic after initial acquisition, and remain quiescent for decades, but these patients may display active melioidosis years later which is often associated with an immune-compromising state. Aerosol infection produced by a biological weapon containing either glanders (B. mallei) or melioidosis (B. pseudomallei) could produce any of these syndromes

Field First Aid:
1. For localized disease, administer **ONE** of the following drugs for a duration of 60 to 150 days. Amoxcillin/clavulanate, 60 mg/kg/day in three divided oral doses.

Tetracycline, 40 mg/kg/day in three divided oral doses. Trimethoprim/sulfa (TMP), four mg per kg per day/sulfa, 20 mg per kg per day in divided oral doses.

2. For localized disease with mild toxicity, the antibiotics should be administered as follows: combine two of the above oral regimens for duration of thirty days, followed by monotherapy with either amoxicillin/clavulanate or TMP/sulfa for 60 to 150 days.

3. For extrapulmonary suppurative disease, the antibiotic therapy should be administered for six to twelve months, and surgical drainage of abscesses is indicated.

4. For severe and/or septicemic disease, administer antibiotics as follows: Ceftazidime, 120 mg/kg/day in three divided doses, combined with TMP/sulfa (TMP, eight mg per kg per day/sulfa, 40 mg per kg per day in four divided doses). Initially, administer parental therapy for two weeks, followed by oral therapy for six months.

5. The addition of streptomycin is indicated if acute pneumonia and sputum studies suggests plague.

Medical Management: Medical staff should apply Standard Precautions in management of victims and contacts. Glanders, melioidosis, and smallpox may have diffuse pustular rashes; **strict isolation and quarantine would be indicated until smallpox can be disregarded as the cause of this disease.** Contact precautions are indicated while caring for victims with skin involvement. Glanders, melioidosis, and smallpox may show as acute pulmonary disease with purulent sputum. Respiratory isolation is recommended pending exclusion of plague if sputum studies disclose gram-positive bacilli with bipolar "safety pin" appearance when using Wright's or methylene blue stains. **Hospital staff should continue using Standard Precautions with glanders and/or melioidosis victims even after smallpox and plague have been excluded from the diagnosis.**

Fire: For an infectious disease or weapon, use available methods and equipment on surrounding fires. Appropriate extinguishing agents: dry chemical, soda ash, sand or lime. Do not use high pressure water streams.

Personal Protective Equipment: At all times wear protective **Level A until you know what you are dealing with;** upon learning the exact type of infectious material(s) you are dealing with, consult proper toxicological levels and downgrade as necessary for safety and comfort of operation.

Spill/Leak Disposal: Outright cleanup or disposal of Infectious Substances (biological agents or weapons) should *only* be handled by trained and experienced personnel. To contact such help, call any or all of the following numbers.

1. SBCCOM Operations — The U.S. Army Soldier and Biological Chemical Command, Edgewood Chemical Biological Center (ECBC), ATTN: AMSSB-RCB-RS, Aberdeen Proving Ground, MD 21010-5424. Tel. 410-436-2148 (twenty-four-hour emergency number).

2. U.S. Army Medical Research Institute if Infectious Diseases, Fort Detrick, Frederick, MD 21702-5011. Tel. 1-888-872-7443.

3. Center for Disease Control and Prevention Emergency Response Line. 1-770-488-7100.

4. ATSDR — Agency for Toxic Substances and Disease Registry, Division of Toxicology, 1600 Clifton Road NE, Mailstop E-29, Atlanta, GA 30333. Tel. 1-888-422-8737. Fax: 404-498-0057.

5. USACHPPM — U.S. Army Center for Health Promotion and Preventive Medicine, Aberdeen Proving Grounds, MD 21010-5422. Tel. 410-671-2208. Email: kwilliam@ aehal.apgea.army.mil
6. National Response Center (for chem/bio hazards & terrorist events); Tel. 1-800-424-8802 or 1-202-267-2675.
7. U.S. Public Health Service: 800-USA-NDMS.
8. National Domestic Preparedness Office (for civilian use). Tel. 202-324-9025.

Symptoms: Onset of symptoms may be either abrupt or gradual. Inhalational exposure produces fever commonly in excess of 102 degrees F, rigors, sweats, myalgias, headache, pleuritic chest pain, cervical adenopathy, hepatosplenomegaly, and generalized papular/ pustular eruptions. Acute pulmonary disease can progress and result in bacteremia and acute septicemic disease.

Vaccines: No vaccine available.

Guides for Emergency Response: Biological Agent or Weapon: Plague

9

AGENT: Plague is a zoonotic (communicable from animals to humans under natural conditions) disease caused by *Yersinia pestis*. Under normal conditions, humans become infected as a result of contact with rodents, and their fleas. The transmission of gram-negative coccobacillus is by the bite of the infected flea, *Xenopsylla cheopis,* the oriental rat flea, or the human flea, *Pulex irritans.* Under normal conditions, three syndromes are recognized: bubonic, primary septicemic, or pneumonic. If a situation develops into a biological warfare or terrorist attack, the plague bacillus would probably be delivered by vectors (fleas) causing the bubonic variety or, more likely, by aerosol causing the pneumonic type. Bubonic plague is the most common form, and has a secondary formation of large regional lymph nodes called buboes. Blood may clot in the vessels, and may show up in blackened fingers and toes. "Natural" plague most often is caused by the bite of a flea that had dined on infected rodents; a secondary source would be by sputum droplets inhaled from coughing victims. There is a limited incidence of plague in the southwestern desert of the United States. Usually, the rodents die of plague, fleas feed on the rodents' bodies, plague multiplies in the flea, flea becomes unable to bite normally, flea gets apprehensive and bites everything, and everything the flea bites gets infected with plague. Some U.S. plague victims have been infected by household cats. "Un-natural" plague, the result of terrorist or enemy action, could possibly be an aerosol, or less likely, a release of plague-carrying fleas, forms of dissemination that may develop into pneumonic plague leading to quick death. Pneumonic plague is an extremely virulent form, can be transferred from person-to-person, and vaccine seems to have little effect on it. During World War II, the Japanese established Unit 731 in Mukden, Manchuria, and carried out experiments in biological warfare on prisoners of war from the United States, Britain, Australia, and New Zealand. They tried aerosolizing plague but were unsuccessful.

History: Late in World War II, the United States researched plague as well as anthrax, brucellosis, encephalitis, glanders, melioidosis, psittacosis, tularemia, typhus, yellow fever, and other biological agents to learn those that would be the most destructive and the least difficult to use in wartime. In 1959, Camp Detrick, located in Frederick, Maryland, produced fleas infected with plague; flies contaminated with anthrax, cholera, and dysentery; mosquitoes carrying yellow fever, malaria, and dengue (den-gee) fever; and ticks doused with tularemia. For testing of biological weapons and other interests, Camp Detrick in the early 1960s became the greatest user of guinea pigs in the entire world.

However, today, aerosols are likely to be used for dissemination of plague in an attack, and that would lead to pneumonic plague because victims would inhale the plague. Pneumonic plague is a very effective biological warfare agent because of the rapid onset of symptoms and the severity of illness. However, "normal" plague is also transmitted through infected rabbits, domestic cats, and dogs. Treatment for pneumonic plague must start within twenty-four hours after onset of the disease or the patient is likely to die. That is, almost all untreated pneumonic plague victims die. Death rate of untreated bubonic plague victims is about 50 percent. Early administration of antibiotics with supportive therapy is effective in treating the disease. If any form of plague is suspected, isolation of the patient is imperative until the victim has been on antibiotic therapy for at least forty hours, and there has been a favorable response to treatment.

Medical Classification: Bacterial

Probable Form of Dissemination: Aerosol

Detection in the Field: None

Infective Dose (Aerosol):   100–500 organisms

Signs and Symptoms:

Pneumonic Plague: Incubation period is two to three days. High fever, chills, headache, hemoptysis (expectoration from some part of the respiratory tract), and toxemia (an abnormal condition associated with the presence of toxic substances in the blood; a generalized intoxication due to absorption and systemic dissemination of bacterial toxins from a focus of infection), progressing rapidly to dyspnea (difficult or labored respiration), stridor (a harsh vibrating sound heard during respiration in cases of obstruction of the air passages) and cyanosis. Death results from respiratory failure, circulatory collapse, and a bleeding diathesis (the constitutional or inborn state disposing to a disease, or metabolic or structural anomaly).

Bubonic Plague: Incubation period is from two to ten days. Malaise, high fever, and tender lymph nodes (buboes: inflammatory swelling of one or more lymph nodes, usually in the groin; the confluent mass of nodes usually suppurates and drains pus); may progress spontaneously to the septicemic form, with spread to the central nervous system (CNS), lungs, and elsewhere.

Incubation Time: One to three days.

Clinical Diagnosis: A presumptive diagnosis can be made from by Gram or Wayson stain of lymph node aspirates, sputum, or cerebrospinal fluid. Plague can also be cultured. As for treatment, early administration of antibiotics is very effective. Supportive therapy for pneumonic and septicemic forms is required.

Differential Diagnosis: In most cases where *bubonic* plague is suspected, tularemia adenitis, staphylococcal or streptococcal adenitis, meningococcemia, enteric gram-negative sepsis, and rickettsiosis need to be ruled out. In *pneumonic* plague, tularemia, anthrax, and SEB agents need to be considered. Continued deterioration without stabilization effectively rules out SEB. The presence of a widened mediastinum on a chest X-ray should alert one to the diagnosis of anthrax. Regarding treatment, care should include strict isolation procedures for cases involving plague. This disease may be spread from person-to-person by droplets.

Persistency: Up to one year in soil; 270 days in bodies.

Personal Protection: Protective clothing must be used as well as protection for the respiratory tract by using a mask with biological filters or self-contained breathing apparatus (SCBA) with positive pressure, at least until you know the specific threat(s).

Routes of Entry to the Body: Inhalation, ingestion, flea bite.

Person-To-Person Transmissible: High.

Duration of Illness: One to six days (usually fatal).

Potential Ability to Kill: High unless treated within twelve to twenty-four hours.

Defensive Measures: Immunizations, good personal hygiene, physical conditioning, use of arthropod repellents, wearing protective mask, and practicing good sanitation. Utilize an insecticide as necessary to kill fleas on victims and their contacts; if local flea and rodent population becomes infected, institute control measures.

Vaccines: Yes (Greer Laboratory Vaccine). A licensed, killed vaccine is available. Initial dose followed by a second smaller dose one to three months later, and a third dose three to six months later. A booster dose is to be given at six, twelve, and eighteen months and then every one to two years. *Please Note:* This vaccine may not protect against aerosol exposure. Live-attenuated vaccines are available outside the United States, but they tend to be highly reactogenic and without proven efficacy when challenged with aerosol dissemination.

Drugs Available: Streptomycin, Doxycycline, and Chloramphenicol.

Decontamination: Soap and water, or diluted sodium hypochlorite solution (0.5 percent). Removal of potentially contaminated clothing should be done by people in full protective clothing in an area away from non-contaminated persons. For victims with bubonic plague, drainage, and secretion procedures need to be employed. Careful treatment of buboes is required to avoid aerosolizing infectious material. For victims with pneumonic plague, *strict isolation is absolutely necessary.* Heat, disinfectants and sunlight renders bacteria harmless.

Ability To Kill: Very high.

Characteristics: Plague, an etiologic agent sometimes referred to as *Yersinia pestis*, is a gram-negative bacillus of the family *Entreobacteriaceae.* The primary reservoir is rodents, but domestic cats and wild carnivores can also transmit plague to humans. In endemic or epidemic plague, the disease is transmitted by infected fleas from rodent to human, dog or cat to human, or person-to-person. Respiratory droplet transmission can happen from person-to-person or from cat-to-person. Plague may also be transmitted by cat bites or scratches. In addition, respiratory transmission is enhanced in humid climates. Bubonic plague features the acute onset of fever and prostration in association with acute, painful lymphadenitis in the lymph node group draining the site of the fleabite. A skin lesion at this point of entry is seen in less than 25 percent of cases; clinically apparent lymphangitis does not occur. The disease progresses with bacteremia resulting in metastatic infection, septic shock, and thrombosis of small arteries, resulting in digital gangrene. Pneumonia due to hematogenous metastasis occurs in approximately 25 percent of cases. The case fatality rate for untreated bubonic plague is about 60 percent; when the patient is given prompt, effective therapy the death rate drops to about 5 percent.

Primary pneumonic plague occurs after inhalation of organisms, which may occur by aerosol transmission from a person or animal with secondary or primary pneumonic plague. Septicemic plague may evolve from any form of plague. It features the acute onset of bacteremia, septic shock, and thrombosis with or without antecedent lymphadenitis. Prognosis for both pneumonic and septicemic pneumonic plague is poor; the fatality rate is 100 percent for untreated cases.

Response on Scene by First Responders

Caution: The death rate from untreated pneumonic plague can reach almost 100 percent. Once a human is infected with plague, a progressive and potentially deadly illness usually results unless antibody therapy is administered. In a progressional sequence, the patient develops blood infection which leads to lung infection.

Field First Aid: Triage categories vary with conditions and available resources. Plague pneumonia is curable if treatment is started at once; therefore, such victims are classified as "Immediate." Victim appearing twenty-four hours after respiratory symptoms have begun are less likely to survive, and such victims should be classed as "Expectant." Supportive care should include IV (intravenous) hydration, supplemental oxygen, and respiratory support as necessary. Early administration of antibiotics is critical, as pneumonic plague is invariably fatal if antibody therapy is delayed more than one day after the onset of symptoms. Select and use *one* of the following treatments using streptomycin, gentamicin, ciprofloxacin, OR doxycycline for ten to fourteen days. Chloramphenicol is the drug of choice for plague meningitis.

Drugs: Use streptomycin at 15 mg/kg lean body mass intra-muscular every twenty-four hours for ten to fourteen days; or use gentamicin at 5 mg/kg lean body mass intra-venous every twenty-four hours for ten to fourteen days; or use gentamicin at 1.75 mg/kg lean body mass intra-venous every eight hours for ten to fourteen days; or use ciprofloxacin at 400 mg intra-venous every twelve hours (oral therapy may be given at 750 mg orally every twelve hours after the patient is clinically improved, for completion of a ten to fourteen-day course of therapy); or use doxycycline at 200 mg intra-venous loading dose followed by 100 mg intra-venous every twelve hours (oral therapy may be given at 100 mg orally every twelve hours after the patient is clinically improved, for completion of a ten to fourteen-day course of therapy.

Prophylaxis: For asymptomatic patients exposed to plague aerosol, or to a patient with suspected pneumonic plague, provide doxycycline at 100 mg orally twice daily for seven days, or for the duration of risk of exposure plus one week. Alternative antibiotics include ciprofloxacin, tetracycline, or chloramphenicol. No vaccine is currently available for plague propylaxis. The previously available licensed, killed vaccine was effective against bubonic plague, but not against aerosol exposure.

Medical Management: Use Standard Precautions for bubonic plague. Suspected pneumonic plague cases require strict isolation with droplet precautions for at least forty-eight hours of antibiotic therapy, or until sputum cultures are negative in confirmed cases. When fleas and rodents are in the area which might prevent a hazard for plague by natural means, they should be eradicated. However in a terrorist attack, dissemination will probably be accomplished by means of aerosol. In addition to the drugs mentioned above, there have been good results reported for tests on laboratory animals, but not on humans, indicating that ciprofloxacin and ofloxacin may also be effective. Usual supportive therapy includes IV crystalloids and hemodynamic monitoring when working with plague. Although low-grade DIC (disseminated intravascular coagulation) may occur, clinically significant hemorrhage is uncommon, as is the need for treatment with heparin. Endotoxin shock is common, but pressor agents are rarely needed. Buboes rarely require any form of local care, but instead recede with systemic antibiotic therapy; incision and drainage poses a risk to others in contact with the patient. Aspiration is recommended for diagnostic purposes.

Isolation Precautions for Bubonic Plague: Standard precautions for all aspects of care including strict hand washing with antimicrobial soap.

Isolation Precautions for Pneumonic Plague: Standard precautions for all aspect of care including droplet precautions and strict hand washing with antimicrobial soap.

Patient Placement for Bubonic Plague: Private room or share with "like" patients if no private rooms are available.

Patient Placement for Pneumonic Plague: Private room or share with "like" patients if no private rooms are available.

Patient Transport with Bubonic Plague: Limit movement for essential purposes only.

Patient Transport with Pneumonic Plague: Limit movement, essential purposes only; mask patient before transport.

Cleaning, Disinfecting Equipment with Bubonic Plague: Terminal cleaning required with Phenolic; disinfect surfaces with 1:9 bleach/water solution (10 percent); linen doubled bagged; and air filter changed before room terminally cleaned.

Cleaning, Disinfecting Equipment with Pneumonic Plague: (The same series of isolation controls as for bubonic plague).

Discharge Management with Bubonic Plague: Teach care givers Standard Precautions.

Discharge Management with Pneumonic Plague: Teach care givers Standard Precautions; do not discharge until no longer infectious; do not discharge until after seventy-two hours of antibiotics.

Post-Mortem Care for Bubonic Plague: Follow Standard Precautions; disinfect surfaces with 1:9 bleach/water solution (10 percent).

Post-Mortem Care with Pneumonic Plague: Follow Standard Precautions; droplet precautions; and disinfect surfaces with 1:9 bleach/water solution (10 percent).

Fire: For an infectious disease or weapon, use available methods and equipment on surrounding fires. Appropriate extinguishing agents: dry chemical, soda ash, sand, or lime. Do not use high pressure water streams.

Personal Protective Equipment: At all times wear protective Level A until you know what you are dealing with; upon learning the exact type of infectious material(s) you are dealing with, consult proper toxicological levels and downgrade as necessary for safety and comfort of operation.

Spill/Leak Disposal:
Outright cleanup or disposal of Infectious Substances (biological agents or weapons) should *only* be handled by trained and experienced personnel. To contact such help, call any or all of the following numbers.
   1. SBCCOM Operations — The U.S. Army Soldier and Biological Chemical Command, Edgewood Chemical Biological Center (ECBC), ATTN: AMSSB-RCB-RS, Aberdeen

Proving Ground, MD 21010-5424. Tel. 410-436-2148 (twenty-four-hour emergency number).

2. U.S. Army Medical Research Institute if Infectious Diseases, Fort Detrick, Frederick, MD 21702-5011. Tel. 1-888-872-7443.

3. Center for Disease Control and Prevention Emergency Response Line. 1-770-488-7100.

4. ATSDR — Agency for Toxic Substances and Disease Registry, Division of Toxicology, 1600 Clifton Road NE, Mailstop E-29, Atlanta, Georgia 30333. Tel. 1-888-422-8737. Fax: 404-498-0057.

5. USACHPPM — U.S. Army Center for Health Promotion and Preventive Medicine, Aberdeen Proving Grounds, Maryland 21010-5422. Tel. 410-671-2208. Email: kwilliam@aehal.apgea.army.mil

6. National Response Center (for chem/bio hazards & terrorist events); Tel. 1-800-424-8802 OR 1-202-267-2675.

7. U.S. Public Health Service: 800-USA-NDMS.

8 National Domestic Preparedness Office (for civilian use). Tel. 202-324-9025.

Symptoms: Following an aerosol release of organisms, unprotected persons will show acute pneumonic plague with high fever, systemic toxicity, productive cough, and hemoptysis. Chest x-rays may show patchy peribronchial infiltrates, cavitation, consolidation, hilar adenopathy, and pleural effusions. Victims may show disseminated intravascular coagulation (DIC) with resultant thrombosis and digital gangrene. However, hemorrhagic complications of DIC in plague are rare. In addition, an aerosol attack of plague could result in an epidemic of bubonic plague if rodent hosts and flea vectors are present in the vicinity of the attack.

Vaccines: The currently available inactivated whole cell vaccine is not recommended for protection from a biological warfare agent since it does *not* protect laboratory animals from aerosolized plague. However, the vaccine is effective in preventing bubonic plague in persons in endemic or epidemic areas.

Guides for Emergency Response: Biological Agent or Weapon: Q Fever

10

AGENT: Q fever is a zoonotic disease caused by a rickettsia, Coxiella burnetii. The most common animal reservoirs are sheep, cattle, and goats. Humans acquire Q fever by inhalation of particles contaminated with the organisms. A biological warfare or terrorist attack would cause disease similar to that occurring naturally. Q fever is a sudden feverish illness involving the respiratory system caused by the Rickettsia burnetii as a result of contact with infected animals, drinking infected raw milk, or ticks. There is a high degree of Q fever in Australia but not in neighboring New Zealand. It makes a good biological weapon because it's a high infectivity agent. Pneumonia is extremely common with Q fever, but patients generally recover.

Medical Classification: Rickettsial.

Probable Form of Dissemination: Aerosol; sabotage of food supply.

Detection in the Field: None.

Infective Dose (Aerosol): 1–10 organisms.

Signs and Symptoms: Fever, cough, and pleuritic chest pain may occur as early as ten days after exposure. Patients are not generally seriously ill, and the illness lasts from two days to two weeks.

Incubation Time: Fourteen to twenty-six days.

Diagnosis: Q fever is not a clinically distinct illness and may resemble a viral illness or other types of atypical pneumonia. The diagnosis is confirmed serologically. As for treatment, Q fever is generally a self-limiting illness even without treatment. Tetracycline or doxycycline are the treatments of choice and are given orally for five to seven days. Q fever endocarditus, which is rare, is much more difficult to treat.

Differential Diagnosis: Q fever usually presents as an undifferentiated febrile illness, or a primary atypical pneumonia, which must be differentiated from pneumonia caused by mycoplasm, Legionnaires disease, psittacosis or Chlamydia pneumoniae. More rapidly progressive forms of pneumonia may look like bacterial pneumonia including tularemia or plague.

Persistency: Months on wood and sand.

Personal Protection: Protective clothing must be used as well as protection for the respiratory tract by using a mask with biological filters or self-contained breathing apparatus (SCBA) with positive pressure, at least until you knows the specific threat(s).

Route of Entry to the Body: Inhalation

Person-To-Person Transmissible: Infrequent or never.

Duration of Illness: Two days to three weeks. A high fever could persist for three weeks or more, but treatment with antibiotics is usually effective within thirty-six to forty-eight hours. With treatment or without treatment, Q fever is generally a self-limiting illness.

Potential Ability to Kill: Very low (estimated at 1–3 percent).

Defensive Measures: Immunizations, good personal hygiene, physical conditioning, use of arthropod repellents, wearing protective mask, and practicing good sanitation. Persons who are regularly exposed to domestic animals should be vaccinated against Q fever.

Vaccines: IND 610 (inactivated whole cell vaccine given as single injection) is available through USAMRIID Fort Detrick, MD 21702; and Q-Vax (CSL Ltd., Parkville, Victoria, AUSTRALIA). This vaccine is effective in eliciting protection against exposure, but severe local reactions to this vaccine may be seen in those persons who already possess immunity.

Drugs Available: Tetracycline, and Doxycycline. Treatment with tetracycline during the incubation period may delay but not prevent the onset of symptoms.

Decontamination: Dermal exposure from any suspected biological agent or weapon, according to *NATO Handbook on the Medical Aspects of NBC Defensive Operations*, AMedP-6(B), Part II – Biological, February 1, 1996, should be managed by decontamination at the earliest possible opportunity. "In the absence of specific guidance, exposed areas should be cleaned using an appropriated diluted sodium hypochlorite solution (0.5 percent) or copious quantity of plain soap and water. This should follow any needed use of decontaminants for chemical agents but should be prompt. Potentially contaminated clothing should be removed as soon as practical by protected personnel (that is, in full individual protective equipment) in an area away from non-contaminated patients."

Symptoms: Symptoms appear about ten to twenty days after the Q fever rickettsia are inhaled. The symptoms resemble flu symptoms and include fever, chills, headache, fatigue and muscle aches. About one half of persons with symptoms will have pneumonia evident on chest X-ray and some of these will have a cough or chest pain. The complications of meningitis or and inflammation of the heart may arise, but these are uncommon. Normally, the duration of Q fever is two days to two weeks at which time the disease resolves without permanent effects on the individual.

Medical Countermeasures: The symptoms of Q fever usually resolve without antibiotic treatment. Antibiotics can be given to shorten the illness. Tetracycline's are the drugs of choice, some recent work indicates ciprofloxacin may also be useful. The timing of antibiotic therapy is important. If given one to eight days after exposure, the antibiotics will merely delay the symptoms for about three weeks. Antibiotics begun eight to twelve days after exposure and continued for ten days will prevent Q fever from occurring. Q fever is *not* usually transmitted directly from person to person, so quarantine of affected individuals is not suggested. Q fever has, however, been transmitted through blood or bone marrow donations, so health providers should be aware of this possibility.

Ability to Kill: Low.

Characteristics: As a natural disease Q fever, a rickettsial illness caused by Coxiella burnetii, is typically spread by inadvertent aerosolisation of organisms from infected animal products. Rickettsia is organisms that have characteristics of both bacteria and viruses. Although highly infectious since just one organism can cause infection, Q fever is rarely, or possibly never, transmissible from person-to-person. Intentional release by terrorists would presumably involve aerosolisation, and Q fever would likely be used as an incapacitating agent since its mortality rate is quite low, about 1 to 3 percent. Spore-like forms of Coxiella burnetii may withstand harsh conditions and thus persist in the environment for prolonged periods of time. Presumably, animals, especially sheep, in such areas would be at risk for acquiring infection, and contact with products from such animals would represent a continuing hazard to human beings. However, little information exists to permit assessment of direct long-term hazards to humans entering an area contaminated by intentional release of aerosolised Q fever.

Response on Scene by First Responders

Caution: Chemoprophylaxis is not effective if given immediately (one to seven days) post-exposure; it merely delays the onset of disease. When doing post-exposure chemoprophylaxis, the drug regimen mentioned below is effective if begun eight to twelve days post-exposure.

Field First Aid: In a terrorist attack with Q fever, the primary threat is dissemination of aerosol, or contamination of food. Acute Q fever can appear to develop as an undifferentiated febrile illness, as an atypical pneumonia, or as a rapidly progressive pneumonia.

Drugs:
1. **Acute Q fever:** Administer doxycycline 100 mg orally every twelvehours for five days after victim is afebrile (free of fever). Administer tetracycline 500 mg every six hours for five days after the victim is afebrile. If a victim appears unable to take tetracycline, try ciprofloxacin and other quinolones, which are active in vitro; the duration of the therapy is usually five to seven days, at least two days after the victim is afebrile.
2. Quinolones **are NOT meant** for children. A word of caution is necessary about prophylaxis: do not begin prophylaxis too early. If it is begun during the incubation period, prophylaxis may delay but not prevent the onset of symptoms. As a result of this, doxycycline or tetracycline should be started eight to twelve days post exposure. This type of regimen has been shown to prevent clinical disease.

Regarding isolation and decontamination, Standard Precautions are recommended for healthcare personnel. Person-to-person transmission is rare. Victims exposed to Q fever by aerosol do not present a risk for secondary contamination or re-aerosolization of the organism. Decontamination can be done with soap and water, or a 0.5 percent chlorine solution on personnel.

Q fever endocarditis, and other firms of chronic Q fever (which is very rare) is much more difficult to treat. Such treatment is very complex, even controversial, and beyond the scope of this volume.

Medical Management: Standard Precautions are recommended for healthcare personnel. Most cases of acute Q fever will eventually resolve without antibiotic treatment, but all suspected cases of Q fever should be treated to reduce the risk of complications.

Tetracycline 500 mg every six hours or doxycycline 100 mg every twelve hours for five to seven days will shorten the duration of illness, and fever usually disappears within one to two days after treatment is begun. Ciprofloxacin and other quinolones are active in vitro and should be considered for victims unable to take tetracycline or doxycycline. Successful treatment of Q fever endocarditis is much more difficult. Tetracycline or doxycycline given in combination with trimethoprim-sulfamethoxazole (TMP-SMX) or rifampin for twelve months or longer has been successful in some cases. However, valve replacement is often required to achieve a cure.

With regard to infection control, Q fever will require Standard Precautions for all aspects of care including strict hand-washing with antimicrobial soap; Patient Placement including placement with "like" patients if there are no private rooms; no restrictions on Patient Transport; Cleaning/Disinfecting Equipment will require terminal cleaning with Phenolic, linen to be double-bagged, and air filter changed before room terminally cleaned; and for Discharge Management no special requirements are necessary. For Post-Mortem Care, the only requirement is to follow Standard Precautions.

Fire: For an infectious disease or weapon, use available methods and equipment on surrounding fires. Appropriate extinguishing agents: dry chemical, soda ash, sand, or lime. Do not use high pressure water streams

Personal Protective Equipment: At all times wear protective Level A until you know what you are dealing with; upon learning the exact type of infectious material(s) you are dealing with, consult proper toxicological levels and downgrade as necessary for safety and comfort of operation.

Spill/Leak Disposal: Outright cleanup or disposal of Infectious Substances (biological agents or weapons) should *only* be handled by trained and experienced personnel. To contact such help, call any or all of the following numbers.

1. CEMCOM Operations — The U.S. Army Soldier and Biological Chemical Command, Edgewood Chemical Biological Center (ECBC), ATTN: AMSSB-RCB-RS, Aberdeen Proving Ground, MD 21010-5424. Tel. 410-436-2148 (twenty-four-hour emergency number).
2. U.S. Army Medical Research Institute if Infectious Diseases, Fort Detrick, Frederick, MD 21702-5011. Tel. 1-888-872-7443.
3. Center for Disease Control and Prevention Emergency Response Line. 1-770-488-7100.
4. ATSDR — Agency for Toxic Substances and Disease Registry, Division of Toxicology, 1600 Clifton Road NE, Mailstop E-29, Atlanta, GA 30333. Tel. 1-888-422-8737. Fax: 404-498-0057.
5. USACHPPM — U.S. Army Center for Health Promotion and Preventive Medicine, Aberdeen Proving Grounds, MD 21010-5422. Tel. 410-671-2208. Email: kwilliam@ aehal.apgea.army.mil
6. National Response Center (for chem/bio hazards & terrorist events); Tel. 1-800-424-8802 or 1-202-267-2675.
7. U.S. Public Health Service: 800-USA-NDMS.
8. National Domestic Preparedness Office (for civilian use). Tel. 202-324-9025.

Symptoms: Q fever typically presents as an undifferentiated illness, with fever, chills, cough, headache, weakness, and chest pain occurring as early as ten days after exposure. Onset may be sudden or insidious. Pneumonia is present in some cases, but pulmonary syndromes are usually not prominent. Victims are not generally critically ill, and the illness last from two days to two weeks. Complications include hepatitis and a peculiar form of chronic endocarditus that may be largely responsible for the few fatal cases that occur.

Vaccines: An inactivated whole cell vaccine (IND 610) given as a single 0.5 mL subcutaneous injection; this vaccine is being tested to determine the necessity of skin testing prior to use. A single dose provides protection for five years, but there are some problems with the presently available vaccine. It is highly reactogenic in immune persons, and can result in induration, sterile abscesses and necrosis at the injection site. The sterile abscesses may spontaneously drain or require surgical drainage. Immunization screening (skin testing and antibody measurements) MUST precede administration of the immunization, and personnel undergoing reaction to the screening must NOT be given this vaccine unless approved by a physician. That is, the vaccine is contraindicated for individuals with positive skin tests and/or antibody titers. A new vaccine prepared by chloroform-methanol extraction is presently being evaluated. This new vaccine seems to be safe, immunogenic in non-sensitized human volunteers, and is not reactogenic in sensitized guinea pigs.

Guides for Emergency Response: Biological Agent or Weapon: Ricin

11

AGENT: Ricin is a glycoprotein toxin (66,000 daltons; unit of mass — $1.657 \times 10 \times 24$) from the seed of the castor plant. It blocks protein synthesis by altering the rRNA (any of the various nucleic acids that contain ribose and uracil as structural components and are associated with the control of cellular chemical activities; also called, ribonucleic acid), thus killing the cell. Ricin is a potential biological warfare or terrorist attack agent since it is available throughout the world, is easy to make, and causes extreme pulmonary toxicity when inhaled. It has so far been used only as a weapon of assassination, but when used as an aerosol it could possibly lead to widespread illness and death among victims. It has been said that one milligram of ricin can kill an adult. Abdominal pain, vomiting, and diarrhea symptoms appear in a few hours. Within a few days, there is severe dehydration and a decrease in urine and blood pressure. The clinical picture observed is completely dependent on the method of exposure to this biological agent. All reported serious or fatal cases of castor bean ingestion have taken approximately the same course: rapid onset of nausea, vomiting, abdominal cramps and severe diarrhea with vascular collapse; death has occurred on the third day or later. With regard to inhalation, one might expect non-specific symptoms of weakness, fever, cough, and hypothermia followed by hypotension and cardiovascular collapse. High dosages in animals receiving inhalation doses appear to produce severe enough pulmonary damage to produce death.

Medical Classification: Toxin

Probable Form of Dissemination: Aerosol (plus many other exposure routes).

Detection in the Field: None.

Infective Dose (Aerosol): 3 to 5 ug/kg.

Signs and Symptoms: Weakness, fever, cough, and hypothermia about thirty-six hours after aerosol exposure, followed in the next twelve hours by hypotension and cardiovascular collapse.

Incubation Time: Hours.

Diagnosis: If signs and symptoms spelled out above are noted in large numbers of geographically clustered patients, exposure to aerosolized ricin is the suggested cause. **The rapid time course of severe symptoms and death would be unusual for infectious agents.** Laboratory findings are nonspecific except for specific serum ELISA (enzyme-linked immunosorbent assay). Acute and convalescent sera (plural of "serum") should be collected.

Differential Diagnosis: In oral intoxication, fever, gastrointestinal involvement, and vascular collapse are prominent, the latter differentiating it from infection with enteric (of or relating to the intestines) pathogens. With regard to inhalation exposure, nonspecific findings of weakness, fever, vomiting, cough, hypothermia, and hypotension in large numbers of patients might suggest several respiratory pathogens. The temporal onset of botulinum intoxication would be similar, but include ptosis and general muscular paralysis with minimal pulmonary effects. Staphylococcal enterotoxin B (SEB) intoxication would likely have a more rapid onset after exposure and a lower mortality rate but could be difficult to distinguish. Nerve agent intoxication is characterized by acute onset of cholinergic crisis with dyspnea (shortness of breath, a subjective difficulty or distress in breathing, usually associated with disease of the heart or lungs occurring normally during intense physical exertion or at high altitude) and profuse secretions.

Persistency: Stable.

Personal Protection: Protective clothing must be used as well as protection for the respiratory tract by using a mask with biological filters, or self-contained breathing apparatus (SCBA) with positive pressure, at least until you know the specific threat(s).

Route of Entry to the Body: Inhalation, Mouth, Sabotage of water supplies and foodstuffs.

Person-To-Person Transmissible: No

Duration of Illness: Days (death within ten to twelve days for ingestion). No specific treatment exists.

Potential Ability to Kill: High.

Defensive Measures: Good personal hygiene, physical conditioning, wearing protective mask, and practicing good sanitation.

Vaccines: For treatment, management is supportive. There is currently no vaccine or prophylactic antitoxin available for human use. The use of a protective mask is currently the best protection against inhalation if an attack is considered likely.

Drugs Available: (No specific anti-toxin) Active immunization and passive antibody prophylaxis are under study, as both are effective in protecting animals from death following exposure by intravenous or respiratory routes. Ricin is not dermally active; therefore, respiratory protection is the most critical means of prevention.

Decontamination: Soap and water, or diluted sodium hypochlorite solution (0.1 percent). Since ricin is not volatile, secondary aerosols are usually not a danger to medical personnel.

Characteristics: Ricin is a glycoprotein toxin from the seed of the castor plant that blocks protein synthesis by altering the rRNA, thus killing the cell. The attractiveness of ricin as a biological agent or weapon hinges on its availability worldwide, its ease of production, and extreme pulmonary toxicity when inhaled. More than one million tons of castor beans are harvested around the world to make castor oil which ends up in all sorts of products ranging from medicines, lubricants, paints to dyes, and other items of interest to the industrial world. Large quantities of waste mash from ricin, which contains about 5 percent ricin by weight, are readily available many places in the world. Ricin itself is a potent inhibitor of DNA (the molecular basis of heredity) replication and protein synthesis. The toxin can be transmitted through contaminated food and water or as a biological weapon aerosol. The primary threat is thought to be dissemination by aerosol release. However, a large quantity would be required to significantly cover even a square mile since ricin is less toxic than botulism. Heretofore, ricin has been used more as an assassination weapon rather than a tactical weapon. It should be understood that naturally-occurring cases of ricin intoxication involve ingestion of castor beans, and are marked by severe gastrointestinal symptoms, vascular collapse, and death. Ricin is toxic by numerous exposure routes.

Response on Scene by First Responders

Caution: Ricin is extremely toxic to cells and acts by inhibiting protein synthesis. After aerosol exposure, signs and symptoms would depend on the dose inhaled. Humans can be expected to develop severe lung inflammation with progressive cough, dyspnea, cyanosis, and pulmonary edema.

Field First Aid: No specific treatment exists for ricin intoxication; therefore, care is supportive but should include treatment for pulmonary edema, use of supplemental oxygen, endotracheal intubation and mechanical ventilation, positive end-expiratory pressure, and hemodynamic monitoring that may be required for respiratory disease. Gastric lavage and cathartics are indicated for ingestion, but activated charcoal is of little value for large molecules such as ricin. Failure to respond to antibiotics helps to differentiate ricin exposure from pulmonary infections produced by bacterial agents. The diagnosis of ricin intoxication is largely clinical and should be suspected in a setting of mass casualties with similar and appropriate clinical picture. In regard to decontamination, use soap and water; ricin can be inactivated with a 0.1 percent solution of hypochlorite. Since ricin is not dermally active and is not volatile, decontamination of victims immediately following exposure should be attempted but may not be as critical as with certain chemical agents. In general, ricin does not pose a risk of secondary aerosolization. At this time, there is no vaccine or prophylactic antitoxin available for human use. Use of a protective mask is currently the best protection against inhalation. Healthcare workers should apply Standard Precautions toward victims. A toxin such as ricin is non-volatile, and secondary aerosols are not expected to be a danger to health care providers. All-in-all, the prognosis for victims of ricin intoxication depends on the route and intensity of exposure.

Medical Management: Pre-exposure prophylaxis and post-exposure prophylaxis are not available for ricin at the current time, but a small number of candidate vaccines are presently under development. Serum and respiratory secretions should be submitted for antigen detection through Enzyme-Linked Immunosorbant Assay (ELISA). Paired acute and convalescent sera for antibody studies could be submitted by survivors. Nonspecific laboratory and radiographic findings may include neutrophilic leukocytosis, bilateral interstitial infiltrates compatible with noncardiogenic pulmonary edema. Differential diagnosis of respiratory disease would include phosgene exposure, SEB (staphylococcal enterotoxin B), Hantavirus pulmonary syndrome, atypical pneumonias including Q fever and tularemia, and diverse causes for adult respiratory disease (ARDS).

Fire: For an infectious disease or weapon, use available methods and equipment on surrounding fires. Appropriate extinguishing agents: dry chemical, soda ash, sand, or lime. Do not use high pressure water streams

Personal Protective Equipment: At all times wear protective Level A until you know what you are dealing with; upon learning the exact type of infectious material(s) you are dealing with, consult proper toxicolical levels and downgrade as necessary for safety and comfort of operation.

Spill/Leak Disposal: Outright cleanup or disposal of Infectious Substances (biological agents or weapons) should *only* be handled by trained and experienced personnel. To contact such help, call any or all of the following numbers.

1. SBCCOM Operations — The U.S. Army Soldier and Biological Chemical Command, Edgewood Chemical Biological Center (ECBC), ATTN: AMSSB-RCB-RS, Aberdeen Proving Ground, MD 21010-5424. Tel. 410-436-2148 (twenty-four-hour emergency number).

2. U.S. Army Medical Research Institute if Infectious Diseases, Fort Detrick, Frederick, MD 21702-5011. Tel. 1-888-872-7443.

3. Center for Disease Control and Prevention Emergency Response Line. 1-770-488-7100.

4. ATSDR — Agency for Toxic Substances and Disease Registry, Division of Toxicology, 1600 Clifton Road NE, Mailstop E-29, Atlanta, GA 30333. Tel. 1-888-422-8737. Fax: 404-498-0057.

5. USACHPPM — U.S. Army Center for Health Promotion and Preventive Medicine, Aberdeen Proving Grounds, MD 21010-5422. Tel. 410-671-2208. Email: kwilliam@ aehal.apgea.army.mil

6. National Response Center (for chem/bio hazards & terrorist events); Tel. 1-800-424-8802 or 1-202-267-2675.

7. U.S. Public Health Service: 800-USA-NDMS.

8. National Domestic Preparedness Office (for civilian use). Tel. 202-324-9025.

Symptoms: Symptoms include acute onset of fever, chest tightness, cough, dyspnea, nausea, and arthralgias which occur four to eight hours after inhalational exposure. Airway necrosis and pulmonary capillary leak resulting in pulmonary edema would likely occur within eighteen to twenty-four hours, followed by severe respiratory distress and death from hypoxemia in thirty-six to seventy-two hours.

Vaccines: At this time, there is no vaccine or prophylactic antitoxin available for use.

Guides for Emergency Response: Biological Agent or Weapon: Staphylococcal Enterotoxin (SEB)

12

AGENT: Staphylococcal Enterotoxin B (SEB): Staphylococcal Enterotoxin B is one of several exotoxins produced by Staphylococcus aureu that causes food poison when ingested. **A biological warfare attack using aerosol delivery of SEB to the respiratory tract produces a distinct syndrome causing significant morbidity and potential mortality.** SEB is a common contributor to staphylococcal food poisoning and could be used by terrorists as an aerosol. **However, since inhalation disease is not a natural phenomena for SEB, intentional aerosolization could be easily recognized.** Its action is rapid but not too serious. Inhalation following aerolization could cause up to 80 percent of those exposed to become ill.

Medical Classification: Toxin.

Probable Form of Dissemination: Aerosol, Sabotage of the food supply.

Detection in the Field: None

Infective Dose (Aerosol): 30 nanograms/person (incapacitating). 1 .7 micrograms/person (lethal).

Signs and Symptoms: From three to six hours after aerosol exposure, sudden onset of chills, fever, headache, pain in one of more muscles, and nonproductive cough. Some patients may develop shortness of breath and retrostenal (situated or occurring behind the sternum) chest pain. Fever may last two to five days, and cough may persist for up to four weeks. Patients may also experience nausea, vomiting, and diarrhea if they swallow toxin. Higher exposure levels can lead to septic shock and death.

Incubation Time: One to six hours

Diagnosis: Diagnosis is clinical. Patients present with a febrile respiratory syndrome without CXR abnormalities. Large numbers of military personnel presenting typical symptoms and signs of staphylococcal enterotoxin B (SEB) pulmonary exposure would suggest an intentional attack with this toxin. Treatment is limited to supportive care. Artificial ventilation might be needed for very severe cases, and attention to fluid management is important.

Differential Diagnosis: Regarding foodborne SEB intoxication, fever and respiratory involvement are not seen, and gastrointestinal symptoms are prominent. The nonspecific findings of fever, nonproductive cough, myalgia (muscular pain), and headache occurring in large numbers of persons in an epidemic setting would suggest any of several infectious respiratory pathogens such as influenza, adenovirus, or mycoplasma. In a biological warfare attack with SEB, cases would likely have their onset within a single day, while naturally occurring outbreaks would make a presentation over longer periods of time. However, naturally occurring outbreaks of Q fever and/or tularemia might cause confusion, but would involve much smaller numbers of individuals, and would more likely be accompanied by pulmonary infiltrates.

Persistency: Stable (resistant to freezing).

Personal Protection: Protective clothing must be used as well as protection for the respiratory tract by using a mask with biological filters, or self-contained breathing apparatus (SCBA) with positive pressure, at least until you know the specific threat(s).

Routes of Entry to the Body: Inhalation, mouth.

Person-to-Person Transmissable: No.

Duration of Illness: Hours, or days to weeks. Treatment is mainly limited to supportive care, but assisted ventilation may be necessary in serious cases, and fluid management is necessary. No antitoxin is available, and antibiotics provide no benefit.

Potential Ability to Kill: Low (less than 1 percent).

Defensive Measures: Good personal hygiene, physical conditioning, wearing protective mask, and practicing good sanitation.

Vaccines: There is currently no human vaccine available to prevent SEB intoxication. One has been under development since 1996. Be certain to wear a protective mask when dealing with patients.

Drugs Available: No specific anti-toxin.

Decontamination: Soap and water, or diluted sodium hypochlorite solution (0.5 percent) for ten-15 minutes. Destroy food that may have become contaminated.

Characteristics: Staphylococcal enterotoxin B (usually called SEB) is an **incapacitating toxin** produced by the bacterium Staphylococcus aureus that is responsible for the fever, chills, and gastrointestinal upsets of so called "food poisoning" from ingestion of improperly prepared food items. **The weaponized form is an aerosol that is a potent incapacitator in small doses that could render up to 80 percent of exposed personnel clinically ill for approximately two weeks.** In normal times, the reservoir of this toxin is human beings, especially food handlers with abscesses, acne eruptions, nasal discharges, and even from normal appearing skin at times; as well as contaminated milk or milk products. **It is believed that terrorists could primarily use an aerosol for dissemination, or might even employ sabotage by contamination of food and/or water supplies.**

Response on Scene by First Responders

Caution: A toxin is a poison produced by a living organism. The middle term of staphylococcal enterotoxin B means a toxin that is produced by microorganisms, such as some staphylococci, and causes gastrointestinal symptoms.

Field First Aid: SEB is considered to be an incapacitating agent where most victims should make a full recovery, although they will be unable to work for a week or two. However, victims who suffer pulmonary edema and respiratory failure may die. At the present time, there is no pre-exposure or post-exposure prophylaxis for staphylococcal entero-toxin B (SEB). Medical care is primarily supportive care; symptomatic relief may be provided by the use of acetaminophen and cough suppressants. There will be some severe cases which will require intensive care such as respiratory support (artificial ventilation may be required in more severe victims), hemodynamic monitoring, and probably diuretics, vasopressors (causing or characterized by vasomotor depression resulting in lowering of the blood pressure), and assistance with fluid management. Antibiotic medicines seems to provide no benefits, and at the present time there is no vaccine for humans. Extensive environmental contamination for any lengthy amount of time due to aerosol of SEB is not foreseen.

Drugs: Acetaminophen. Drug Classes = antipyretic (reduces fever by acting directly on the hypothalamic heat-regulation center to cause vasodilation and sweating, which helps dissipates heat), and analgesic-non-narcotic (site and mechanism of action unclear).

Medical Management: Victims may have a feverish respiratory syndrome without chest x-ray abnormalities, and diagnosis is usually clinical. Medical and emergency medical services personnel should be aware of and should report any number of victims showing up within a limited amount of time presenting typical symptoms and instances of SEB pulmonary exposure as being indicative of being an intentional attack with SEB toxin.

Fire: For an infectious disease or weapon, use available methods and equipment on surrounding fires. Appropriate extinguishing agents: dry chemical, soda ash, sand, or lime. Do not use high pressure water streams.

Personal Protective Equipment: At all times wear protective Level A until you know what you are dealing with; upon learning the exact type of infectious material(s) you are dealing with, consult proper toxicolical levels and downgrade as necessary for safety and comfort of operation.

Spill/Leak Disposal: Outright cleanup or disposal of Infectious Substances (biological agents or weapons) should only be handled by trained and experienced personnel. To contact such help, call any or all of the following numbers.
  1. SBCCOM Operations — The U.S. Army Soldier and Biological Chemical Command, Edgewood Chemical Biological Center (ECBC), ATTN: AMSSB-RCB-RS, Aberdeen Proving Ground, MD 21010-5424. Tel. 410-436-2148 (twenty-four-hour emergency number).
  2. U.S. Army Medical Research Institute if Infectious Diseases, Fort Detrick, Frederick, MD 21702-5011. Tel. 1-888-872-7443.
  3. Center for Disease Control and Prevention Emergency Response Line. 1-770-488-7100.
  4. ATSDR — Agency for Toxic Substances and Disease Registry, Division of Toxicology, 1600 Clifton Road NE, Mailstop E-29, Atlanta, GA 30333. Tel. 1-888-422-8737. Fax: 404-498-0057.
  5. USACHPPM — U.S. Army Center for Health Promotion and Preventive Medicine, Aberdeen Proving Grounds, MD 21010-5422. Tel. 410-671-2208. Email: kwilliam@aehal.apgea.army.mil
  6. National Response Center (for chem/bio hazards & terrorist events). Tel. 1-800-424-8802 or 1-202-267-2675.

7. U.S. Public Health Service: 800-USA-NDMS.

8. National Domestic Preparedness Office (for civilian use). Tel. 202-324-9025.

Symptoms: SEB food poisoning introduces the acute onset of fever, nausea, vomiting, and diarrhea within hours of intoxication. The toxin increases intestinal peristalsis; severe nausea and vomiting may be due to stimulation of the central nervous system (CNS). The staphylococcal enterotoxins belong to a class of potent immune stimulants known as bacterial superantigens. Superantigens bind to monocytes at major histcompatibility complex type II receptors rather than the normal antigen binding receptors. This leads to the direct stimulation of a large population of T-helper cells while bypassing the usual antigen processing and presentation. This results in a brisk cascade of pro-inflammatory cytokines and recruitment of other immune effector cells, and a relatively deficient activation of counter-regulatory negative feedback loops. The end result is an intense inflammatory response that can injure host tissues. However, illness due to inhalation, the expected form of dissemination to be used in a terrorist attack, will result in respiratory tract disease not encountered in the endemic disease due to the activation of pro-inflammatory cytochylema cascades in the lungs, leading to pulmonary capillary leak and pulmonary edema. Symptoms include fever, headache, myalgia, nonproductive cough, and in cases that are more severe, dyspnea (difficult or labored respiration). Gastrointestinal symptoms may also appear due to inadvertent swallowing of SEB delivered via aerosol and deposited in the upper aero-digestive tract, or coughed following mucociliary clearance. More severe victims could face pulmonary edema and respiratory failure.

Vaccines: No human vaccine available.

Guides for Emergency Response: Biological Agent or Weapon: Smallpox (Variola) **13**

AGENT: Smallpox virus, an orthopoxvirus with a narrow host range confined to humans, was an important cause of morbidity and mortality in the developing world until recent times. Eradication of the natural disease was completed in 1977 and the last human cases from laboratory infections occurred in 1978. Appearances of human cases outside the laboratory would signal use of the virus as a biological weapon. Under normal conditions, the virus is transmitted by direct (face-to-face) contact with an infected case by formites (objects, such as clothing, towels, and utensils that possibly harbor a disease agent and are capable of transmitting it), and occasionally by aerosols. Smallpox virus is highly stable and retains infectivity for long periods of time outside the host. A related virus, monkeypox, clinically resembles smallpox and causes sporadic human disease in West and Central Africa. Smallpox is a highly contagious virus with fever and a blister-like rash. Smallpox can survive for centuries, and antibiological drugs are not able to handle viruses. It is caused by two species of pox-virus, variola major or variola minor; the disease is only carried by humans. Since vaccination with vaccinia throughout the world, not one case of natural smallpox has occurred. Only two samples of this virus are known to still exist in the world; one at the Center for Disease Control and Prevention in Atlanta, Georgia, and one at the Vector laboratory in Novizbersk in the former Russia. Smallpox has a death rate of about 30 percent for unvaccinated persons and three percent for those who have been vaccinated. Since smallpox has been exterminated in the world, civilians and military personnel have not been vaccinated for smallpox since the 1980s. You are unprotected if your vaccination is three or more years old. If you are old enough, you probably remember the smallpox vaccine scarification method; paint the vaccine on your arm and then get jabbed sixteen times with a sharp needle. Pregnant woman, and carriers of HIV, or other immunity diseases should not be vaccinated for smallpox. As the skin becomes pock-marked and sloughs off, smallpox can easily be confused with chicken pox although lesions are a smooth, orderly progression in contrast to chicken pox.

Medical Classification: Virus.

Probable Form of Dissemination: Aerosol.

Detection in the Field: None.

Infective Dose: (Aerosol): 10–100 organisms (assumed low).

Signs and Symptoms: Clinical manifestations begin acutely with malaise, fever, rigors, vomiting, headache, and backache. Two to three days later lesions appear which quickly progress from macules (a small, discolored patch or spot on the skin, neither elevated above nor depressed below the skin's surface) to papules (a small, circumscribed, solid elevation on the skin), and eventually to pustular vesicles. They are more abundant on the extremities and face, and develop synchronously.

Incubation Time: Ten to seventeen days (average equals twelve days).

Diagnosis: Electron and light microscopy are not capable of discriminating variola (a synonym for smallpox) from vaccinia (an infection, primarily local and limited to the site of inoculation induced in man by inoculation with the vaccinia, cowpox, virus in order to confer resistance to smallpox, variola. On about the third day after vaccination, papules form at the site of inoculation which become transformed into umbilicated vesicle and later pustules; they then dry up, and the scab falls of on about the twenty-first day, leaving a pittted scar; in some cases there are more or less marked constitutional disturbances, monkey-pox or cowpox. The new PCR (polymerase chain reaction) diagnostic techniques is an in vitro method for enzymatically synthesizing and amplifying defined sequences of DNA in molecular biology. This method can be used for improving DNA-based diagnostic procedures for identifying unknown biowarfare agents) diagnostic techniques may be more accurate in discriminating between variola and other Orthopoxviruses. As far as treatment is concerned, at present there is no effective chemotherapy, and treatment of a clinical case remains supportive. Patients with smallpox should be treated by vaccinated medical personnel using universal precautions. Objects in contact with the patient, including bed linens, clothing, ambulance and the like require disinfection by fire, steam, or sodium hypochlorite solution.

Differential Diagnosis: The eruption of chickenpox (varicella) is typically centripetal (passing inward, as from a sense organ to the brain or spinal column) in distribution (worse on trunk than face and extremities) and characterized by crops of lesions in different stages of development. Chickenpox papules (small, circumscribed, solid elevation on the skin) are soft and superficial, compared to the firm, shotty, and deep papules of smallpox. Chickenpox crusts fall off rapidly and usually leave no scar. Monkeypox cannot be easily distinguished from smallpox clinically, although generalized lymphadenopathy (abnormal enlargement of the lymph nodes) is a more common feature of the disease. Monkeypox occurs only in forested areas of West and Central Africa as a sporadic, zoonotic infection transmitted to humans from wild squirrels. Person-to-person spread is rare and ceases after one to two generations. Mortality is 15 percent. Some other diseases are confused with smallpox, including typhus, secondary syphilis, and malignant measles.

Persistency: Very stable.

Personal Protection: Protective clothing must be used as well as protection for the respiratory tract by using a mask with biological filters or self-contained breathing apparatus (SCBA) with positive pressure, at least until you know the specific threat(s).

Routes of Entry to the Body: Inhalation by direct, face-to-face contact with an infected case, by fomites (objects, such as clothing, towels, and utensils that possibly harbor a disease agent and are capable of transmitting it), and by aerosols.

Person-To-Person Transmissible: High.

Duration of Illness: One to two weeks. Smallpox therapy is mainly supportive care, and no specific antiviral therapy exists.

Potential Ability to Kill: High (20–40 percent for Variole major, and less than 1 percent for Variole minor).

Defensive Measures: Immunizations, good personal hygiene, physical conditioning, wearing protective mask, and practicing good sanitation.

Vaccines: Yes. Wyeth Vaccine, called VIG or vaccinia immune globulin, one dose by scarification. (Pre- and post-exposure vaccination recommended if greater than three years passed since last vaccine.) Persons who are pregnant, have clinical immunosuppression (such as persons who have undergone heart and other transplant operations and take medicines to control rejection of their new, transplanted organ), eczema, and/or leukemia/lymphoma would be contraindicated for such vaccine. As one example, the fatality rate for immunosuppressed persons taking such vaccine could be as high as 75 percent or more.

Drugs Available: Vaccinia Immune Globulin (VIG). The U.S. Army maintains a supply of VIG, and as of November of 2001 the U.S. government is attempting to buy VIG in large amounts. Cidofovir, another prescription drug, is effective in vitro.

Decontamination: Soap and water, or diluted sodium hypochlorite solution (0.5 percent). Smallpox has great potential for person-to-person exposure. Removal of potentially contaminated clothing should be done by people in full protective clothing in an area away from non-contaminated persons. All infectious cases should be quarantined for seventeen days following exposure for all contacts, and strict isolation would apply to any victims. All material used to treat victims or coming in contact with victims should be autoclaved, boiled or burned. Patients should be considered infectious until all scabs separate.

Ability to Kill: High.

Characteristics: Smallpox (variola) is an infection by a virus called variola which occurs in at least two strains, variola major and the milder disease, variola minor. Viruses are organisms which require living cells in which to replicate. They are therefore intimately dependent upon the cells of the host that they infect. They produce diseases that generally do not respond to antibiotics but which may be responsive to antiviral compounds, of which there are few available, and those that are available are of limited use. In spite of the global eradication of smallpox and availability of a vaccine, the potential of smallpox being used as a weapon in the age of terrorism continues to pose a serious threat to the entire world. The reason for such a threat is quite simple; smallpox has great aerosol infectivity, and it can be easily manufactured on a large scale. It is highly transmissible from person-to-person, its infective dose is assumed to be low, and it has an incubation period of from ten to twelve days. It is extremely contagious, particularly in modern society where air travel is so often used by people all over the world. This infective disease lasts about four weeks, lethality is high to moderate, and is very stable and therefore persistent.

Response on Scene by First Responders

Caution: No specific antiviral therapy exists for smallpox, so supportive care remains the mainstay for smallpox therapy. Victims who do acquire this infective disease will probably be quarantined, or isolated, for at least two weeks following exposure. Animals do not catch smallpox, but humans do; that is, smallpox has a narrow host range confined strictly to human beings. Smallpox virus is highly stable and retains infectivity for long periods outside the host. Victims with smallpox should be treated by vaccinated personnel using universal precautions. Anything in contact with a person having smallpox, including bed linens, clothing, ambulance, equipment or the like will require disinfection by fire, steam, or sodium hypochlorite solution.

Field First Aid: Monitor the incident scene, and victims, to attempt to identify the chemical or biological agent that is creating the problem. Separate the walking wounded from the prone (e.g., "All those who can walk, stand over there"). Assess the number of prone and walking wounded. If Emergency Medical Service personnel are not on the scene, call them. If it appears to be a mass casualty incident, call for help immediately. As soon as a person trained in triage arrives on scene (physician, physicians assistant, nurse, paramedic or EMS person) have triage instituted.

There is no proven treatment for smallpox, but in persons exposed to smallpox who do not show symptoms as yet, the vaccine — if given within four days after exposure — can lessen the severity of or even prevent illness. However, once a patient shows symptom, treatment is limited to supportive therapy and antibiotics to treat bacterial infections. Patients with smallpox can benefit from supportive therapy such as intravenous fluids, and medicines to control fever or pain.

Drugs: The drugs available to treat smallpox victims include vaccines and an antiviral drug.

Vaccinia Virus: This is a live poxvirus vaccine that induces strong cross-protection against smallpox for at least five years and partial protection for ten years of more. The vaccine is administered by dermal scarification or intradermal jet injection. The appearance of a vesicle or pustule within a few days is an indication that the vaccine has been effective. Complication are infrequent but include: 1. Progressive vaccinia in immunosuppressed persons with a case fatality of more than 75 percent; 2. Eczema vaccinatum in persons with eczema or a history of eczema, or in contact with eczema with a case fatality of ten to 15 percent; 3. Postvaccinal encephalitus, almost exclusively seen after primary vaccination, occurring at an incidence of about one in 500,000, with a case fatality rate of 25 percent; 4. Generalized vaccinia, seen in immunocompetent individuals and having a good prognosis; and 5. Autoinnoculation of the eye or genital area, with a secondary lesion.

Vaccinia-immune human globulin at a dose of 0.3 mg/kg body weight provides 79 percent more or less protection against naturally occurring smallpox if given during the early incubation period. Administration immediately after or within the first twenty-four hours of exposure would provide the highest level of protection, especially in unvaccinated persons. If vaccinia-immune globulin is unavailable, vaccination or revaccination should be performed as early as possible after, and within twenty-four hours of exposure, with careful surveillance for signs of illness.

The anti-viral drug, n-methylisatin B-thiosemicarbazone, commonly called by its registered trade name, Marboran, afforded protection in some early trials, but not others, possibly because of noncompliance due to unpleasant gastrointestinal side effects Drugs under Consideration for Smallpox: Cidofovir (Vistide) is an antiviral drug that is being considered to inhibit poxvirus replication and cell disintegration. The therapeutic actions include antiviral activity in which the drug selectively inhibits CMV (cytomegalovirus) replication of viral DNA synthesis. Also under investigation for smallpox, are adefovir and ribavirin.

Medical Management: Any confirmed case of smallpox should be considered an international emergency with an immediate report made to public health authorities. Droplet and airborne protection for a minimum of seventeen days following exposure for *all* persons in direct contact with the beginning case should be ensured, particularly for the unvaccinated persons. In a civilian setting, strict quarantine of asymptomic contacts may prove to be impractical and impossible to enforce. A reasonable alternative would be to require contacts to check their temperatures daily. Any fever above 101 F (or 83 C) during a seventeen-day period following exposure to a confirmed case would suggest the development of smallpox. If such a temperature develops, the contact should then be isolated immediately, preferably at home, until smallpox is either confirmed or ruled out; and remain in isolation until all scabs separate. Patients should be considered infectious until all scabs separate! Immediate vaccination or revaccination should also be undertaken for all personnel exposed to either weaponized variola virus or a clinical case of smallpox. If an actual quarantine were decried by government sources, this would be required of all emergency medical services personnel who may have treated or transported known or suspected persons having smallpox.

Decontamination plays a very important role in the approach to chemical casualty management. However, the incubation period of biological agents makes it unlikely that victims of a biological agent attack will even know he or she has been contaminated until days after an attack. At this point, the need for decontamination is minimal or non-existent. In the rare cases where decontamination is warranted, simple soap and water bathing will generally suffice. Certainly, standard decontamination solutions, such as hypochlorite, typically employed in cases of chemical agent contamination, would be effective against all biological agents. In fact, even 0.1 percent bleach reliably kills anthrax spores, the most stable of biological agents. However, routine use of caustic substances, particularly on human skin, is rarely warranted following a biological attack.

Fire: (Not normally applicable with an infectious disease) However, surrounding fire can be fought with dry chemical, soda ash, lime, or sand.

Personal Protective Equipment: At all times wear protective Level A until you know what you are dealing with; upon learning the exact type of infectious material(s) you are dealing with, consult proper toxicological levels and downgrade as necessary for safety and comfort of operation.

Spill/Leak Disposal: Outright cleanup or disposal of Infectious Substances (biological agents or weapons) should ONLY be handled by trained and experienced personnel. To contact such help, call any or all of the following numbers.

1. SBCCOM Operations — The U.S. Army Soldier and Biological Chemical Command, Edgewood Chemical Biological Center (ECBC), ATTN: AMSSB-RCB-RS, Aberdeen Proving Ground, MD 21010-5424. Tel. 410-436-2148 (twenty-four-hour emergency number).
2. U.S. Army Medical Research Institute if Infectious Diseases, Fort Detrick, Frederick, MD 21702-5011. Tel. 1-888-872-7443.
3. Center for Disease Control and Prevention Emergency Response Line. 1-770-488-7100.
4. ATSDR — Agency for Toxic Substances and Disease Registry, Division of Toxicology, 1600 Clifton Road NE, Mailstop E-29, Atlanta, GA 30333. Tel. 1-888-422-8737. Fax: 404-498-0057.
5. USACHPPM — U.S. Army Center for Health Promotion and Preventive Medicine, Aberdeen Proving Grounds, MD 21010-5422. Tel. 410-671-2208. Email: kwilliam@ aehal.apgea.army.mil

6. National Response Center (for chem/bio hazards & terrorist events). Tel. 1-800-424-8802 or 1-202-267-2675.
7. U.S. Public Health Service: 800-USA-NDMS.
8. National Domestic Preparedness Office (for civilian use). Tel. 202-324-9025.

Symptoms: Malaise, rigors, vomiting, fever, headache, and backache, followed in two to three days by lesions that speedily become macules (an anatomical structure having a form of a spot differential from the surrounding area tissues), progress to papules (a small solid, usually conical, elevation of the skin caused by inflation), and end up as pustular vesicles.

Vaccines: Vaccinia virus is a live poxvirus that can lead to strong cross-protection to smallpox for about five years, plus partial protection for ten or more years. If your are old enough, or have been in military service in recent years, you may have been vaccinated with this vaccine administered by dermal scarification, or intra-dermal jet injection. However, certain persons should NOT receive smallpox vaccine, including persons who are pregnant, persons who underwent a clinical immunosuppresion, persons with eczema, or persons with leukemia or lymphoma.

Guides for Emergency Response: Biological Agent or Weapon: Mycotoxins/T-2

14

AGENT: Trichothecene mycotoxins are a large group of toxins produced by several species of fungi. T-2 is one of the most stable of these toxins and therefore the most likely to be used in terrorist actions. This toxin was allegedly used as an aerosol, popularly known as "yellow rain," in Laos (1975–1981), Kampuchea (1979–1981), and Afghanistan (1979–1981). The trichothecene mycotoxins belong to a group of more than forty compounds produced by fungi. They tend to be strong inhibitors of protein synthesis, interfere with DNA synthesis, limit mitochondrial respiration, and change cell membrane structure and function. Secondary metabolites of fungi, such as T-2 toxin, as well as others, produce toxic reactions called mycotoxicoses upon inhalation or consumption of contaminated food products by humans or animals. Naturally occurring trichothecenes have been identified in agricultural products and have been implicated in a disease of animals known as moldy corn poisoning.

Medical Classification: Toxin

Probable Form of Dissemination: Aerosol; sabotage

Detection in the Field: None

Infective Dose: Aerosol

Signs and Symptoms: Exposure causes skin pain, pruritus (localized or generalized itching due to irritation of sensory nerve endings from organic or psychological causes), redness, vesicles, necrosis (pathological death of one or more cells, or of a portion of tissue or organ, resulting from irreversible damage) and sloughing of epidermis. Effects on the airway include nose and throat pain, nasal discharge, itching and sneezing, cough, dyspnea, wheezing, chest pain, and hemoptysis (the spitting of blood derived from the lungs or bronchial tubes as a result of pulmonary or bronchial hemorrhage). Toxin also produces effects after ingestion or eye contact. Severe poisoning may result in prostration, weakness, ataxia (inability to coordinate muscle activity during voluntary movement, so that smooth movements occur), collapse, shock, and death.

Incubation Time: Two to four hours

Diagnosis: Be suspicious if an aerosol attack occurs in the form of "yellow rain" with droplets of yellow fluid contaminating clothing and the environment. Confirmation requires testing of blood, tissue and environmental samples. As for treatment, there is no specific antidote; however, super-activated charcoal should be given orally if the toxin is swallowed. The only defense is to wear a protective mask and clothing during an attack. No specific immunotherapy or chemotherapy is available for use in the field.

Differential Diagnosis: Other diagnoses to consider include radiation toxicity, and plant or chemical toxicity.

Persistency: Stable (for years at room temperature).

Personal Protection: Protective clothing should be used as well as protection for the respiratory tract by using a mask with biological filters, or self-contained breathing apparatus (SCBA) with positive pressure, at least until you know the specific threats.

Route of Entry to the Body: Inhalation, mouth, and skin

Person-To-Person Transmissable: No

Duration of Illness: Days to Months. Therapy is mainly supportive.

Defensive Measures: Good personal hygiene, physical conditioning, wearing protective mask, and practicing good sanitation. Mycotoxin-induced disease is not contagious, but the stability of this toxin in the environment is quite persistent.

Vaccines: None (vaccines and chemoprotective pretreatment are being studied in animal models, but are not available for field use).

Drugs Available: (No specific anti-toxin)

Decontamination: Soap and water, or diluted sodium hypochlorite solution (0.5 percent). Clothing of T-2 victims should be removed and exposed to a 5 percent solution of hypochlorite for six to ten hours, or destroyed. Skin may be cleaned with soap and water, and eye exposure should undergo saline irrigation. Regular disinfectants useful against most other biological agents are often inadequate against the very stable mycotoxins. After decontamination, isolation in not required. Ensure standard precautions for healthcare personnel.

Characteristics: Trichothecene mycotoxins are a group of more than forty compounds produced by fungi. They act to inhibit protein synthesis, impair DNA synthesis, alter cell membrane structure and function, and inhibit mitochondrial respiration. Secondary metabolites of fungi, like T-2 toxin and others produce toxic reactions called mycotoxicoses upon inhalation or consumption of contaminated food. T-2 toxins are not very soluble in water, but they tend to be soluble in organic solvents; they are very stable and persistent. In those purported incidents where T-2 was reportedly used as a biological warfare agent, victims had acute exposure via inhalation and/or dermal routes as well as oral exposure from consuming contaminated food products and water. Symptoms reported during those reported incidents included painful skin lesions, lightheadedness, dyspnea, and rapid onset of hemorrhage, incapacitation and death. People who lived through such incidents suffered a radiation-like illness, including fever, nausea, vomiting, diarrhea, leukopenia, bleeding, and sepsis. The mode of action with T-2 mycotoxins is unknown. What is known is the toxin inhibits protein synthesis, and effects clotting factors in the blood leading to hemorrhage. The most pronounced effects occur in rapidly dividing cells (blood and bone marrow cells).

Response on Scene by First Responders

Caution: The T-2 mycotoxins are the only potential biological warfare agents that can harm and be absorbed through intact skin. Aerosol doses of T-2 toxins may be ten times more potent than parenteral doses.

Field First Aid: Victims require general supportive care. No antidote or antitoxin is available at this time. For ingested toxin, repeated doses of oral charcoal may be helpful. For pre-exposure propylaxis, first responders could use a topical antivesicant cream or ointment that may provide limited protection on skin surfaces. There is no post-exposure for T-2 toxins at the present time. Mycotoxins are highly cytotoxic (toxic to cells), tend to mimic radioactive illness, and have effects similar to vesicants, especially with respect to mustard agents. The skin offers a ready entryway to T-2, and respiratory exposure may result in nasal itching with pain, rhinorrhea, sneezing, epistaxis (a nosebleed), dyspnea, wheezing and cough. Exposure in the conjunctive and other mucosal surfaces may result in local burning pain and redness, followed by necrosis (death of living tissue). Systemic absorption may follow delivery by any route and result in late toxicity of decreased blood cell counts thus predisposing the victim to bleeding and sepsis. Standard Precautions should be used throughout the victim(s) care. Use soap and water for decontamination of victims; even after four to six hours after exposure, decontamination can significantly reduce dermal toxicity; washing within one hour may entirely prevent toxicity. Outer clothing should be removed and exposed skin decontaminated with soap and water. Eye exposure should be treated with saline irrigation. Secondary aerosols are not a problem with T-2 mycotoxins, but contact with contaminated skin and clothing can produce secondary dermal exposures. Contact precautions should be required until decontamination is done. At that time, Standard Precautions are recommended for healthcare workers. Environmental decontamination requires the use of a hypochlorite solution under alkaline conditions (1 percent sodium hypochlorite and 0.1M NaOH with a one hour contact time).

Medical Management: There is no specific antidote or therapeutic regimen at the present time. All therapy used is supportive. If a victim is unprotected during the attack, the outer clothing shall be removed within four hours, and the clothing will be discarded OR decontaminated by exposure or soaking in 5 percent hypochlorite solution for six to ten hours. The victim's skin shall be thoroughly washed with soap and uncontaminated water, if available. This action can reduce dermal toxicity, even if such action is delayed four to six hours after exposure. Standard burn care should be provided for cutaneous damage. Also, standard therapy for poison ingestion, including the use of superactivated charcoal to absorb swallowed T-2, should be administered to victims of an unprotected aerosol attack. Respiratory support could be necessary. The eyes should be irrigated with normal saline or water to remove toxin. Physical protection of the skin, mucous membranes, and airway (use of chemical protective mask and clothing) are the only proven effective methods of protection during a T-2 attack.

Fire: For an infectious disease, or a mycotoxins weapon, use available methods and equipment on surrounding fires. Appropriate extinguishing agents: dry chemical, soda ash, sand, or lime. Do not use high pressure water streams.

Personal Protective Equipment: At all times wear protective **Level A until you know what you are dealing with;** upon learning the exact type of infectious material(s) you are dealing with, consult proper toxicological levels and downgrade as necessary for safety and comfort of operation.

Spill/Leak Disposal: Outright cleanup or disposal of Infectious Substances, or biological agents or weapons including mycotoxins, should *only* be handled by trained and experienced personnel. To contact such help, call any or all of the following numbers.

1. SBCCOM Operations — The U.S. Army Soldier and Biological Chemical Command, Edgewood Chemical Biological Center (ECBC), ATTN: AMSSB-RCB-RS, Aberdeen Proving Ground, MD 21010-5424. Tel. 410-436-2148 (twenty-four-hour emergency number).
2. U.S. Army Medical Research Institute if Infectious Diseases, Fort Detrick, Frederick, MD 21702-5011. Tel. 1-888-872-7443.
3. Center for Disease Control and Prevention Emergency Response Line. 1-770-488-7100.
4. ATSDR — Agency for Toxic Substances and Disease Registry, Division of Toxicology, 1600 Clifton Road NE, Mailstop E-29, Atlanta, GA 30333. Tel. 1-888-422-8737. Fax: 404-498-0057.
5. USACHPPM — U.S. Army Center for Health Promotion and Preventive Medicine, Aberdeen Proving Grounds, MD 21010-5422. Tel. 410-671-2208. Email: kwilliam@aehal.apgea.army.mil
6. National Response Center (for chem/bio hazards & terrorist events). Tel. 1-800-424-8802 or 1-202-267-2675.
7. U.S. Public Health Service: 800-USA-NDMS.
8. National Domestic Preparedness Office (for civilian use). Tel. 202-324-9025.

Symptoms: Low-dose symptoms include nausea, shortness of breath, dizzziness, eye and skin irritation; as well as formation of small, hard blisters; and chest pains. Trichothecene symptoms may resemble those of blister agents, whereas nausea commonly occurs with exposure to trichothecene toxins. These toxins mimic the effects of ionizing radiation. High doses can include bloody vomit or diarrhea and blistering of the skin within hours. Massive hemorrhaging and shock can bring rapid death, or death may be postponed until weeks later. The delayed death would result from bone marrow suppression which lead to anemia and reduction in immunity, liver failure, and/or internal bleeding.

Vaccines: Immunological and chemoprotective pre-treatments are being studied in models, but are not available for first responders as yet.

Guides for Emergency Response: Biological Agent or Weapon: Tularemia 15

AGENT: Tularemia is a zoonotic disease caused by Francisella tularesis, a gram-negative bacillus. Humans acquire the disease under natural conditions through inoculation of skin or mucous membranes with blood or tissue fluids of infected animals, or bites of infected deerflies, mosquitoes, or ticks. Less commonly, inhalation of contaminated dusts or ingestion of contaminated food or water may produce clinical disease. A biological warfare or terrorist attack with tularemia disseminated by aerosol would primarily cause typhoidal turaremia, a syndrome expected to have a case fatality rate that may be higher than the 5 to 10 percent when the disease is acquired naturally. It takes only a very small aerosol dose to make people ill. Any attack with dissemination by aerosol would be likely to spread typhoidal tularemia. Strict isolation of patients is not required.

Medical Classification: Bacteria.

Probable Form of Dissemination: Aerosol

Detection in the Field: Limited.

Infective Dose (Aerosol): 10–50 organisms.

Signs and Symptoms: Ulceroglandular tularemia presents a local ulcer and regional lymphadenopathy (any disease process affecting a lymph node or lymph nodes), fever, chills, headache, and malaise. Typhoidal or septicemic tularemia presents fever, headache, malaise, substernal discomfort, prostration, weight loss and a non-productive cough.

Incubation Time: One to ten days

Diagnosis: Clinical diagnosis. Physical findings are usually non-specific. Chest X-ray may reveal a pneumonic process, mediastinal lymphadenopathy or plural effusion. Routine culture is possible but difficult. The diagnosis can be established retrospectively by serology. As for treatment, administration of antibiotics such as streptomycin or gentamicin with early treatment can be very effective.

Differential Diagnosis: The clinical presentation of tularemia may be severe, yet non-specific. Differential diagnosis includes typhoidal syndromes (e.g., slamonella, rickettsia, malaria) or pneumonic processes (e.g., plague, mycoplasma, SEB). A clue to the diagnosis of tularemia delivered in a biological warfare of terrorist strike might be a large number of temporally clustered patients presenting similar systemic illnesses, a proportion of which will have a nonproductive pneumonia.

Persistency: Not very stable (but can last for months in moist soil or other media).

Personal Protection: Protective clothing must be used as well as protection for the respiratory tract by using a mask with biological filters, or self-contained breathing apparatus (SCBA) with positive pressure, at least until you know the specific threat(s).

Routes of Entry to the Body: Inhalation, mouth, and skin.

Person-To-Person Transmissible: No.

Duration of Illness: Two or more weeks.

Potential Ability to Kill: Moderate if untreated. If left untreated, the death rate from tularemia can reach five to 15 percent. **There would be an expected death rate of 35 percent if survivors were not treated with antibiotics.**

Defensive Measures: Immunizations, good personal hygiene, physical conditioning, use of arthropod repellents, wearing protective mask, and practicing good sanitation.

Vaccines: Yes. LVS - Live Attenuated Vaccine as an investigational new drug (IND) administered once by scarification. A two week course of tetracycline is effective as prophylaxis when given after exposure.

Drugs Available: Streptomycin, Gentamicin, and Doxycycline.

Decontamination: Institute standard precautions for healthcare workers. Organisms are relatively easy to render harmless by mild heat (55 degrees Celcius for ten minutes) and standard disinfectants. Secretion and lesion precautions should be practiced, **although strict isolation of patients is not required.** Soap and water, or diluted sodium hypochlorite solution (0.5 percent). **Secretion and lesion precautions are necessary, but strict isolation of victims is *not* required.**

Characteristics: Tularemia is an illness that can be varied in different forms; including typhoidal, pneumonic, intestinal, oropharyngeal, oculoglandular, glandular, or ulceroglandular. Clinical diagnosis is supported by evidence or history of a tick or deerfly bite, exposure to tissues of a mammalian host of Francisella tulareensis, or exposure to potentially contaminated water. **Naturally acquired tularemia often has a glandular component, although a minority of cases have the systemic or pneumonic forms; but when using tularemia as a biological weapon, this infectious disease would likely lead to systemic and pneumonic cases, with a shortened incubation period.**

Response on Scene by First Responders

Caution: Tularemia has long been a weapon of war used and/or studied by the Japanese, the Russians, and the United States. Chemoprophylaxis is not recommended following potential natural exposures (tick bite, and/or rabbit or other animal exposures). Inhalation tularemia can lead to fulminant pneumonia with a case fatality rate of 30 to 60 percent without treatment.

Field First Aid: The way to cure humans who contact tularemia is the administration of the antibiotics as an early treatment. Streptomycin has been the choice drug for treatment of tularemia for a long time, but it might not be readily available at once after a mass casualty attack therefore four drugs regimens are provided below. Medical or EMS personnel should select *one* of the plans:
- Gentamicin three to 5 mg/kg intravenous daily for ten to furteen days.
- Ciprofloxacin 400 mg intravenous every twelve hours, switch to oral ciprofloxacin (500 mg every twelvehours) after the victim is clinically improved; continue for completion of a ten to fourteen day course of therapy.

- Ciprofloxacin 750 mg orally every twelve hours for ten to fourteen days.
- Streptomycin 7.5–10 mg/kg intramuscular every twelve hours for ten to fourteen days.

The U. S, government is worried that a fully virulent streptomycin-resistant strain for tularemia was produced in the 1950s and other governments may have obtained that same strain. The same strain was sensitive to gentamicin, and gentamicin also offers an advantage of providing broader coverage for gram-negative bacteria and may be useful when the diagnosis of tularemia may be in doubt. **Apply Standard Precautions throughout treatment.**

Medical Management: Following aerosol exposure, an undifferentiated febrile illness called typhoidal tularemia, or an acute pneumonia featuring fever, coughing, substernal chest tightness, and pleuritic chest pain may be present. Usually, coughing is non-productive; hemoptysis is rarely evident. In a biological warfare situation, the primary threat for type of dissemination is considered to be by aerosol, or by contamination of food or water supplies (all food must be thoroughly heated before use, and water must be thoroughly disinfected. With regard **to pre-exposure prophylaxis,** a live attenuated vaccine is available as an IND (Investigational New Drug) that is given by scarification which has been tested on animals and human volunteers. For **post-exposure prophylaxis** following a tularemia biological agent attack, administer doxycycline 100 mg orally every twelve hours for two weeks, or tetracycline 500 mg orally every six hours for two weeks, or ciprofloxacin 500 orally every twelve hours for two weeks.

Serologic testing is the preferred procedure for laboratory testing of tularemia. Confirmation of diagnosis requires a four-fold increase in titer (the strength of a solution or the concentration of a substance in solution as determined by titration); serologies may need to be repeated at seven and ten day intervals. Agglutination tests and ELISA (enzyme-linked immunosorbant assay) are also available. The organism does not grow on standard bacteriologic growth media, but Francicella tularensis can be cultured on special supportive media containing cystine or another sulfhydryl source. **Whenever cultures for tularemia are submitted to a laboratory, staff employed there should be warned that such cultures** *MUST* **be tested and processed at Biosafety Level Three.** Since there is no known human-to-human transmission with tularemia, **neither isolation nor quarantine are required as Standard Precautions are appropriate for care of victims with draining lesions or pneumonia. Strict adherence to the drainage/secretion recommendations of Standard Precautions is demanded for the safety of medical personnel.**

Fire: For an infectious disease or weapon, use available methods and equipment on surrounding fires. Appropriate extinguishing agents: dry chemical, soda ash, sand. or lime. Do not use high pressure water streams

Personal Protective Equipment: At all times wear protective **Level A until you know what you are dealing with;** upon learning the exact type of infectious material(s) you are dealing with, consult proper toxicological levels and downgrade as necessary for safety and comfort of operation.

Spill/Leak Disposal: Outright cleanup or disposal of Infectious Substances (biological agents or weapons) should *only* be handled by trained and experienced personnel. To contact such help, call any or all of the following numbers:

1. SBCCOM Operations — The U.S. Army Soldier and Biological Chemical Command, Edgewood Chemical Biological Center (ECBC), ATTN: AMSSB-RCB-RS, Aberdeen Proving Ground, MD 21010-5424. Tel. 410-436-2148 (twenty-four-hour emergency number).
2. U.S. Army Medical Research Institute if Infectious Diseases, Fort Detrick, Frederick, MD 21702-5011. Tel. 1-888-872-7443.
3. Center for Disease Control and Prevention Emergency Response Line. 1-770-488-7100.
4. ATSDR — Agency for Toxic Substances and Disease Registry, Division of Toxicology, 1600 Clifton Road NE, Mailstop E-29, Atlanta, GA 30333. Tel. 1-888-422-8737. Fax: 404-498-0057.
5. USACHPPM — U.S. Army Center for Health Promotion and Preventive Medicine, Aberdeen Proving Grounds, MD 21010-5422. Tel. 410-671-2208. Email: kwilliam@aehal.apgea.army.mil
6. National Response Center (for chem/bio hazards & terrorist events). Tel. 1-800-424-8802 or 1-202-267-2675.
7. U.S. Public Health Service: 800-USA-NDMS.
8. National Domestic Preparedness Office (for civilian use). Tel. 202-324-9025.

Symptoms: Ulceroglandular disease involves a necrotic , tender ulcer at the site of inoculation, along with tender, enlarged regional lymph nodes. Fever, chills, headache, and malaise often accompany such symptoms. Systemic and pneumonic forms usually involve serious unproductive cough, abdominal pain, generalized muscle pain as well as prolonged fever, chills, and headache.

Vaccines: An investigational live-attenuated vaccine administered by scarification has been given to more that five thousand persons without significant adverse reactions and prevents typhoidal and ameliorates ulceroglandular forms of laboratory-acquired tularemia. Aerosol challenge tests in laboratory animal and human volunteers have demonstrated significant protection. **As with all vaccines, the degree of protection depends upon the magnitude of the dose; so vaccine-induced protection could be overwhelmed by extremely high doses of tularemia.**

Guides for Emergency Response: Biological Agent or Weapon: Viral Encephalitus

16

AGENT: Venezuelan Equine Encephalitis (VEE) also includes Western and Eastern Equine Encephalitis (WEE and EEE). VEE consists of eight serologically distinct viruses belonging to the Venezuelan equine encephalitis complex that have been associated with human disease. The most important of these pathogens are designated subtype 1, variants A, B, and C. Such agents also cause severe disease in horses, mules, and donkeys. Natural infections are acquired by the bites of a wide variety of mosquitoes, and such animals serve as viremic (the presence of virus in the blood of a host) hosts and source of mosquito infection. In natural human epidemics, severe and often fatal encephalitis in horses, mules and donkeys always precedes that in humans. A biological warfare or terrorist attack with virus disseminated as an aerosol would cause human disease as a primary event. If horses, mules, and donkeys were present in the area, disease in these animals would occur simultaneously with human disease, while secondary spread by person-to-person contact occurs at a negligible rate. However, in an attack in an area populated by horses, mules, and donkeys with appropriate mosquito vectors (an organism that transmits a pathogen from one organism to another, as fleas carry plague, mosquitoes carry VEE) could initiate an epizootic (an outbreak of a disease affecting many animals of one kind at the same time) epidemic. Children and the elderly are more susceptible to the VEE virus than the average adult; however, pregnant women might develop serious problems although most victims live through this disease.

Medical Classification: Virus.

Probable Form of Dissemination: Aerosol; infected vectors (carrier organism).

Detection in the Field: None.

Infective Dose (Aerosol):   10–100 organisms.

Sign and Symptoms: Sudden onset of illness with general malaise, spiking fevers, rigors, severe headache, photophobia (intolerance to light), and myalgias (muscular pain). Nausea, vomiting, cough, sore throat, and diarrhea may follow. Full recovery takes one to two weeks.

Incubation Time: One to five days.

Diagnosis: Clinical diagnosis. Physical findings are usually non-specific. The white blood cell count often shows a striking leukopenia (the antithesis of leukocytosis; any situation in which the total number of leukocytes in the circulating blood is less than normal, the lower limit of which is generally regarded as 4,000–5,000 per cubic mm) and lymphopenia (a reduction, relative or absolute, in the number of lymphrocytes in the circulating blood). Virus isolation may be made from serum, and in some cases throat swab specimens. Both neutralizing or lgG antibody in paired sera or VEE specific lgM present in a single serum sample indicate recent infection. Treatment is supportive only.

Differential Diagnosis:　An outbreak of VEE may be difficult to distinguish from influenza on clinical grounds. Clues to the diagnosis are the appearance of a small proportion of neurological cases or disease in horses, mules and donkeys, but these might be absent in a biological warfare or terrorist attack.

Persistency:　Relatively unstable.

Personal Protection:　Protective clothing must be used as well as protection for the respiratory tract by using a mask with biological filters, or self-contained breathing apparatus (SCBA) with positive pressure, at least until you know the specific threat(s).

Routes of Entry to the Body:　Skin.

Person-To-Person Transmissable:　Low.

Duration of Illness:　Days to weeks.

Potential Ability to Kill:　Low; less than 1 percent for naturally occurring disease.

Defensive Measures:　Immunizations, good personal hygiene, physical conditioning, use of arthropod repellents, wearing protective mask, and practicing good sanitation.

Vaccines:　Yes. TC-83 Live Attenuated Vaccine (IND) and C-84: Formalin inactivation of TC-83 (IND) is available for boosting antibody titers in those initially receiving the live vaccine.

Drugs Available:　No specific anti-viral drugs are available, but supportive therapy might call for analgesics and anticonvulsants.

Decontamination:　**Standard precautions for healthcare workers.** Human cases are infectious for mosquitoes for at least seventy-two hours. The virus can be destroyed by heat (80 degrees centigrade for thirty minutes), soap and water, or diluted sodium hypochlorite solution (0.5 percent). **Blood and body fluid precautions are necessary.**

Characteristics:　Venezuelan equine encephalitis is a member of the Alphavirus family transmitted by mosquitoes that generally infects horses but can cause epidemics in humans. It can also cause infections if aerosols containing the virus are inhaled. Infection is manifested by fever, headache, sore throat, vomiting, and muscle aches. Eight serologically distinct viruses belonging to the Venezuelan equine encephalitis (VEE) complex have been associated with human disease; the most important of these pathogens are designated subtype 1, variants A, B, and C. The same agents cause severe disease in Equidae (horses, mules, and donkeys). So-called "natural" infections are acquired from bites from a wide selection of different types of mosquitoes. In natural human epidemics, severe and often fatal encephalitis in horses, mules or donkeys always precedes that in humans. **A biological warfare or terrorist attack using aerosol for dissemination of this virus would cause human disease as a primary event; but if Equine were in the area, disease in such animals would occur at the same time as human disease. Such an attack in a region populated by Equidae and appropriate mosquito vectors, however, could initiate an epizootic/epidemic.** Western and Eastern Equine Encephalitis (WEE and EEE) are similar to the Venezuelan Equine Encephalitis (VEE) complex, and are often confused because clinical differentiation is difficult and all three have many similarities. In the final analysis, the question always comes down to which one of the three would make the best virus for an attack. **The human infective dose for VEE is thought to be 10 to 100 organisms, and that tends to be the principal why it wins out over WEE and EEE.** Another point is noted **by various governments: neither the population density of infected mosquitoes nor the aerosol concentration of the virus particles has to be great to allow transmission of VEE during an attack.** A key point for other countries is that VEE is very infectious, although it does have some weaknesses as an agent.

VEE particles are not considered stable in the environment, and they are less persistent than the bacteria responsible for anthrax, tularemia or Q fever. Heat and standard disinfectants can easily kill the VEE virus complex. It is true that VEE is better understood than either WEE or EEE because back in the 1950s and 1960s the United States studied, weaponized and stored VEE as a component in our offensive biological weapons program. In 1969 all biological weapons were destroyed within the United States, and the in-stock supply of VEE was done away with. VEE is not really a killer, but rather more of an **incapacitating agent** with a mortality rate of less than one percent as a naturally occurring disease. However, susceptibility is nearly 100 percent.

Response on Scene by First Responders

Caution: **Humans are infectious for mosquitoes for at least seventy-two hours after the onset of symptoms, so mosquito control can provide an important measure of control to ensure prevention of secondary Venuzuelan equine encephalitis following either intentional or natural outbreaks of VEE.**

Field First Aid: Treatment of VEE is largely supportive, and the diagnosis is usually due to clinical factors. **Human cases are infectious for mosquitoes for seventy-two hours, and mosquitoes are infectious for life. No specific first aid treatment is available for VEE, and there is no specific antiviral therapy.** First responders should administer anticonvulsive therapy for victims with seizures and provide other supportive care as necessary. Remember, the VEE virus is an incapacitating agent; while acute morbidity can be severe, most victims will recover, at least from naturally occurring VEE. **However, animal studies seem to indicate that aerosol exposure leads to viral attachment to olfactory nerve endings and the direct invasion of the central nervous system (CNS) via the olfactory nerve, thus resulting in a high incidence of CNS disease.** This factor seems to suggest that unlike the mosquito-borne disease, VEE disseminated as an aerosol in an attack would be more likely to bring about CNS involvement that could lead to higher morbidity and mortality.

Medical Management: No specific viral therapy exists so treatment is supportive only. Treat patients with uncomplicated VEE infection with analgesics to relieve headache and myalgia. Patients who develop encephalitis could require anticonvulsants and intensive care to maintain fluid and electrolyte balance, ensure adequate ventilation, and avoid complicating secondary bacterial infections. **Patients should be treated in a screened room or in quarters treated with residual insecticide for at least five days after onset, or until afebrile (without fever) to foil mosquitoes since humans may remain infectious for mosquitoes for at least seventy-two hours. Isolation and qaurantine is *not* required. Standard Precautions should be practiced when dealing with infection control for VEE victims as shown below:**

Infection Control when Dealing with Victims of VEE

Isolation Precautions: Use Standard Precautions for all aspects of care. (airborne, contact, HEPA hood by all entering room, and strict hand washing with antimicrobial soap.

Patient Placement: Private room, negative pressure room, and door closed at all times.

Patient Transport: Limit movement for essential purposes only, and mask patient before transport.

Cleaning/Disinfecting Equipment: Terminal cleaning required with Phenolic, linen doubled bagged, disinfect equipment before taking it from room, and change air filter before having room terminally cleaned.

Discharge Management: Do not discharge until no longer infectious.

Post-mortem Care: Follow Standard Precautions, airborne precautions, contact precautions, use of HEPA hood by all entering room, and negative pressure required. **At the present time, there is no pre-exposure or post-exposure immunoprophylaxis available for humans.**

Fire: For an infectious disease or weapon, use available methods and equipment on surrounding fires. Appropriate extinguishing agents: dry chemical, soda ash, sand, or lime. Do not use high pressure water streams

Personal Protective Equipment: At all times wear protective **Level A until you know what you are dealing with;** upon learning the exact type of infectious material(s) you are dealing with, consult proper toxicological levels and downgrade as necessary for safety and comfort of operation.

Spill/Leak Disposal: Outright cleanup or disposal of Infectious Substances (biological agents or weapons) should *only* be handled by trained and experienced personnel. To contact such help, call any or all of the following numbers.

1. SBCCOM Operations — The U.S. Army Soldier and Biological Chemical Command, Edgewood Chemical Biological Center (ECBC), ATTN: AMSSB-RCB-RS, Aberdeen Proving Ground, MD 21010-5424. Tel. 410-436-2148 (twenty-four-hour emergency number).
2. U.S. Army Medical Research Institute if Infectious Diseases, Fort Detrick, Frederick, MD 21702-5011. Tel. 1-888-872-7443.
3. Center for Disease Control and Prevention Emergency Response Line. 1-770-488-7100.
4. ATSDR — Agency for Toxic Substances and Disease Registry, Division of Toxicology, 1600 Clifton Road NE, Mailstop E-29, Atlanta, GA 30333. Tel. 1-888-422-8737. Fax: 404-498-0057.
5. USACHPPM — U.S. Army Center for Health Promotion and Preventive Medicine, Aberdeen Proving Grounds, MD 21010-5422. Tel. 410-671-2208. Email: kwilliam@ aehal.apgea.army.mil
6. National Response Center (for chem/bio hazards & terrorist events). Tel. 1-800-424-8802 or 1-202-267-2675.
7. U.S. Public Health Service: 800-USA-NDMS.

8. National Domestic Preparedness Office (for civilian use). Tel. 202-324-9025.

Symptoms: Acute systemic febrile illness with encephalitis developing in a small percentage of victims (4 percent for children, and less than 1 percent for adults). Symptoms include generalized malaise, spiking fevers, rigors, severe headache, photophobia, and myalgias for twenty-four to seventy-two hours. This may be followed by nausea, vomiting, cough, sore throat and diarrhea. Full recovery will take one to two weeks. *Please note: The incidence of central nervous system disease and associated morbidity and mortality will be much higher after an aerosol attack.*

Vaccines: There are two IND (investigative new drug) human unlicensed VEE vaccines available. The first, named TC-83, developed in the 1960s is a live, attenuated cell-culture-propagated vaccine produced by the Salk Institute, and is not effective against all the serotypes in the VEE complex. It is used in a single subcutaneous dose of 0.5 mL.

The second IND vaccine, C-85, has been tested but not licensed for humans. It is prepared by formalin-inactivation of the TC-83 strain (see above). The vaccine is not used for the primary immunization, but rather to boost non-responders to TC-83. The required regimen for this vaccine is to administer 0.5 mL subcutaneously at two to four week intervals for up to three inoculations or until an antibody response is measured. Booster shots are required for the C-84 vaccine, and it does not protect rodents from an aerosol challenge.

Guides for Emergency Response: Biological Agent or Weapon: Viral Hemorrhagic Fevers (VHFs)

17

Viral Hemorrhagic Fevers (VHF): Viral hemorrhagic fevers (VHFs), including Crimean-Congo fever, Rift Valley fever, Lassa fever, Ebola fever, and others, are a diverse group of human viral illnesses characterized by acute febrile onset accompanied by headache and complicated by increased vascular permeability, damage, and bleeding; mortality is high. Viral pathogenesis is complex, incompletely understood and varies among specific viruses. Some infections result in immune complex deposition that activates complement and other inflammatory results. Such a process damages vascular endothelium, results in capillary leak, and deregulates vascular smooth muscle tone. They lead to hypotension, shock, and end-organ failures. Some of these diseases activate coagulation cascades and result in disseminated intravascular coagulation (DIC). Hemorrhage can also be enhanced by specific end-organ failures. As an example, yellow fever can cause massive hepatic (relating to, affecting, or associated with the liver) necrosis resulting in a deficiency of vitamin K-dependent clotting factors. The uremia complicating the acute renal failure of the hemorrhagic fever syndrome (HFRS) leads to platelet dysfunction, further promoting hemorrhage. The end result is that viral hemorrhagic fever is highly contagious, but some VHF viruses can cause relatively mild illnesses, while many others can cause severe, life-threatening disease. With a few exceptions, there is no cure or an established drug treatment for viral hemorrhagic fevers.

Viral Hemorrhagic Fevers that Can Infect Humans

Disease	Geography	Reservoir	Transmission
Lassa Fever	W. Africa	Rodent	Aerosol, Fomites*
Argentine VHF	S. America	Rodent	Aerosol, Formites
Bolivian VHF	S. America	Rodent	Aerosol, Formites
Venezuelan VHF	S. America	Rodent	Aerosol, Formites
Brazilian VHF	S. America	Unknown	Aerosol, Formites
Rift Valley Fever	Africa	Mosquito	Mosquito, Aerosol, Formites
Crimean-Congo Fever	Africa, Middle East, & Eastern Europe	Ticks	Mosquito, Aerosol, Formites
VHF with Renal Syndrome	Asia, Europe, & Worldwide	Rodent	Aerosol, Fomites
Ebola VHF	Africa, Asia	Unknown	Unknown
Marburg VHF	Africa	Unknown	Unknown
Yellow Fever	S.America, Africa	Mosquito, Primate	Mosquito
Dengue VHF	Tropics/Subtropics	Mosquito, Humans	Mosquito
Kyasanur Forest Disease	India	Rodent, Monkey	Tick Disease:
Omsk VHF	Iberia	Rodent	Tick

* Formites: Objects such as clothing, towels, and utensils that possibly harbor a disease agent and are capable of transmitting it.

Crimean-Congo Hemorrhagic Fever

AGENT: Crimean-Congo hemorrhagic fever is one of the viral hemorrhagic fevers, a varied group of human illnesses due to RNA (any of various nucleic acids that contain ribose and uracil as structural components and are associated with the control of cellular chemical activities — also called *ribonucleic acid*) viruses from several different viral families: the *Filoviridae,* including the Ebola and Marburg viruses; the *Arenaviridae,* such as Lassa fever, Argentine and Bolivian hemorrhagic fever viruses; the *Bunyaviridae,* including various members from the Hantavirus genus, Congo-Crimean hemorrhagic fever virus from the Nairovirus genus, and Rift Valley fever from the *Phlebovirus* genus; and *Flavivirida,* including Yellow fever, Dengue hemorrhagic fever virus, and others. The viruses may be spread in a variety of ways, and for some there is the possibility that humans could be infected through a respiratory portal of entry. Although evidence for weaponization does not exist for many of these viruses, some are included here in this manual because of *potential of aerosol dissemination or weaponization.* **Crimean-Congo hemorrhagic fever (CCHF) is caused by transmission of a virus by ticks with intermediate hosts varying with the tick species.** Humans become infected by tick bites, at the slaughter of viremic (the presence of virus in the blood of a host), or the crushing of a tick. The spread of this disease within hospitals poses a potentially significant problem. In a biological warfare or terrorist attack, CCHF would probably be delivered by aerosol.

Medical Classification: Virus.

Probable Form of Dissemination: Aerosol.

Detection in the Field: None.

Infective Dose (Aerosol): One to ten organisms.

Signs and Symptoms: VHFs are febrile illnesses which can be complicated by easy bleeding, petechiae (minute hemorrhagic spots, of pinpoint to pinhead size, in the skin, which are not blanched by pressure), hypotension and even shock, flushing of the face and chest, and edema (an accumulation of an excessive amount of watery fluid in cells, tissues, or serous cavities). Constitutional symptoms such as malaise, myalgias (a pain in one or more muscles), headaches, vomiting, and diarrhea may occur in any of the hemorrhagic fevers.

Incubation Time: Three to twelve days.

Diagnosis: Definitive diagnosis rests on specific virologic techniques. Significant numbers of military troops with a hemorrhagic fever syndrome should suggest the diagnosis of viral hemorrhagic fever. For treatment, intensive supporting care may be required. Antiviral therapy with ribavirin may be useful in several of these infections. Also, convalescent plasma may be effective in Argentine hemorrhagic fever.

Differential Diagnosis: Thrombocytopenia (a condition in which there is an abnormally small number of platelets in the circulating blood) and an elevated aspartate aminotransferase (AST) may provide a clue to suggest CCHF is the culprit in the febrile (denoting or relating to fever) patient seen early in the course of infection. Other viral hemorrhagic fevers, meningococcemia, rickettsial diseases, and similar conditions may resemble full-blown CCHF. Particular care must be taken in the case of massive gastrointestinal bleeding not to confuse CCHF with surgical conditions.

Persistency: Relatively Stable.

Personal Protection: Protective clothing must be used as well as protection for the respiratory tract by using a mask with biological filters or self-contained breathing apparatus (SCBA) with positive pressure, at least until you know the specific threat(s).

Routes of Entry to the Body: Inhalation of aerosol, tick bites, crushing an infected tick, or at the slaughter of viremic livestock.

Person-To-Person Transmissable: Moderate.

Duration of Illness: Days to weeks.

Potential Ability to Kill: High

Defensive Measures: Good personal hygiene, physical conditioning, use of arthropod repellents, wearing protective mask, and practicing good sanitation.

Vaccines: The only licensed vaccine for VHF is yellow fever vaccine.

Drugs Available: Prophylactic ribavirin may be effective for Crimean-Congo hemorrhagic fever, Lassa fever, and Rift Valley fever.

Decontamination: Contact precautions for healthcare workers. Diluted sodium hypochlorite solution (0.5 percent). Isolation measures and barrier nursing procedures are necessary.

Rift Valley Fever

AGENT: Rift Valley Fever is one of the viral hemorrhagic fevers, a varied group of human illnesses due to RNA (any of various nucleic acids that contain ribose and uracil as structural components and are associated with the control of cellular chemical activities — also called ribonucleic acid) viruses from several different viral families: the Filoviridae, including the Ebola and Marburg viruses; the Arenaviridae, such as Lassa fever, Argentine and Bolivian hemorrhagic fever viruses; the Bunyaviridae, including various members from the Hantavirus genus, Congo-Crimean hemorrhagic fever virus from the Nairovirus genus, and Rift Valley fever from the Phlebovirus genus; and Flavivirida, including Yellow fever, Dengue hemorrhagic fever virus, and others. The viruses may be spread in a variety of ways, and for some there is the possibility that humans could be infected through a respiratory portal of entry. Although evidence for weaponization does not exist for many of these viruses, some are included here in this manual because of potential of aerosol dissemination or weaponization. The natural path of Rift Valley Fever is through mosquitoes that feed on infected animals and then bite humans, but aerosols or virus-laden droplets could also be used. The disease tends to be similar whether acquired through aerosol or mosquito bites.

Medical Classification: Virus.

Probable Form of Dissemination: Aerosol,

Detection in the Field: None.

Infective Dose (Aerosol):   One to ten organisms.

Signs and Symptoms: VHFs are febrile illnesses which can be complicated by easy bleeding, petechiae (minute hemorrhagic spots, of pinpoint to pinhead size, in the skin, which are not blanched by pressure), hypotension and even shock, flushing of the face and chest, and edema (an accumulation of an excessive amount of watery fluid in cells, tissues, or serous cavities). Constitutional symptoms such as malaise, myalgias (a pain in one or more muscles), headaches, vomiting, and diarrhea may occur in any of the hemorrhagic fevers.

Incubation Time: Three to twenty-one days.

Diagnosis: Definitive diagnosis rests on specific virologic techniques. Significant numbers of military troops with a hemorrhagic fever syndrome should suggest the diagnosis of viral hemorrhagic fever. For treatment, intensive supporting care may be required. Antiviral therapy with ribavirin may be useful in several of these infections. Also, convalescent plasma may be effective in Argentine hemorrhagic fever.

Differential Diagnosis: The clinical syndrome in an individual is not pathognomonic (distinctively characteristic of a particular disease or condition); but the occurrence of an epidemic with febrile disease, hemorrhagic fever, eye lesions, and encephalitus in different patients would be characteristics of Rift Valley fever.

Persistency: Relatively Stable.

Personal Protection: Protective clothing must be used as well as protection for the respiratory tract by using a mask with biological filters or self-contained breathing apparatus (SCBA) with positive pressure, at least until you know the specific threat(s).

Routes of Entry to the Body: Inhalation of aerosol, mosquito bites, crushing an infected mosquito, or at the slaughter of viremic livestock.

Person-To-Person Transmissible: Moderate.

Duration of Illness: Days to Weeks.

Potential Ability to Kill: Low to very high depending on specific virus.

Defensive Measures: Immunizations, good personal hygiene, physical conditioning, use of arthropod repellents, wearing protective mask, and practicing good sanitation

Vaccines: The only licensed vaccine for VHF is yellow fever vaccine.

Drugs Available: Prophylactic ribavirin may be effective for Crimean-Congo hemorrhagic fever, Lassa fever, and Rift Valley fever.

Decontamination: Contact precautions for healthcare workers; diluted sodium hypochlorite solution (0.5 percent), and isolation measures and barrier nursing procedures are necessary.

Response on Scene by First Responders

Caution: With some noteworthy exceptions, there is no cure or established drug treatment for viral hemorrhagic fever. All the VHF agents, except for dengue fever, are infectious by aerosol in the laboratory.

Field First Aid: Intensive supportive care may be required. Antiviral therapy may be useful in several of the VHF viruses, while convalescent plasma may be effective in Argentine hemorrhagic fever; but both are available only as IND (investigational new drug) under protocol. The only licensed vaccine is yellow fever vaccine. **Prophylactic ribavirin may be effective for Lassa fever, Rift Valley fever, Crimean-Congo hemorrhagic fever, and possibly hemorrhagic fever with renal syndrome (HFRS) (available only as IND under protocol). Contact isolation, with the addition of a surgical mask and eye protection for those coming within three feet of a patient, is indicated for suspected or proven Lassa fever, Crimean-Congo hemorrhagic fever, Ebola hemorrhagic fever, or Marburg hemorrhagic fever. Respiratory protection should be upgraded to airborne isolation, including the use of a fit-tested HEPA filtered respirator, a battery powered air purifying respirator, or a positive pressure supplied air respirator, if patients with the above conditions have prominent cough, vomiting, diarrhea, or hemorrhage.** Decontamination can be accomplished with hypochlorite or phenolic disinfectants.

Medical Management: Viral hemorrhagic fevers is a group of illnesses that are caused by several families of viruses (arenaviruses, filoviruses, bunyaviruses, and flaviviruses), some of which can cause relatively mild illnesses while many of these viruses can cause severe, life-threatening disease. Each of these four families share some traits: they are all RNA (ribonucleic acid) viruses, their survival depends on an animal or insect host, the specific viruses are geographically restricted to where their hosts live, and human beings are *not* a natural host reservoir (an organism in which a parasite that is pathogenic for some other species lives and multiplies without damaging its host) for any of these viruses. Humans are infected when they come in contact with infected hosts. In the case of some viruses, after the accidental transmission from the host, humans can transmit the virus to another human. Viruses that can cause VHF are initially transmitted in a natural fashion to humans when the activities of infected reservoir hosts or vectors (an organism that transmits a pathogen from one organism to another as rodents or ticks can be vectors for VHF) and humans overlap. As an example, the viruses carried in rodent reservoirs are transmitted when humans have contact with urine, fecal matter, saliva, or other body excretions from infected rodents. In like manner, the viruses associated with arthropod vectors are spread when the vector mosquito or tick bites a human, or when a person crushes a tick. However some vectors may spread virus to animals, and then humans who care for or slaughter the animals. **Some viruses like Crimean-Congo VHF, Ebola VHF, Marburg VHF, and Lassa fever can spread from one person to another person.** This secondary transmission of virus can occur directly, through close contact with infected people or their body fluids; or indirectly, through contact with objects contaminated with infected body fluids.

For prevention of some types of viral hemorrhagic fevers, there is pre-exposure prophylaxis in the form of yellow fever vaccinations, and some vaccines are available as IND (investigation new drug) including IND for Argentine Hemorrhagic Fever (an attenuated vaccine, that might also provide protection against Bolivian VHF) and Rift Valley Hemorrhagic Fever (an inactivated vaccine and a live attenuated vaccine). Also, there is post-exposure chemoprophylaxis under IND status that is protective against Crimean-Congo VHF and Lassa fever. It requires the administration of ribavirin at a dose of 500 mg orally every six hours for seven days. Also, ribavirin, both intravenous and oral, is available under IND status and available only through a human use protocol.

For diagnosis, serologic methods do include enzyme-linked immunosorbent assay (ELISA) and IgM antibody capture (a class of antibodies of high molecular weight including those appearing early in the immune response to be replaced later by IgG of lower molecular weight) systems to detect the antigen. Also, tissue can be processed through immunohistochemical staining, electron microscopy, or genetic typing. Serum or other clinical specimens can be sent for viral culture **under maximum containment of Biosafety Level 4.** Laboratory findings tend to be nonspecific and variable; it is normal for these to result in thrombocytopenia and leukopenia. Elevated liver function tests and other nonspecific laboratory findings may be present. The blood urea nitrogen in the BUN test will be related to the circulatory status, with the exception of the Hemorrhagic Fever with Renal Syndrome (HFRS) in which the kidneys are target organs of the Hantaviruses.

Viral hemorrhagic fever treatment in medical care facilities is primarily supportive to be assisted by procedures designed to control fluid and electrolyte balance; and treat shock, blood loss, renal failure, seizures, and coma. Treatment may also include intensive care such as mechanical ventilation, dialysis, or neurological assistance. As an example, Heparin therapy may improve the disseminated intravascular coagulation (DIC) but such therapy may be controversial in some quarters and should be used only with victims having significant hemorrhage and laboratory evidence of DIC. **Aspirin and other drugs that might impair platelet function are contraindicated; intramuscular injections are also contraindicated.**

In treating VHF, specific antiviral therapy is pretty much limited to one choice, at least in the cases or Lassa fever and hemorrhagic fever with renal syndrome (HFRS), intravenous ribavirin. The use of ribavirin therapy for Crimean-Congo VHF is supported by in vitro studies but there is no clinical experience. **Ribavirin**, both intravenous and oral, is available under investigational new drug (IND) procedures and is available through human use protocol only. For Lassa fever and Crimean-Congo VHF, medical personnel should administer a loading dose of ribavirin of 30 mg/kg intravenous, followed by 16 mg/kg intravenous every six hours for four days; then eight mg/kg intravenous every eight hours for six days to complete a ten-day course of therapy. For hemorrhagic fever with renal syndrome (HFRS), therapy may benefit victims who have been febrile for six days or less. Medical personnel should administer a loading dose of ribavirin 33 mg/kg intravenous, followed by 16 mg/kg intravenous every six hours for four days, then eight mg/kg intravenous every eight hours for three days, to complete a seven day course of therapy.

For those VHF viruses that can be transmitted from one victim to another person, response and medical personnel should avoid close physical contact with the victim and his or her body fluids. Barrier nursing or infection control techniques include isolating infected persons, wearing protective clothing, disinfection, and disposal of instruments and equipment used for the treatment of VHF, including needles and thermometers. Special Precautions are a "must do" factor when responding to or caring for viral hemorrhagic fever victims who were infected by "natural" means or dissemination of aerosol organisms.

Isolation Control: The U.S. Centers for Disease Control and Prevention (CDC), and the World Health Organization, have produced a manual, *Infection Control for Viral Hemorrhagic Fevers in the African Health Care Setting*. The manual is available at the CDC Web site http://www.cdc.gov/ncidod/dvrd/spb/mnpages/vhfmanual.htm and can be downloaded. Factors covered in this lengthy manual are include:

Section 1: Use standard precautions for all patients
Section 2: Identify suspected cases of VHF
Section 3: Isolate the patient
Section 4: Wear protective clothing
Section 5: Disinfect reusable supplies
Section 6: Dispose of waste safely
Section 7: Use safe burial practices
Section 8: Mobilize community resources and conduct community education
Section 9: Make advance preparations to use VHF isolation precautions

There are also sixteen annexes identified:

Annex 1: Standard precautions for hospital infection control
Annex 2: Specific features of VHFs
Annex 3: Planning and setting up the isolation area
Annex 4: Adapting VHF isolation precautions for a large number of patients
Annex 5: Making protective clothing
Annex 6: Requirement for purchasing protective clothing
Annex 7: Disinfecting water for drinking, cooking and cleaning
Annex 8: Preparing disinfection solutions
Annex 9: Making supplies; sharps containers, incinerator, and boot remover
Annex 10: Sample job-aids and posters for use in the health care facility
Annex 11: Laboratory testing for VHFs
Annex 12: Skin biopsy on fatal cases for diagnosis of Ebola
Annex 13: Community education materials
Annex 14: Conducting in-service training for VHF isolation precautions
Annex 15: Local resources for community mobilization and education
Annex 16: International and regional contacts

In 1995, an outbreak of Ebola VHF affected more than three hundred people in and around the city of Kikwit in the Democratic Republic of the Congo (the former Zaire) and approximately 80 percent of the victims died. An international investigation team worked with local authorities to introduce VHF isolation precautions as well as standard precautions. When the types of precaution featured in the manual mentioned above were installed in Kikwit, no further nosocomial (hospital) transmission of the Ebola virus was documented.

Infection Control in the United States: For any VHF viruses in the United States, it is recommended that the following protocols be implemented:

Isolation Precautions:
  Standard Precautions for all aspects of care
  Airborne Precautions
  Contact Precautions
  Use of HEPA hood by all entering room
  Strict hand washing with antimicrobial soap

Patient Placement:
  Private room
  Negative pressure room
  Door closed at all times

Patient Transport:
  Limit movement, essential purposes only
  Mask patient before transport

Cleaning, Disinfecting Equipment:
  Terminal cleaning required with Phenolic
  Linen double bagged
  Disinfect equipment before taking it from room
  Air filter changed before room terminally cleaned

Discharge Management:
  Do not discharge until no longer infectious

Post-Mortem Care:
  Follow Standard Precautions
  Airborne Precautions
  Contact Precautions
  Use of HEPA hood by all entering room
  Negative pressure required
  Disinfect surfaces with 1:9 bleach/water solution (10 percent)

One family of VHF viruses, the filoviruses from the filoviridae family, has both the Marburg virus and the Ebola virus, both of which caused great fear and concern in Africa in 1976 through 1995. Both viruses have shown themselves to be highly lethal since 90 percent of the Congo and 50 percent of the Sudanese cases resulted in death. Even with such death rates, VHF viruses — including Marburg virus and Ebola virus — can be conquered by using strict special precautions, isolation procedures, and so-called "barrier nursing."

The U. S. Army Medical Research Institute of Infectious Diseases (USAMRIID) points out in their extremely good manual, *Medical Management of Biological Casualties Handbook*, (4th ed., February 2001) some clinical factors that stand out for a particular VHF virus may point toward a specific etiologic agent. Although hepatic involvement is common among VHF diseases, a clinical picture actually dominated by jaundice and other features of hepatitis is only seen in some cases of Rift Valley fever, Crimean-Congo VHF, Marburg fever, Ebola fever, and yellow fever. Kyanasur Forest disease and Omsk hemorrhagic fever are notable for pulmonary involvement, and a biphasic illness with subsequent central nervous system (CNS) manifestations. Among the arenavirus infections, Lassa fever can cause severe peripheral edema due to capillary leak, but hemorrhage is uncommon, while hemorrhage is commonly caused by the South American areanviruses. Severe hemorrhage and nosocomial (hospital) are typical for Crimean-Congo VHF. Retinitis is commonly seen in Rift Valley fever, and hearing loss is common among Lassa fever survivors.

Fire: For an infectious disease or weapon, use available methods and equipment on surrounding fires. Appropriate extinguishing agents: dry chemical, soda ash, sand, or lime. Do not use high pressure water streams

Personal Protective Equipment: At all times, wear protective Level A until you know what you are dealing with; upon learning the exact type of infectious material(s) you are dealing with, consult proper toxicological levels and downgrade as required.

Spill/Leak Disposal: Outright cleanup or disposal of Infectious Substances (biological agents or weapons) should *only* be handled by trained and experienced personnel. To contact such help, call any or all of the following numbers.

1. SBCCOM Operations — The U.S. Army Soldier and Biological Chemical Command, Edgewood Chemical Biological Center (ECBC), ATTN: AMSSB-RCB-RS, Aberdeen Proving Ground, MD 21010-5424. Tel. 410-436-2148 (twenty-four-hour emergency number).
2. U.S. Army Medical Research Institute if Infectious Diseases, Fort Detrick, Frederick, MD 21702-5011. Tel. 1-888-872-7443.
3. Center for Disease Control and Prevention Emergency Response Line. 1-770-488-7100.
4. ATSDR — Agency for Toxic Substances and Disease Registry, Division of Toxicology, 1600 Clifton Road NE, Mailstop E-29, Atlanta, GA 30333. Tel. 1-888-422-8737. Fax: 404-498-0057.
5. USACHPPM — U.S. Army Center for Health Promotion and Preventive Medicine, Aberdeen Proving Grounds, MD 21010-5422. Tel. 410-671-2208. Email: kwilliam@aehal.apgea.army.mil
6. National Response Center (for chem/bio hazards & terrorist events). Tel. 1-800-424-8802 or 1-202-267-2675.
7. U.S. Public Health Service: 800-USA-NDMS.
8. National Domestic Preparedness Office (for civilian use). Tel. 202-324-9025.

Symptoms: Viral hemorrhagic fevers are febrile illnesses which can feature flushing of the face and chest, petechiae (a minute reddish or purplish spot containing blood that appears in skin or mucous membranes, especially in some infectious diseases), bleeding, edema, hypotension, and shock. Malaise, myalgias, headache, vomiting, and diarrhea may occur in any of the hemorrhagic fevers. Illness may progress with prostration, fatigue, and hemorrhage. The most dreaded complications are shock, multiple organ system failure, and death.

Vaccines: With the exception of yellow fever and Argentine hemorrhagic fever for which vaccines have been developed, there are presently no vaccines is existence that can protect humans from VHF diseases. As a result, prevention efforts must concentrate on avoiding contact with host species.

Guides for Emergency Response: Chemical Agents and Weapons

18

Because they can be easily manufactured and used, chemical agents and weapons are the most likely weapons of mass destruction (WMD) to be used in any wars of the immediate future. They have been widely used in the past. The U.S. Marine Corps has estimated that chemical agents are stockpiled, or under development, in twenty to twenty-five countries. "Many of the same countries associated with nuclear or biological programs have established, or are seeking to establish, chemical weapons programs. Egypt, France, Iran, Libya, Iraq, People's Republic of China, Syria, Russia, and many of the countries belonging to the former Warsaw Pact (Bulgaria, Czech Republic, Poland, Romania, Slovakia, and the former Yugoslav republics) have, or are suspected of having, chemical weapons programs" (*Biological, and Chemical Defense Operations*, pp. 1–5, MCWP 3-37 MAGTF Nuclear).

In the standard format used for each Chemical agent on the following pages, the following definitions apply.

Formula: An expression of the constituents of a compound by symbols and figures.

Vapor Density: The ratio of the density of any gas or vapor to the density of air, under the same conditions of temperature and pressure; that is, a measure of how heavy the vapor is in relation to the same volume of air. Air is considered to have a molecular weight of 1. As an example, the higher the vapor density is in relation to 1.0, the longer it will persist in low-lying areas such as valleys, trenches, and cellar holes. If the vapor pressure is less than 1, a chemical agent would will likely be non-persistent and dissipate quickly into the atmosphere.

Vapor Pressure: A measure of the tendency of a liquid to become a gas at a given temperature. Chemical agents with a high vapor pressure evaporate rapidly, while those with a low vapor pressure evaporate more slowly.

Molecular Weight: The sum of the atomic weights of all the atoms in a molecule. In regard to chemical agents, the molecular weight is a guide to persistency; an agent with a high molecular weight would tend to have a lower rate of evaporation and greater persistency.

Liquid Density: Liquid density of a chemical agent measures the weight of the agent compared to water which has a density of 1.00. Specific gravity, the ratio of the mass of a unit volume of a substance to the mass of the same volume of a standard substance, usually water, at a standard temperature, is the result of this comparison. Liquid agent forms layers in water; the greater the density, the more it will sink, while less dense agents will float. Nerve agents are roughly the same density as water so they tend to mix throughout the depth of water.

Volatility: Passing off rapidly in the form of vapor; evaporating rapidly. It provides a measure of how much material evaporates under given conditions and varies directly with temperature. The volatility depends on vapor pressure.

Median Lethal Dosage: (LD50) The lethal dose of a poison, when taken orally or absorbed through the skin, which is lethal to 50 percent of the exposed laboratory animal population.

Physical State: Condition with respect to structure, form, phase, etc..

Odor: Smell, scent, aroma, fragrance.

Freezing/Melting Point: Freezing point is the temperature at which a liquid changes to a solid.Melting point is the temperature at which a solid changes to a liquid. Since dissemination characteristics vary greatly with physical state, the freezing/melting point can be an important piece of information in using poison gases.

Boiling Point: The temperature at which the vapor pressure of a liquid equals the atmospheric pressure and the liquid becomes vapor. Knowing the boiling point, you can estimate the persistency of a chemical agent under a given set of conditions because the vapor pressure and the evaporating tendency vary inversely with its boiling point. Chemical agents with high boiling points tend to be persistent, while agents with low boiling points are normally nonpersistent.

Action Rate: Effect or influence.

Physiological Action: Of or pertaining to physiology, the science dealing with the functions of living organisms or their parts.

Required level of protection: Chemical protective clothing/equipment required to work safely.

Decontamination: The physical and/or chemical process of reducing and preventing the spread of contamination from persons and equipment used at a hazardous materials incident.

Detection in the field: Methods and apparatus for detecting and/or identifying chemical agents in the field.

Use: The primary use of this chemical agent.

CAS registry number: An identifying numbering system for chemicals called the Chemical Abstract Service (CAS) with individual numbers assigned by the American Chemical Society. CAS numbers identify specific chemicals and are assigned sequentially. Such a number is a concise and unique means of material identification.

RTECS number: Registry of Toxic Effects of Chemical Substances number is published by NIOSH and present basic toxicity data on thousands of materials.

In like manner, the standard introductory format for each **Biological** agent discussed will be as follows:

Medical Classification, Probable Form of Dissemination, Detection in the Field, Infective Dose, Incubation Time, Persistence, Personal Protection, Routes of Entry to the Body, Person-to-Person Transmissible, Duration of Illness, Potential Ability to Kill, Defensive Measures, Vaccines, Drugs Available, and Decontamination. In each case, for both **Chemical** and **Biological** agents, each agent will have guidelines laid out within the book.

Keeping Local First Responders Viable to Respond to Chemical and Biological Agents

Terrorism is merely another responsibility for local fire departments and other first responders including police officers, emergency medical service personnel, civil defense employees, emergency management officials, emergency room physicians and other staff, health regulatory staff, the military, commercial response contractors, industrial hazardous materials response teams and fire brigades, and county, local, state, federal first responders to chemical and biological agents of all types. Attaining the demands and protocols for required training, and certification required by law, can be a serious drain on first responder staff and management. Local personnel get to a chemical or biological agent incident first, stay the longest, make the necessary decisions, and have limited money available for specialized equipment and training. Federal agencies and the military, on the other hand, have spent a whooping sum in each of the last two years for terrorist acts and response alone. A fire chief in Oklahoma City, Oklahoma, pointed out a few years ago that 3 percent of federal money goes to local fire departments while police forces obtain $7 billion a year in federal funds for local police departments. **However, during the past four years, this practice has changed drastically, as Department of Homeland Security (DHS) has been handing respective amounts of money to fire departments to respond to chemical and biological agents and weapons, terrorism, and other threats within the United States.** The local fire department is one of the best protections available against anarchy in the United States. It is the seeming belief of citizens that local firefighters can do anything and/or everything to protect their community. Firefighters and other first responders have also become targets of terrorist attacks.

Living In a Time of Aggression within a Rage of Communication

Terrorism is getting easier every month. The seemingly long-ago domestic terrorist incident at the Alfred P. Murrah federal building in Oklahoma City, Oklahoma, that happened at 9:02 a.m. on April 19, 1995, resulting in 169 deaths and 475 injuries was an example of basic terrorist procedure, action and tactics. The terrorist(s) used a ready-made "weapon" available to almost anyone, a fertilizer known as ammonium nitrate with a fuel oil igniter to set if off that can be obtained without undo notice, and be transformed by any person of average intelligence into a devastating bomb that can be exploded by very ordinary means and comes dirt cheap. The absolute power of ammonium nitrate was known for many years in the explosives industry, ever since the French ship *Grandcamp* exploded in the harbor at Texas City, Texas, on April 16, 1947, killing 510 persons and injuring thousands. Almost the entire local fire department of Texas City was obliterated by the *Grandcamp* explosion and fire.

On September 11, 2001, nineteen Arab terrorists stole four airliners from three different airports and killed over three thousand victims at the World Trade Center in the financial district of New York City, and the U.S. Military Headquarters at the Pentagon just outside our nation's Capital. Among the dead, were over 400 firefighters, police officers, and emergency medical service technicians and paramedics. Our first responders have definitely become terrorist targets. These Americans and others were basically killed by aviation fuel, a cheap method of mass killing if ever there was one. Within slightly more than a month later, on October 15, 2001, the United States underwent another first class tragedy when a "super" anthrax shut down government buildings in Washington, D.C. Let's face it, our federal government was totally unable to handle this particular mass attack with the Ames strain of anthrax, and, as of this writing, does *not* admit to knowing who was responsible for this sensational act of terrorism._

Over recent years, advances in science and society have made hazardous materials, including chemical and biological agents, much more widely available to a wider range of people throughout the world and thus made WMD accessible to terrorist groups or solitary loners. There are also many possible victims because innocent people are prime targets the terrorists aim for to get their messages across. A policy of "no innocents to be spared," means there are an infinite number of targets. Ten years ago most terrorist incidents were politically motivated. At the present time, they can be motivated by ethnicity, fraternal, nationalistic, abortion and/or anti-abortion, racial, cult, poverty, environmental, geographic, sexual, and medical reasons. Terrorism has been increasingly adopted by religious groups and cults. Everybody has a viewpoint, and with chemical and biological agents, one man or woman can make a strong point alone. You no longer have to have a country, a group, a state, an army, or a follower. You can work alone to make a violent statement on any issue, and your statement can be very bloody if you can draw from people who believe in suicidal attacks for their cause.

Terrorism can be very cheap and can be very successful in attracting attention while carrying low-risk to the terrorist organization. Right now, we are living in a time of maximum aggression nestled within a rage of communication. It seems like every person in the entire world has a message they want to tell. This is fairly new. Great increases in technological, economical, and political segments are altering the status of violence and crime throughout the entire world. Terrorists do not recognize any rules, laws, or statutes; and no locations, persons, or objects of social prestige or value can be immune from terrorist incidents. The primary goal of a terrorist attack is to cause panic among innocent and defenseless persons in order to generate political pressure against an enemy, government, organization, or group. Terrorists seek to control the news media to provide them with status they may never obtain by other means. With a terrorist incident, governments have to expect mass panic, large numbers of dead and injured, severe stress for those who live through the event, and perhaps delayed-action explosives designed to eliminate responding firefighters, police officers, EMS personnel and health workers.

A chemical or biological agent attack will produce unfamiliar situations for all first responders who are unfamiliar with such agents, sheer panic, confusion, psychological reactions, as well as death and physical injury. There could be a mass of unharmed persons who are *not* afflicted with chemical/biological aerosol or skin absorption who try to overwhelm area hospitals seeking help. First responders may have to install quarantine that can be very unpopular with both injured and non-injured persons who might rebel and fight such action. Medical triage is the sorting of and allocation of treatment to patients, especially with battle and disaster victims, according to a system of priorities designed to maximize the number of survivors. The U.S. Army uses the following four categories of triage for potential victims who have been involved with chemical agents: immediate, delayed, minimal, and expectant (interested readers should obtain *Field Management of Chemical Casualties Handbook* (2nd ed.), published July 2000 by the U.S. Army Medical Research Institute of Chemical Defense (USAMRICD), Chemical Casualty Care Division at Aberdeen Proving Ground in Maryland. The book is available at http://www.ccc@apq.amedd.army.mil or by contacting Linda Harris at EAI Corporation, 410-676-1449).

IMMEDIATE: A casualty classified as **immediate** has an injury that will be fatal of he does not receive immediate care. In a non-mass casualty situation, he would be the first casualty to receive care. However, in a mass casualty situation, particularly in a far-forward medical treatment facility, he may not receive this care. The required care may not be available at that echelon (e.g., a casualty may need major chest surgery)or the time needed to provide the care may be so prolonged that other casualties would suffer.

DELAYED: A **delayed** casualty is one who needs further medical care but can wait for that care without risk of compromising successful recovery. That person may require extensive surgical procedures and long-term hospitalization, but is presently stable and requires no immediate care. A casualty with a leg wound or fracture is an example of a conventional casualty who would be delayed. A casualty recovering from severe nerve agent poisoning will be delayed. Most casualties with vesicant burns will be delayed.

MINIMAL: A casualty would be classified as **minimal** is one who (1) can be treated by a medic and does not need to see a physicians or physician's assistant, (2) will not be evacuated, and (3) will return to duty within a day or so.

EXPECTANT: The **expectant** casualty is one for whom medical care cannot be provided at the medical treatment facility and cannot be evacuated for more advanced care in time to save his life. This category is used only during mass casualty situations. This category does not mean that these casualties will receive medical care.

Medical personnel will then transfer casualties for treatment/evacuation based on established priorities for treatment. "Casualties who have been classified as **immediate** are transferred to the contaminated medical treatment area for stabilization. After stabilization, these casualties are taken to the litter patient decontamination area. Casualties who have been classified as **delayed** may or may not require treatment in the collective protection treatment area before evacuation. If they need to enter this clean area for treatment, they are sent to the ambulatory or litter decontamination line, whichever is appropriate. If they do not need treatment in this area, they are sent directly to the evacuation holding area. Casualties who have been categorized as **minimal** may receive treatment in the collective protection shelter or the contaminated emergency treatment area. If they can be treated in the contaminated emergency treatment area and they have no break in their BDO (Battle Dress Over-Garment), they will be returned to duty from this area. If they require treatment in the clean treatment area, they will need to be sent to one of the decontamination areas before entry before entry into the area. If there is a break in their BDO, they will need re-supply. They must go through decontamination to the clean treatment area for re-supply. Casualties who have been categorized as **expectant** will be transferred to designated contaminated holding areas. These casualties will be constantly monitored while in this area and provided with available comfort measures."

Chemical Agents

Attackers with a chemical agent may have more problems than the defenders have under a chemical agent attack. If you chose to use a chemical agent to attack a military objective, or to stage a terrorist raid, your success ratio may change a great deal based on the weather report and how you use such information. Success, or failure, of either the attackers or potential victims may depend as much on weather and terrain conditions as any other factor in determining where the chemical agent can be an effective weapon. The wind may carry vapors more or less than you might have intended. Choosing vapors, aerosols, liquids, or other delivery systems will affect your intended outcome. Atmospheric stability, dispersion, temperature, humidity, and precipitation could all be involved.

High wind speed decreases effective coverage of an area, and higher temperatures equal greater vaporization. Precipitation in the form of rain, sleet, snow, and hail can wash chemical agents from vegetation, air, and material. Ground levels can create problems; higher concentration of released agents can be obtained in narrow valleys. Trees, grass, and brush can cause a gas cloud to dissipate more rapidly.

Chemical agents can basically be described as "persistent" or "non-persistent." Persistent chemical agents can poison people or animals for a long period of time by remaining a contact hazard or vaporizing to form an inhalation hazard. Conversely, non-persistent chemical agents disperse quickly as airborne particles, gases or liquids to provide an inhalation threat of short duration. Generally, chemical agents are cumulative in their effects. The human body can detoxify them to a limited extent, but an individual who suffers a thirty-minute exposure to distilled mustard or phosgene in two increments at two different times undergoes the same effect as in a one-hour exposure. However, hydrogen chloride or cyanogen chloride can be detoxified by the human body to a significant degree rather quickly compared to other chemical agents. Therefore, it would take a higher concentration of these two agents to result in maximum casualties.

Nerve agents such as tabun (GA), sarin (GB), and soman (GD) have a persistence that lasts from ten minutes to twenty-four hours in summer and from two hours to three days during winter. Aerosol is most likely to enter the eyes or lungs, while vapor may enter the eyes, skin, or mouth. VX is more persistent than other nerve agents having a persistence of two days to one week in summer and two days to some weeks in winter. Possible points of entry for VX are the same as those listed for other nerve agents. Blood agents such as hydrogen cyanide (AC) or cyanogen chloride (CK) persist for one to ten minutes during warm weather and ten minutes to one hour during cold weather. Possible entry routes for AC and CK via vapor or aerosol are through the eyes and/or lungs, while liquid could enter through the eyes, skin, and mouth.

Blister agents such as distilled mustard (HD) and nitrogen mustard (HN) have a slow rate of action and might be persistence for three days to one week in warm weather and for some weeks in cold weather, while lewisite (L) and mustard/lewisite (HL) have a quick rate of action and would be persistent for one to three days during summer and for weeks during winter. Route of entry for these blister agents would be nearly the same: eyes, lungs, and skin for vapor or aerosol, but for liquid, the points of entry would be the eyes and skin for HD and HN and eyes, skin, and mouth for L and HL.

The U.S. Marine Corps Biological Incident Response Force (CBIRF)

The U.S. Marine Corps Chemical Biological Incident Response Force (CBIRF), 350 Marines and Navy personnel of both sexes drawn from forty-four military occupational specialties, uses new tactics, procedures, and equipment to provide mass decontamination to local

U.S. civilian populations who may be contaminated by nuclear, biological, or chemical (NBC) agents. In a worst case terrorist, criminal, or accidental event, the CBIRF provides a standing, highly trained consequence management force tailored for short notice response to civilian victims of NBC materials or weapons of mass destruction.

In June of 1995, The White House issued Presidential Decision Directive 39 (PDD-39), "United States Policy of Counterterrorism." PDD-39 directed a number of measures to reduce the nation's vulnerability to terrorism, to deter and respond to terrorist acts, and to strengthen capabilities to prevent and manage the consequences of terrorist use of NBC weapons. The laws of the United States assign primary authority to the *states* to respond to the consequences of terrorism; the federal government provides assistance as required.

Within the military, in addition to the CBIRF, the U.S. Army Soldier and Biological Chemical Command through its "Domestic Preparedness" program has trained local emergency responders in 120 large population centers. The trainees are EMT/Paramedics/medical personnel, firefighters, law enforcement officers, and other local emergency personnel. There are also a number "Joint Task Force-Civil Support" (JTF-CS) teams integrating National Guard and military reserve components dedicated to assisting local civilian authorities in the event of an attack or accident within the United States; most states will use such teams in-state, guided by a headquarters component located at Fort Monroe, Virginia (for more information on such JTF-CS teams, see http://www.jtfcs.northcom.mil, or contact the Public Affairs Directorate at 757-788-6631). The U.S. Department of Defense also has a Chemical Biological Rapid Response Team (CB-RRT) which, under certain circumstances, will assist state and local responders in the detection, neutralization, containment, dismantlement and disposal of weapons of mass destruction containing chemical, biological or related hazardous materials.

The commanding officer of the CBIRF knows who he works for: "The 350 members of the CRIRF know our focal point, the person we support, is the civilian, local incident commander," he says. "He or she is the constitutionally mandated authority; and all local, state, and federal assistance should be focused on helping that civilian incident commander accomplish his or her tasks." In every training exercise undertaken by the CBIRF, there is one person in civilian clothes equipped with a radio who is required to be present at a chemical/biological incident: the local, civilian incident commander.

CBIRF's operational mission is to turn victims into patients who can be treated at civilian medical care facilities. Marine and Navy personnel have eight basic tasks. They determine the chemical or biological agent present, the threat it poses to potential victims, mark the location of victims, and secure the site. Another group will locate casualties, triage and tag them, and perform evacuation. A casualty clearing crew will perform another triage, stabilize casualties, and record information on charts. A hot zone coordinator will monitor casualty movement, determine resource needs, and maintain a hot zone perimeter. A separate group will perform a further triage and log-in casualties, stabilize and decontaminate injured, and confirm decontamination with a CAM (the chemical agent monitor is a hand-held device for monitoring chemical agent contamination of personnel and equipment; it detects G-series nerve agents and H-series blister agent vapors by sensing molecular ions of specific mobility's, or time of flight, and uses timing and microprocessor techniques to reject inferences and displays the relative concentration). A medical team will triage and stabilize casualties, record casualty information, and prioritize evacuation status. At the command and operations center, the incident is monitored and managed, control and coordination of CBIRF activities is undertaken, the casualty flow is tracked, and, through a "Reach Back" Web capability, experts can be contacted to provide pertinent information. Finally, a medical regulation team transmits casualty evacuation data and tracks evacuation resources.

The CBIRF has a Command Element; a Reconnaissance Element that handles agent detection and identification, a chemical/biological sample collection group, hazard area identification, and down-wind hazardous area determination staff; a Medical Element that advises and assists local medical authorities, does triage and emergency treatment, provides organic medical support, and does initial epidemiological investigation; a Security Element responsible for incident area security and isolation, site evacuation, critical personnel/government property security, crowd control, and other operations as required; and a Support Service Element that handles transport, embarkation, engineering, and supply sections. However, the Decontamination Element plays a central role in the operational mission of the CBIRF.

The primary mission of the Decontamination Element is to turn chemical/biological victims into patients through mass decontamination procedures by establishing a site capable of providing initial and sustained operational decontamination of Force personnel (rescue workers), ambulatory, and non-ambulatory patients. The Decontamination Element also handles decontamination of CBIRF members, attachments, vehicles, and equipment that have entered the incident site; controls access into and out of the incident site; handles processing of surety material and evidence while maintaining chain of custody through the site; and handles limited area decontamination of the incident site.

The Decontamination Element is staffed by one officer and thirty-one enlisted personnel, and can be set up in seven to eight minutes. Using just one decontamination site, about twenty-five to thirty non-ambulatory patients, about seventy to one hundred ambulatory, and a larger number of Force personnel can be handled in one hour. Most of this unit is cross-trained, so with the available two decontamination sites, such counts can be doubled. There are also five "hospital sites," each staffed by five Marines prepared to do ambulatory and non-ambulatory decontamination right at the hospital(s) for people who might have made it to a hospital on their own and have not undergone decontamination. Marines and Navy personnel who staff the Decontamination Element are trained at Fort McClellan, Alabama, at the Nuclear/Biological/Chemical Defense Specialist School and have a good background in weapons of mass destruction and decontamination. When on site, they provide assistance to the local, civilian incident commander, and do not take over the site in any manner.

Ambulatory Decontamination Line

Ambulatory decontamination involves moving mostly civilians going through the decontamination site. The Team Leader will sort casualties into ambulatory and non-ambulatory patients. Ambulatory patients will proceed through the site following directions of each station attendant. At a valuable registration point, all personal items (i.e., watches, wallets, jewelry, radios, etc.) will be dropped off, a receipt given, inserted in zip-locked bags, and an inventory tag prepared for collection at a later date. The bagged items will be decontaminated in a bucket containing a 5 percent HTH (calcium hypochlorite) solution or another appropriate decontamination solution, and placed in a crate to be transferred to the rear of the site. After processing through a shuffle pit containing decontamination solution, ambulatory patients step into the tent through the "Male" or "Female" side and walk through a shower to reduce airborne contamination (this procedure will be used for biological contamination only).

Patients will remove all remaining clothing. Force personnel who have been sent to ambulatory decon will remove the filter on their M40 mask once all clothing is removed, being sure to retain their mask for future decontaminations. At this station, all personnel will take a three-minute shower containing 0.5 percent HTH solution or an appropriate solution to decontaminate the entire body. Soap is provided to physically remove contamination only. Force personnel with masks will thoroughly wash the mask in the shower at this time. After the decontamination shower, personnel will move into the rinse shower. They will then dry off with the towels provided which are then discarded into trash chutes. Patients will be issued a hospital gown with slippers, and be monitored with appropriate detection equipment. Any person still contaminated will be processed through the site again. Patients will be logged in using a casualty tag as a tracking number. In a biological decontamination, procedures will remain the same except the door shower will be added at the beginning of the ambulatory tent. For a radiological decon, procedures remain the same except for using warm, soapy water instead of HTH.

Non-Ambulatory Decontamination Line

Non-ambulatory patients will be removed from the incident area by the casualty clearing team, and given to the triage team for treatment of wounds and categorized according to their injuries. The Decontamination Team Leader will sort patients where a determination is made; based on wounds and triage tags, on what order or priority each casualty will proceed. At the first station, all clothing will be removed; a system of cutting and rolling will be performed to safely remove all clothing. Bandages and splints will *not* be removed during clothing removal or at any other station; clothing will be cut off around the wound or split. An emphasis on safety and basic principles of decontamination is paramount at this station to prevent further injury to the patient and to avoid contaminating the casualty.

Closing the Decontamination Site

After all casualties and Force personnel are decontaminated, the site will either be torn down or converted to an equipment decontamination site. All items that cannot be decontaminated will be bagged and placed aside for future disposal. Starting at the dirty end of the site, all items to be decontaminated will be sprayed down with 5 percent HTH decontamination solution. Working from the dirty end towards the clean end, items will be passed down the line to the contamination control line (CCL) and be rinsed with clean water. Then, the item will be checked with detection equipment for contamination. If the equipment is clean, it will be passed across the CCL.

When all equipment is removed from the decontamination site, the containment pits will be sprayed with decontamination solution, rinsed, and checked for contamination. All bag items meant for disposal will be picked up by another agency. Once the site has been closed, Force personnel will perform MOPP (Mission-Oriented Protective Posture, the protective clothing used by members of the U.S. military who engage in nuclear, biological, or chemical warfare) dress-down. Once undressed, Force personnel will move towards the rear of the site area and be monitored for contamination. Area decontamination can be accomplished by spraying the decontamination area with 5 percent decontamination solution.

Lone Individuals as Terrorists

Identification of freelancer terrorists who operate alone, like those at Oklahoma City, are extremely difficult to spot by governmental agencies more accustomed to identifying anti-government *groups* rather than lone individuals. One or more individuals with few organizational attachments who follow standard "cell" practices and are not known by any other group that can identify cell members can be very difficult to identify. Agencies tend to have no knowledge of their existence until they stage an attack.

The Washington Post noted in early March of 2002 that a previously unknown report done by the U.S. Army Surgeon General after the September 11, 2001, attacks seems to conclude that as many as 2.4 million American people could be killed or injured in a terrorist attack against a toxic chemical plant within a densely populated area in the United States. The study seems to suggest that terrorist assaults on chemical industrial complexes could result in twice as many casualties as by other "worst case" situations as assumed by government. At the time, the federal government was asking for a budget of $38 billion for the next fiscal year (2002/2003), approximately double the amount of the previous year's budget, to spend strictly for domestic security.

On September 11, 2001, six of the nineteen hijackers, participants in possibly the greatest tragedy to ever befall United States citizens, were recognized by airline passenger profiling systems and received special handling. Two more were recognized because of problems with their identification, while another one was keyed for special evaluation because he was traveling with a man who had questionable personal identification. That is, 47.4 % were recognized as needing special evaluation and scrutiny as a result of a profiling test. The end result was that the nine men, who would go down in history as heroes in Moslem countries, had their checked baggage examined for explosives (all passengers have their carry-on baggage and themselves checked). They took over the four planes with box cutters and small knives, which we not considered "weapons" on September 11, 2001. Two of the hijackers were also on a FBI watch list of potential terrorists, but nobody told the airlines of this fact.

On Monday, March 11, 2002, six months to the day of the September 11, 2001, terrorist attack on America, a flight school in Venice, Florida (Huffman Aviation), had received student visa approval forms for Mohamed Atta and Marwan Al-Shehhi from the U.S. Immigration and Naturalization Service, a part of the United States Justice Department. Our country's "justice" department had granted two of the most abhorred names in the United States student visas to live and study at a flight school six months after they had participated in the most broad-based terrorist attack to ever hit the United States—Atta the chief honcho, pilot, and field planner, and Al-Shehhi , possibly the pilot of the second plane to hit the World Trade Center. They were together in Germany planning the attack and died in separate airliners used for the attack.

Dedicated Services of U.S. Federal Agencies and the Threat of Weapons of Mass Destruction

Specific capabilities of federal agencies, bureaus, and offices in dealing with WMD are stated below. Fire and police chiefs in the United States are often frustrated in not knowing exactly what response services are available from the federal government to help local officials when a terrorist incident occurs with the United States.

Locate and Examine an Unknown WMD Device

Army Technical Escort Unit; 52nd Explosives Ordnance Disposal Unit and other selected Department of Defense (DOD) units: DOE Joint Technical Operations Team, DOE Nuclear-Radiological Advisory Team, DOE Nuclear Emergency Search Team, DOE Lincoln Gold Emergency Team.

Render Safe an Armed WMD Device

Army Technical Escort Unit; 52nd Explosives Ordnance Disposal Unit and other selected DOD unit:; DOE Joint operations Team; Nuclear-Radiological Advisory Team; Nuclear Emergency Search Team; Lincoln Gold Augmentation Team.

Identify or Evaluate WMD Agents

Army Technical Escort Unit, Marine Corps Chemical Biological Incident Response Force, Army Medical Research Institute for Infectious Diseases, Naval Medical Institute, HHS Center for Disease Control and Prevention, HHS National Institutes of Health, HHS Agency for Toxic Substance Registry, HHS Food and Drug Administration, EPA Radiological Emergency Response Team, EPA Environmental Response Team; Environmental Radiation Ambient Monitoring System, EPA National Enforcement Investigations Center, EPA Research Laboratories, EPA Contract Laboratories, DOE Joint Technical Operations Team, Nuclear-Radiological Advisory Team, DOE Nuclear Emergency Search Team, DOE Lincoln Gold Augmentation Team.

Project Dispersion of WMD Agents

DOD Defense Special Weapons Agency, HHS Center for Disease Control and Prevention, HHS Agency for Toxic Substance Registry, EPA Radiological Emergency Response Team, EPA Environmental Response Team, EPA Environmental Radiation Ambient Monitoring System, DOE Federal Radiological Monitoring and Assessment Center (before handoff to EPA), DOE Atmospheric Release Advisory Capability.

Track Dispersion of WMD Agents

DOD Defense Special Weapons Agency, EPA Radiological Emergency Response Team, Environmental Radiation Ambient Monitoring System, EPA Environmental Response Teams, Federal Radiological Monitoring and Assessment Center (after handoff from

DOE), Federal Radiological Monitoring and Assessment Center (before handoff to EPA), DOE Aerial Measuring Systems, Atmospheric Release Advisory Capability.

Provide Medical Advice on Health Impact of WMD

DOD Army Medical Research Institute for Infectious Diseases, DOD Naval Research Institute, HHS Center for Disease Control and Prevention, HHS National Institutes of Health, Agency for Toxic Substance Registry, HHS Food and Drug Registry, EPA Radiological Emergency Response Team, EPA Environmental Radiation Ambient Monitoring System, EPA Environmental Response Team, EPA Federal Radiological Monitoring Center (after handoff from DOE); DOE Federal Radiological Monitoring and Assessment Center (before handoff to EPA), DOE Atmospheric Release Advisory Capability, DOE Radiation Emergency Assistance Center and Training Site.

Provide Triage and Medical Treatment

DOD Marine Corps Chemical Biological Incident Response Force, DOD Army Medical Research Institute for Infectious Diseases, DOD Naval Medical Research Institute, HHS National Medical Response Teams, HHS Disaster Medical Assistance Teams, HHS Metropolitan Medical Strike Teams, HHS Experts from Public Health Safety agencies, DOE Radiation Emergency Assistance Center and Training Site.

Administer Antidotes, Vaccines, and Chelating Agents

DOD Army Medical Research Institute for Infectious Diseases, DOD Naval Medical Research Institute, Variety of Potential HHS Units, DOE Radiation Emergency Assistance Center and Training Site.

Decontaminate Victims

DOD Marine Corps Chemical Biological Incident Response Force, variety of potential HHS units.

Decontaminate Equipment and Other Materials

DOD Marine Corps Chemical Biological Incident Response Force, EPA Environmental Response Team, EPA Emergency Response Team, EPA Emergency Response Contract Services, DOE Radiological Assistance Program.

Package and Transport WMD Devices and Agents

DOD Army Technical Escort Unit, DOD 52nd Explosives Ordnance Disposal Unit, EPA Environmental Response Team, EPA Emergency Response Contract Services, DOE Joint Technical Operations Team.

Other Departments

U.S. State Department Foreign Emergency Response Team (FEST)
Federal Bureau of Investigation Hazardous Materials Response Unit (HMRU)
Hostage Rescue Team (ninety-one agents)
Enhanced SWAT Teams (nine teams, 355 agents)
SWAT Teams (forty-seven teams, 706 agents)
Domestic Emergency Response Team (DEST)
Joint Operations Center
U.S. Army Reserve Chemical Biological Rapid Response Teams (C/B RRTs).
U.S. State Department, Bureau of Diplomatic Security (investigations).
U.S. Bureau of Alcohol, Tobacco, and Firearms (investigations, training).
U.S. Secret Service (investigations, training)
U.S. Customs Service (investigations)
U.S. Department of the Treasury (Financial Crimes Enforcement Network)
U.S. Immigration and Naturalization Services (investigations)
U.S. Environmental Protection Agency (investigations)
Postal Inspection Service (investigations)
U.S. State Department (administers the Antiterrorism Assistance Program)
U.S. Immigration and Naturalization Services (investigations)
U.S. Environmental Protection Agency (investigations)
Postal Inspection Service (investigations)
Federal Aviation Administration (training)
U.S. Department of Justice (counter-proliferation)
F.B.I. and U.S. Department of Defense (counter-proliferation)

Guides for Emergency Response: Chemical Agent or Weapon: Arsenical Vesicants including Ethyldichloroarsine (ED), Methyldichloroarsine (MD), Phenyldichloroarsine (PD)

19

Arsenical Vesicants: Such chemical agents are organic dichloroarsines. The main ones that could become chemical agents or weapons are ethyldichloroarsine, methyldichloroarsine, and phenyldichloroarsine. All arsenical vesicants are colorless to brown liquids, soluble in most organic solvents but poorly soluble in water. **They are more volatile than mustard agents and have fruity to geranium-like odors.** They react rapidly with water to yield the corresponding solid arsenoxides, with concurrent loss of volatility and most of their vesicant properties. As liquids, they gradually penetrate rubber and most impermeable fabrics. Arsenical vesicants are much more dangerous as liquids than as vapors. The liquids do cause hazardous burns on the eyes and skin, while field concentrations of vapors are **not** likely to cause permanent significant injuries. Immediate decontamination is required to remove the liquid agents in time to prevent severe burns, but decontamination is not required for vapor exposure unless there is pain to the victim. When inhaled, the vapors cause sneezing and may instill irritation of the upper respiratory tract.

BLISTER AGENT (Vesicant): Ethyldichloroarsine (ED). ED is a liquid with a fruity but biting and irritating odor first produced by the Germans in 1918 while they were attempting to come up with a volatile chemical agent with a short duration of effectiveness that would act more quickly than diphosgene or mustard, and would last longer in its effects than phenyldichloroarsine (PD). Resembling other chemical agents that contain arsenic, ED is irritating to the respiratory tract and will cause lung injury after sufficient exposure. The vapor is irritating to the eyes as well, while the liquid can produce serious eye injury. ED may also absorb either liquid or vapor through the skin in sufficient amounts to cause systemic poisoning or death, and prolonged contact with either liquid or vapor will blister the skin. Its flash point is high enough not to interfere with military use, and its rapid hydrolysis produces hydrochloric acid as well as ethylarsenious oxide. It is similar to other arsenicals in that sub-lethal amounts also detoxify rapidly. The following personal protective equipment (PPE) is required to deal with ED: protective mask and permeable protective clothing for ED vapor and small droplets; impermeable protective clothing for protection against large droplets, splashes, and smears. **Decontamination** is not often necessary in the field, but it may be in enclosed areas.

Formula: C2H5AsC12

Vapor Density: (Air = 1) 6.0

Vapor Pressure: (mm^Hg) 2.09 at 20 degrees C.

Molecular Weight: 174.88

Liquid Density: (g/cc) 1.66 at 20 degrees C.

Volatility: (mg/m3) 20,000 at 20 degrees C.

Median Lethal Dosage: (mg-min/m3) 3,000–5,000 by inhalation; 100,000 by skin exposure.

Physical State: (at 20 degrees C) Colorless liquid.

Odor: Fruity, but biting; irritating.

Freezing/melting point: (at degrees C) –65 C.

Boiling point: (at degrees C) 156 C.

Action rate: Irritating effect on nose and throat is intolerable after one minute at moderate concentrations, delayed blistering .

Physiological action: Damages respiratory tract; affects eyes; blisters; can cause systemic poisoning.

Required level of protection: Protective mask and clothing.

Decontamination: Not usually necessary in the field. If necessary for enclosed areas, use HTH, STB, household bleach, caustic soda, or DS2. (Decontamination liquid agent on skin with M258A1, M258, or M259 skin decontamination kit. Decontaminate individual equipment with M280 decontamination kit).

Detection in the field: M18A2

Use: Delayed-action casualty agent.

CAS registry number: 598-14-1.

RTECS number: CH3500000.

LCt50:   (percutaneous) 3,000–5,000 mg-min/m3, depending on the period of exposure.

BLISTER AGENT (VESICANT):   Methyldichloroarsine (MD). MD is quite similar to ED and, like many other arsenicals, it causes immediate irritation of the eyes and nose but with blistering effects being delayed for hours. It also is irritating to the respiratory tract and can produce lung damage if given a high exposure. The **vapor** irritates the eyes, and the **liquid** could seriously injure the eyes. **The absorption of either liquid or vapor through the skin in sufficient amounts may lead to systemic poisoning and lead to death.** Prolonged contact with either liquid or vapor produces blistering of the skin, but vapor concentrations that present a blistering effect are quite difficult to obtain in the field. Although there is reported to be no accurate data, the LCt/50 is probably similar to ethyldichloroarsine (ED) at 3,000–5,000 mg-min/m3. **MD's persistency is relatively short, but its intended use is as a delayed-action casualty agent; that is, it has immediate irritation but the blistering effect, if present, is delayed for several hours.** The human body is able to detoxify MD at an appreciable rate.

Formula: CH3AsCl2

Vapor Density: (Air = 1) 5.5

Vapor Pressure: (mm^Hg) 7.76 at 20 degrees C

Molecular Weight: 160.86

Liquid Density: (g/cc) 1.836 at 20 degrees C

Volatility: (mg/m3) 74,900 at 20 degrees C

Median Lethal Dosage: (mg-min/m3) Estimated at 3,000–5,000

Physical State: (at 20 degrees C) Colorless liquid

Odor: None

Freezing/melting point: (at degrees C) –55 C

Boiling point: (at degrees C) 133 C

Action rate: Immediate irritation of eyes and nose; delayed blistering

Physiological action: Irritates respiratory tract; injures lungs and eyes; causes systemic poisoning

Required level of protection: Protective mask and clothing

Decontamination: Bleach, DS2, caustic soda, M258A1, M280

Detection in the field: M18A2

Use: Delayed-action casualty agent

CAS registry number: 593-89-5

RTECS number: CH4375000

LCt50:   No accurate data; probably similar to ED, 3,000 to 5,000 mg-min/m3

BLISTER AGENT (VESICANT):   Phenyldichloroarsine (PD). PD is a colorless liquid that is used as a delayed-action chemical agent, and its persistency depends on the manner of dissemination and the weather. **It is classed as a blister agent but it behaves like a vomiting agent, although its use during World War I did not tend to be superior to other vomiting agents.** Phenyldichloroarsine does have an immediate effect on eyes but a delayed effect of thirty minutes to one hour on skin. PD blisters bare skin but wet clothing decomposes it at once. A protective mask and protective clothing provide adequate protection, but **protection against large droplets, splashes, and smears requires impermeable clothing.** The rate of hydrolysis of PD is very rapid producing hydrochloric acid and phenylarsenious oxide. **When, and if, it gets dispersed as an aerosol, its volatility (390 mg/m3 at 25 degrees C) is such that this agent would be effective against unprotected people, although only as a chemical agent with a short duration of effectiveness.** It is classed as a delayed-action casualty agent.

Formula: C6H5AsCl2

Vapor Density: (Air = 1) 7.7

Vapor Pressure: (mm^Hg) 0.033 at 25 degrees C

Molecular Weight: 222.91

Liquid Density: (g/cc) 1.65 at 20 degrees C

Volatility: (mg/m3) 390 at 25 degrees C

Median Lethal Dosage: (mg-min/m3) 2,600 by inhalation

Physical State: (at 20 degrees C/68 degrees F).

Odor: None

Freezing/melting point: (at degrees C): –20 C

Boiling point: (at degrees C): 252 to 255

Action rate: Immediate eye effect, with skin effects appearing in thirty to sixty minutes.

Physiological action: Irritates; causes nausea, vomiting, and blisters.

Required level of protection: Protective mask and clothing.

Decontamination: Bleach, DS2, caustic soda, M258A1, M280.

Detection in the field: M18A2, and Draeger Tube Organic Arsenic Compounds and Arsine.

Use: Delayed-action casualty agent.

CAS registry number: 696-28-6

RTECS number: CH5425000

LCt50:   (respiratory) 2,600 mg-min/m3

Ability to Kill: Limited.

Characteristics: Arsenical vesicants cause severe damage to the eye, where upon contact pain and blepharospasm (a spasm of the eyelid) occur at once. Edema of the conjunctivae and lids follow in a rapid fashion and tend to close the eye within an hour. Inflammation of the iris can be evident by this time, but after a few hours the edema of the lids starts to subside, while the haziness of the cornea develops and iritis increases. The corneal injury, that varies with the amount of exposure, may heal without residuals, induce pannus (a vascular tissue causing a superficial opacity of the cornea) formation, or progress to massive necrosis. However, the iritis may subside without permanent impairment of vision if the amount of exposure was mild; but after heavy exposure, hypopyon (an accumulation of white blood cells in the anterior chamber of the eye) may begin which terminates in necrosis, depigmentation of the iris, and synechia (adhesion of the iris to the cornea or the crystalline lens) formation. Arsenical vesicants instantly produce a gray scarring of the cornea, rather like an acid burn, at the point of contact. Necrosis and sloughing of both bulbar and palpebral conjunctivae may follow very heavy exposure. All injuries to eyes are susceptible to secondary infections. Mild conjunctivitis due to arsenical vesicants heals in a few days without specific treatment, but severe exposure may cause permanent injury or blindness.

Caution: Some serious burns from arsenical vesicants can cause shock and systemic poisoning and are life-threatening; even when the victim survives the acute effects the

prognosis must remain guarded for several weeks. Decontamination must be done at once to get rid of liquid agents to prevent serious burns, but decontamination is usually not required for vapor contamination unless the victim experiences pain. In severe cases, the systemic use of morphine has cause to be used for the control of pain. **Arsenical vesicants such as phenyldichloroarsine (PD) are often mixed with mustard agent; such use does not produce more severe lesions than either agent used alone, but is meant to confound and confuse the "enemy" and make diagnosis more difficult to define correctly.**

Field First Aid: Remove victims from the scene and source of contamination; immediately establish a decontamination procedure for persons exposed, or possibly exposed, to **liquid** arsenical vesicants. For victims potentially exposed to **vapor,** no decontamination is normally required **unless a victim complains of having pain.** Medical treatment is largely symptomatic, **but liquid arsenical vesicants produce more severe lesions of the skin and eyes than liquid mustard.** Arsenical vesicants cause serious damage to the eye; both pain and blepharospasm can occur at once, and edema of the conjunctivae and lids follows rapidly forcing the eye to close within one hour or less. The iris becomes inflamed usually, but after a few hours the edema of the lids starts to lessen, and haziness in the cornea develops and iritis increases. The corneal injury which is varied by the severity of exposure may heal without residual injury, induce pannus (a vascular tissue causing a superficial opacity of the cornea) formation, or lead to massive necrosis. The iritis might fail to provide permanent impairment of vision if the exposure received was small; whereas after a heavy exposure, hypopyon (an accumulation of white blood cells in the anterior chamber of the eye) may be present which can terminate in necrosis, depigmentation of the iris, and synechia (an adhesion of the iris to the cornea) formation. Arsenical vesicants immediately cause a gray scarring of the cornea similar to an acid burn at the position of contact. Necrosis and sloughing of the conjunctivae may be a result of heavy exposure. **All injured eyes can suffer secondary infection. Mild conjunctivitis can actually heal in about a week, while major exposures can change the end-result to permanent injury and/or blindness.**

Burns serious enough to cause shock and systemic poisoning are life-threatening. On scene medical personnel need to understand indications for treatment of victim(s) contaminated with arsenical vesicants by any route: look for a cough with dyspnea and frothy sputum that may be blood tinged and other signs of pulmonary edema; search for a skin burn the size of the palm of a hand or larger caused by **liquid** arsenical vesicants which was not decontaminated within the first fifteen minutes; or seek out skin contamination by an arsenical vesicant covering 5 percent, or about one square foot of skin, in which there is evidence of immediate skin damage such as gray or white blanching of the skin, or in which erythema develops over the surface within the first fifteen minutes. On scene medical personnel, after sorting and arranging decontamination of exposed victims, should begin two treatments immediately. Dimercaprol (British anti-lewisite, or BAL) ointment should be applied to victims of skin exposure which are found before actual vesication has begun. First, for local neutralization on and within the dermal surface, begin by applying a liberal portion of BAL ointment. Any and all protective ointment already on a victim's skin **must be removed before any application of the BAL ointment because BAL will destroy the other ointment.** Spread BAL ointment in a thin film that should be rubbed in with the medical person's fingers, left on the victim(s) skin for at least five minutes, and later washed off with water. A modicum of mild dermatitis may occur if BAL ointment is frequently administered to the same area of skin. BAL should *not* be used as a protective barrier ointment on unaffected skin because of its dermatitis properties. The second phase is the immediate intramuscular injection of "BAL in Oil" (10 percent) given deep into the muscles of the buttocks taking every possible precaution against injecting into a blood vessel. The amount of dosage must be nearly approximate to the weight of the victim allowing 0.5 mL per every 25 pounds, up to a maximum of 4 mL (i.e., 125 pounds for 2.5 mL; or 200 pounds and over for 4 mL). Repeat the intramuscular injection of BAL in Oil at different locations in the buttocks at four, eight, and twelve hours after the initial injection for a total of four equal doses. If other signs or symptoms or serious arsenical poisoning seem to be present, the interval between the first and second dose may be shortened to two hours. In severe cases, half doses could be given at the rate of one injection per day for three to four days. Use of BAL in Oil does have its own symptoms that arrive in fifteen to thirty minutes, and may end about thirty minutes after each intramuscular injection. These mild symptoms can be disconcerting to victims, but with the exception of symptoms that appear to be unduly serious or prolonged, medical staffs are cautioned to provide the full regimen of treatment. In toxic victims, liberal fluids by mouth or intravenous, and high-vitamin, high-protein, high-carbohydrate diets could be indicated. For those victims where shock is in evidence, provide the usual supportive measures such as intravenous administration, blood transfusions, or other vascular volume expanders should be indicated.

Drugs: Dimercaprol (trade name, BAL in Oil) is both an antidote and a chelating agent (any of various compounds that combine with metals to form chelates and that include some used medically in the treatment of metal poisoning). The product is used for treatment or arsenic, gold, and mercury poisoning; treatment of acute lead poisoning with calcium edetate disodium; and treatment of acute mercury poisoning if used within the first one or two hours. BAL in Oil can cause severe renal damage; alkalinize urine to increase chelated complex excretion; use extreme caution with children.

Medical Management: **Immediate decontamination after exposure is the only way to prevent damage to victims, followed by symptomatic management of lesions.** Hospital care tends to be supportive. It should be repeated that **liquid** arsenical vesicants produce more serious lesions on dermal surfaces than do liquid mustard. In toxic victims, liberal fluids by mouth or intravenous, and high-vitamin, high-protein, high-carbohydrate diets could be indicated. For those victims where shock is in evidence, provide the usual supportive measures such as intravenous administration, blood transfusions, or other vascular volume expanders should be indicated.

Fire: Extinguishing media for arsenal vesicants are water, fog, foam, or CO2. Avoid use of extinguishing methods that will cause splashing or spreading of chemical product or containers.

Personal Protective Equipment: For arsenical vesicants wear full protective clothing consisting of Level A: Level A should be worn when the highest level of respiratory, skin, eye, and mucous membrane protection is needed. This level consists of a fully-encapsulated, vapor-tight, chemical-resistant suit, chemical-resistant boots with steel toe and shank, chemical-resistant inner/outer gloves, coveralls, hard hat, and self-contained (positive pressure) breathing apparatus (SCBA). Both the Environmental Protection Agency and NIOSH recommend that initial entry into unknown environments or into confined space that has not been chemically characterized be conducted wearing at least Level B, if not Level A, protection. For protective gloves, it is mandatory to wear Butyl toxicological protective gloves such as M3, M4, or glove set. Once on the scene, monitor the area paying close attention to low-lying areas to ascertain if you can downgrade your level of personal protective equipment (PPE) for ease of operation within a guaranteed safety level.

Spill/Leak Disposal: Arsenical vesicant agents or weapons should be contained using vermiculite, diatomaceous earth, clay, or fine sand and neutralized as soon as possible using copious amounts of 5.25 percent sodium hypochlorite solution. Scoop up all material and place it in an approved Department of Transportation container. Cover the contents with decontaminating solution as mentioned above. The exterior of the salvage container shall be decontaminated and labeled according to EPA and DOT regulations. All leaking containers will be over packed with sorbent (e.g., vermiculite) placed between the interior and exterior containers. Decontaminate and label according to EPA and DOT regulations. Dispose of the material in accordance with waste disposal methods provided below. Conduct general area monitoring with an approved monitor to confirm that the atmospheric concentrations do not exceed the airborne exposure limits. If 5.25 percent sodium hypochlorite solution is not available, the following decontaminants may be used instead and are listed in order of preference: Calcium Hypochlorite, Contamination Solution No. 2 (DS2), and Super Tropical Bleach Slurry (STB).

Approved Methods of Waste Disposal: Decontamination of waste or excess material shall be accomplished according to procedures outlined above and can be destroyed by incineration in EPA approved incinerators according to appropriate provisions of federal, state, and local Resource Conservation Act (RCRA) regulations. Note: Some decontamination solutions are hazardous waste according to RCRA regulations and must be disposed of according to these regulations.

Symptoms: Symptoms include stinging pain felt usually in ten to twenty seconds after contact with liquid arsenical vesicants; the pain continues to get more serious with penetration that results in a level of deep, aching pain that signals a need for immediate decontamination to save victim(s) from lesion damage. After about five minutes of contact, a gray area of dead epithelium appearing the same as seen in corrosive burns. Erythema is like that caused by mustard agent but is accompanied by more pain. Itching and irritation will last for roughly twenty-four hours whether or not a blister develops. Blisters are often developed in twelve hours and are painful at first which is a contrast to the relatively painless mustard blister. The erythema of arsenical vesicants normally recedes more rapidly than the erythema of mustard agent and with less pigmentation. Small blisters heal in about the same time as those from mustard agent, while large lesions may well involve serious injuries which heal slowly and may require skin grafts.

Vaccine: None.

Guides for Emergency Response: Chemical Agent or Weapon: Arsine 20

Agent: Blood agents are substances that injure a person by interfering with cell respiration (the exchange of oxygen and carbon dioxide between blood and tissues). Potential blood agents that might be used as weapons of mass destruction (WMD) are listed in a standard format. Two agents are of most concern, hydrogen cyanide (AC) and cyanogen chloride (CK). These blood agents have a very high vapor pressure which causes the rapid evaporation of the liquid after exposure. **Such vaporization greatly reduces the chance of a liquid exposure.** The high vapor pressure of AC and/or CK distributed will quickly expand outward and up, and high volatility will quickly cause the vapor to lose its lethal concentration near the delivery point. Because of quick dissipation of the vapor, AC and/or CK are called non-persistent agents. Cyanide causes few signs and symptoms in humans. Death occurs within a few minutes upon inhalation of a large amount, while inhalation of low concentration will be less severe. **The only aid for a person exposed to either AC or CK is to don a protective mask.** Rapid onset of symptoms may make medical assistance of no valid use since they occur in a very short time period and patients could rapidly succumb to death.

Another blood agent is arsine (SA) which is a colorless, flammable, extremely poisonous gas.

Formula: AsH3

Vapor Density: (Air = 1) 2.69

Vapor Pressure: (mm^Hg) 11,100 at 20 degrees C

Molecular Weight: 77.93

Liquid Density: (g/cc) 1.34 at 20 degrees C

Volatility: (mg/m3) 30,900,000 at zero degrees C

Median Lethal Dosage: (mg-min/m3) 5,000

Physical State: (at 20 degrees C) Colorless gas

Odor: Mild, garlic-like

Freezing/melting point: (at degrees C) –116 C

Boiling point: (at degrees C) –62.5 C

Action rate: Delayed action, two hours to eleven days.

Physiological action: Damages blood, liver, and kidneys.

Required level of protection: Protective mask.

Decontamination: None needed.

Detection in the field: Draeger Tube Arsine 0.05/a.

Use: Delayed-action casualty agent

CAS registry number: 7784-42-1

RTECS number: CG6475000

LCt50: 5000 mg-min/m3

Arsine Patient Information

What Is Arsine?

Arsine is a colorless gas that does not burn the eyes, nose, or throat like some other dangerous gases. It has a garlic-like or fishy smell, but only at relatively high concentrations. A person can be exposed to a high concentration of arsine and not be able to smell it.

Certain ores or metals may contain traces of arsenic. If water or acid contacts these ores or metals, they may release small amounts of arsine gas. Arsine is widely used in manufacturing of fiberoptic equipment and computer microchips. It is sometimes used in galvanizing, soldering, etching, and lead plating.

What Immediate Health Effects May Result from Arsine Exposure?

Besides the odor, there may be no other immediate sign that a person is breathing arsine. Its main effect is to destroy red blood cells, causing anemia (destruction of red blood cells) and kidney damage (from red blood cell debris). Within hours after a serious exposure, the victim may develop dark red or brown urine, back pain or belly pain, weakness, or shortness of breath. The skin or eyes may become yellow or bronze in color. Although arsine is related to arsenic, it does not produce the usual signs of arsenic poisoning.

What Is the Treatment for Arsine Poisoning?

There is no antidote for arsine poisoning, but its effects can be treated. The doctor may give the exposed patient fluids through a vein to protect the kidneys from damage. For severe poisoning, blood transfusions and cleansing of the blood (hemodialysis) may be needed.

Are Any Future Health Effects Likely to Occur?

After a serious exposure, symptoms usually begin within two to 24 hours. People who have no signs of poisoning during this time probably have not breathed a large amount of arsine and may be sent home with instructions for follow-up medical care. Most people do not have long term effects from a single, small exposure to arsine. In rare cases, permanent kidney damage or nerve damage has developed after a severe exposure.

Repeated exposure to arsine may cause skin and lung cancer.

What Tests Can Be Done if a Person Has Been Exposed to Arsine?

There are no specific tests for arsine exposure. However, blood, urine and other tests may show if there has been any serious injury to the lungs, blood cells, kidneys, or nerves.

Where Can More Information about Arsine be Obtained?

More information about arsine can be obtained from your regional poison control center, your state, county, or local health department; the Agency for Toxic Substances and Disease Registry (ATSDR); your doctor; or a clinic in your area that specializes in occupational and environmental health. If the exposure happened at work, you may wish to contact the Occupational Safety and Health Administration (OSHA) or the National Institute for Occupational Safety and Health (NIOSH).

Ability to Kill: Moderate.

Characteristics: Arsine is a highly toxic gas that can destroy red blood cells and create organ injuries leading to death. It is flammable, a potential cancer causing and teratogenic agent, and a systemic toxin. It reacts with strong oxidizers, chlorine, and nitric acid. This possible agent or weapon, which has a number of uses in industry, has an identifying number of UN 2188. For this agent, inhalation is the major route of exposure, and its odor threshold is ten times greater that the OSHA permissible exposure limit. **Odor is not an adequate indicator of arsine's presence and does not provide reliable warning of hazardous concentrations.** This chemical gas is heavier than air so hazardous concentrations may develop quickly in closed or poorly ventilated areas, or in low areas. After absorption by the lungs, arsine enters red blood cells where hemolysis (a process of disintegration or dissolution of red blood cells with liberation of hemoglobin) and impairment of oxygen transport. Inhibition of catalase (an enzyme) may lead to accumulation of hydrogen peroxide (an oxidizer) which destroys red cell membranes. Acute intravascular hemolysis develops with hours and may continue for up to ninety-six hours and anemia develops. Difficult breathing is an early symptom of arsine exposure, and kidney failure is fairly common. The skin of a victim beset by arsine toxicity is induced by hemolysis and may be caused by hemoglobin deposits; however, this is not true jaundice which can occur in severe cases.

Caution: Contact with arsine vapor can be fatal. *Do not breathe the fumes!* If inhalation is a possibility, hold breath until respiratory mask is donned. Firefighters should wear full protective clothing and respiratory protection during both fire-fighting and rescue (positive pressure, full face, NIOSH-approved self-contained breathing apparatus/SCBA shall be worn).

Field First Aid: Monitor the incident scene, and victims, to attempt to identify the chemical or biological agent that is creating the problem. Separate the walking wounded from the prone (e.g., "All those who can walk, stand over there"). Assess the number of prone and walking wounded. If Emergency Medical Service personnel are not on the scene, call them. If it appears to be a mass casualty incident, call for help immediately. As soon as a person trained in triage arrives on scene (physician, physician's assistant, nurse, paramedic or EMS person) have triage instituted. For inhalation exposure, remove victims from the source **immediately**; give artificial respiration if breathing has stopped (refrain from mouth-to-mouth attempts to provide artificial respiration; administer oxygen if breathing is difficult; *seek medical attention at once!* If there is contact with the eyes, speed in decontaminating the eyes is absolutely essential; remove victim(s) from the liquid or inhalation source; if necessary, flush the eyes immediately with water for ten to fifteen minutes by tilting the head to the side, pulling the eyelids apart with fingers, and pouring water slowly into the eyes; do not cover eyes with bandages, but if necessary, protect eyes by means of dark glasses or opaque goggles; seek medical help immediately! Time is of the essence. For skin contact, remove patient from the source at once; remove contaminated clothing; decontaminate affected areas by flushing with ten percent sodium carbonate solution; wash off with soap and water after three to four minutes to protect against erythema; seek medical attention at once. Isolate contaminated items such as clothing, shoes, personal items, and equipment.

No one enters the Hot Zone or Warm Zone without appropriate personal protective equipment. Ensure that responding agencies know the danger(s) they face by responding to this incident.

Quickly ensure ABC (airway/breathing/circulation). See that the victim has a clear airway. If trauma is suspected, maintain cervical immobilization manually and apply a cervical collar and a backboard if feasible. Check for adequate respiration, and administer supplemental oxygen if cardiopulmonary compromise is suspected. Maintain adequate circulation. Establish intravenous access when necessary and warranted. Use a cardiac monitor as necessary. Stop any bleeding when necessary. When ingestion is evident, **do not induce emesis** (vomiting). For victims who are alert and able to swallow, give four to eight ounces of milk to drink. Administration of activated charcoal is **not** viewed as beneficial.

Treatment consists primarily of supportive care.

Response on Scene by First Responders

Caution: Arsine is a flammable and highly toxic gas that does not provide adequate warning of hazardous levels. Inhalation is the major route of arsine exposure, although there is little information about absorption through the skin or toxic effects on the skin or eyes. **Contact with liquid arsine may result in frostbite.**

Guidelines for arsine are as follows: the OSHA PEL (permissible exposure limit) equals 0.05 ppm (averaged over a eight-hour workshift); the NIOSH IDLH (immediately dangerous to life or health) equals three ppm; and the AIHA ERPG-2 (emergency response planning guideline) equals 0.5 ppm (maximum airborne concentration below which it is believed that nearly all persons could be exposed for up to one hour without experiencing or developing irreversible or other serious health effects or symptoms that could impair their abilities to take protective action). **Arsine is an extremely flammable, toxic gas and may be fatal if inhaled in sufficient quantities; it may be ignited by heat, sparks or flames; vapors may travel to a source of ignition and flash back.** Initially, some patients may look relatively well. Common initial symptoms of exposure include malaise, headache, thirst, shivering, abdominal pain and dyspnea. Such symptoms usually occur within thirty to sixty minutes with heavy exposure, but can be delayed for two to twenty-four hours. Hemoglobinuria (the presence of free hemoglobin in the urine) usually appears within hours, and jaundice within one to two days. Chronic arsine exposure can result in gastrointestinal upset, anemia, and damage to lungs, kidneys, liver, nervous system, heart, and blood-forming organs. Little information is provided on health effects of chronic low-level exposures to arsine. **Arsine has not been classified as yet for carcinogenic effects, but arsenic compounds and metabolites have been classified as known human carcinogens by IARC and EPA. In addition, arsine should be treated as a *potential* teratogenic** (related to, or causing, developmental malformations; abnormal in growth or structure) agent.

Field First Aid: Evacuate the Hot Zone at once when there is any release of arsine; consider any victims who may have inhaled arsine to have suffered a potentially toxic dose. Although small amounts of arsine can be trapped in the victim's clothing or hair, these quantities are not likely to cause a danger for first response personnel outside the Hot Zone. Toxic effects could be delayed for up to two to twenty-four hours after exposure; arsine exposure victims should all be evaluated at a medical facility. **There is no specific antidote for arsine; treatment is symptomatic and consists of actions to support respiratory, vascular, and renal functions.**

Since arsine is a highly toxic poison, rescuers should be properly dressed in personal protective clothing (PPC) and equipment. If proper equipment is not available, or rescue personnel have not been trained in the use of PPC, call for aid from a properly equipped and trained hazardous materials response team (HMRT). Chemical-protective clothing is usually not necessary since arsine gas is not absorbed through the skin and fails to cause skin irritation. However, **respiratory protection is certainly required with positive-pressure, self-contained breathing apparatus (SCBA) with a full face-piece for entering any potential arsine exposure level.** Remember, contact with arsine liquid (compressed gas) can cause frostbite burns to eyes and skin. In such situations, wash frosted skin with water, gently remove clothing from affected area, dry with clean towels, and keep patient warm and quiet.

Rescued persons who have been exposed _only to arsine gas_ do NOT need decontamination, and rescue personnel in the Support Zone or other "clean" areas do not need any specialized protection if everybody has already been decontaminated or all victims now in the support areas have been exposed _only to arsine gas._

With arsine exposure, there may be potential severe hemolysis (the breakdown of red blood cells and the release of hemoglobin). Ensure adequate oxygenation by arterial blood measurement or pulse oxygenation monitoring. Use diuretics to maintain urinary flow.

Drugs: Diuretics (to promote flow of urine).

Antidotes: No specific antidote available for arsine.

Medical Management: Arsine poisoning causes acute intravascular hemolysis that may lead to renal failure. Evaluate and support airway, breathing, and circulation; establish intravenous access in symptomatic patients; monitor cardiac rhythm; monitor fluid balance, plasma electrolytes, and hematocrit. Observe victims who have inhaled arsine for up to twenty-four hours. For inhalation exposure, give supplemental oxygen, and treat patients who have bronchospasm with aerolized bronchodiators which may present additional risks. If hemolysis develops, initiate urinary alkalinization. Consider hemodialysis if renal failure is severe. **Do not administer arsenic chelating drugs.** Although BAL/dimercaprol and other chelating agents are acceptable for arsenic poisoning, **they are not effective antidotes for arsine poisoning** and are not recommended! Patients who do not develop hemolysis after twenty-four hours of close monitoring can be discharged from hospital care when given instructions to seek medical care again if symptoms return, and told to get plenty of rest and drinks lots of fluids.

Fire: Arsine is flammable in air and has a lower explosion limit of 5.8 percent. Immediately evacuate everybody from the area of the spill/leak/toxic gas exposure. Stop the flow of gas if possible and feasible without needless exposure. Supply assisted respiration if victim(s) have stopped breathing. Isolate spill or leak area for 350 to 700 feet for 360 degrees. Completely eliminate all ignition sources. When entering closed spaces, you should ventilate them first. Only if leaks or spills can be stopped, extinguish fire(s); do not extinguish fire(s) if leaks or spills are still occurring. Use water spray or fog, alcohol resistant foam, dry chemical, or CO_2 to fight fire(s).

Personal Protective Equipment: Respiratory protection is required (positive pressure, full face piece, NIOSH-approved SCBA will be worn). When response personnel respond to handle rescue or reconnaissance, they will wear Level A protection that should be worn when the highest level of respiratory, skin, eye, and mucous membrane protection is needed. This level consists of a fully-encapsulated, vapor-tight, chemical-resistant suit, chemical-resistant boots with steel toe and shank, chemical-resistant inner/outer gloves (butyl rubber glove M3 and M4 Norton, chemical protective glove set), coveralls, hard hat, and self-contained (positive pressure) breathing apparatus (SCBA).

Do not breathe fumes; contact with arsine gas can kill responders.

Spill/Leak Disposal: Evacuate/Isolate the area, ensure control of ignition sources, stop leak or spill if possible, control/contain/ventilate the gas dispersion area if possible, ground all equipment, attend to security of the site, and wear proper protective clothing and equipment.

Symptoms: Initial symptoms include malaise, dizziness, nausea, abdominal pain, and dyspnea which may develop within several hours of exposure to three parts-per-million to arsine. **Children exposed to the same levels as adults may get a higher dose because they have a greater lung surface area and they are closer to ground level.**

Vaccines: None.

Note: Information on arsine can be found in Guide Number 119 of the *Emergency Response Guidebook* (2004 ed.), and identified as UN 2188 (United Nation 2188). This book is basically for first responders of all types and sources during the initial phase of a dangerous goods/hazardous materials incident. Guide 119 provides pointed information and facts for handling health, fires, explosions, public safety, protective clothing, evacuation, fire, spills and leaks, or first aid providers. Since every fire department in the United States, Canada, and Mexico, and many businesses and companies, have the 2004 *Guidebook* and should know the following data and advice from experts.

> **Page 354:** Covers Criminal/Terrorist Use of Chemical/Biological/Radiological agents, and differences between a Chemical, Biological and/or a Radiological agent, and indicators of a possible chemical incident.
> **Page 355:** Treats indicators of a potential Biological incident.
> **Pages 356 to 357:** Deals with personal safety considerations.
> **Page 358 to 365:** Provides a Glossary of agent-specific definitions.

Guides for Emergency Response: Chemical Agent or Weapon: Cyanogen Chloride (CK)

21

BLOOD AGENT: **Cyanogen Chloride (CK).** Blood agents such as cyanogen chloride are readily absorbed by the mucous membranes and dermal surfaces. CK has an odor like bitter almonds although many people cannot detect such odor because of irritating and lacrimatory effects of this chemical agent. **Cyanogen chloride is not absorbed very well by the metallic salt-impregnated charcoal filters in the common protective mask.** The basic practice of CK is to inhibit certain enzymes, including cytochrome oxidase, which are very important for oxidation-reduction in the cells. The end result is that cell respiration is inhibited and oxygen carried by the hemoglobin is not consumed thus causing the venous blood to remain bright red. The symptoms of this process can be quite striking: violent convulsions, increased deep respiratory movements, which are followed by cessation of respiration within one minute. The final results are rather two-sided depending on the actual dose the victim received. **With high concentrations, the victim could be dead very quickly; with lower doses and concentrations, the victim will probably survive spontaneously.**

Formula: CNCl

Vapor Density: (Air = 1) 2.1

Vapor Pressure: (mm^Hg) 1,000 at 25 degrees C

Molecular Weight: 61.48

Liquid Density: (g/cc) 1.18 at 20 degrees C

Volatility: (mg/m3) 2,600,000 at 12.8 degrees C; 6,132,000 at 25 degrees C.

Median Lethal Dosage: (mg-min/m3) 11,000

Physical State: (at 20 degrees C) Colorless liquid or gas.

Odor: Pungent, biting; can go unnoticed.

Freezing/melting point: (at degrees C) –6.9

Boiling point: (at degrees C) 12.8

Action rate: Very rapid.

Physiological action: Chokes, irritates, causes slow breathing rate

Required level of protection: Protective mask.

Decontamination: Decontaminate eyes and skin. Even though this chemical agent is highly volatile, pay particular attention to the eyes. Remove wet, contaminated clothing and the underlying skin; decontaminated with water.

Detection in the field: M18A2, M256, M256A1, M8 alarm, and Draeger Tube Cyanogen Chloride 0.25/a.

Use: Quick-action casualty agent.

CAS registry number: 506-77-4

RTECS number: GT2275000

LCt50:   11,000 mg-min/m3

Ability to Kill: High.

Characteristics: Blood agents, including cyanogen chloride (CK), interfere with oxygen utilization at the cellular level, **and all blood agents are non-persistent.** The vapor of CK is heavier than air and very irritating to the eyes and mucous membrane surfaces, and inhalation presents the primary route of entry. **With cyanogen chloride, recovery from the systemic effects of this chemical agent is prompt as it is with hydrogen cyanide poisoning, meaning that usually death occurs rapidly or the victim suddenly recovers.** The difference between the two agents (CK and AC) is that with cyanogen chloride, there is a higher residual damage to the central nervous system expected for the victim. **Once again, the dose and exposure concentration of CK a victim suffered will be a factor in determining if pulmonary effects may develop at once or may be delayed until the systemic effects have subsided.**

Response on Scene by First Responders

Caution: **Rescue personnel should wear Level A personal protective equipment when dealing with cyanogen chloride.** Even in low concentrations, cyanogen chloride irritates the eyes and respiratory tract. Acute exposure produces great irritation of the lungs signaled by coughing and breathing problems which may rapidly lead to a pulmonary edema. Inside the human body, cyanogen chloride converts to hydrogen cyanide; this *inactivates* the enzyme cytochrome oxidase which prohibits the utilization of oxygen by the cells. **The toxic hazard is high for inhalation, ingestion, and skin and eye exposure, but CK is basically an inhalation hazard because of its high volatility. Blood agent cyanogen chloride becomes a liquid at temperatures below 55 degrees F.**

Field First Aid: Cyanogen chloride acts in two ways in that its systemic effects are similar to hydrogen cyanide, but CK also has an irritant effect on the eyes, upper respiratory tract, and lungs. It injures the respiratory tract, resulting in severe inflammatory changes in the bronchioles, as well as congestion and edema in the lungs. It should be understood that very low concentration, as low as ten to 20 mg.min.m-3, can produce eye irritation and lachrymation. Remove victims from contaminated area. **If a victim's respiration is feeble or has ceased, medical personnel on scene should immediately apply assisted ventilation with oxygen, and treatment with sodium nitrite and sodium thiosulfate.** Continue with assisted ventilation until the victim's spontaneous breathing returns, or until ten minutes after no normal heart action has occurred and been observed. Sodium nitrite and sodium thiosulfate must ONLY be administered by intravenous injection. Administer ten mL of a 3 percent solution (300 mg) intravenously of sodium nitrite over a period of three minutes. Then intravenously inject 50 mL of a 25 percent solution (12.5 g) of sodium thiosulfate over a ten minute period. The sodium nitrite is given to produce methemoglobin in a manner to sequester the cyanide on the methemoglobin. The sodium thiosulfate combines with the sequestered cyanide to form thiocyanate that is then excreted from the body. The development of a slight degree of cyanosis provides evidence of a desirable degree of methemoglobin formation; it is not anticipated that the doses mentioned above will develop an excess or injurious amount of methemoglobin formation. If that does happen, administer treatment by 100 percent oxygen inhalation.

There are four entry points in the human body for cyanogen chloride to contaminate a victim: inhalation, eye contact, skin contact, and ingestion. First aid procedures require directing first efforts dealing with inhalation. If a victim is conscious, direct first aid and medical treatment towards the relief of any pulmonary symptoms. First, put the victim immediately at bed rest with head slightly elevated, and seek medical attention at once. Next, administer oxygen if there seems to be dyspnea or evidence of pulmonary edema. Combined therapy, in the case of long exposures, should include oxygen plus amyl nitrite inhalations and artificial respiration, or managed breathing. For **eye contact,** hold eyes open while flushing and flush affected areas with large amount of water at once. For **skin contact,** without delay remove all contaminated clothing including boots or shoes, then wash skin areas with water immediately to clean the victim's body of contamination of CK. In an **ingestion** episode, do not induce vomiting, but rather give the victim water or milk to drink.

Antidotes: Intravenous sodium nitrite and sodium thiosulfate, and/or amyl nitrite perles are packaged in the cyanide antidote kit. When possible, treatment with cyanide antidotes should be given under medical supervision to **unconscious** victims who are known or suspected of having been exposed to cyanide poison. Amyl nitrite perles should be broken onto a gauze pad, and held under the nose, over the Ambu-valve intake, or placed under the lip of the face mask. Inhale for thirty seconds every minute, and use a new perle every three minutes if sodium nitrate infusions will be delayed. If a patient has not responded to oxygen and amyl nitrite treatment, infuse sodium nitrite intravenously as soon as possible. The usual adult dose is ten mL (milliliter) of a 3 percent solution (300 mg) infused **over absolutely no less than five minutes;** the average pediatric dose is 0.12 to 0.33 mL/kg body weight up to ten mL infused as above. Medical personnel should monitor blood pressure during sodium nitrite administration, and slow the rate of infusion if hypotension develops. The next step is to infuse sodium thiosulfate intravenously. The usual adult dose is mL of a 25 percent solution (12.5 g) infused over ten to twenty minutes. The average pediatric for this step is 1.65 mL/kg of a 25 percent solution. Repeat one-half of the initial dose thirty minutes later if there is an inadequate clinical response.

Cyanogen chloride is absorbed through the skin and mucosal surfaces, and is quite dangerous when inhaled since toxic amounts are absorbed through bronchial mucosa and alveoli. This process is similar in toxicity and mode of action to the function seen with hydrogen cyanide (AC); the difference is that cyanogen chloride (CK) is far more irritating. CK may cause serious irritation of the respiratory tract, hemorrhagic exudate of the bronchi and trachea, and pulmonary edema. The liquid form will burn skin and eyes; while lengthy exposure can cause dermatitis, loss of appetite, headache, and upper respiratory irritation in victims.

Medical Management: Cyanogen chloride poisoning should be treated in the same manner as hydrogen cyanide poisoning in regards to its cyanide-like effects, while pulmonary irritation should be treated in the same way as phosgene poisoning. In casualties exposed to CK, the diagnosis is suggested by the intense irritation and the rapid onset of symptoms. Two major approaches are involved in the treatment of cyanide poisoning: provision of binding sites for cyanide ions, and provision of additional sulfur groups. The first approach provides alternative sites to those of cytochrome oxidase and essentially reactivates that enzyme. The binding site may be provided by drugs such as dicobalt edetate and by hydroxocobalamin or by production of methaemoglobin in the blood. Methaemoglobin binds avidly to cyanide ions and may be produced by compounds such as **sodium nitrite** and **amyl nitrite** as well as dimethylaminophenol. Methaemoglobin forming compounds should be used cautiously in victims suffering from concurrent carbon monoxide poisoning or hypoxia. The second approach calls for provision of additional sulfur groups to enhance the detoxification of cyanide and thiocyanate by endogenous rhodanese; this comes about by giving **sodium thiosulphate**.

Fire: No flash point.

Personal Protective Equipment: Laboratory personnel should wear appropriate chemical cartridge respirator, Butyl or Neoprene rubber gloves, and full-length faces shields with forehead protection depending on the amount of exposure. However, rescue personnel should be equipped with self-contained breathing apparatus (SCBA); and have available and use as appropriate Level A personal protective equipment (PPE). When you do *not* know the degree of hazard, use Level A personal **protective equipment (PPE) as follows**:

Level A protection should be worn when the highest level of respiratory, skin, eye, and mucous membrane protection is needed. This level consists of a fully-encapsulated, vapor-tight, chemical-resistant suit, chemical-resistant boots with steel toe and shank, chemical-resistant inner/outer gloves, coveralls, hard hat, and self-contained (positive pressure) and self-contained breathing apparatus (SCBA).

Spill/Leak Disposal: Evacuate/Isolate the area, ensure control of ignition sources, stop leak or spill if possible, control/ contain/ventilate the dispersion area if possible, ground all equipment, attend to security of the site, and wear proper protective clothing and equipment.

Symptoms: The signs and symptoms of cyanogen chloride (CK) are like a combination of those produced by hydrogen cyanide and a lung irritant. **At first, CK stimulates the respiratory center and then quickly paralyzes it.** However, with high concentrations its local irritant action may be so great that dyspnea is produced. Exposure is followed at once by severe irritation of the eyes, throat and nose along with coughing, lacrimation, and tightness in the chest. Thereafter, the exposed victim may become dizzy and increasingly dyspneic (difficult or labored breathing). **Unconsciousness come first and then is joined by failing respiration and death within a few minutes. Convulsion, retching, and involuntary urination and defecation may occur.** If such effects are not fatal, the sign and symptoms of pulmonary edema can develop. There could be persistent cough with much frothy sputum, rales in the chest, severe dyspnea, and marked cyanosis.

Vaccines: None

Guides for Emergency Response: Chemical Agent or Weapon: Diphosgene and Phosgene

22

AGENT: Pulmonary (choking) Agents. Pulmonary (choking) agents cause physical injury to the lungs through inhalation. Membranes may swell and lungs become filled with liquid in serious cases causing death from lack of oxygen. Potential pulmonary (choking) agents that might be used as weapons of mass destruction (WMD) are listed in a standard format below.

PULMONARY (Choking) AGENT: Diphosgene (DP) is a colorless liquid which smells like either new-mown hay, grain, or green corn and has a higher boiling point than phosgene (CG). Diphosgene (DP) causes a tearing effect in victim's eyes when used in an attack, so it comes as less of a surprise than phosgene would have on the same victim. In addition, diphosgene (DP) has a lower volatility (vapor pressure) than does phosgene (CG), so phosgene (CG) would be the better surprise weapon. **Because a victim's body converts DP (diphosgene) to CG (phosgene), the physical effects are the same for both agents. That is, the human body does not detoxify either diphosgene or phosgene; each dose is cumulative.** The rate of action for diphosgene is delayed; immediate symptoms may follow high doses of agent, but lower doses may take three hours or more to become apparent to the victim or to medical personnel.

Formula: CLCOOCCL3

Vapor Density: (Air = 1) 6.8

Vapor Pressure: (mm^hg) 4.2 at 29 degrees C

Molecular Weight: 197.85

Liquid Density: (g/cc) 1.65 at 20 degrees C

Volatility: (mg/m3) 45,000 at 20 degrees C

Median Lethal Dosage: (mg-min/m3) 3,200

Physical State: (at 20 degrees C) Colorless oily liquid

Odor: New-mown hay; green corn

Freezing/melting point: (at degrees C) –57 C

Boiling point: (at degrees C) 127 C to 128 C

Action rate: Immediate to 3 hours, depending on concentration

Physiological action: Damages and floods lungs

Required level of protection: Protective mask

Decontamination: None needed in field; aeration in closed spaces

Detection in the field: Odor (but *no* Draeger Tube).

Use: Delayed—or immediate—action casualty agent

CAS registry number: 503-38-8

RTECS number: LQ7350000

LCt50:   3,000 mg-min/3 for resting persons. Since the effects of diphosgene are cumulative, the Ct does not significantly change with variations in time of exposure.

PULMONARY (Choking) AGENT: **Phosgene (CG)** has plenty of experience! In World War I, more than 80 percent of the chemical agent fatalities where due to the use of phosgene. Phosgene tends to hug low-lying areas and stay close to the ground; it was a perfect chemical agent for World War I where enemies fought from trenches. It condenses to a liquid below 46 degrees F (7.8 degrees C). **However, phosgene reacts rapidly with moisture so rain, fog, and thick vegetation all tend to reduce its concentration in the air.** If you become a victim of phosgene, you may die of what is commonly called "dry-land drowning." During exposure it is likely that coughing and wheezing may be self-evident but exposure to a low dose causes no ill effects for three hours or more. **The severity of poisoning cannot be estimated from immediate symptoms since the full effect of this cumulative agent is not usually evident until three to four hours after exposure.** During the World War I, phosgene with its rapid vaporization from liquid to gas and its fast reaction to moisture often made an instant white cloud. **Phosgene, with a density four times that of air, at-once converted to low-lying gas hugging the trenches of the enemy and became a much feared symbol to don gas masks, the only real defense for both phosgene and diphosgene.** Phosgene rapidly hydrolyzes to form carbon dioxide and hydrochloric acid, and that acid causes ocular, nasal, and central airway irritation from high doses of CG. The odor thresh-hold for CG is roughly 1.5 mg/m3.

Formula: COCl2

Vapor Density: (Air = 1) This information is not available, *but* it is heavier than air

Vapor Pressure: (mm^Hg) 1.173 at 20 degrees C

Molecular Weight: 98.92

Liquid Density: (g/cc) 1.37 at 20 degrees C

Volatility: (mg/m3) 4,300,000 at 7.6 degrees C

Median Lethal Dosage: (mg-min/m3) 1,600

Physical State: (at 20 degrees C) Colorless gas

Odor: New-mown hay; green corn

Freezing/melting point: (at degrees C) –128 C

Boiling point: (at degrees C) 7.6 C

Action rate: Immediate to three hours, depending on concentration

Physiological action: Damages and floods lungs

Required level of protection: Protective mask

Decontamination: One needed in field; aeration in closed spaces

Detection in the field: M18A2, odor, and Draeger Tube Phosgene 0.25/b

Use: Delayed-action casualty agent

CAS registry number: 75-44-5

RTECS number: SY5600000

LCt50:   3,200 mg-min/m3

Ability to Kill: High

Characteristics: Phosgene directly reacts with amine, sulfhydryl, and alcohol groups in cells, thereby adversely affecting cell macromolecules as well as cell metabolism. Direct toxicity to the cells leads to an increase in capillary permeability, resulting in large shifts of body fluid, thus decreasing plasma volume. Also, when phosgene hydrolyzes, it forms hydrochloric acid, which can also damage surface cells and cause cell death in the alveoli and bronchioles. Hydrochloric acid release into the mucosa triggers a systemic inflammatory response.

Response on Scene by First Responders

Caution: Product is a Poison A; response personnel need to wear proper personal protective equipment (PPE).
   (Please refer to the section below on PPE.)

Field First Aid: First responder rescue personnel should use breathing apparatus and chemical protective clothing if there is any possibility of exposure to unsafe levels of phosgene contamination. Victims exposed only to phosgene do NOT pose a substantial risk of secondary contamination. Persons whose clothing or skin is contaminated with liquid phosgene with the ambient temperature below 47 degrees F. can secondarily contaminate response personnel through direct contact or off-gassing vapor (below 47 degrees F., phosgene is a fuming liquid and contact can cause frostbite). When moisture is in the area including saliva, sweat, and tears, the liquid or gas slowly hydrolyzes to hydrochloric acid, which can irritate or damage cells. Phosgene is absorbed to some extent by the lungs, but not intact skin. Phosgene is corrosive to the lungs and intact skin. In industry, phosgene is shipped as a liquefied, compressed gas; and is a combustion product of many household products that contain volatile organochlorine compounds which may contribute to hazards of smoke inhalation in fire victims and firefighters.
   Inhalation is the major route of phosgene exposure with an odor threshold five times higher than the OSHA PEL (public exposure limit); thus, odor provides insufficient warning of hazardous concentrations. The irritating quality of CG can be mild or delayed, which may result in a lack of avoidance leading to exposure for prolonged periods. **Response personnel should remember that phosgene (CG) is a cumulative poison, piling one dose on top of another without end. It requires good monitoring practices to constantly test for the actual level of contamination and the use of personal chemical protective**

equipment and self-contained breathing apparatus (please see "Personal Protective Equipment" in this guide for phosgene). Phosgene is heavier than air and could cause asphyxiation in poorly ventilated, low-lying, or enclosed spaces. The OSHA PEL (permissible exposure limit) is 0.1 parts-per-million (ppm) averaged over an eight-hour day, while the NIOSH IDLH (immediately dangerous to life and health) standard for phosgene is 2 parts-per-million.

Phosgene poisoning may cause respiratory and cardiovascular failure which results in low plasma volume, increased hemoglobin concentration, low blood pressure, and an accumulation of fluid in the lungs. There is no antidote for phosgene, and treatment consists of support of respiratory and cardiovascular functions.

For inhalation, escort victim(s) to fresh air; keep person calm and avoid any necessary exertion or movement; maintain airway and blood pressure; medical personnel should administer oxygen if breath is hindered; and give artificial respiration if victim is not breathing. For eye contact, flush eyes at once with running water or normal saline solution for at least fifteen minutes, and hold eyelids apart during irrigation. To avoid permanent eye injury, do not delay rinsing the eyes. For **dermal contact**, it is unlikely that emergency treatment will be required, but gently wrap the affected portion in blankets if warm water is not available, then allow circulation to return naturally. For **ingestion**, handle symptomatically and supportively, keep head lower than hips if vomiting occurs in order to prevent aspiration. In all cases (inhalation, eye contact, dermal contact, and ingestion), after providing first aid, seek qualified medical attention at once.

Quickly access for a victim airway, and ensure adequate respiration and pulse. If trauma is suspected, maintain cervical immobilization manually and apply cervical collar and a backboard if feasible. Victims should be kept warm and quiet as any activity subsequent to phosgene exposure may lead to death. If exposure levels are determined to be safe, decontamination should be conducted by personnel wearing a lower level of protection than that worn in the hot zone. If the exposure involved liquid phosgene, and the temperature is less than 47 degrees F., and the victim's clothing has been contaminated, remove and double-bag the clothing.

Medical Management: Chemical agents that attack lung tissue, primarily causing pulmonary edema, are classified as lung-damaging agents, or choking agents, because irritation of the bronchi, trachea, larynx, and nose may occur and, with pulmonary edema, contribute to the sensation of choking. **Aside from mild conjunctival irritation, the direct effects of exposure to phosgene are confined to the lungs, and the outstanding feature of phosgene poisoning is massive pulmonary edema.** After cessation of the exposure, medical management requires treatment of the ABCs, enforced rest and observation, oxygen with or without positive airway pressure for signs of respiratory distress, and other supportive therapy as needed. Tightness of the chest and coughing should be treated with immediate rest and comfortable warmth. The victim should be evacuated in a semiseated position if dyspnea or orthopnea make a supine posture impractical. Sedation should be used only in a sparing fashion. Codeine in doses of 30 to 60 mg may be effective for cough. Restlessness may be a manifestation of hypoxia, and only cautious use of sedatives is advised. Use of sedatives should be withheld until adequate oxygenation is assured and facilities for possible respiratory assistance are available. Contraindicated are atropine, barbituates, analeptics, and antihistamines. Hypoxemia may be controlled by oxygen supplementation. Early administration of positive airway pressure (intermittent positive pressure breath or IPPB, positive end-expiratory pressure "peep mask," or, if necessary, intubation with or without a ventilator) may delay and/or minimize the pulmonary edema and reduce the degree of hypoxemia.

Fire: Phosgene is a **Poison A** and an environmentally hazardous substance, but it is not flammable. However, serious exposure from flames or intense heat of other burning objects or containers may result in violent rupturing and rocketing. First responders should alert people on scene of Poison A being present at the incident scene, cool exposed containers of cylinders, move any victims to fresh air, and evacuate the incident scene if the leak cannot be controlled or the spill is sizable. Rescue personnel should shun any contact with either liquid or vapor, wear proper personal protective equipment (PPE) for a Class A poison, and remain upwind using water spray to control vapor. Fight fire in surrounding materials or packaging, possibly using secured hose nozzle locations. **All first responders or firefighters should wear personal protective equipment as indicated below, or wait for assistance from responders or teams who have such equipment.**

Personal Protective Equipment: Phosgene is a severe respiratory tract and skin irritant, and contact with the liquid will cause frostbite. Respiratory protection requires positive-pressure-demand, self-contained breathing apparatus (SCBA), while skin protection requires chemical-protective clothing since phosgene gas can cause skin irritation and burns. NIOSH recommends protective suits from Responder™ (Kappler Co.), Tychem 10000 TM (DuPont Co.) or Teflon™ (DuPont Co.).

Spill/Leak Disposal: Isolate the incident scene; dress in proper personal protective equipment (**see above**); do not allow contact with any materials, liquid or gas; stop and/or control leak or hazard if possible to do so; and control water – use water spray to control vapor and any vapor cloud. Contain product and keep phosgene from entering sewers, streams, or water intakes. Dike surface flow, and depending on the temperature, try to neutralize the product for disposal using agricultural lime (slaked lime), crushed limestone, or soda ash, or sodium bicarbonate.

Symptoms: Phosgene produces a pulmonary edema that evidences itself after a varied period of time judged by the intensity of exposure, or the Ct. Following a period of latency, the victim may present evidence of worsening respiratory distress that could progress to pulmonary edema and death. Exposure to large doses of phosgene may irritate moist mucous membranes, possibly because of the generation of hydrochloric acid; **CG (phosgene) is a severe mucous membrane irritant. Phosgene is a severe eye, mucous membrane, and skin irritant; but it is primarily a toxic hazard by inhalation exposure. Two parts per million in air is immediately dangerous to life and health.** Acute inhalation may cause respiratory and circulatory failure with symptoms of chills, dizziness, thirst, burning of eyes, cough, viscous sputum, dyspnea, feeling of suffocation, tracheal rhonchi (a whistling or snoring sound heard on auscultation of the chest when air channels are partly obstructed), burning in the throat, vomiting, pain in the chest, and cyanosis. Chronic inhalation may cause irreversible pulmonary changes resulting in emphysema and fibrosis. Acute dermal contact can result in lesions similar to those caused by frostbite and burns, and chronic skin contact may lead to dermatitis. Acute eye contact may cause conjunctivitis, lacrimation, lesions, and burns.

Vaccines: Not available

Guides for Emergency Response: Chemical Agent or Weapon: Distilled Mustard (H, HD)

23

Blister Agents: The vesicants have been widely used in different types of combat since introduced during World War I. They are vapor and liquid hazards to skin and mucous membranes. Mustard's effects appear hours after the actual exposure, and most commonly affect skin, eyes, and airways. **With exposure to large amounts of mustard, cells in the bone marrow are damaged.** Within the military, blister agents are of the second level of concern just behind nerve agents that are of utmost concern. **The primary threat blister agents are sulfur mustard (H/HD), Lewisite (L), and a mixture of mustard and Lewisite (HL).** Mustard assumes major significance because it is *both* lethal and incapacitating. That is, there were many casualties from mustard during World War I but only 3 percent of such casualties died, even though there were no antibiotics available at that time. The true danger of mustard is that a person exposed to a significant amount of vapor or liquid mustard faces total systemic assault from failure of the body's immune system, along with sepsis, infection, and pulmonary injury.

Environmental conditions at the time of exposure tend to help control, or increase, the severity of blister agent damage. Cold weather retards the time of symptom onset; if the skin remains cool, it will lessen the severity of blister agent damage. On the other hand, warm and/or humid condition, increase the severity of blister agent damage and decrease the time required for symptoms onset. Blister agents also have a relatively high vapor density when compared to air given that mustard is 5.4 times greater than the density of air, Lewisite is 7.1 greater than the density of air, and a mixture of mustard and Lewisite is 6.5 times heavier than air. A clue is provided by vapor density. Air is considered to have a molecular weight of 1. The higher the density is to 1, the longer it will persist in low-lying areas such as valleys, trenches, and cellar holes. In other words, the denser a vapor may be, the more it will settle. On the other hand, if the vapor pressure is less than 1, a chemical agent will likely be non-persistent and dissipate quickly into the atmosphere.

Clinical signs and symptoms from mustard (H/HD) **are *not*** apparent until hours later; skin blisters might not appear for up to twenty-four hours, but tissue damage occurs *within two minutes*. **If decontamination is not done within the first two minutes after exposure, nothing can be done to prevent a mustard injury.** Since mustard (H/HD) can be detected by human beings by smell in concentrations of 0.6 to 1.0 mg/m3 as a garlic, horseradish, or mustard odor; an alert person will most likely smell the mustard vapor before encountering the liquid.

The hazard of a mixture of mustard and Lewisite (HL) on the eyes and skin, or vapor in the eyes or respiratory tract, is immediate. Within an hour, edema of the conjunctivae and lids begin and soon results in eye closure. Any casualty experiences serious pain seconds after contact with HL liquid, but such extreme pain makes the injured person decontaminate at once. Rapid decontamination is the sole manner to avoid severe burns; since after a few minutes of contact with a mixture of mustard and Lewisite, the upper layer of skin will die and appear gray, painful erythema will be apparent shortly thereafter, and painful blisters may appear within twelve hours. The pain from contact with HL vapor is so irritating that persons who are able will immediately leave the area or don a protective mask. **Blister agents are so persistent that any first responder, or witness, who attempts to decontaminate an exposed person should quickly don a mask as well as protective clothing to provide that aid.**

BLISTER AGENT (VESICANT):   Distilled Mustard (HD)

Formula: ClCH2CH2-S-CH2CH2Cl

Vapor Density: (Air = 1) 5.4

Vapor Pressure: (mm^Hg) 0.072 at 20 degrees C.

Molecular Weight: 159.08

Liquid Density: (g/cc) 1.27 at 20 degrees C

Volatility: (Mg/M3) 610 at 20 degrees C

Median Lethal Dosage: (mg-min/m3) 1,500 by inhalation; 10,000 by skin exposure

Physical State: (at 20 degrees C): Oily, colorless to pale yellow liquid

Odor: Similar to garlic or horseradish

Freezing/melting point: (at degrees C) 14.45 C

Boiling point: (at degrees C) 217 C

Action rate: Delayed—usually four to six hours until first symptoms appear

Physiological action: Blisters; destroys tissue; injures blood cells

Required level of protection: Protective mask and clothing

Decontamination: Bleach, fire, DS2, M258A1, M280

Detection in the field: M18A2, M256, M256A1, M8 and M9 paper, and Draeger Tube

Thioether.   Mustard agent received its name because of garlic, horseradish, or mustard odor that can be detected by smell. The human nose can detect mustard (H, HD) in concentrations of 0.6 to 1.0 mg/m3. It must be understood that until recently, the U.S. military had no automatic vapor/liquid detection capability. Therefore, alert soldiers would most likely smell the agent vapor before encountering the liquid (after release, H or HD appears as a thick, colorless or pale yellow liquid; HL, or mustard/lewisite mixture, appears as a dark oily cloud that can be detected visually).

Use: Delayed-action casualty agent

CAS registry number: 505-60-2

RTECS number: WQ0900000

LD50:   (oral) 0.7 mg/kg

Ability to Kill: Moderate

Characteristics: Sulfur mustards are vesicants and alkylating agents. Agent H (sulfur mustard) contains about 20 to 30 percent impurities which are mostly sulfur, while HD (distilled mustard) is nearly pure. Agent HT, on the other hand, is a mixture of 60 percent HD (distilled mustard) and 40 percent agent T, which is a closely related vesicant with a lower freezing point. Sulfur mustards evaporate slowly, and are sparingly soluble in water although they are soluble in oil, fats, and organic solvents. They tend to be stable at ambient temperatures but decompose at temperatures greater than 149 degrees C. **People whose skin or clothing is contaminated by sulfur mustard can contaminate rescuers by direct contact or through off-gassing vapor. Although volatility is low, vapors can reach hazardous levels during warm weather.** Sulfur mustards are absorbed by the skin, causing erythema and lesions (blisters). Eye exposure to such agents may cause incapacitating damage to the cornea and conjunctiva. Inhalation contact to the respiratory tract epithelium may cause death. **The only way to prevent damage to victims who have had contact with sulfur mustard agents is to thoroughly and immediately decontaminate them at once, just seconds or minutes after exposure.**

Response on Scene by First Responders

Caution: The Airborne Exposure Limit for sulfur mustard agents is 0.003 mg/m3 as a time-weighted average (TWA) for the workplace. Ingestion may cause chemical burns in the gastrointestinal (GI) tract and cholinergic stimulation, and nausea and vomiting may occur following ingestion or inhalation. High doses of sulfur mustard to the central nervous system (CNS) can cause hyperexcitability, convulsions, and insomnia. Additional health effects of this agent (H, HD, or HT) can include systemic absorption which may induce bone marrow suppression and an increased risk for fatal complicating infection, hemorrhage, and anemia. They might be delayed effect years later after apparent healing of severe eye lesions when relapsing keratitis (inflammation of the cornea or the eye) or keratopathy (any noninflammatory disease of the eye) may develop. Persistent eye conditions, loss of taste and smell, and chronic respiratory illness may remain following exposure to sulfur mustards. Prolonged or repeated acute exposure to such agents may cause cutaneous (affecting the skin) sensitization and chronic respiratory disease. **Repeated exposures result in cumulative effects because mustards are not normally detoxified by the body.** The International Agency for Research on Cancer (IARC) has classified sulfur mustard as carcinogenic to humans (Group 1). Evidence tends to indicate that repeated exposures to sulfur mustard may lead to cancers of the upper airways. Also, there are at least two reproductive and developmental effects alleged for sulfur mustard, but data tends to be inconclusive. The effects of exposure to any hazardous substance depend upon the dose, the duration, how you are exposed, personal traits and habits, and whether other chemicals are present. Spills over one pound of mustard agent must be reported to the National Response Center.

Field First Aid: Decontaminate At Once for All Exposed Victims! Although sulfur mustards cause cellular changes within minutes of contact, the onset of pain and other clinical effects are delayed for one to twenty-four hours. Sulfur mustards are alkylating agents that may cause bone marrow suppression and neurologic and gastrointestinal toxicity. However, the biochemical mechanisms of action are not clearly understood by anyone. The death rate from exposure to sulfur mustard during World War I was 2–3 percent, with death usually occurring between the fifth and the tenth day due to pulmonary insufficiency complicated by infection due to immune system compromise. The eye is the most sensitive tissue to sulfur mustard effects. Vapor or liquid may cause intense conjunctival and scleral pain, swelling, lacrimation, blepharospasm, and photophobia although these effects do not appear for a hour or more. High concentrations of liquid or vapor can cause corneal edema, perforation, blindness, and later scarring.

Quickly ensure ABC (airway/breathing/circulation). See that the victim has a clear airway. If trauma is suspected, maintain cervical immobilization manually and apply a cervical collar and a backboard if feasible. Check for adequate respiration, and administer supplemental oxygen if cardiopulmonary compromise is suspected. Maintain adequate circulation. Establish intravenous access when necessary and warranted. Use a cardiac monitor as necessary. Stop any bleeding when necessary. When ingestion is evident, **do not induce emesis (vomiting).** For victims who are alert and able to swallow, give four to eight ounces of milk to drink. Administration of activated charcoal is not viewed as beneficial.

For inhalation contact, remove from source at once, and provide artificial respiration if breathing is not apparent. Provide oxygen if breathing is difficult. For eye contact, **flush eyes immediately with water for 10 to 15 minutes** by pulling eyelids apart with fingers, and pouring water into the eyes. Do not cover eyes with bandages, but protect eyes with dark or opaque glasses after flushing the eyes. For skin contact, don respiratory mask and gloves and **remove victim from the source of contact**, then remove the patient's clothing and decontaminate the skin immediately by flushing with five percent solution of liquid household bleach followed in three to four minutes by a wash with soap and water to remove the five percent bleach solution used to decontaminate the victim's skin and to protect against erythema (to prevent systemic toxicity, decontamination can be done as late as two to three hours after exposure even if it increases the severity of the local reaction; you are dealing with a chemical reaction, and you have to deal with it. Speed is definitely needed, but endurance may save the victim much suffering even if applied late). As stated before, for ingestion contact do not induce vomiting, provide milk to drink if victim is capable. **For all types of contact introduced above, seek medical attention immediately.**

Dermal (skin) contact with sulfur mustard agents causes erythema and lesions (blistering), while contact with vapor may result in first and second degree burns; contact with liquid typically produces second and third degree chemical burns. Any burn area covering 25 percent or more of the body surface area **may be fatal.** Respiratory contact is a dose-related factor in the sense that inflammatory reactions in the upper and lower airway begin to develop several hours after exposure and progress over several days.

Decontamination immediately after exposure is the only way to present damage, with symptomatic management of lesion/blisters thereafter. All victims and first responders shall be decontaminated when leaving the Hot Zone. Clothes should be removed if at all possible, and no one should be transported to a hospital until he of she has been thoroughly decontaminated. Care in a hospital is strictly supportive. First responders have to ensure that every one contaminated goes through an efficient decontamination procedure.

Medical Management: There is no known antidote for mustard exposure, and the process of cellular destruction is irreversible. It is essential to remove the mustard agent as quickly as possible. Vesicants rapidly penetrate the skin causing both localized cellular damage and systemic damage. The deadly nature of such agents' effect is that a person exposed to a large amount of liquid or vapor faces total systemic assault. The reasons for this are (a) the failure of the body's immune system, with sepsis and infection as the major contributing causes of death, and (b) pulmonary damage which wins out as the major contributory factor in death. **Rapid decontamination** is the only way to avoid severe burns. Medical management of a victim who has undergone exposure to mustard agent can be as simple as providing of symptomatic care for a sunburn-like erythema, or vastly complex as in providing long-term services to fight immunosuppression, caring for serious burns, and/or dealing with other multi-system factors. A patient severely ill with mustard agent poisoning requires the general supportive care provided for any severely ill patient as well as specific care given to a burn patient. Liberal use of systemic analgesics and antipruritics, as needed, maintenance of fluid and electrolyte balance, and other supportive measures are necessary. Food supplements, including vitamins, may also be helpful. Some medical management schemes do not work out so well. Activated charcoal given orally does not seem to provide better care. Hemodialysis has been ineffective and even harmful to several patients. The rapid transformation of the mustard molecule suggests that few methods are actually available and beneficial hours or days after exposure with a mustard agent. **Mustard agents are classed as a mutagen and a carcinogen based on laboratory studies.**

Fire: The flash point for agents H and HD is 221 degrees F or 105 degrees C; for agent HT, the flash point is 212 degrees F or 100 degrees C. For fire, use the following extinguishing agents: Dry chemical, CO/2, alcohol-resistant foam, and/or water spray. Attempt to move containers to a safe area; contain run-off; use Kelly coils or monitor nozzles or other unmanned means to fight fire from a distance; dam or cover sewer openings; do not splatter or scatter this agent.

Personal Protective Equipment: For mustard agents (H, HD, HL, HT) wear full protective clothing consisting of **Level A:** Level A should be worn when the highest level of respiratory, skin, eye, and mucous membrane protection is needed. This level consists of a fully-encapsulated, vapor-tight, chemical-resistant suit, chemical-resistant boots with steel toe and shank, chemical-resistant inner/outer gloves, coveralls, hard hat, and self-contained (positive pressure) breathing apparatus (SCBA). Both the Environmental Protection Agency and NIOSH recommend that initial entry into unknown environments or into confined space that has not been chemically characterized be conducted wearing at least Level B, if not Level A, protection. For protective gloves, it is mandatory to wear Butyl toxicological protective gloves such as M3, M4, or glove set.

Spill/Leak Disposal: Mustard agents or weapons should be contained using vermiculite, diatomaceous earth, clay, or fine sand and neutralized as soon as possible using copious amounts of 5.25 percent sodium hypochlorite solution. Scoop up all material and place it in an approved Department of Transportation container. Cover the contents with decontaminating solution as mentioned above. The exterior of the salvage container shall be decontaminated and labeled according to EPA and DOT regulations. All leaking containers will be over packed with sorbent (e.g., vermiculite) placed between the interior and exterior containers. Decontaminate and label according to EPA and DOT regulations. Dispose of the material in accordance with waste disposal methods provided below. Conduct general area monitoring with an approved monitor to confirm that the atmospheric concentrations do not exceed the airborne exposure limits. If 5.25 percent sodium hypochlorite solution is not available, the following decontaminants may be used instead and are listed in order of preference: Calcium Hypochlorite, Contamination Solution No. 2 (DS2), and Super Tropical Bleach Slurry (STB).

Approved Methods of Waste Disposal: **(Open pit burning or burying of HD or items containing or contaminated with HD in any quantity is *prohibited!*)** Decontamination of waste or excess material shall be accomplished according to procedures outlined above and can be destroyed by incineration in EPA approved incinerators according to appropriate provisions of Federal, State, and local Resource Conservation Act (RCRA) regulations. *Note*: Some decontamination solutions are hazardous waste according to RCRA regulations and must be disposed of according to these regulations.

Symptoms: Erythema, blisters, irritation of the eyes, cough, dyspnea, asymtomatic latent period (hours). Also, mild upper respiratory signs to marked airway damage, GI effects and bone marrow stem cell suppression possible. Mustard is a blister agent that affects the eyes, lungs, and skin. A person exposed to mustard will feel very little pain and will *not* notice symptoms for quite some time. However, the longer the exposure without removal of the mustard agent, the more severe will be the damage to the affected areas of the body.

Vaccines: None

Special Section on Thickened HD

Styrene-butyl acrylate copolymer is used to "thicken" HD and make THD. A thickened agent is one to which a polymer or plastic has been added to retard evaporation and cause it to adhere to surfaces. Essentially, THD is the same as HD except for the following factors: Fire, Health Hazard, Spill/Leak/Disposal Factors, and Special Precautions. THD has a slight fire hazard when exposed to heat or flame. The health hazard is the same for HD except for skin contact. For skin contact don respiratory protective mask and remove contaminated clothing **IMMEDIATELY. IMMEDIATELY** scrape the THD from the skin surface, then wash the contaminated surface with acetone. Seek medical attention **IMMEDIATELY.** If spills or leaks occur, handle the same as HD but dissolve THD in acetone before introducing and decontaminating solution. Containment of THD is generally not necessary. Spilled THD can be carefully scrapped off the contaminated surface and placed in a fully removable head drum with a high density, polyethylene lining. THD then can be decontaminated, after it has been dissolved in acetone with the same procedures used for HD. Contaminated surfaces should be treated with acetone then decontaminated using the same procedures as those used for HD. *Note:* Surfaces contaminated with THD and then rinsed-decontaminated may evolve sufficient HD vapor to produce a physiological response. As far as special precautions are concerned, watch for "stringers" (elastic, thread like attachments) formed when agents are transferred or dispensed. These stringers **must be broken cleanly** before moving the contaminating device or dispensing device to another location, or unwanted contamination of a working surface will result. Avoid contact with strong oxidizers, excessive heat, sparks, or open flame.

Guides for Emergency Response: Chemical Agent or Weapon: Blood Agent Hydrogen Cyanide (AC)

24

Introduction: Hydrogen Cyanide (AC) is a so-called blood agent (this is actually a historical and long-lasting misnomer; this agent has little or nothing to do with movement of blood) that is fast acting, highly poisonous, and may be fatal if inhaled. It has been called the perfect weapon if you want to kill yourself; but with speedy action and treatment, recovery after overexposure can be quick and complete. AC is a non-persistent, colorless liquid that is highly volatile yet dissipates quickly in the air. The United Nation number is 1051, and the Emergency Response Guidebook guide number is 117 if the solution is greater than 20 percent. If it is less than 20 percent solution, the United Nation number becomes 1613, and the Emergency Response Guidebook guide number becomes 154.

Formula: HCN

Vapor Density: (Air = 1) 0.990 at 20 degrees C

Vapor Pressure: (mm^Hg) 742 at 25 degrees C; 612 at 20 degrees C

Molecular Weight: 27.02

Liquid Density: (g/cc) 0.687 at 20 degrees C

Volatility: (mg/m3) 1,080,000 at 25 degree C

Median Lethal Dosage: (mg-min/m3) Varies widely with concentration

Physical State: (at 20 degrees C) Colorless gas or liquid

Odor: Bitter almonds

Freezing/melting point: (at degrees C) –13.3 C

Boiling point: (at degrees C) 25.7

Action rate: Very rapid.

Physiological action: Interferes with body tissues, oxygen use and rate of breathing

Required level of protection: Protective mask; protective clothing in unusual situations

Decontamination: None needed in field

Detection in the field: M18A2, M256, M256A1, M8 alarm, and Draeger Tube Hydrocyanic Acid 2/a.

Use: Quick-action casualty agent

CAS registry number: 74-90-8

RTECS number: MW6825000

LCt50: Varies widely with concentration because of the rather high rate at which the human body detoxifies hydrogen cyanide.

Ability to Kill: High.

Characteristics: Cyanide is a rapidly acting lethal agent where death can occur in six to eight minutes after inhalation, but both sodium nitrite and sodium thiosulfate can be effective antidotes. It is a non-persistent blood agent that has an affinity for oxygen and is flammable. It is also an extremely volatile liquid. Because of its physical properties, hydrogen cyanide will not remain for long in its liquid state so decontamination should not be necessary. Hydrogen cyanide is a highly poisonous material, and an extremely hazardous liquid and/or vapor under pressure. **It is a prompt killer that requires speed in responding to personal or victim exposure. Hydrogen cyanide, other than being a substance that has been studied as a chemical agent or weapon, is an important industrial chemical as well.** Yet, with prompt treatment and effective antidotes following exposure, recovery is normally quick and complete. Hydrogen cyanide inactivates the enzyme cytochrome oxidase, preventing the use of oxygen by the cells. **The toxic hazard is high for inhalation, ingestion, and skin and eye exposure, but this chemical is primarily an inhalation hazard because of its high volatility. It hinders the vital oxidation-reduction reactions in the body resulting in anoxia affecting the central nervous system (CNS) resulting in respiratory paralysis.** Hydrogen cyanide is lighter than air, and the evaporation rate of its vapor in open-air can be only a short time period. The OSHA-Personal Exposure Limit equals ten parts-per-million on skin averaged over 15 minutes. The NIOSH-Immediately-Dangerous-to-Life-or-Health equals 50 parts-per-million. Dermal absorption can occur, leading to systemic toxicity, and absorption occurs more readily at high ambient temperature and relative humidity. **Hydrogen cyanide (AC) has not been classified for carcinogenic effects, and no carcinogenic effects have been reported for it.**

Response on Scene by First Responders

Caution: Hydrogen cyanide is unstable with heat, alkaline materials, and water. Do NOT store wet hydrogen cyanide as it may react strongly with mineral acids; experience shows mixtures of 20 percent or more sulfuric acid will explode; effects of other acids are not quantified, but strong acids such as hydrocloric or nitric acid would probably react in a similar fashion. **Polymerization may occur in the presence of heat, alkaline materials, or moisture.** When initiated, polymerization becomes uncontrollable since the reaction is autocatalytic, producing heat and alkalinity; confined polymerization can cause a violent explosion. Hydrogen cyanide is stabilized with small amount of acid to prevent polymerization, and should not be stored for extended lengths of time unless routine. Protective charcoal filters tend to clog when saturated with cyanide.

Field First Aid: Death may come very quickly to anyone who inhales cyanide. Roughly 15 seconds after inhalation of a high concentration of cyanide vapor there is a transient hyperpnea (rapid or deep breathing) followed in 15 to 30 seconds by the onset of convulsions. Respiratory activity stops two or three minutes later, and cardiac activity ceases several minutes after that, or about six to minutes after exposure. **With that process provided above, it really is amazing that so many people actually survive.** Successful treatment for acute cyanide poisoning, or "intoxication," depends upon quick fixation for the cyanide ion, either by methaemoglobin (metHB) formation or by fixation with cobalt compounds. Victims who may be fully conscious and breathing normally more than five minutes after presumed exposure to cyanide agents has ceased will recover spontaneously and does not need or require treatment since cyanide can be very rapidly detoxified in the body. Artificial resuscitation is NOT likely to be helpful without use of the drugs considered below.

Inhalation: Remove the victim to fresh air and allow the victim to recline. Give oxygen and amyl nitrate; keep the victim quiet and warm. With inhalation poisoning, thoroughly check clothing and skin to assure no cyanide is present. **Get medical assistance at once.**

Eye Contact: Flush eyes immediately with plenty of water; remove contaminated clothing; keep victim quiet and warm; **seek medical assistance at once.**

Skin Contact: Wash skin at once to remove cyanide while removing all contaminated clothing including shoes; **do not delay!** Skin absorption can take place from cyanide dust, solutions, or hydrogen cyanide vapor. Absorption is slower than with inhalation, often measured in minutes rather than seconds (AC or HCN is absorbed much faster than metal cyanides from solutions such as sodium, potassium or copper cyanide solutions). After going though decontamination on the victim(s), watch him or her for at least one to two hours, if possible, since absorbed cyanide can continue to work into the blood stream. As a final note, wash clothing before reuse, and **destroy contaminated shoes.**

Ingestion: Give victim 1 percent of sodium thiosulfate solution (or plain water) immediately by mouth and induce vomiting; repeat as needed until all vomit fluid is clear. Never give anything by mouth to an unconscious victim; give oxygen, and seek medical help immediately.

Drugs: Amyl nitrite and sodium nitrite are used to produce methaemoglobin. Amyl nitrite is useful only in a closed positive pressure respiratory system. Crushing the ampoule around the face or even inside the facepiece of the respirator is not really useful. It should not be used with concurrent oxygen administration due to the risk of explosion. Treatment with amyl nitrite should be followed by sodium thiosulphate. Sodium nitrite should be given intravenously at ten milliliters of a three percent solution (300 mg) over a period of three minutes. **WARNING: The therapeutic index of sodium nitrate is very low; the dose given is meant for adults and is reported to have killed children.** The sodium nitrite is given to produce methaemoglobin, thus sequestering the cyanide on the methaemoglobin. **The cyanide is then removed from the body as thiocyanate after administration of sodium thiosulphate.**

   Skin decontamination is usually **NOT** necessary since cyanide agents are highly volatile. Wet, contaminated clothing should be removed and the underlying skin decontaminated with water or other standard decontaminates. **However, persons whose skin or clothing is contaminated with cyanide-containing solutions can secondarily contaminate responder personnel by direct contact or off-gassing vapor.** Hydrogen cyanide (AC) is very **volatile**, and readily produces flammable and toxic concentrations at room temperatures. Hydrogen cyanide gas mixes well with air, and explosive mixtures are easily formed.

Antidotes: Intravenous sodium nitrite and sodium thiosulfate, and/or amyl nitrite perles are packaged in the cyanide antidote kit. When possible, treatment with cyanide antidotes should be given under medical supervision to **unconscious victims** who are known or suspected of having been exposed to cyanide poison. Amyl nitrite perles should be broken onto a gauze pad and held under the nose, over the Ambu-valve intake, or placed under the lip of the face mask. Inhale for thirty seconds every minute, and use a new perle every three minutes if sodium nitrate infusions will be delayed. If a patient has not responded to oxygen and amyl nitrite treatment, infuse sodium nitrite intravenously as soon as possible. The usual adult dose is ten mL (milliliter) of a three percent solution (300 mg) infused over **absolutely no less than five minutes**; the average pediatric dose is 0.12 to 0.33 mL/kg body weight up to ten mL infused as above. Medical personnel should monitor blood pressure during sodium nitrite administration, and slow the rate of infusion if hypotension develops. The next step is to infuse sodium thiosulfate intravenously. The usual adult dose is mL of a 25 percent solution (12.5 g) infused over ten to twenty minutes. The average pediatric for this step is 1.65 mL/kg of a 25 percent solution. Repeat one-half of the initial dose thirty minutes later if there is an inadequate clinical response.

Hospital staff working in an enclosed area can be secondarily contaminated by cyanide vapor off-gassing from heavily soaked clothing or skin, or from toxic vomit. Avoid dermal contact with cyanide-contaminated victims or with gastric contents of persons who may have ingested cyanide-containing materials (patients do not usually pose secondary contamination risks after contaminated clothing is removed and the skin is washed.

Medical Management: The treatment given to victims of hydrogen cyanide exposure is based on encouraging and speeding-up the body's own ability to excrete cyanide and to bind cyanide in the blood. Two major approaches are involved in the treatment of cyanide poisoning: provision of binding sites for cyanide ions, and provision of additional sulfur groups. The first approach provides alternative sites to those of cytochrome oxidase, and essentially reactivates that enzyme. The binding site may be provided by drugs such as dicobalt edetate and by hydroxocobalamin or by production of methaemoglobin in the blood. Methaemoglobin binds avidly to cyanide ions and may be produced by compounds such as **sodium nitrite** and **amyl nitrite** as well as **dimethylaminophenol**.

Methaemoglobin forming compounds should be used cautiously in victims suffering from concurrent carbon monoxide poisoning or hypoxia. The second approach calls for provision of additional sulfur groups to enhance the detoxification of cyanide and thiocyanate by endogenous rhodanese; this comes about by giving **sodium thiosulphate**.

Fire: Hydrogen cyanide (AC) is a highly flammable liquid. The liquid contains an acid stabilizer; aging of this liquid may cause an explosion of the mixture due to lessening of the acid stabilizer. AC can also polymerize in an explosive fashion. Use the buddy system whenever working with AC, only work in areas where all ignition sources have been controlled, and wear proper personal protective clothing and equipment. Hydrogen cyanide is flammable in air and a poison. Immediately evacuate everybody from the area of the spill/leak/toxic gas exposure. Stop the flow of gas if possible and feasible without needless exposure. Supply assisted respiration if victim(s) have stopped breathing. Isolate spill or leak area for 350 to 700 feet for 360 degrees. Completely eliminate all ignition sources. When entering closed spaces, you should ventilate them first. Only if leaks or spills can be stopped, extinguish fire(s); do not extinguish fire(s) if leaks or spills are still occurring. Use water spray or fog, alcohol resistant foam, dry chemical, or CO2 to fight fire(s).

Personal Protective Equipment: Respiratory protection is required (positive pressure, full face piece, NIOSH-approved SCBA will be worn). When response personnel respond to handle rescue or reconnaissance, they will wear **Level A** protection that should be worn when the highest level of respiratory, skin, eye, and mucous membrane protection is needed. This level consists of a fully-encapsulated, vapor-tight, chemical-resistant suit, chemical-resistant boots with steel toe and shank, chemical-resistant inner/outer gloves (butyl rubber glove M3 and M4 Norton, chemical protective glove set), coveralls, hard hat, and self-contained (positive pressure) breathing apparatus (SCBA).

Do not breathe fumes; contact with hydrogen cyanide can kill responders.

Spill/Leak Disposal: Evacuate/Isolate the area, ensure control of ignition sources, stop leak or spill if possible, control/ contain/ventilate the dispersion area if possible, ground all equipment, attend to security of the site, and wear proper protective clothing and equipment.

Symptoms: The more rapidly hydrogen cyanide levels build up, the more acute are the signs and symptoms of poisoning and the smaller is the total absorbed dose required to produce a given effect. In **high concentrations** there is an increase in the depth of respiration within a few seconds. Such stimulation may be so powerful that a victim cannot voluntarily hold his or her own breath. Violent convulsions occur after twenty to thirty seconds with cessation of respiration within one minute. Cardiac failure follows within a few minutes. With **lower concentrations**, the early symptoms are a weakness of the legs, vertigo, nausea, and headache. These may be followed by convulsions and coma which may last for hours or days depending on the length of exposure to the hydrogen cyanide. If the coma is lengthy, recovery may disclose residual damage to the central nervous system evidenced by irrationality, altered reflexes, and unsteady gait which may last for several weeks or longer. In mild cases there may be headache, vertigo, and nausea for several hours before complete recovery.

Vaccines: Not applicable

Guides for Emergency Response: Chemical Agent or Weapon: Blister Agent Lewisite (L), Blister Agent Mustard-Lewisite Mixture (HL)

25

Lewisite is an organic arsenical known for its vesicant properties. Blister agent Lewisite (L) acts as a systemic poison causing pulmonary edema, diarrhea, restlessness, weakness, subnormal temperature, and low blood pressure. In order of severity and appearance of symptoms it is a blister agent, a toxic lung irritant, absorbed in tissues, and a systemic poison. When inhaled in high concentrations, it may be fatal in as short a time as ten minutes. Lewisite is **not** detoxified by the human body. Common routes of entry into the human body include eyes, skin, and inhalation. L presents both a vapor and a liquid hazard and may damage the eyes, skin, respiratory tract, and the circulatory system. Exposure to L causes **immediate pain or at least irritation. Vapor may be inhaled into the respiratory tract which will cause onset of burning pain, irritation of the nose, and reflex coughing and chest tightness. Vapor also affects the eyes with the immediate onset of pain, redness, uncontrollable blinking, and swelling of the eyelids. Vapor or a liquid splash of Lewisite on the skin will be followed immediately by stinging pain and destruction of tissue followed by blistering within twelve hours.**

Mustard-Lewisite mixture is a liquid mixture of distilled mustard (HD) and Lewisite. Due to its low freezing point, the mixture remains a liquid in cold weather and at high altitudes; its low freezing point make it better for ground dispersal and aerial spraying. The mixture with the lowest freezing point consists of 63 percent Lewisite and 37 percent mustard.

BLISTER AGENT (VESICANT):   Lewisite (L)

Formula: C2H2AsC13

Description: Oily, colorless liquid

Warning: Odor like geraniums

Molecular weight: 207.32.

Boiling point: (760 mm Hg) = 374 Degrees F (190 Degrees C).

Freezing point: 0.4 F (–18 C)

Specific gravity: 1.888 at 68 Degrees F (20 Degrees C) (water = 1.0)

Vapor pressure: 0.394 mm Hg at 69 Degrees F (20 Degrees C)

Vapor Density: 7.1 (air = 1.0)

Liquid density: 1.89 g/cm3 at 77 Degrees F (25 Degrees C)

Flash point: Does not burn easily. When heated, toxic fumes of hydrogen chloride and arsenic are emitted.

Solubility in water: Negligible

Volatility: 4,480 mg/me (20 Degrees C)

BLISTER AGENT (VESICANT):   Mustard/Lewisite Mix (HL)

Formula: (None, a variable mix of HD and L for use in cold weather

Vapor Density: (Air = 1) 6.5

Vapor Pressure: (mm^Hg) 0.248 at 20 degrees C

Molecular Weight: 186.4

Liquid Density: (g/cc) 1.66 at 20 degrees C

Volatility: (mg/m3) 2,730 at 20 degrees C

Median Lethal Dosage: (mg-min/m3) 1,500 by inhalation; over 10,000 by skin exposure

Physical State: (at 20 degrees C) Dark, oily liquid

Odor: Garlic

Freezing/melting point: (at degrees C) –42 degrees C for plant purity; –25.4 degrees C when pure

Boiling point: (at degrees C: Less than 190 C

Action rate: Immediate stinging of skin and redness within thirty minutes; blistering delayed about thirteen hours.

Physiological action: Similar to HD; may cause systemic poisoning

Required level of protection: Protective mask and clothing

Decontamination: Bleach, fire, DS2, caustic soda, M258A1, M280

Detection in the field: M18A2, M256, M256A1

Use: Delayed-action casualty agent

CAS registry number: (mixture)

RTECS number: (mixture)

LCt50:   (respiratory) About 1,500 mg-min/m3

Ability to Kill: Moderate

Characteristics: Persons whose skin or clothing is contaminated by liquid Lewisite or mustard-Lewisite mixture can contaminate rescuers by direct contact or through off-gassing vapor. Lewisite is an oily, colorless liquid with an odor like geraniums. Mustard-Lewisite mixture is a liquid with a garlic-like odor. Volatility of both agents is significant at high ambient temperatures.

Response on Scene by First Responders

Caution: Contact with Lewisite vapor or liquid can be fatal. **Do not breathe the fumes! Skin contact must be avoided at all times.** If inhalation is a possibility, hold breath until respiratory mask is donned. Firefighters should wear full protective clothing and respiratory protection during both fire-fighting and rescue (positive pressure, full face, NIOSH-approved self-contained breathing apparatus/SCBA shall be worn).

Field First Aid: Monitor the incident scene, and victims, to attempt to identify the chemical or biological agent that is creating the problem. Separate the walking wounded from the prone (e.g., "All those who can walk, stand over there"). Assess the number of prone and walking wounded. If Emergency Medical Service personnel are not on the scene, call them. If it appears to be a mass casualty incident, call for help **immediately!** As soon as a person trained in triage arrives on scene (physician, physicians assistant, nurse, paramedic, or EMS person) have triage instituted. For inhalation exposure, remove victims from the source **immediately**; give artificial respiration if breathing has stopped (refrain from mouth-to-mouth attempts to provide artificial respiration); administer oxygen if breathing is difficult, **seek medical attention at once!** If there is contact with the eyes, speed in decontaminating the eyes is absolutely essential; remove victim(s) from the liquid source; flush the eyes immediately with water for ten to fifteen minutes by tilting the head to the side, pulling the eyelids apart with fingers, and pouring water slowly into the eyes; do not cover eyes with bandages, but if necessary, protect eyes by means of dark glasses or opaque goggles; seek medical help immediately! Time is of the essence. For skin contact, remove patient from the source at once; remove contaminated clothing; decontaminate affected areas by flushing with 10 percent sodium carbonate solution; wash off with soap and water after three to four minutes to protect against erythema; seek medical attention at once. Regarding ingestion of blister agent Lewisite, **do not induce vomiting**; give the injured person milk to drink if able; seek medical attention immediately. Isolate contaminated items such as clothing, shoes, personal items, and equipment.

No one enters the Hot Zone or Warm Zone without appropriate personal protective equipment. Ensure that responding agencies know the danger(s) they face by responding to this incident.

Quickly ensure ABC (airway/breathing/circulation). See that the victim has a clear airway. If trauma is suspected, maintain cervical immobilization manually and apply a cervical collar and a backboard if feasible. Check for adequate respiration, and administer supplemental oxygen if cardiopulmonary compromise is suspected. Maintain adequate circulation. Establish intravenous access when necessary and warranted. Use a cardiac monitor as necessary. Stop any bleeding when necessary. When ingestion is evident, **do not induce emesis** (vomiting). For victims who are alert and able to swallow, give four to eight ounces of milk to drink. Administration of activated charcoal is **not** viewed as beneficial.

Antidotes: British Anti-Lewisite (BAL) can be given by intramuscular injection as an antidote for systemic effects but has no effect on the local lesions of the skin, eyes, or airways. **Treatment consists primarily of supportive care.**

Medical Management: It is essential to remove the mustard and/or Lewisite agents as quickly as possible. Vesicants rapidly penetrate the skin causing both localized cellular damage and systemic damage. The deadly nature of such agents' effect is that a person exposed to a large amount of liquid or vapor faces total systemic assault. The reasons for this are (a) the failure of the body's immune system, with sepsis and infection as the major contributing causes of death, and (b) pulmonary damage which wins out as the major contributory factor in death. Rapid decontamination is the only way to avoid severe burns. Medical management of a victim who has undergone exposure to mustard agent can be very simple, as providing of symptomatic care for a sunburn-like erythema; or vastly complex as in providing long-term services to fight immunosuppression, caring for serious burns, and/or dealing with other multi-system factors. A patient severely ill with mustard agent poisoning requires the general supportive care provided for any severely ill patient as well as specific care given to a burn patient. Liberal use of systemic analgesics and antipruritics, as needed, maintenance of fluid and electrolyte balance, and other supportive measures are necessary. Food supplements, including vitamins, may also be helpful. Some medical management schemes do not work out so well. Activated charcoal given orally does not seem to provide better care. Hemodialysis has been ineffective and even harmful to several patients. The rapid transformation of the mustard molecule suggests that few methods are actually available and beneficial hours or days after exposure with a mustard agent. **Mustard agents are classed as a mutagen and a carcinogen based on laboratory studies. Lewisite is generally considered a suspect carcinogen because of its arsenic content.**

Fire: The flash point for agents H and HD is 221 degrees F or 105 degrees C; for agent HT, the flash point is 212 degrees F or 100 degrees C. There is no flash point for Lewisite, so that information is not applicable. Extinguishing media for Lewisite is water, fog, foam, or CO2. Avoid use of extinguishing methods that will cause splashing or spreading of Lewisite.

Personal Protective Equipment: For mustard agents (H, HD, HL, HT) wear full protective clothing consisting of Level A: **Level A** should be worn when the highest level of respiratory, skin, eye, and mucous membrane protection is needed. This lever consists of a fully-encapsulated, vapor-tight, chemical-resistant suit, chemical-resistant boots with steel toe and shank, chemical-resistant inner/outer gloves, coveralls, hard hat, and self-contained (positive pressure) breathing apparatus (SCBA). Both the Environmental Protection Agency and NIOSH recommend that initial entry into unknown environments or into confined space that has not been chemically characterized be conducted wearing at least Level B, if not Level A, protection. For protective gloves, it is good practice to wear Butyl toxicological protective gloves such as M3, M4, or glove set.

Spill/Leak Disposal: If leaks or spills of Lewisite or mustard occur, only personnel in full protective clothing will be allowed in the area. **Mustard agents** or weapons should be contained using vermiculite, diatomaceous earth, clay, or fine sand and neutralized as soon as possible using copious amounts of 5.25 percent sodium hypochlorite solution. Scoop up all material and place it in an approved Department of Transportation container. Cover the contents with decontaminating solution as mentioned above. The exterior of the salvage container shall be decontaminated and labeled according to EPA and DOT regulations. All leaking containers will be over packed with sorbent (e.g., vermiculite) placed between the interior and exterior containers. Decontaminate and label according to EPA and DOT regulations. Dispose of the material in accordance with waste disposal methods provided below. Conduct general area monitoring with an approved monitor to confirm that the atmospheric concentrations do not exceed the airborne exposure limits. If 5.25 percent sodium hypochlorite solution is not available, the following decontaminates may be used instead and are listed in order of preference: Calcium Hypochlorite, Contamination Solution No. 2 (DS2), and Super Tropical Bleach Slurry (STB).

For **Lewisite**, it should be contained using vermiculite, diatomaceous earth, clay, or fine sand and neutralized as soon as possible using copious amounts of alcoholic caustic, carbonate, or Decontaminating Agent (DS2). Caution should be exercised when using these decontaminates since acetylene will be given off. Household bleach can also be used if accompanied by stirring to allow contact. With the differences expressed for Lewisite, now proceed as you would for mustard agent.

Special Section on Thickened HD

Styrene-butyl acrylate copolymer is used to "thicken" HD and make THD. A thickened agent is one to which a polymer or plastic has been added to retard evaporation and cause it to adhere to surfaces. Essentially, THD is the same as HD except for the following factors: Fire, Health Hazard, Spill/Leak/Disposal Factors, and Special Precautions. THD has a slight fire hazard when exposed to heat or flame. The health hazard is the same for HD except for skin contact. For skin contact don respiratory protective mask and remove contaminated clothing **IMMEDIATELY. IMMEDIATELY** scrape the THD from the skin surface, then wash the contaminated surface with acetone. Seek medical attention **IMMEDIATELY.** If spills or leaks occur, handle the same as HD but dissolve THD in acetone before introducing and decontaminating solution. Containment of THD is generally not necessary. Spilled THD can be carefully scrapped off the contaminated surface and placed in a fully removable head drum with a high density, polyethylene lining. THD then can be decontaminated, after it has been dissolved in acetone with the same procedures used for HD. Contaminated surfaces should be treated with acetone then decontaminated using the same procedures as those used for HD. Note: Surfaces contaminated with THD and then rinsed-decontaminated may evolve sufficient HD vapor to produce a physiological response. As far as special precautions are concerned, watch for "stringers" (elastic, thread like attachments) formed when agents are transferred or dispensed. These stringers **must be broken cleanly** before moving the contaminating device or dispensing device to another location, or unwanted contamination of a working surface will result. Avoid contact with strong oxidizers, excessive heat, sparks, or open flame.

Approved Methods of Waste Disposal: (**Open pit burning or burying of HD or items containing or contaminated with HD in any quantity is** *prohibited***!**) Decontamination of waste or excess material shall be accomplished according to procedures outlined above and can be destroyed by incineration in EPA approved incinerators according to appropriate provisions of Federal, State, and local Resource Conservation Act (RCRA) regulations. *Note*: Some decontamination solutions are hazardous waste according to RCRA regulations and must be disposed of according to these regulations.

Symptoms: Erythema, Blisters, Irritation of the Eyes, Cough, Dyspnea, Asymtomatic Latent Period (hours). Also, mild upper respiratory signs to marked airway damage, GI effects and bone marrow stem cell suppression possible. Mustard is a blister agent that affects the eyes, lungs, and skin. A person exposed to mustard will feel very little pain and will NOT notice symptoms for quite some time. However, the longer the exposure without removal of the mustard agent, the more severe will be the damage to the affected areas of the body

Vaccines: None

Guides for Emergency Response: Chemical Agent or Weapon: Nerve Agent GF (GF) 26

NERVE AGENT: GF (GF)

Introduction: GF is a chemical nerve agent that is a fluoride-containing organophosphate. It is a slightly volatile liquid that is almost insoluble in water. GF enters the body primarily through the respiratory tract, but is also highly toxic through the skin and digestive track. It is a strong cholinesterase inhibitor. **It is approximately twenty times more persistent than sarin (CB).** GF is about as persistent as tabun, and evaporates about twenty times more slowly than water. Heavily splashed liquids persist one to two days under average weather conditions. It is a quick action casualty agent, and is very stable.

Nerve Agent: GF (GF)

Formula: CH3PO(F)OC6H11

Vapor Density: (Air = 1) 6.2

Vapor Pressure: (mm^Hg) 0.044 at 20 degrees C

Molecular Weight: 180.2

Liquid Density: (g/cc) 1.1327 at 20 degrees C

Volatility: (mg/m3) 438 at 20 degrees C

Median Lethal Dosage: (mg-min/m3) unknown

Physical State: (at 20 degrees C) Liquid

Odor: Sweet, musty, peaches, shellac

Freezing/melting point: (at degrees C) –30 C

Boiling point: (at degrees C) 239 C

Action rate: Very rapid

Physiological action: Cessation of breathing—death may follow

Required level of protection: Protective mask and clothing

Decontamination: Bleach slurry, dilute alkali, or DS2; steam or ammonia in confined area; M258A1, M280

Detection in the field: M18A2, M256, M256A1, M8 and M8A1 alarms, M8 and M9 paper

Use: Quick-action casualty agent

CAS registry number: 329-99-7

RTECS number: TA822500050: (subcutaneous) Values are reported from 16ug/kg to 400 ug/kg for mice.

Ability to Kill: HIGH

Characteristics: Nerve agents are liquid under temperate conditions, but, when dispersed, the more volatile ones constitute **both** a vapor and a liquid hazard. However, the less volatile nerve agents represent primarily a liquid hazard (mainly, the G-agents are more volatile than the nerve agent VX, while sarin (GB) is the most volatile and nerve agent GF is the least volatile of the so-called "G-agents." GB has an LCt-50 of 100 (vapor toxicity of mg-min/m3), an ICt-50 of 75 (vapor toxicity of mg-min/m30, and an MCt-50 of 3 (vapor toxicity of mg-min/m3). The LD-50 on skin is 1700mg.

Response on Scene by First Responders

Caution: Higher temperatures on site can increase the danger to sarin contamination particularly with cutaneous exposure. GF cannot be transported legally except by the military (U.S. Army Technical Escort Unit) according to 49 CFR 172. Nerve agents are potent acetylcholinesterase inhibitors causing the same signs and symptoms regardless of the exposure route, although the initial effects depend on the dose and route of exposure.

Skin or clothing contaminated with liquid nerve agent can contaminate rescuers by direct contact or through off-gassing vapor. Nerve agents are extremely toxic and can cause loss of consciousness and convulsions within seconds and death from respiratory failure within minutes of exposure. Atropine and pralidoxime chloride (2-PAM Cl) are antidotes for nerve agent toxicity but pralidoxime must be administered within minutes to a few hours following exposure (depending on the specific agent) to be effective. Basically, treatment consists of supportive measures and repeated administration of antidotes.

Field First Aid: Nerve agents are the most toxic of the known chemical warfare agents. Chemically similar to organophosphate pesticides, their method of acting is to inhibit acetylcholinesterase enzymes. Individuals whose skin or clothing is contaminated with nerve agent can contaminate rescuers by direct contact or through off-gassing vapor; but persons whose skin is exposed only to nerve agent vapor pose no risk of secondhand contamination. However, it is important to remember that clothing can trap vapor. Nerve agents can be readily absorbed by ingestion, skin, and inhalation that can lead to fatal systemic effects. Most of the nerve agents were failed pesticides that tended to be so toxic they were referred to armies around the world, mainly in Germany, and made into chemical nerve agents. Nerve agents cause the same signs and symptoms regardless of the exposure route. Chemical agent casualty triage is based on walking feasibility (if they can walk, their contamination is slight, or maybe, moderate), respiratory status, age, and additional conventional injuries. Any person appointed to handle triage must understand, at a minimum, the natural course of a given injury, the medical resources immediately available, the current and likely casualty flow, and the medical evacuation capabilities in order to decide Immediate, Delayed, Minimal, and Expectant priority assignments for the injured.

Nerve agent intoxication requires rapid decontamination to prevent further absorption by the patient and to prevent exposure to others; ventilation when necessary, as well as administration of antidotes and supportive therapy. Skin decontamination is not necessary with exposure to vapor alone, but clothing should be removed to get rid of any trapped vapor. With nerve agents, there can be high airway resistance due to bronchoconstriction and secretions, and initial ventilation is often difficult. The restriction will decrease with atropine administration. Copious secretions, which may be thickened by atropine, also impede ventilatory actions and will require frequent suctioning. For inhalation exposure to nerve agents, ventilation support is essential

Inhalation: Hold breath until respiratory protective mask is donned. If severe signs of agent exposure appear (chest tightens, pupil constriction, etc.) immediately administer, in rapid succession, all three; Nerve Agent Antidote Kit(s), Mark I kit injectors (or atropine if directed to by a physician). Injections using the Mark I kit injectors may be repeated at five to twenty minute intervals if signs and symptoms are progressing until all three series of injections have been administered. No more injections will be administered unless indicated by medical personnel. **In addition, a record will be maintained of all injections given. If breathing has stopped, give artificial respiration**. Mouth-to-mouth resuscitation should be used when mask-bag or oxygen delivery systems are not available. **Do not use mouth-to-mouth resuscitation** when facial contamination exists. If breathing is difficult, administer oxygen. Seek medical attention **immediately.**

Eye Contact: **Immediately** flush eyes with water for ten to fifteen minutes, then don respiratory protective mask. Although miosis (pinpointing of the pupils) may be an early sign of agent exposure, an injection will not be administered when miosis is the only sign present. Instead, the individual will be taken **immediately** to a medical treatment facility for observation.

Skin Contact: Don respiratory protective mask and remove contaminated clothing. Immediately wash contaminated skin with copious amounts of soap and water, 10 percent sodium carbonate solution, or 5 percent liquid household bleach. Rinse well with water to remove excess decontaminant. Administer Nerve Agent Antidote Kit, Mark I, only if local sweating and muscular twitching symptoms are observed. Seek medical attention immediately.

Ingestion: Do not induce vomiting. First symptoms are likely to be gastrointestinal. **Immediately** administer Nerve Agent Antidote Kit, Mark I. Seek medical attention **immediately.**

Drugs: For mild or moderate nerve agent contamination in an adult, the standard first dose for atropine is 2 to 4 mg IM, (intramuscular) and for 2-PAM Cl is 15 mg/kg (1g) IV (intravenous) slowly; for an adult victim with severe symptoms, a first dose would be atropine at 6 mg IM (intramuscular), and for 2-PAM Cl the dose should be 15 mg/kg(1g) IV (intravenous) slowly. There are different recommended doses for Infants, Children, Adolescent, and Elderly/Frail.

Atropine: Atropine is a cholinergic blocking, or anticholinergic, compound that is very effective in blocking the effects of excess acetylcholine at peripheral muscarinic sites. The amount of atropine in three MARK I kits may cause adverse effects on military performance in a normal person. In people NOT exposed to nerve agents, amounts of 10 mg or higher may cause delirium. However, potentially the most hazardous effect of inadvertent use of atropine (two mg, intramuscular) to a young person **not exposed to nerve agents**

in a warm or hot atmosphere is inhibition of the sweating function which may lead to heat injury. In the military, atropine is packaged in auto-injectors with each injector containing 2 mg.

Pralidoxime chloride: Pralidoxime chloride (Protopam chloride; 2-PAM Cl) is an oxime that attaches to the nerve agent that is inhibiting the cholinesterase and breaks the agent-enzyme bond to restore the normal activity of the enzyme. This drug at a dosage of 600 mg is in an auto-injector for self- or buddy use that is included in the military MARK I kit along with the atropine injector, and each military person in an area where nerve agents might be used is issued three MARK I kits. MARK I kits are also being provided to civilian first responders at the present time.

Diazepam: Diazepam is an anticonvulsant drug used to decrease convulsive activity in order to reduce brain damage caused by prolonged seizure activity. Without the use of pyridostigmine pretreatment, experimental animals died quickly after super-lethal doses of nerve agents despite conventional therapy. With pyridostigmine pretreatment which is followed by conventional therapy, animals survive super-lethal doses of soman, but had prolonged periods of seizure activity before recovery. The administration of diazepam with other standard therapy to soman-poisoned animals pretreated with pyridostigmine reduced the seizure activity. Current military doctrine (also being used by local responders) is to administer diazepam with other therapy (three MARK I's) at the onset of severe effects from a nerve agent, whether or not seizure activity is among these effects. Each military person or responders should carry one auto-injector of 10 mg of diazepam for his buddy to administer to him; if he could administer it to himself, he would not need it! Diazepam should be administered with the three MARK I's when the casualty's condition warrants the use of three MARK I's at the same time. Medical personnel can administer more diazepam to a casualty when necessary. A paramedic should carry more diazepam injectors, and be authorized to administer two additional injectors at ten minute intervals to a convulsing victim. Paramedics might make a mistake in giving too much atropine to a mild to moderate casualty, but more importantly he or she might err by giving too little atropine to a severe victim. The casualty with skin exposure to liquid is more difficult to evaluate and manage than is a victim contaminated with exposure to vapor. Agent on the skin can be decontaminated, but agent absorbed into the skin cannot be removed. **A casualty from liquid exposure on the skin may continue to worsen because of continued absorption of the agent.**

Pyridostigmine: The U.S. military about fifteen years ago provided pyridostigmone bromide as a pretreatment for nerve agent exposure. Each trooper received a blister pack containing twenty-one 30-mg tablets for a dose regimen of one 30-mg tablet every eight hours. When given before **soman** exposure and when that exposure is followed by the standard MARK I therapy, the use of this pretreatment will increase the LD-50 several fold over the LD-50 obtained without the use of the pretreatment. When soman is the nerve agent used, the use of pyridostigmine increases survival. When the agent used is sarin (GB) or VX (VX), survival after standard MARK I therapy is essentially the same whether or not pyridostigmine pretreatment is used, i.e., **pyridostigmine use provides no benefit in sarin or VX (VX) poisoning.** Pyridostigmine is a **pretreatment** rather than an antidote and **should not be given after soman exposure. Its use will not decrease the effects of soman.** One consequence of the greater survival from the use of pyridostigmine is prolonged seizure activity and subsequent brain damage in the survivors. The early administration of diazepam will decrease these effects.

Fire: Sarin failed to flash to 280 degrees F. Flammability limits, lower explosive limit, and upper explosive limits are not applicable for this product. Sarin reacts with steam and water to produce toxic and corrosive vapors. Any personnel not fighting a fire of sarin should immediately be evacuated from the area. Respiratory protection is required (positive pressure, full face piece, NIOSH-approved SCBA will be worn). When response personnel respond to handle rescue or reconnaissance, they will wear **Level A** protection that should be worn when the highest level of respiratory, skin, eye, and mucous membrane protection is needed. This level consists of a fully-encapsulated, vapor-tight, chemical-resistant suit, chemical-resistant boots with steel toe and shank, chemical-resistant inner/outer gloves (butyl rubber glove M3 and M4 Norton, chemical protective glove set), coveralls, hard hat, and self-contained (positive pressure) breathing apparatus (SCBA).

Do not breath fumes; skin contact with nerve agents should always be avoided; contact with liquid sarin or vapors can kill responders. Hydrogen may be present! Use water mist, fog, foam, or CO-2 to fight the fire; do NOT splash or spread sarin.

National Fire Protection Association—704 System Warning

The NFPA—704 marking system was created to give firefighters an indication as to the degree of hazard a chemical presents in a fire situation. The system was designed for use at fixed facilities rather than on transport vehicles. These markings are also found on individual containers. The top three sections of the marking uses numbers from 0 to 4 to identify the degree of hazard presented for Health Hazards (blue), Flammability (red), Reactivity (yellow), reading from left to right across the top, and Special Hazards (white) across the bottom of the markings. The number 0 represents no significant hazard, while the number 4 indicates a severe hazard. The bottom section (Special Hazards/White) indicates, if marked at all, special hazards such as radioactive, oxidizer, or water-reactive. The NFPA—704 system for GF chemical nerve agent is HEALTH—4, FLAMMABILIY—1, REACTIVITY—1, and SPECIAL HAZARD—0.

Personal Protective Equipment: Standard turnout gear with SCBA provides a first responder with sufficient protection from nerve agent vapor hazards inside interior or downwind areas of the hot zone to allow *thirty minutes rescue* time for known live victims. Self-taped turnout gear with SCBA provides sufficient protection in an unknown nerve agent environment for a *three-minute reconnaissance to search for living victims,* (or a two-minute reconnaissance if mustard, blister agent (HD) is suspected).

When you do NOT know the degree of hazard, use Level A personal protective equipment (PPE) as follows: **Level A** protection should be worn when the highest level of respiratory, skin, eye, and mucous membrane protection is needed. This level consist of a fully-encapsulated, vapor-tight, chemical-resistant suit, chemical-resistant boots with steel toe and shank, chemical-resistant inner/outer gloves, coveralls, hard hat, and self-contained (positive pressure) and self-contained breathing apparatus (SCBA).

Spill/Leak Disposal: If leaks or spills of nerve agents occur, only personnel in full protective clothing will remain in the area. Spills must be contained by covering with vermiculite, diatomaceous earth, clay, fine sand, sponges, and paper or cloth towels. Decontaminate with copious amounts of aqueous sodium hydroxide solution (a minimum of 10 wt.%). Scoop up all material and place it in a DOT approved container. Cover the contents with decontamination solution as above. After sealing, the exterior will be decontaminated and labeled according to EPA and DOT regulations. All leaking containers will be over packed with sorbent (e.g., vermiculite) placed between the interior and exterior containers. Decontaminate and label according to EPA and DOT regulations. Dispose of decontaminate according to federal, state, and local laws. Conduct general area monitoring to confirm that the atmospheric concentrations do not exceed the airborne exposure limits. If 10 wt.% aqueous sodium hydroxide is not available, then the following decontaminants may be used instead (listed in order of preference): Decontaminating Agent (DS2), Sodium Carbonate, and Supertropical Bleach Slurry (STB).

Symptoms: Nerve agents exposures results in rhinorrhea, chest tightness, pinpoint pupils, shortness of breath, excessive salivation and sweating, nausea, vomiting, abdominal cramps, involuntary defecation and urination, muscle twitching, confusion, seizures, flaccid paralysis, coma, respiratory failure, and death.

Vaccine(s): No

Guides for Emergency Response: Chemical Agent or Weapon: Nerve Agent Sarin (GB)

27

Introduction: Nerve agents are organophosphate ester derivatives of phosphoric acid. Some general practices, procedures, and methods can be applied to a number of different toxic chemical agents. Chemical weapons can work at maximum effectiveness when used against untrained people and unprotected targets. Chemical warfare attacks, by their very nature and history, can seriously make psychological warfare a reality through combat stress, poor morale, and general inefficiency. *In this chapter we deal only with chemical agents designed to kill or seriously injure;* that is, we do not deal with riot control agents, incendiary agents, noxious chemicals, temporary incapacitating weapons, or smoke agents. Chemical agents can be inhaled, absorbed through the skin/wounds/abrasions/eyes, or consumed through food and/or drink. They includes blister agents, nerve agents, pulmonary (choking) agents, and blood agents; they can be distributed by spray devices, bombs, aircraft, rockets, missiles, mines, water supplies/reservoirs, and other methods. Some signs that a toxic chemical agent has been released might include an unexplained runny nose, an obvious attack using spray from an aircraft, dead birds/insects/animals/fish, mass casualties, distribution of victims in a pattern suggesting a particular plan for dissemination, similar symptoms among casualties, smoke/mist/fumes/clouds of unknown origin, laughter or strange behavior in other persons, slurred speech, difficulty in breathing, eyesight problems, an oily film on ground or leaves, or unexplained vectors (hosts), or discolored grass or bushes.

NERVE AGENT: Sarin

Formula: CH3PO(F)OCH(GH3)2

Vapor Density: (Air = 1) 4.86

Vapor Pressure: (mm^Hg) 2.9 at 25 degrees C; 2.10 at 20 degrees C

Molecular Weight: 140.1

Liquid Density: (g/cc) 0.4 at 25 degrees C

Volatility: (mg/m3) 22,000 at 25 degrees C; 16,090 at 20 degrees C

Median Lethal Dosage: (mg-min/m3) 100 for a resting person

Physical State: (at 20 degrees C) Colorless liquid

Odor: Odorless

Freezing/melting point: (at degrees C) –56 C

Boiling point: (at degrees C): 158

Action rate: Very rapid

Physiological action: Cessation of breathing—death may follow

Required level of protection: Protective mask and clothing

Decontamination: Steam and ammonia in confined area; hot soapy water; M258A1, M280

Detection in the field: M18A2, M256, M256A1, M8 and M8A1 alarms, M8 and M9 paper, and Draeger Tube Phosphoric Acid Ester 0.05/a

Use: Quick-action casualty agent

CAS registry number: 107-44-8

RTECS number: TA8400000

LCt50: (inhalation): 70 mg-min/m3

Ability to Kill: HIGH

Characteristics: Nerve agents are liquid under temperate conditions; but when dispersed the more volatile ones constitute **both** a vapor and a liquid hazard. However, the less volatile nerve agents represent primarily a liquid hazard (mainly, the G-agents are more volatile than the nerve agent VX, while GB is the most volatile and nerve agent GF is the least volatile of the so-called "G-agents". GB has an LCt-50 of 100 (vapor toxicity of mg-min/m3), an ICt-50 of 75 (vapor toxicity of mg-min/m30, and an MCt-50 of 3 (vapor toxicity of mg-min/m3). The LD-50 on skin is 1700mg.

Response on Scene by First Responders

Caution: Higher temperatures on site can increase the danger to sarin contamination particularly with cutaneous exposure. Sarin cannot be transported legally except by the military (U.S. Army Technical Escort Unit) according to 49 CFR 172. Nerve agents are potent acetylcholinesterase inhibitors causing the same signs and symptoms regardless of the exposure route, although the initial effects depend on the dose and route of exposure. **Casualties whose skin or clothing is contaminated with liquid nerve agent can contaminate rescuers by direct contact or through off-gassing vapor.** Nerve agents are extremely toxic and can cause loss of consciousness and convulsions within seconds and death from respiratory failure within minutes of exposure. Atropine and pralidoxime chloride (2-PAM Cl) are antidotes for nerve agent toxicity but pralidoxime must be administered within minutes to a few hours following exposure (depending on the specific agent) to be effective. Basically, treatment consists of supportive measures and repeated administration of antidotes.

Field First Aid: Nerve agents are the most toxic of the known chemical warfare agents. Chemically similar to organophosphate pesticides, their method of acting is to inhibit acetylcholinesterase enzymes. Individuals whose skin or clothing is contaminated with nerve agent can contaminate rescuers by direct contact or through off-gassing vapor; but persons whose skin is exposed only to nerve agent vapor pose no risk of secondhand contamination. However, it is important to remember that clothing can trap vapor. Nerve agents can be readily absorbed by ingestion, skin, and inhalation that can lead to fatal systemic effects. Most of the nerve agents were "failed pesticides" that tended to be so toxic they were referred to armies around the world, mainly in Germany, and made into chemical nerve agents. Sarin's (GB) pure liquid is clear, colorless, and tasteless, and discolors with aging to dark brown. Nerve agents cause the same signs and symptoms regardless of the exposure route. Chemical agent casualty triage is based on walking feasibility (if they can walk, their contamination is slight, or maybe, moderate) respiratory status, age, and additional conventional injuries. Any person appointed to handle triage must understand, at a minimum, the natural course of a given injury, the medical resources immediately available, the current and likely casualty flow, and the medical evacuation capabilities **in order to decide Immediate, Delayed, Minimal, and Expectant priority assignments for the injured.**

Nerve agent intoxication requires rapid decontamination to prevent further absorption by the patient and to prevent exposure to others, ventilation when necessary, administration of antidotes, as well as supportive therapy. Skin decontamination is not necessary with exposure to vapor alone, but **clothing should be removed to get rid of any trapped vapor.** With nerve agents, there can be high airway resistance due to bronchoconstriction and secretions, and initial ventilation is often difficult. The restriction will decrease with atropine administration. Copious secretions which may be thickened by atropine also impede ventilatory actions and will require frequent suctioning. **For inhalation exposure to nerve agents, ventilation support is essential.**

Inhalation: Hold breath until respiratory protective mask is donned. If severe signs of agent exposure appear (chest tightens, pupil constriction, etc.), immediately administer, in rapid succession, all three Nerve Agent Antidote Kit(s), Mark I kit injectors (or atropine if directed to by a physician). Injections using the Mark I kit injectors may be repeated at five to twenty minute intervals if sign and symptom are progressing until all three series of injections have been administered. No more injections will be administered unless indicated by medical personnel. In addition, a record will be maintained of all injections given. If breathing has stopped, give artificial respiration. Mouth-to-mouth resuscitation should be used when mask-bag or oxygen delivery systems are **not** available. **Do not use mouth-to-mouth resuscitation when facial contamination exists.** If breathing is difficult, administer oxygen. Seek medical attention **immediately.**

Eye Contact: **Immediately** flush eyes with water for ten to fifteen minutes, then don respiratory protective mask. Although miosis (pinpointing of the pupils) may be an early sign of agent exposure, an injection will not be administered when miosis is the only sign present. Instead, the individual will be taken **immediately** to a medical treatment facility for observation.

Skin Contact: Don respiratory protective mask and remove contaminated clothing. **Immediately** wash contaminated skin with copious amounts of soap and water, ten percent sodium carbonate solution, or five percent liquid household bleach. Rinse well with water to remove excess decontaminant. Administer Nerve Agent Antidote Kit, Mark I, only if local sweating and muscular twitching symptoms are observed. Seek medical attention **immediately.**

Ingestion: Do not induce vomiting. First symptoms are likely to be gastrointestinal. **Immediately** administer Nerve Agent Antidote Kit, Mark I. Seek medical attention **immediately.**

Drugs: For mild or moderate nerve agent contamination in an adult, the standard first dose for atropine is 2 to 4 mg IM, (intramuscular) and for 2-PAM Cl is 15 mg/kg (1g) IV (intravenous) slowly; for an adult victim with severe symptoms, a first dose would be atropine at 6 mg IM, (intramuscular) and for 2-PAM Cl the dose should be 15 mg/kg(1g) IV (intravenous) slowly. There are different recommended doses for Infants, Children, Adolescent, and Elderly/Frail.

Atropine: Atropine is a cholinergic blocking, or anticholinergic, compound that is very effective in blocking the effects of excess acetylcholine at peripheral muscarinic sites. The amount of atropine in three MARK I kits may cause adverse effects on military performance in a normal person. In people **NOT exposed to nerve agents**, amounts of 10 mg or higher may cause delirium. However, potentially the most hazardous effect of inadvertent use of atropine (2 mg, intramuscular) young person **not** exposed to nerve agents in a warm or hot atmosphere is inhibition of the sweating function which may lead to heat injury. In the military, atropine is packaged in auto-injectors with each injector containing 2 mg.

Pralidoxime chloride: Pralidoxime chloride (Protopam chrolide; 2-PAM Cl) is an oxime that attaches to the nerve agent that is inhibiting the cholinesterase and breaks the agent-enzyme bond to restore the normal activity of the enzyme. This drug at a dosage of 600 mg is in an auto-injector for self or buddy use is included in the military MARK I kit along with the atropine injector, and each military person in an area where nerve agents might be used is issued three MARKI kits.

Diazepam: Diazepam is an anticonvulsant drug used to decrease convulsive activity in order to reduce brain damage caused by prolonged seizure activity. Without the use of pyridostigmine pretreatment, experimental animals died quickly after super-lethal doses of nerve agents despite conventional therapy. With pyridostigmine pretreatment which is followed by conventional therapy animal survives super-lethal doses of soman, but had prolonged periods of seizure activity before recovery. The administration of diazepam with other standard therapy to soman-poisoned animals pretreated with pyridostigmine reduced the seizure activity. Current military doctrine (also used by local responders) is to administer diazepam with other therapy (three MARK I's) at the onset of severe effects from a nerve agent, **whether or not seizure activity is among these effects.** Each military person or responders should carry one auto-injector of 10 mg of diazepam for his buddy to administer to him; **if he could administer it to himself, he would not need it! Diazepam should be administered with the three MARK I's when the casualty's condition warrants the use of three MARK I's at the same time.** Medical personnel can administer more diazepam to a casualty when necessary. A paramedic should carry more diazepam injectors, and be authorized to administer two additional injectors at ten minute intervals to a convulsing victim. Paramedics might make a mistake in giving too much atropine to a mild to moderate casualty, but more importantly he or she might err by giving too little atropine to a severe victim. The casualty with skin exposure to liquid is more difficult to evaluate and manage than is a victim contaminated with exposure to vapor. Agent on the skin can be decontaminated, but agent absorbed into the skin cannot be removed. A casualty from liquid exposure on the skin may continue to worsen because of continued absorption of the agent.

Pyridostigmine: About twelve years ago, the U. S. military provided pyridostigmone bromide as a pretreatment for nerve agent exposure. Each trooper received a blister pack containing twenty-one 30-mg tablets for a dose regimen of one 30-mg tablet every eight hours. When given before **soman** exposure and when that exposure is followed by the standard MARK I therapy, the use of this pretreatment will increase the LD-50 several fold over the LD-50 obtained without the use of the pretreatment. When **soman** is the nerve agent used, the use of pyridostigmine increases survival. When the agent used is sarin (GB) or VX (VX), survival after standard MARK I therapy is essentially the same whether or not pyridostigmine pretreatment is used, i.e., pyridostigmine use provides no benefit in sarin or VX (VX) poisoning. Pyridostigmine is a **pretreatment** rather than an antidote and should not be given **after** soman exposure. Its use will not decrease the effects of soman. One consequence of the greater survival from the use of pyridostigmine is prolonged seizure activity and subsequent brain damage in the survivors. The early administration of diazepam will decrease these effects.

Fire: Sarin failed to flash to 280 degrees F. Flammability limits, lower explosive limit, and upper explosive limits are not applicable for this product. Sarin reacts with steam and water to produce toxic and corrosive vapors. Any personnel not fighting a fire of sarin should immediately be evacuated from the area. Respiratory protection is required (positive pressure, full face piece, NIOSH-approved SCBA will be worn). When response personnel respond to handle rescue or reconnaissance, they will wear Level A protection that should be worn when the highest level of respiratory, skin, eye, and mucous membrane protection is needed. This level consists of a fully-encapsulated, vapor-tight, chemical-resistant suit, chemical-resistant boots with steel toe and shank, chemical-resistant inner/outer gloves (butyl rubber glove M3 and M4 Norton, chemical protective glove set), coveralls, hard hat, and self-contained (positive pressure) breathing apparatus (SCBA).

   Do not breath fumes; skin contact with nerve agents should always be avoided; contact with liquid sarin or vapors can kill responders. Hydrogen may be present! Use water mist, fog, foam, or CO-2 to fight the fire; do NOT splash or spread sarin.

   Standard turnout gear with SCBA provides a first responder with sufficient protection from nerve agent vapor hazards inside interior or downwind areas of the hot zone to allow *thirty minutes rescue* time for known live victims.

   Self-taped turnout gear with SCBA provides sufficient protection in an unknown nerve agent environment for a three-minute reconnaissance to search for living victims, (or a two-minute reconnaissance if HD (mustard, blister agent) is suspected.

   When you do **NOT** know the degree of hazard, use **Level A personal protective equipment (PPE) as follows: Level A** protection should be worn when the highest level of respiratory, skin, eye, and mucous membrane protection is needed. This level consist of a fully-encapsulated, vapor-tight, chemical-resistant suit, chemical-resistant boots with steel toe and shank, chemical-resistant inner/outer gloves, coveralls, hard hat, and self-contained (positive pressure) and self-contained breathing apparatus (SCBA).

Spill/Leak Disposal: If leaks or spills of Sarin occur, only personnel in full protective clothing will remain in the area. Spills must be contained by covering with vermiculite, diatomaceous earth, clay, fine sand, sponges, and paper or cloth towels. Decontaminate with copious amounts of aqueous sodium hydroxide solution (a minimum of 10 wt.%). Scoop up all material and place it in a DOT approved container. Cover the contents with decontamination solution as above. After sealing, the exterior will be decontaminated and labeled according to EPA and DOT regulations. All leaking containers will be over packed with sorbent (e.g., vermiculite) placed between the interior and exterior containers.

Decontaminate and label according to EPA and DOT regulations. Dispose of decontaminate according to federal, state, and local laws. Conduct general area monitoring to confirm that the atmospheric concentrations do not exceed the airborne exposure limits. If 10 wt.% aqueous sodium hydroxide is not available, then the following decontaminants may be used instead (which are listed in order of preference): Decontaminating Agent (DS2), Sodium Carbonate, and Supertropical Bleach Slurry (STB).

Symptoms: Sarin exposure includes rhinorrhea, chest tightness, pinpoint pupils, shortness of breath, excessive salivation and sweating, nausea, vomiting, abdominal cramps, involuntary defecation and urination, muscle twitching, confusion, seizures, flaccid paralysis, coma, respiratory failure, and death.

Vaccine(s):   No

Guides for Emergency Response: Chemical Agent or Weapon: Nerve Agent Soman (GD)

28

Introduction: Soman nerve gas (GD) is defined as Pinacolyl methyl phosphonofluoridate, methyl-1, 2, 2-trimethyyl ester. It is a colorless liquid with a fruity or camphor odor. It undergoes "aging" very quickly, rendering oxime therapy and making poisoning with this agent more difficult to treat. Like most chemical nerve agents, soman's course of death may be caused by anoxia resulting from airway obstruction, weakness of the muscle of respiration and central depression of respiration. Airway obstruction is due to pharyngeal muscular collapse, upper airway and bronchial secretions, bronchial constriction, and, occasionally, laryngospasm and paralysis of the respiratory muscles. Respiration is shallow, labored, and rapid and the casualty may gasp and struggle for air. Cyanosis increases. Respiration becomes slow and then ceases. Unconsciousness ensures. The blood pressure falls, and cardiac rhythm may become irregular and death may be the result without assisted ventilation.

NERVE AGENT: Soman (GD)

Formula: CH3PO(F)OCH(CH3)C(CH3)3

Vapor Density: (Air = 1) 6.33

Vapor Pressure: (mm^Hg) 0.4 at 25 degrees C

Molecular Weight: 182.178

Liquid Density: (g/cc) 1.0222 at 25 degrees C

Volatility: (mg/m3) 3,900 at 25 degrees C

Median Lethal Dosage: (mg-min/m3) Approximately GB, GA range

Physical State: (at 20 degrees C) Colorless liquid

Odor: Fruity; camphor when impure

Freezing/melting point: (at degrees C) –42 C

Boiling point: (at degrees C) 198 C

Action rate: Very rapid. Death usually occurs within fifteen minutes after absorption of fatal dose.

Physiological action: Cessation of breathing — death may follow

Required level of protection: Protective mask and clothing

Decontamination: Bleach slurry, dilute alkali; in confined area, hot soapy water; M285A1, M280.

Detection in the field: M18A2, M256, M256A1, M8 and M8A1 alarms, M8 and M9 paper, and Draeger Tube Phosphoric Acid Ester 0.05/a

Use: Quick-action casualty agent

CAS registry number: 96-64-0

RTECS number: TA8750000

LCt50: (respiratory) 70 mg-min/m3

Ability to Kill: HIGH

Characteristics: Nerve agents are liquid under temperate conditions; but when dispersed the more volatile ones constitute both a vapor and a liquid hazard. However, the less volatile nerve agents represent primarily a liquid hazard (mainly, the G-agents are more volatile than the nerve agent VX, while sarin (GB) is the most volatile and nerve agent GF is the least volatile of the so-called "G-agents." Sarin (GB) has an LCt-50 of 100 (vapor toxicity of mg-min/m3), an ICt-50 of 75 (vapor toxicity of mg-min/m30, and an MCt-50 of 3 (vapor toxicity of mg-min/m3). The LD-50 on skin is 1700mg.

Response on Scene by First Responders

Caution: Higher temperatures on site can increase the danger to soman contamination particularly with cutaneous exposure. Soman cannot be transported legally except by the military (U.S. Army Technical Escort Unit) according to 49 CFR 172. Nerve agents are potent acetylcholinesterase inhibitors causing the same signs and symptoms regardless of the exposure route, although the initial effects depend on the dose and route of exposure. Casualties whose skin or clothing is contaminated with liquid nerve agent can contaminate rescuers by direct contact or through off-gassing vapor. Nerve agents are extremely toxic and can cause loss of consciousness and convulsions within seconds and death from respiratory failure within minutes of exposure. Atropine and pralidoxime chloride (2-PAM Cl) are antidotes for nerve agent toxicity but pralidoxime must be administered within minutes to a few hours following exposure (depending on the specific agent) to be effective. Basically, treatment consists of supportive measures and repeated administration of antidotes.

Field First Aid: Nerve agents are the most toxic of the known chemical warfare agents. Chemically similar to organophosphate pesticides, their method of acting is to inhibit acetylcholinesterase enzymes. Individuals whose skin or clothing is contaminated with nerve agent can contaminate rescuers by direct contact or through off-gassing vapor; but persons whose skin is exposed only to nerve agent vapor pose no risk of secondhand contamination. However, it is important to remember that clothing can trap vapor. Nerve agents can be readily absorbed by ingestion, skin and inhalation that can lead to fatal systemic effects. Most of the nerve agents were "failed pesticides" that tended to be so toxic they were referred to armies around the world, mainly in Germany, and made into chemical nerve agents. Nerve agents cause the same signs and symptoms regardless of the exposure route. Chemical agent casualty triage is based on walking feasibility (if they can walk, their contamination is slight, or maybe, moderate), respiratory status, age, and additional conventional injuries. Any person appointed to handle triage must understand, at a minimum, the natural course of a given injury, the medical resources immediately available, the current and likely casualty flow, and the medical evacuation capabilities in order to decide Immediate, Delayed, Minimal, and Expectant priority assignments for the injured.

Nerve agent intoxication requires rapid decontamination to prevent further absorption by the patient and to prevent exposure to others, ventilation when necessary, administration of antidotes, as well as supportive therapy. Skin decontamination is not necessary with exposure to vapor alone, but clothing should be removed to get rid of any trapped vapor. With nerve agents, there can be high airway resistance due to bronchoconstriction and secretions, and initial ventilation is often difficult. The restriction will decrease with atropine administration. Copious secretions which may be thickened by atropine also impede ventilatory actions and will require frequent suctioning. For inhalation exposure to nerve agents, ventilation support is essential.

Inhalation: Hold breath until respiratory protective mask is donned. If severe signs of agent exposure appear (chest tightens, pupil constriction, etc.), immediately administer, in rapid succession, all three Nerve Agent Antidote Kit(s), Mark I kit injectors (or atropine if directed to by a physician). Injections using the Mark I kit injectors may be repeated at five to twenty minute intervals if sign and symptom are progressing until all three series of injections have been administered. No more injections will be given unless indicated by medical personnel. In addition, a record will be maintained of all injections given. If breathing has stopped, give artificial respiration. Mouth-to-mouth resuscitation should be used when mask-bag or oxygen delivery systems are not available. Do not use mouth-to-mouth resuscitation when facial contamination exists. If breathing is difficult, administer oxygen. Seek medical attention immediately.

Eye Contact: Immediately flush eyes with water for ten to fifteen minutes, then don respiratory protective mask. Although miosis (pinpointing of the pupils) may be an early sign of agent exposure, an injection will not be administered when miosis is the only sign present. Instead, the individual will be taken immediately to a medical treatment facility for observation.

Skin Contact: Don respiratory protective mask and remove contaminated clothing. Immediately wash contaminated skin with copious amounts of soap and water, 10 percent sodium carbonate solution, or 5 percent liquid household bleach. Rinse well with water to remove excess decontaminant. Administer Nerve Agent Antidote Kit, Mark I, only if local sweating and muscular twitching symptoms are observed. Seek medical attention immediately.

Ingestion: Do not induce vomiting. First symptoms are likely to be gastrointestinal. Immediately administer Nerve Agent Antidote Kit, Mark I. Seek medical attention immediately.

Drugs: For mild or moderate nerve agent contamination in an adult, the standard first dose for atropine is 2 to 4 mg IM, (intramuscular) and for 2-PAM Cl is 15 mg/kg (1g) IV (intravenous) slowly; for an adult victim with severe symptoms, a first dose would be atropine at 6 mg IM (intramuscular), and for 2-PAM Cl the dose should be 15 mg/kg(1g) IV (intravenous) slowly. There are different recommended doses for Infants, Children, Adolescent, and Elderly/Frail.

Atropine: Atropine is a cholinergic blocking, or anticholinergic, compound; which is very effective in blocking the effects of excess acetylcholine at peripheral muscarinic sites. The amount of atropine in three MARK I kits may cause adverse effects on military performance in a normal person. In people NOT exposed to nerve agents, amounts of 10 mg or higher may cause delirium. However, potentially the most hazardous effect of inadvertent use of atropine (2 mg, intramuscular) to a young person not exposed to nerve agents in a warm or hot atmosphere is inhibition of the sweating function which may lead to heat injury. In the military, atropine is packaged in auto-injectors with each injector containing 2 mg.

Pralidoxime chloride: Pralidoxime chloride (Protopam chloride; 2-PAM Cl) is an oxime that attaches to the nerve agent that is inhibiting the cholinesterase and breaks the agent-enzyme bond to restore the normal activity of the enzyme. This drug at a dosage of 600 mg is in an auto-injector for self or buddy use that is included in the military MARK I kit along with the atropine injector, and each military person in an area where nerve agents might be used is issued three MARK I kits. MARK I kits are also being provided to civilian first responders at the present time.

Diazepam: Diazepam is an anticonvulsant drug used to decrease convulsive activity in order to reduce brain damage caused by prolonged seizure activity. Without the use of pyridostigmine pretreatment, experimental animals died quickly after super-lethal doses of nerve agents despite conventional therapy. With pyridostigmine pretreatment which is followed by conventional therapy animal survives super-lethal doses of soman, but had prolonged periods of seizure activity before recovery. The administration of diazepam with other standard therapy to soman-poisoned animals pretreated with pyridostigmine reduced the seizure activity. Current military doctrine (also used by local responders) is to administer diazepam with other therapy (three MARK I's) at the onset of severe effects from a nerve agent, whether or not seizure activity is among these effects. Each military person or responders should carry one auto-injector of 10 mg of diazepam for his buddy to administer to him; if he could administer it to himself, he would not need it! Diazepam should be administered with the three MARK I's when the casualty's condition warrants the use of three MARK I's at the same time. Medical personnel can administer more diazepam to a casualty when necessary. A paramedic should carry more diazepam injectors, and be authorized to administer two additional injectors at ten minute intervals to a convulsing victim. Paramedics might make a mistake in giving too much atropine to a mild to moderate casualty, but more importantly he or she might err by giving too little atropine to a severe victim. The casualty with skin exposure to liquid is more difficult to evaluate and manage than is a victim contaminated with exposure to vapor. Agent on the skin can be decontaminated, but agent absorbed into the skin cannot be removed. A casualty from liquid exposure on the skin may continue to worsen because of continued absorption of the agent.

Pyridostigmine: About twelve years ago, the U.S. military provided pyridostigmone bromide as a pretreatment for nerve agent exposure. Each trooper received a blister pack containing twenty-one 30-mg tablets for a dose regimen of one 30-mg tablet every eight hours. When given before soman exposure and when that exposure is followed by the standard MARK I therapy, the use of this pretreatment will increase the LD-50 several fold over the LD-50 obtained without the use of the pretreatment. When soman is the nerve agent used, the use of pyridostigmine increases survival. When the agent used is sarin (GB) or VX, survival after standard MARK I therapy is essentially the same whether or not pyridostigmine pretreatment is used, i.e., pyridostigmine use provides no benefit in sarin or VX poisoning. Pyridostigmine is a pretreatment rather than an antidote and should not be given after soman exposure. Its use will not decrease the effects of soman. One consequence of the greater survival from the use of pyridostigmine is prolonged seizure activity and subsequent brain damage in the survivors. The early administration of diazepam will decrease these effects.

Fire: Soman failed to flash to 280 degrees F. Flammability limits, lower explosive limit, and upper explosive limits are not applicable for this product. Soman reacts with steam and water to produce toxic and corrosive vapors. Any personnel not fighting a fire of soman should immediately be evacuated from the area. Respiratory protection is required (positive pressure, full face piece, NIOSH-approved SCBA will be worn). When response personnel respond to handle rescue or reconnaissance, they will wear Level A protection that should be worn when the highest level of respiratory, skin, eye, and mucous membrane protection is needed. This level consists of a fully-encapsulated, vapor-tight, chemical-resistant suit, chemical-resistant boots with steel toe and shank, chemical-resistant inner/outer gloves (butyl rubber glove M3 and M4 Norton, chemical protective glove set), coveralls, hard hat, and self-contained (positive pressure) breathing apparatus (SCBA).

Do not breath fumes; skin contact with nerve agents should always be avoided; contact with liquid sarin or vapors can kill responders. Hydrogen may be present! Use water mist, fog, foam, or CO-2 to fight the fire; do NOT splash or spread any chemical nerve agent.

National Fire Protection Association—704 System Warning: The NFPA—704 marking system was created to give firefighters an indication as to the degree of hazard a chemical presents in a fire situation. The system was designed for use at fixed facilities rather than on transport vehicles. These markings are also found on individual containers. The top three sections of the marking uses numbers from 0 to 4 to identify the degree of hazard presented for Health Hazards (blue), Flammability (red), Reactivity (yellow), reading from left to right across the top, and Special Hazards (white) across the bottom of the markings. The number 0 represents no significant hazard, while the number 4 indicates a severe hazard. The bottom section (Special Hazards/white) indicates, if marked at all, special hazards such as radioactive, oxidizer, or water-reactive. The NFPA—704 system for soman chemical nerve agent is HEALTH— 4;   FLAMMABILITY—1; REACTIVITY—1; and SPECIAL HAZARD—0.

Personal Protective Equipment: "Standard turnout gear with SCBA provides a first responder with sufficient protection from nerve agent vapor hazards inside interior or downwind areas of the hot zone to allow thirty minutes rescue time for known live victims.

Self-taped turnout gear with SCBA provides sufficient protection in an unknown nerve agent environment for a three-minute reconnaissance to search for living victims, (or a two-minute reconnaissance if HD (mustard, blister agent) is suspected.

When you do NOT know the degree of hazard, use Level A personal protective equipment (PPE) as follows: Level A protection should be worn when the highest level of respiratory, skin, eye, and mucous membrane protection is needed. This level consist of a fully-encapsulated, vapor-tight, chemical-resistant suit, chemical-resistant boots with steel toe and shank, chemical-resistant inner/outer gloves, coveralls, hard hat, and self-contained (positive pressure) and self-contained breathing apparatus (SCBA).

Spill/Leak Disposal: If leaks or spills of soman occur, only personnel in full protective clothing will remain in the area. Spills must be contained by covering with vermiculite, diatomaceous earth, clay, fine sand, sponges, and paper or cloth towels. Decontaminate surfaces with copious amounts of aqueous sodium hydroxide solution (a minimum of 10 wt.%). Scoop up all material and place it in a DOT approved container. Cover the contents with decontamination solution as above. After sealing, the exterior will be decontaminated and labeled according to EPA and DOT regulations. All leaking containers will be over packed with sorbent (e.g., vermiculite) placed between the interior and exterior containers. Decontaminate and label according to EPA and DOT regulations. Dispose of decontaminate according to Federal, state, and local laws. Conduct general area monitoring to confirm that the atmospheric concentrations do not exceed the airborne exposure limits. If 10 wt.% aqueous sodium hydroxide is not available, then the following decontaminants may be used instead (which are listed in order of preference): Decontaminating Agent (DS2), Sodium Carbonate, and Supertropical Bleach Slurry (STB).

Symptoms: Soman exposure results in pupil constriction, blurred and dimmed vision, pain in the eyeballs; chest tightness, difficulty in breathing; sweating, salivation, increased bronchial secretions, bradycardia, hypotension, vomiting and diarrhea, bronchoconstriction, and urinary and fecal incontinence.

Vaccine(s): No

Guides for Emergency Response: Chemical Agent or Weapon: Nerve Agent Tabun (GA)

29

Introduction: In yesterday's world, there were literally thousands of poisonous substances but less than sixty-five to seventy have actually been made and stored as chemical weapons during the entire twentieth century. In the world of today as we wait for the beginning of the year 2008, the United States considers five nerve agents that make a credible argument in terms of what we as citizens may have to defend ourselves against being used by attacking countries or groups (as well as blister/blood/pulmonary chemical agents, biological agents and incapacitating agents). To be considered suitable for use as a nerve agent today, the agent must be highly toxic but in a "suitable" fashion so it as not too difficult to handle. It must be capable of being stored for long periods in containers without degradation and without corroding the container, relatively resistant to atmospheric water and oxygen so it does not lose effect when dispersed, and able to withstand the heat developed when dispersed. Those that graduated to the new grade are listed below:

Tabun (O-ethyl dimethyamidophosphorylcyanide), called GA in the United States, is the easiest to manufacture and is therefore likely to be used by backward countries while more industrialized countries view it as out-of-date and of limited use.

Sarin (isopropyl methylphosphonofluoridate), called GB in the United States, is a volatile substance taken up mainly through inhalation.

Soman (pinacolyl methylphosphonofluoridate), called GD in the United States, is a moderately volatile substance that can be taken by inhalation or skin contact.

GF (cyclohexyl methyylphosphonofluorridate) is a low volatility nerve agent taken by inhalation either as a gas or aerosol.

VX (O-ethyl S-diisopropylaminomethyl methylphosphonothiolate) is a persistent substance with the ability to remain on material, terrain, and equipment for long periods of time. Exposure is mainly through the skin but also by inhalation of VX as a gas or aerosol.

Nerve agents interfere with the central nervous system and are called cholinesterase inhibitors. They affect the transmission of nerve impulses by reacting with the enzyme acetylcholinesterase, thus creating an excess of acetylcholine. Nerve agents tend to be the most toxic of known chemical agents. Sample nerve agents include GB (sarin), GA (tabun), GD (soman), GF, and VX. Potential nerve agents could be used as weapons of mass destruction (WMD) are listed in a standard format below. Nerve agents can be absorbed through the eyes, skin, and respiratory tract. Such multiple routes of entry, and because they have a high toxicity and effectiveness, nerve agents are a major threat to the United States. Their vapor toxicity can be figured as mg-min/m3 in terms of Ct (the product of the **Concentration** of the vapor or aerosol to which one is exposed and the **Time** that one is exposed to that concentration), and LCt50 (the Ct of the agent vapor that will be **Lethal** to half the population exposed to it).

Body components that can probably be affected by the excessive acetylcholine accumulation include the eyes, the nose (glands), the mouth (glands), the respiratory tract, the gastrointestinal tract, the cardiac muscle, sweat glands, skeletal muscle, and the central nervous system. Care, provided at once (provided within several minutes after exposure to a nerve agent), including the administration of antidotes, means the difference between life and death for the casualty. That is, remedial service at the site of the nerve agent exposure must be done immediately. Nerve agents can be encountered in both liquid and vapor forms, and first responders or any aid workers should don protective masks and clothing at once. When exposure effects progress to more than one organ system, the effective situation moves quickly from mild to a severe exposure.

The administration of nerve agent antidotes must be done as soon as possible, generally at the site of an exposure, but the antidotes exhibit problems if used incorrectly and care should be done by emergency medical service personnel if possible. The usual antidotes for nerve agent poisoning include atropine sulfate, 2 PAM Cl, and diazepam. Atropine blocks the effects of acetylcholine at muscarinic receptors and produces relief from many symptoms. A combination of adequate atropine usage plus assisted ventilation is several times more effective in saving lives than assisted ventilation alone. A victim's ability to perspire, however, is reduced due to atropine usage; doses of atropine that can be tolerated well in colder climates could be seriously incapacitating in hot or tropical climates due to inability to sweat. It should be noted that in hot climates or with heat-stressed persons, one dose (2 mg) can reduce efficiency, 4 mg can sharply reduce efficiency, and 6 mg will incapacitate persons for several hours. 2 PAM Cl (trade names protopam chloride or pralidoxime chloride) is an oxime that increases the effectiveness of drug therapy in poisoning by some, but not all, cholinesterase inhibitors. 2 PAM Cl thus relieves the skeletal neuromuscular block and also reactivates the acetylcholinesterase clinically at muscarinic sites. 2 PAM Cl is least effective in dealing with soman (GD) nerve agent. Diazepam blocks the effects of acetylcholine on the central nervous system; that is, diazepam added to the basic antidote schedule prevents or ameliorates convulsions in moderate to severe nerve agent poisoning. Diazepam is also called CANA (convulsive antidote for nerve agent). CANA is **NOT** to be self-injected. The military has a caveat: "If you know who you are, where you are, and what you are doing, you do *NOT* need CANA!"

NERVE AGENT: Tabun (GA)

Formula: $C_2H_5OPO(CN)n(CH_3)_2$

Vapor Density: (Air = 1) 5.63

Vapor Pressure: (mm^Hg) 0.037 at 20 degrees C

Molecular Weight: 162.3

Liquid Density: (g/cc) 1.073 at 25 degrees C

Volatility: (mg/m3) 610 at 25 degrees C

Median Lethal Dosage: (mg-min/m3) 400 for a resting person

Physical State: (at 20 degrees C) Colorless to brown liquid

Odor: Faintly fruity; none when pure

Freezing/melting point: (at degrees C –5 C

Boiling point: (at degrees C) 240

Action rate: Very rapid

Physiological action: Cessation of breathing — death may follow

Required level of protection: Protective mask and clothing

Decontamination: Bleach, slurry, dilute alkali, or DS2; steam and ammonia in confined area; M258A1, M280

Detection in the field: M18A2, M256A1, M8 and M8A1 alarms, M8 and M9 paper, and Draeger Tube phosphoric Acid Ester 0.05/a

Use: Quick-action casualty agent

CAS registry number: 77-81-6

RTECS number: TB4550000

LCt50:   (respiratory) Approximately 400 mg-min/m3

Ability To Kill: HIGH

Characteristics: Nerve agents are liquid under temperate conditions, but, when dispersed, the more volatile ones constitute both a vapor and a liquid hazard. However, the less volatile nerve agents represent primarily a liquid hazard (mainly, the G-agents are more volatile than the nerve agent VX, while sarin (GB) is the most volatile and nerve agent GF is the least volatile of the G-agents. Sarin (GB) has an LCt-50 of 100 (vapor toxicity of mg-min/m3), an ICt-50 of 75 (vapor toxicity of mg-min/m30, and an MCt-50 of 3 (vapor toxicity of mg-min/m3). The LD-50 on skin is 1700mg.

Response on Scene by First Responders

Caution: Higher temperatures on site can increase the danger to tabun contamination particularly with cutaneous exposure. Tabun cannot be transported legally except by the military (U.S. Army Technical Escort Unit) according to 49 CFR 172. Nerve agents are potent acetylcholinesterase inhibitors causing the same signs and symptoms regardless of the exposure route, although the initial effects depend on the dose and route of exposure. **Casualties whose skin or clothing is contaminated with liquid nerve agent can contaminate rescuers by direct contact or through off-gassing vapor.** Nerve agents are extremely toxic and can cause loss of consciousness and convulsions within seconds and death from respiratory failure within minutes of exposure. Atropine and pralidoxime chloride (2-PAM Cl) are antidotes for nerve agent toxicity but pralidoxime must be administered within minutes to a few hours following exposure (depending on the specific agent) to be effective. Basically, treatment consists of supportive measures and repeated administration

Field First Aid: Nerve agents are the most toxic of the known chemical warfare agents. Chemically similar to organophosphate pesticides, their method of acting is to inhibit acetylcholinesterase enzymes. **Individuals whose skin or clothing is contaminated with**

nerve agent can contaminate rescuers by direct contact or through off-gassing vapor; but persons whose skin is exposed only to nerve agent vapor pose no risk of second-hand contamination. However, it is important to remember that clothing can trap vapor. Nerve agents can be readily absorbed by **ingestion, skin, and inhalation** that can lead to fatal systemic effects. Most of the nerve agents were failed pesticides that tended to be so toxic they were referred to armies around the world, mainly in Germany, and made into chemical nerve agents. Nerve agents cause the same signs and symptoms **regardless** of the exposure route. Chemical agent casualty triage is based on walking feasibility (if they can walk, their contamination is slight, or maybe, moderate) respiratory status, age, and additional conventional injuries. Any person appointed to handle triage must understand, at a minimum, the natural course of a given injury, the medical resources immediately available, the current and likely casualty flow, and the medical evacuation capabilities in order to decide Immediate, Delayed, Minimal, and Expectant priority assignments for the injured.

Nerve agent intoxication requires rapid decontamination to prevent further absorption by the patient and to prevent exposure to others, ventilation when necessary, administration of antidotes, as well as supportive therapy. Skin decontamination is not necessary with exposure to vapor alone, but **clothing should be removed to get rid of any trapped vapor.** With nerve agents, there can be high airway resistance due to bronchoconstriction and secretions, and initial ventilation is often difficult. The restriction will decrease with atropine administration. Copious secretions which may be thickened by atropine also impede ventilatory actions and will require frequent suctioning. For inhalation exposure to nerve agents, ventilation support is essential.

Inhalation: Hold breath until respiratory protective mask is donned. If severe signs of agent exposure appear (chest tightens, pupil constriction, etc.) immediately administer, in rapid succession, all three Nerve Agent Antidote Kit(s), Mark I kit injectors (or atropine if directed to by a physician). Injections using the Mark I kit injectors may be repeated at five to twenty minute intervals if sign and symptom are progressing until all three series of injections have been administered. No more injections will be given unless indicated by medical personnel. **In addition, a record will be maintained of all injections given. If breathing has stopped, give artificial respiration.** Mouth-to-mouth resuscitation should be used when mask-bag or oxygen delivery systems are not available. **Do not use mouth-to-mouth resuscitation** when facial contamination exists. If breathing is difficult, administer oxygen. Seek medical attention **immediately.**

Eye Contact: **Immediately** flush eyes with water for ten to fifteen minutes, then don respiratory protective mask. Although miosis (pinpointing of the pupils) may be an early sign of agent exposure, an injection will not be administered when miosis is the only sign present. Instead, the individual will be taken **immediately** to a medical treatment facility for observation.

Skin Contact: Don respiratory protective mask and remove contaminated clothing. Immediately wash contaminated skin with copious amounts of soap and water, ten percent sodium carbonate solution, or five percent liquid household bleach. Rinse well with water to remove excess decontaminant. Administer Nerve Agent Antidote Kit, Mark I, only if local sweating and muscular twitching symptoms are observed. Seek medical attention **immediately.**

Ingestion: Do not induce vomiting. First symptoms are likely to be gastrointestinal. Immediately administer Nerve Agent Antidote Kit, Mark I. Seek medical attention **immediately**.

Drugs: For mild or moderate nerve agent contamination in an adult, the standard first dose for atropine is 2 to 4 mg IM (intramuscular), and for 2-PAM Cl is 15 mg/kg (1g) IV (intravenous) slowly; for an adult victim with severe symptoms, a first dose would be atropine at 6 mg IM (intramuscular), and for 2-PAM Cl the dose should be 15 mg/kg(1g) IV (intravenous) slowly. There are different recommended doses for Infants, Children, Adolescents, and Elderly/Frail.

Atropine: Atropine is a cholinergic blocking, or anticholinergic, compound that very useful in blocking the effects of excess acetylcholine at peripheral muscarinic sites. The amount of atropine in three MARK I kits may cause adverse effects on military performance in a normal person. In people **NOT exposed to nerve agents**, amounts of 10 mg or higher may cause delirium. However, potentially the most hazardous effect of inadvertent use of atropine (two mg, intramuscular) young person not exposed to nerve agents in a warm or hot atmosphere is inhibition of the sweating function which may lead to heat injury. In the military, atropine is packaged in auto-injectors with each injector containing 2 mg.

Pralidoxime chloride: Pralidoxime chloride (Protopam chloride; 2-PAM Cl) is an oxime that attaches to the nerve agent that is inhibiting the cholinesterase and breaks the agent-enzyme bond to restore the normal activity of the enzyme. This drug at a dosage of 600 mg is in an auto-injector for self or buddy use is included in the military MARK I kit along with the atropine injector, and each military person in an area where nerve agents might be used is issued three MARK I kits. MARK I kits are also being provided to civilian first responders at the present time.

Diazepam: Diazepam is an anticonvulsant drug used to decrease convulsive activity in order to reduce brain damage caused by prolonged seizure activity. Without the use of pyridostigmine pretreatment, experimental animals died quickly after super-lethal doses of nerve agents despite conventional therapy. With pyridostigmine pretreatment which is followed by conventional therapy animal survives super-lethal doses of soman, but had prolonged periods of seizure activity before recovery. The administration of diazepam with other standard therapy to soman-poisoned animals pretreated with pyridostigmine reduced the seizure activity. Current military doctrine (also being used by local responders) is to administer diazepam with other therapy (three MARK I's) at the onset of sever effects from a nerve agent, **whether or not seizure activity is among these effects.** Each military person or responders should carry one auto-injector of 10 mg of diazepam for his buddy to administer to him; **if he could administer it to himself, he would not need it! Diazepam should be administered with the three MARK I's when the casualty's condition warrants the use of three MARK I's at the same time.** Medical personnel can administer more diazepam to a casualty when necessary. A paramedic should carry more diazepam injectors, and be authorized to administer two additional injectors at ten minute intervals to a convulsing victim. Paramedics might make a mistake in giving too much atropine to a mild to moderate casualty, but more importantly he or she might err by giving too little atropine to a severe victim. The casualty with skin exposure to liquid is more difficult to evaluate and manage than is a victim contaminated with exposure to vapor.

Agent on the skin can be decontaminated, but agent absorbed into the skin cannot be removed. A casualty from liquid exposure on the skin may continue to worsen because of continued absorption of the liquid agent.

Pyridostigmine: About twelve years ago, the U.S. military provided pyridostigmone bromide as a pretreatment for nerve agent exposure. Each trooper received a blister pack containing twenty-one 30-mg tablets for a dose regimen of one 30-mg tablet every eight hours. When given before **soman** exposure and when that exposure is followed by the standard MARK I therapy, the use of this pretreatment will increase the LD-50 several fold over the LD-50 obtained without the use of the pretreatment. When soman is the nerve agent used, the use of pyridostigmine increases survival. When the agent used is sarin (GB) or VX (VX), survival after standard MARK I therapy is essentially the same whether or not pyridostigmine pretreatment is used, i.e., pyridostigmine use provides no benefit in sarin or VX (VX) poisoning. Pyridostigmine is a **pretreatment** rather than an antidote and should not be given **after** soman exposure. Its use will not decrease the effects of soman. One consequence of the greater survival from the use of pyridostigmine is prolonged seizure activity and subsequent brain damage in the survivors. The early administration of diazepam will decrease these effects.

Fire: Tabun reacts with steam and water to produce toxic and corrosive vapors. Any personnel not fighting a fire of sarin should immediately be evacuated from the area. Respiratory protection is required (positive pressure, full face piece, NIOSH-approved SCBA will be worn). When response personnel respond to handle rescue or reconnaissance, they will wear Level A protection that should be worn when the highest level of respiratory, skin, eye, and mucous membrane protection is needed. This level consists of a fully-encapsulated, vapor-tight, chemical-resistant suit, chemical-resistant boots with steel toe and shank, chemical-resistant inner/outer gloves (butyl rubber glove M3 and M4 Norton, chemical protective glove set), coveralls, hard hat, and self-contained (positive pressure) breathing apparatus (SCBA).

Do not breath fumes; skin contact with nerve agents should always be avoided; contact with liquid tabun or vapors can kill responders. Use water mist, fog, foam, or CO-2 to fight the fire; do **NOT** splash or spread sarin.

National Fire Protection Association—704 System Warning

The NFPA—704 marking system was created to give firefighters an indication as to the degree of hazard a chemical presents in a fire situation. The system was designed for use at fixed facilities rather than on transport vehicles. These markings are also found on individual containers. The top three sections of the marking uses numbers from 0 to 4 to identify the degree of hazard presented for Health Hazards (blue), Flammability (red), Reactivity (yellow), reading from left to right across the top, and Special Hazards (white) across the bottom of the markings. The number 0 represents no significant hazard, while the number 4 indicates a severe hazard. The bottom section (Special Hazards/white) indicates, if marked at all, special hazards such as radioactive, oxidizer, or water-reactive. The NFPA—704 system for tabun chemical agent is HEALTH— 4; FLAMMABILITY—2; REACTIVITY—1; and SPECIAL HAZARD—0.

Personal Protective Equipment: Standard turnout gear with SCBA provides a first responder with sufficient protection from nerve agent vapor hazards inside interior or downwind areas of the hot zone to allow *thirty minutes rescue-*time for known live victims.

Self-taped turnout gear with SCBA provides sufficient protection in an unknown nerve agent environment for a *three-minute reconnaissance to search for living victims,* (or a two-minute reconnaissance if HD (mustard, blister agent) is suspected.

When you do **NOT** know the degree of hazard, use **Level A personal protective equipment (PPE) as follows: Level A** protection should be worn when the highest level of respiratory, skin, eye, and mucous membrane protection is needed. This level consist of a fully-encapsulated, vapor-tight, chemical-resistant suit, chemical-resistant boots with steel toe and shank, chemical-resistant inner/outer gloves, coveralls, hard hat, and self-contained (positive pressure) and self-contained breathing apparatus (SCBA).

Spill/Leak Disposal: If leaks or spills of tabun occur, only personnel in full protective clothing will remain in the area. Spills must be contained by covering with vermiculite, diatomaceous earth, clay, fine sand, sponges, and paper or cloth towels. Decontaminate with copious amounts of aqueous sodium hydroxide solution (a minimum of 10 wt.%). Scoop up all material and place it in a DOT approved container. Cover the contents with decontamination solution as above. After sealing, the exterior will be decontaminated and labeled according to EPA and DOT regulations. All leaking containers will be over packed with sorbent (e.g., vermiculite) placed between the interior and exterior containers. Decontaminate and label according to EPA and DOT regulations. Dispose of decontaminate according to federal, state, and local laws. Conduct general area monitoring to confirm that the atmospheric concentrations do not exceed the airborne exposure limits. If 10 wt.% aqueous sodium hydroxide is not available, then the following decontaminants may be used instead (which are listed in order of preference): Decontaminating Agent (DS2), Sodium Carbonate, and Supertropical Bleach Slurry (STB).

Symptoms: Tabun has the following signs and symptoms. For a small exposure of vapor the victim might experience miosis, rhinorrhea, and mild difficulty in breathing. For a large exposure, he or she might suffer sudden loss of consciousness, convulsions, apnea, flaccid paralysis, copious secretions, and miosis. For liquid on the skin, the victim might have small to moderate exposure and feel localized sweating; nausea, vomiting, and a feeling of weakness. For a large exposure, the victim may experience a sudden loss of consciousness, convulsions, apnea, flaccid paralysis, and copious secretions.

Vaccine(s):   No

Guides for Emergency Response: Chemical Agent or Weapon: Nerve Agent VX (VX)

30

NERVE AGENT: VX (VX). The U.S. standard V-agent is a very persistent nerve agent called VX (VX). A few drops of VX, the most deadly nerve agent produced today, can kill a person in minutes. Death is often by suffocation; the nerves that control breathing are disrupted and the diaphragm fails to expand and contract. Even small doses can cause blindness, hallucinations, convulsions, and death; VX vapors linger on the ground for weeks before dispersing. Although VX is many times more persistent than G-agents, it is very similar to sarin (GB) in mechanisms of action and effects. Since VX has low volatility, liquid droplets on the skin do not evaporate quickly, thereby increasing absorption. VX by the percutaneous (skin) route is estimated to be more than 100 times as toxic as GB. VX by inhalation is thought to be twice as toxic as sarin (GB). It is extremely toxic by skin and eye absorption, although VX liquid does not injure the skin or eyes but penetrates rapidly.

Immediate decontamination is required for the smallest drop. VX can be very rapid, and death can occur within fifteen minutes after absorption of a fatal dosage. Heavily splashed liquid persists for long periods under average weather conditions, and VX can persist for months in cold weather. For instance, VX is calculated to be approximately 1,500 times slower in evaporation than sarin (GB). As for LD-50 dose on the skin, sarin (GB) requires a dose of 1700 mg while VX requires only ten mg (tabun requires 1000 mg, soman requires 50 mg, and GF requires 30 mg).

Formula: (C2H5O)(CH3O)P(O)S(C2H4)N{C2H2(CH3)2}2

Vapor Density: (Air = 1) 9.2

Vapor Pressure: (mm^Hg) 0.0007 at 20 degrees C.

Molecular Weight: 267.38

Liquid Density: (g/cc) 1.0083 at 20 degrees C

Volatility: (mg/m3) 10.5 at 25 degrees C

Median Lethal dosage: (mg-min/m3) 100

Physical State: (at 20 degrees C): Colorless to amber, oily liquid.

Odor: None.

Freezing/melting point: (at degrees C) Below −51 C

Boiling point: (at degrees C) 298

Action rate: Very rapid

Physiological action: Produces casualties when inhaled or absorbed

Required level of protection: Protective mask and clothing

Decontamination: STB slurry or DS2 solution; hot soapy water; M258A1, M280

Detection in the field: M18A2, M256, M256A1, M8 and M8A1 alarms, M8 and M9 paper, and Draeger Tube Phosphoric Acid Ester 0.05/a

Use: Quick action casualty agent.

CAS registry number: 50782-69-9

RTECS number: TB1090000 LCt50: (respiratory) 100 mg-min/m3 (resting); 30 mg-min/m3 (mild activity)

Ability to Kill: HIGH

Characteristics: Nerve agents are liquid under temperate conditions, but, when dispersed, the more volatile ones constitute **both** a vapor and a liquid hazard. However, the less volatile nerve agents represent primarily a liquid hazard (mainly, the G-agents are more volatile than the nerve agent VX, while sarin (GB) is the most volatile and nerve agent GF is the least volatile of the G-agents. VX (VX) has an LCt-50 of 100 (vapor toxicity of mg-min/m3 while resting, and 30 mg-min/m3 under mild activity), an ICt-50 of 35 (vapor toxicity of mg-min/m30, and an MCt-50 of 0.04 (vapor toxicity of mg-min/m3).

Response on Scene by First Responders

Caution: VX is *not* thought of as a prime vapor hazard except on very warm days. Irregardless of warm temperatures, the prime concern for VX is as a **liquid** hazard; VX can be a particular danger through skin contact since this nerve agent evaporates quite slowly, much more slowly than the G-agents, thus giving it much more time to be absorbed through the skin. Like any nerve or mustard agent, **immediate decontamination** is one of the first tasks to be performed by first responders. VX cannot be transported legally except by the military (U.S. Army Technical Escort Unit) according to 49 CFR 172. Nerve agents are potent acetylcholinesterase inhibitors causing the same signs and symptoms regardless of the exposure route, although the initial effects depend on the dose and route of exposure. **Casualties whose skin or clothing is contaminated with liquid nerve agent can contaminate rescuers by direct contact or through off-gassing vapor. Nerve agents are extremely toxic and can cause loss of consciousness and convulsions within seconds and death from respiratory failure within minutes of exposure.** Atropine and pralidoxime chloride (2-PAM Cl) are antidotes for nerve agent toxicity but pralidoxime must be administered within minutes to a few hours following exposure (depending on the specific agent) to be effective. Basically, treatment consists of supportive measures and repeated administration of antidotes.

Field First Aid: Nerve agents are the most toxic of the known chemical warfare agents. Chemically similar to organophosphate pesticides, their method of acting is to inhibit acetylcholinesterase enzymes. Individuals whose **skin or clothing is contaminated with nerve agent can contaminate rescuers by direct contact or through off-gassing vapor**; but persons whose skin is exposed **only** to nerve agent vapor pose no risk of secondhand contamination. However, it is important to remember that clothing can trap vapor. Nerve agents can be readily absorbed by ingestion, skin and inhalation that can lead to fatal systemic effects. Most of the nerve agents were failed pesticides that tended to be so toxic they were referred to armies around the world, mainly in Germany, and made into chemical nerve agents. Nerve agents cause the same signs and symptoms regardless of the exposure route. Chemical agent casualty triage is based on walking feasibility (if they can walk, their contamination is slight, or maybe, moderate) respiratory status, age, and additional conventional injuries. Any person appointed to handle triage must understand, at a minimum, the natural course of a given injury, the medical resources immediately available, the current and likely casualty flow, and the medical evacuation capabilities in order to decide Immediate, Delayed, Minimal, and Expectant priority assignments for the injured.

Nerve agent intoxication requires rapid decontamination to prevent further absorption by the patient and to prevent exposure to others, ventilation when necessary, administration of antidotes, as well as supportive therapy. Skin decontamination is not necessary with exposure to vapor alone, but **clothing should be removed to get rid of any trapped vapor.** With nerve agents, there can be high airway resistance due to bronchoconstriction and secretions, and initial ventilation is often difficult. The restriction will decrease with atropine administration. Copious secretions which may be thickened by atropine also impede ventilatory actions and will require frequent suctioning. For inhalation exposure to nerve agents, ventilation support is essential.

Inhalation: Hold breath until respiratory protective mask is donned. If severe signs of agent exposure appear (chest tightens, pupil constriction, etc.), immediately administer, in rapid succession, all three Nerve Agent Antidote Kit(s), Mark I kit injectors (or atropine if directed to by a physician). Injections using the Mark I kit injectors may be repeated at five- to twenty-minute intervals if sign and symptom are progressing until all three series of injections have been administered. No more injections will be administered will be given unless indicated by medical personnel. **In addition, a record will be maintained of all injections given.** If breathing has stopped, give artificial respiration. Mouth-to-mouth resuscitation should be used when mask-bag or oxygen delivery systems are not available. **Do not use mouth-to-mouth resuscitation when facial contamination exists.** If breathing is difficult, administer oxygen. Seek medical attention immediately.

Eye Contact: **Immediately** flush eyes with water for ten to fifteen minutes after donning respiratory protective mask. Although miosis (pinpointing of the pupils) may be an early sign of agent exposure, an injection will not be administered when miosis is the only sign present. Instead, the individual will be taken **immediately** to a medical treatment facility for observation.

Skin Contact: Don respiratory protective mask and remove contaminated clothing. **Immediately wash contaminated skin** with copious amounts of soap and water, 10 percent sodium carbonate solution, or 5 percent liquid household bleach. Rinse well with water to remove excess decontaminant. Administer Nerve Agent Antidote Kit, Mark I, only if local sweating and muscular twitching symptoms are observed. Seek medical attention **immediately.**

Ingestion: Do not induce vomiting. First symptoms are likely to be gastrointestinal. **Immediately** administer Nerve Agent Antidote Kit, Mark I. Seek medical attention immediately.

Drugs: For mild or moderate nerve agent contamination in an adult, the standard first dose for atropine is 2 to 4 mg IM, (intramuscular) and for 2-PAM Cl is 15 mg/kg (1g) IV (intravenous) slowly; for an adult victim with severe symptoms, a first dose would be atropine at 6 mg IM (intramuscular), and for 2-PAM Cl the dose should be 15 mg/kg(1g) IV (intravenous) slowly. There are different recommended doses for Infants, Children, Adolescent, and Elderly/Frail.

Atropine: Atropine is a cholinergic blocking, or anticholinergic, compound; which is very effective in blocking the effects of excess acetylcholine at peripheral muscarinic sites. The amount of atropine in three MARK I kits may cause adverse effects on military performance in a normal person. In people NOT exposed to nerve agents, amounts of 10 mg or higher may cause delirium. However, potentially the most hazardous effect of inadvertent use of atropine (two mg, intramuscular) to a young person not exposed to nerve agents in a warm or hot atmosphere is inhibition of the sweating function which may lead to heat injury. In the military, atropine is packaged in auto-injectors with each injector containing 2 mg.

Pralidoxime chloride: Pralidoxime chloride (Protopam chloride; 2-PAM Cl) is an oxime that attaches to the nerve agent that is inhibiting the cholinesterase and breaks the agent-enzyme bond to restore the normal activity of the enzyme. This drug at a dosage of 600 mg is in an auto-injector for self or buddy use that is included in the military MARK I kit along with the atropine injector, and each military person in an area where nerve agents might be used is issued three MARK I kits. MARK I kits are also being provided to civilian first responders at the present time.

Diazepam: Diazepam is an anticonvulsant drug used to decrease convulsive activity in order to reduce brain damage caused by prolonged seizure activity. Without the use of pyridostigmine pretreatment, experimental animals died quickly after super-lethal doses of nerve agents despite conventional therapy. With pyridostigmine pretreatment, which is followed by conventional therapy, animals survive super-lethal doses of soman, but had prolonged periods of seizure activity before recovery. The administration of diazepam with other standard therapy to soman-poisoned animals pretreated with pyridostigmine reduced the seizure activity. Current military doctrine, that is also being used by local responders, is to administer diazepam with other therapy (three MARK Is) at the onset of severe effects from a nerve agent, **whether or not seizure activity is among these effects**. Each military person or responders should carry one auto-injector of 10 mg of diazepam for his buddy to administer to him; if he could administer it to himself, he would not need it! Diazepam should be administered with the three MARK Is when the casualty's condition warrants the use of three MARK Is at the same time. Medical personnel can administer more diazepam to a casualty when necessary. A paramedic should carry more diazepam injectors, and be authorized to administer two additional injectors at 10-minute intervals to a convulsing victim. Paramedics might make a mistake in giving too much atropine to a mild to moderate casualty, but more importantly he or she might err by giving too little atropine to a severe victim. The casualty with skin exposure to liquid is more difficult to evaluate and manage than is a victim contaminated with exposure to vapor. Agent on the skin can be decontaminated, but agent absorbed into the skin cannot be removed. A casualty from liquid exposure on the skin may continue to worsen because of continued absorption of the agent.

Pyridostigmine: The U.S. military about twelve years ago provided pyridostigmone bromide as a pretreatment for nerve agent exposure. Each trooper received a blister pack containing twenty-one 30-mg tablets for a dose regimen of one 30-mg tablet every eight hours.

When given before **soman** exposure and when that exposure is followed by the standard MARK I therapy, the use of this pretreatment will increase the LD-50 several fold over the LD-50 obtained without the use of the pretreatment. When **soman** is the nerve agent used, the use of pyridostigmine increases survival. When the agent used is sarin (GB) or VX (VX), survival after standard MARK I therapy is essentially the same whether or not pyridostigmine pretreatment is used, i.e., **pyridostigmine use provides no benefit in sarin or VX (VX) poisoning.** Pyridostigmine is a **pretreatment** rather than an antidote and should not be given **after** *soman* exposure. Its use will not decrease the effects of **soman.** One consequence of the greater survival from the use of pyridostigmine is prolonged seizure activity and subsequent brain damage in the survivors. The early administration of diazepam will decrease these effects.

Fire: VX (VX) nerve agent reacts with steam and water to produce toxic and corrosive vapors. Any personnel not fighting a fire of VX should immediately be evacuated from the area. Respiratory protection is required (positive pressure, full face piece, NIOSH-approved SCBA will be worn). When response personnel respond to handle rescue or reconnaissance, they will wear Level A protection that should be worn when the highest level of respiratory, skin, eye, and mucous membrane protection is needed. This level consists of a fully-encapsulated, vapor-tight, chemical-resistant suit, chemical-resistant boots with steel toe and shank, chemical-resistant inner/outer gloves (butyl rubber glove M3 and M4 Norton, chemical protective glove set), coveralls, hard hat, and self-contained (positive pressure) breathing apparatus (SCBA).

Do not breath fumes; skin contact with nerve agents should always be avoided; contact with liquid VX or vapors can kill responders. Use water mist, fog, foam or CO-2 to fight the fire; do NOT splash or spread VX.

National Fire Protection Association — 704 System Warning: The NFPA—704 marking system was created to give firefighters an indication as to the degree of hazard a chemical presents in a fire situation. The system was designed for use at fixed facilities rather than on transport vehicles. These markings are also found on individual containers. The top three sections of the marking uses numbers from 0 to 4 to identify the degree of hazard presented for Health Hazards (Blue), Flammability (Red), Reactivity (Yellow), reading from left to right across the top, and Special Hazards (White) across the bottom of the markings. The number 0 represents no significant hazard, while the number 4 indicates severe hazard. The bottom section (Special Hazards/White) indicates, if marked at all, special hazards such as radioactive, oxidizer, or water-reactive. The NFPA—704 system for VX chemical nerve agent is HEALTH—4; FLAMMABILIY—1; REACTIVITY—1; and SPECIAL HAZARD—0.

Personal Protective Equipment: Standard turnout gear with SCBA provides a first responder with sufficient protection from nerve agent vapor hazards inside interior or downwind areas of the hot zone to allow *30 minutes rescue* time for known live victims.

Self-taped turnout gear with SCBA provides sufficient protection in an unknown nerve agent environment for a *three-minute reconnaissance to search for living victims,* (or a two-minute reconnaissance if HD (mustard, blister agent) is suspected.

When you do **NOT** know the degree of hazard, use **Level A personal protective equipment (PPE) as follows: Level A** protection should be worn when the highest level of respiratory, skin, eye, and mucous membrane protection is needed. This level consist of a fully-encapsulated, vapor-tight, chemical-resistant suit, chemical-resistant boots with steel toe and shank, chemical-resistant inner/outer gloves, coveralls, hard hat, and self-contained (positive pressure) and self-contained breathing apparatus (SCBA).

Spill/Leak Disposal: If leaks or spills of VX occur, only personnel in full protective clothing will remain in the area. Spills must be contained by covering with vermiculite, diatomaceous earth, clay, fine sand, sponges, and paper or cloth towels. Decontaminate with copious amounts of aqueous sodium hydroxide solution (a minimum of 10 wt.%). Scoop up all material and place it in a DOT approved container. Cover the contents with decontamination solution as above. After sealing, the exterior will be decontaminated and labeled according to EPA and DOT regulations. All leaking containers will be over packed with sorbent (e.g., vermiculite) placed between the interior and exterior containers. Decontaminate and label according to EPA and DOT regulations. Dispose of decontaminate according to federal, state, and local laws. Conduct general area monitoring to confirm that the atmospheric concentrations do not exceed the airborne exposure limits. If 10 wt.% aqueous sodium hydroxide is not available, then the following decontaminants may be used instead (which are listed in order of preference): Decontaminating Agent (DS2), Sodium Carbonate, and Supertropical Bleach Slurry (STB).

Symptoms: VX for **vapor** contamination: Miosis, rhinorrhea, dyspnea, convulsions, and apnea. VX for **liquid** contamination: Sweating, vomiting, convulsions, and apnea.

Vaccine(s): No

Guides for Emergency Response: Chemical Agent or Weapon: Blister Agent Nitrogen Mustard (HN-1), Nitrogen Mustard (HN-2), Nitrogen Mustard (HN-3)

31

Introduction to Nitrogen Mustard(s): The blister agent Nitrogen Mustard (HN-1) can be identified as United Nation number 2810, and Guide 153 in the 2004 Emergency Response Guide available to all fire departments in the United States, Canada, and Mexico. HN-1 was the first of the nitrogen mustard agents produced in the late 1920s and early 1930s. It was meant to be a drug to remove warts, but found its true mission as a chemical agent. HN-2 rather reversed that procedure, being designed as a chemical agent but being used as a pharmaceutical drug. HN-3 could possibly be the only nitrogen mustards used or stored today as a military agent; it is the principal representative of the nitrogen mustards since its vesicant properties are almost equal to HD (distilled mustard). HN-3 is a cumulative poison highly irritating to the eyes and throat. The median incapacitating dose for eyes is 200 mg-min/m3. Nitrogen Mustard (HN-3) will interfere with hemoglobin functioning in the blood, hindering the production of new blood cells and destroying white blood cells. The persistency of HN-3 is two to three times that of HD (distilled mustard), so it can endure longer as a terrain denial, and adheres well to both equipment as well as personnel.

There is no antidote for nitrogen mustard toxicity, and decontamination of potentially exposed persons must be done within minutes to avoid tissue damage. Victims should be moved out of the hot zone, administered oxygen and/or assisted ventilation, and seek medical attention at once. The nitrogen mustards are delayed chemical agents, and may delay for up to twenty-four hours to show symptoms. They are poisons, and contact with vapor or liquids can be fatal. Do **not** eat, drink, or smoke during response to a nitrogen mustard incident or criminal event.

Blister Agent (Vesicant): Nitrogen Mustard (HN-1)

Formula: (ClCH2CH2)2NC2H5

Vapor Density: (Air = 1) 5.9

Vapor Pressure: (mm^Hg) 0.24 at 25 degrees C

Molecular Weight: 170.08

Liquid Density: (g/cc) 1.09 at 25 degrees C

Volatility: (mg/m3) 1,520 at 20 degrees C

Median Lethal Dosage: (mg-min/m3) 1,500 by inhalation; 20,000 by skin exposure.

Physical State: (at 20 degrees C) Oily, colorless to pale yellow liquid

Odor: Faint, fishy or musty

Freezing/Melting Point: (at degrees C) –34 C

Boiling Point: (at degrees C): 194 C. At atmospheric pressure, HN-1 decomposes below the boiling point.

Action Rate: Delayed twelve hours or longer

Physiological Action: Blisters; affects respiratory tract, destroys tissue, injures blood cells.

Required Level of Protection: Protective mask and clothing

Decontamination: Bleach, fire, DS2, M258A1. M280

Detection in the Field: M18A2, M256, M256A1, M8 and M9 paper, and Draeger Tube Organic Basic Nitrogen Compounds

Use: Delayed-action casualty agent.

CAS registry number: 538-07-8

RTECS Number: YE1225000

LCt50:   (respiratory) 1,500 mg-min/m3

Blister Agent (Vesicant):   Nitrogen Mustard (HN-2)

Formula: (ClCH2CH2)2NCH3

Vapor Density: (Air = 1) 5.4

Vapor Pressure: (mm^Hg) 0.29 at 20 degrees C

Molecular Weight: 156.07

Liquid Density: (g/cc) 1.15 at 20 degrees C

Volatility: (mg/m3) 3,580 at 25 degrees C

Median Lethal Dosage: (mg-min/m3) 3,000 by inhalation

Physical State: (at 20 degrees C): Dark liquid

Odor: Soapy in low concentrations; fruity in high concentrations

Freezing/Melting Point: (at degrees C): –65 to -60 C

Boiling Point: (at degrees C): 75 at 15 mm Hg

Action Rate: On the skin, delayed twelve hours or more; on the eyes, faster than HD

Physiological Action: Similar to HD; bronchopneumonia possible after twenty-four hours

Required Level of Protection: Similar to HD. (Protective mask and permeable protective clothing for vapor and small droplets; impermeable clothing for protection against large droplets)

Decontamination: Bleach, fire, DS2, M258A1, M280

Detection in the Field: M18A2, M256, M256A1, M8 and M9 paper, and Draeger Tube Organic Basic Nitrogen Compounds.

Use: Delayed-action casualty agent.

CAS Registry Number: 51-75-2

RTECS Number: 1A1750000

LCt: (respiratory) 3,000 mg/min/m3

Blister Agent: Nitrogen Mustard (HN-3)

Formula: C6H12Cl3N

Vapor Density: (Air = 1) 204.54

Vapor Pressure: (mm^Hg) 0.0109 at 25 degrees C

Molecular Weight: 204.54

Liquid Density: (g/cc) 1.24 at 25 degrees C

Volatility: (mm.m3) 121 at 25 degrees C

Median Lethal Dosage: (mg-min/m3) 1,500 by inhalation; about 10,000 by skin exposure

Physical State: (at 20 degrees C): Dark, oily liquid

Odor: None if pure

Freezing/Melting Point: (at degrees C): −3.7 C

Boiling Point: (at degrees C): 256 C

Action Rate: Serious effects same as for HD (four–six hours); minor effects sooner (such as eye irritation, tearing, and light sensitivity).

Physiological Action: Similar to HN-2.

Required Level of Protection: Protective mask and clothing.

Decontamination: Bleach, fire, DS2, M258A1, M280

Detection in the Field: M18A2, M256, M256A1, M8 and M9 paper, and Draeger Tube Organic Basic Nitrogen Compounds

Use: Delayed-action casualty agent

CAS Registry Number: 555-77-1

RTECS Number: YE2625000

LCt50:   (respiratory) 1,500 mg-min/m3

Ability to Kill: Moderate (delayed casualty agent)

Characteristics: People whose skin or clothing is contaminated with nitrogen mustard can contaminate rescue personnel by direct contact or through off-gassing vapor. Nitrogen mustards are absorbed by the skin causing erythema and blisters. Ocular exposure to such agents may cause very serious injury to the cornea and conjunctiva. When inhaled, nitrogen mustards 1, 2, and 3 can damage the respiratory tract epithelium **and may cause death**. Nitrogen mustards are vesicant and alkylating agents (as alkylating agents **they may cause bone marrow suppression, and an increased risk of fatal complicating infections, hemorrhage, and anemia; as well as neurologic toxicity**). HN-1 has a faint, fishy or musty odor, and is slightly soluble in water but miscible in acetone and other organic solvents. HN-2 smells "fruity" at high concentrations, but has a soapy odor at low concentrations; its solubility is similar to HN-1. HN-3 seems to be odorless when pure, and is the most stable of the nitrogen mustards. It has a much lower vapor pressure than HN-1 or HN-2 and is insoluble in water. Nitrogen mustard vapors are heavier than air. The Airborne Exposure Limit of HN-1 equals 0.003 mg/m3 as a time weighted average (TWA) for the workplace. No such standards exist for HN-2 or HN-3. **Nitrogen mustards may decrease fertility, they are probably carcinogenic to humans (Group 2A), and there is some evidence they may cause leukemia in humans (nitrogen mustards have been shown to cause leukemia and cancers of the lung, liver, uterus, and large intestine in test animals).**

Response on Scene by First Responders

Caution: There is no antidote for nitrogen mustard toxicity: **decontamination of all potentially exposed areas within minutes after exposure is the only effective method to decrease tissue damage.** Other than that, treatment is mainly supportive.

Field First Aid: Basically, there is no field first aid for vesicants such as HN-1, HN-2, or HN-3; but first responders can and should provide the following supportive services. **Victims whose skin or clothing is contaminated with liquid nitrogen mustard can contaminate rescue personnel by direct contact or through off-gassing vapor.** HN-1, HN-2, and HN-3 are extremely toxic and may damage the eyes, skin, and respiratory tract **and suppress the immune system**. These agents cause cellular changes within minutes of contact, but your body does not react with pain or to other symptoms to warn you in sufficient time to protect yourself. **Rescuers must be trained and dressed in appropriate protective clothing before entering the hot zone.** Before transport to hospitals or other medical-related facilities, all casualties must be decontaminated. Victims who have been decontaminated pose no serious risk of secondary contamination to rescue personnel, and support zone personnel require no specialized protective gear. Because most signs and symptoms of nitrogen mustard exposure occur some time after exposure, **patients should be observed for at least six hours, or sent home with instructions to return immediately if symptoms do develop.** It is important to remember that symptoms of exposure to nitrogen mustard may not appear for up to twenty-four hours. The sooner after exposure the symptoms appear, the more likely they are to progress and become severe.

Quickly ensure ABC (airway/breathing/circulation). See that the victim has a clear airway. If trauma is suspected, maintain cervical immobilization manually and apply a cervical collar and a backboard if feasible. Check for adequate respiration, and administer supplemental oxygen if cardiopulmonary compromise is suspected. Maintain adequate circulation. Establish intravenous access when necessary and warranted. Use a cardiac monitor as necessary. Stop any bleeding when necessary. When ingestion is evident, **do not induce emesis** (vomiting). For victims who are alert and able to swallow, give four to eight ounces of milk to drink. **Administration of activated charcoal is not viewed as beneficial.**

For inhalation contact, **remove from the source at once**, and provide artificial respiration if breathing is not apparent. Provide oxygen if breathing is difficult. For eye contact, flush eyes **immediately with water for ten to fifteen minutes** by pulling eyelids apart with fingers, and pouring water into the eyes. **Do not cover eyes with bandages**, but protect eyes with dark or opaque glasses after flushing the eyes. For skin contact, don respiratory mask and gloves and remove victim from the source of contact, then remove the patient's clothing and decontaminate the skin immediately by flushing with 5 percent solution of liquid household bleach followed in three to four minutes by a wash with soap and water to remove the 5 percent bleach solution used to decontaminate the victim's skin and to protect against erythema (to prevent systemic toxicity, decontamination can be done as late as two to three hours after exposure even if it increases the severity of the local reaction; you are dealing with a chemical reaction, and you have to deal with it. Speed is definitely needed, but endurance may save the victim much suffering even if applied late). As stated before, for ingestion contact do not induce vomiting, provide milk to drink if victim is capable. For all types of contact introduced above, seek medical attention immediately.

Medical Management: Vesicants rapidly penetrate the skin causing both localized cellular damage and systemic damage. **The deadly nature of such agents' effect is that a person exposed to a large amount of liquid or vapor faces total systemic assault. The reasons for this are (a) the failure of the body's immune system, with sepsis and infection as the major contributing causes of death, and (b) pulmonary damage which wins out as the major contributory factor in death. Rapid decontamination is the only way to avoid severe burns.** Medical management of a victim who has undergone exposure to mustard agent can be very simple, as providing of symptomatic care for a sunburn-like erythema; or vastly complex as in providing long-term services to fight immunosuppression, caring for serious burns, and/or dealing with other multi-system factors. A patient severely ill with mustard agent poisoning requires the general supportive care provided for any severely ill patient as well as specific care given to a burn patient. Liberal use of systemic analgesics and antipruritics, as needed, maintenance of fluid and electrolyte balance, and other supportive measures are necessary. Food supplements, including vitamins, may also be helpful. Some medical management schemes do not work out so well. Activated charcoal given orally does not seem to provide better care. Hemodialysis has been ineffective and even harmful to several patients. **The rapid transformation of the mustard molecule suggests that few methods are actually available and beneficial hours or days after exposure with a mustard agent. Mustard agents are classed as a mutagen and a carcinogen based on laboratory studies.**

Fire: For HN-1 and HN-2 there is no immediate danger of fire or explosion. For HN-3, the flash point is high enough not to interfere with military use of this chemical agent.

Personal Protective Equipment: For HN-1, HN-2, as well as HN-3, wear full protective clothing consisting of Level A: Level A should be worn when the highest level of respiratory, skin, eye, and mucous membrane protection is needed. This level consists of a fully-encapsulated, vapor-tight, chemical-resistant suit, chemical-resistant boots with steel toe and shank, chemical-resistant inner/outer gloves, coveralls, hard hat, and self-contained (positive pressure) breathing apparatus (SCBA). Both the Environmental Protection Agency and NIOSH recommend that initial entry into unknown environments or into confined space that has not been chemically characterized be conducted wearing at least Level B, if not Level A, protection. For protective gloves, it is mandatory to wear Butyl toxicological protective gloves such as M3, M4 or glove set.

Spill/Leak Disposal: Mustard agents or weapons should be contained using vermiculite, diatomaceous earth, clay, or fine sand and neutralized as soon as possible using copious amounts of 5.25 percent sodium hypochlorite solution. Scoop up all material and place it in an approved Department of Transportation container. Cover the contents with decontaminating solution as mentioned above. The exterior of the salvage container shall be decontaminated and labeled according to EPA and DOT regulations. All leaking containers will be over packed with sorbent (e.g., vermiculite) placed between the interior and exterior containers. Decontaminate and label according to EPA and DOT regulations. Dispose of the material in accordance with waste disposal methods provided below. Conduct general area monitoring with an approved monitor to confirm that the atmospheric concentrations do not exceed the airborne exposure limits. If 5.25 percent sodium hypochlorite solution is not available, the following decontaminants may be used instead and are listed in order of preference: Calcium Hypochlorite, Contamination Solution No. 2 (DS2), and Super Tropical Bleach Slurry (STB).

Approved Methods of Waste Disposal: **(Open pit burning or burying of HD or items containing or contaminated with HD in any quantity is prohibited!)** Decontamination of waste or excess material shall be accomplished according to procedures outlined above and can be destroyed by incineration in EPA approved incinerators according to appropriate provisions of Federal, State, and local Resource Conservation Act (RCRA) regulations. *Note:* Some decontamination solutions are hazardous waste according to RCRA regulations and must be disposed of according to these regulations.

Symptoms: Erythema, blisters, irritation of the eyes, cough, dyspnea

Vaccines: None

Glossary: Quick Guide to Agents, Drugs, Equipment, Gear, Programs, and Terminology

Vocabulary is defined as the stock of words used or known to a particular people or group of persons. People in technical or professional fields of work have to "talk the talk" and tend to develop their own vocabulary of special words, terms, phrases and definitions. Physicians have developed a vocabulary made of words and terms consistent with the work they do. Person's who work, respond to, control, manufacture, or provide medical services to victims of chemical and biological agents and weapons also have to understand a special trade or business vocabulary. What follows is a quick guide to chemical and biological agents, drugs, specialized equipment, general gear, programs, and terminology.

ug/g: Microgram per gram, one part per million (ppm).

ug/l: Microgram per liter, one part per billion (ppb).

Absorbent Materials: An absorbing material designed to pick and hold liquid hazardous materials to prevent a spread of contamination.

Acetylchorine (ACH, ACh):   The neurotransmitter substance at cholinergic synapse that causes cardiac inhibition, vasodilation, gastrointestinal peristalsis, and other parasympathetic effects. It is liberated from preganglionic and post-ganglionic endings of parasympathetic fibers and from pre-ganglionic fibers of the sympathetic nervous system as a result of nerve injuries, whereupon it acts as a transmitter on the effective organ; it is hydrolyzed into choline and acetic acid by acetylcholinesterase before a second impulse may be transmitted.

ACGIH: The American Conference of Governmental Industrial Hygienists is a voluntary membership organization of professional industrial hygiene personnel in governmental or educational institutions. The ACGIH develops and publishes recommended occupational exposure limits each year called Threshold Limit Values (TLV's) for hundreds of chemicals, physical agents, and biological exposure indices.

Acetylcholinesterase: An enzyme that hydrolyzes the neurotransmitter acetylcholine. The action of this enzyme is inhibited by nerve agents.

Active Immunization: The act of artificially stimulating the body to develop antibodies against infectious disease by the administration of vaccines or toxoids.

Acute: Health effect that occurs over a short term; brief and severe as opposed to chronic.

Acute Exposure: A single encounter to toxic concentrations of a hazardous material or multiple encounters over a short period of time.

Adenopathy: Swelling or morbid enlargement of the lymph nodes.

Adsorption: The attraction and accumulation of one substance on the surface of another.

Aerosol: A fine aerial suspension of liquid, fog or mist; or solid, dust, fume or smoke; particles sufficiently small in size to be stable.

After Action Report: A post-incident analysis report gathered by a responsible party or responding agency after termination of a hazardous materials incident, describing actions taken, materials involved, impacts and similar information.

Air Monitoring : To observe, record, and/or detect pollutants in ambient air.

Aleukia: Absence or extremely decreased number of leukocytes in the circulating blood.

Analgesic: A substance used by medical personnel to relieve pain.

Anhydrous: Without water, dry. Describes a substance in which no water molecules are present.

Antibiotics: Substances produced by and obtained from living cells, such as bacteria or molds. Examples of antibiotics would include penicillin and streptomycin.

Anticonvulsant: An agent which prevents or arrests seizures.

Antidote: A remedy to relieve, prevent, or counteract the effects of a poison.

Antigen: A molecule capable of eliciting a specific antibody or T-cell response.

Antiserum: A serum containing an antibody or antibodies produced from animals or humans that have survived exposure to an antigen.

Antitoxin: An antibody formed in response to and capable of neutralizing a biological poison, an animal serum containing antitoxins, or a solution of antibodies (e.g., diphtheria antitoxin and botulinum antitoxin) derived from the serum of animal immunized with specific antigens. Antitoxins are used to confer passive immunity and for treatment.

ANSI: American National Standards Institute, a private organization that is engaged in creating voluntary standards or characteristics and performance of materials, products, systems, and services.

Aquifer: A permeable geologic unit with the ability to store, transmit, and yield fresh water in usable quantities.

Area Plan: A plan established for emergency response to a release or threatened release of a hazardous material.

Arsenicals: A category of blister agents in which arsenic is the central atom. Although more volatile than mustard agents, they are much more dangerous as liquids than as vapors.

Assessment: To determine the nature and degree of hazard of a hazardous materials or a hazardous materials incident from a safe vantage point.

Asthenia: Weakness or debility.

ASTM: American Society for Testing and Materials, a voluntary group in which members devise consensus standards for materials characterization and use.

Ataxia: An inability to coordinate muscle activity during voluntary movement, so that smooth movements occur. Most often due to disorders of the cerebellum or the posterior columns of the spinal cord; may involve the limbs, head, or trunk.

Atropine: Sometimes used as an antidote for nerve agents. It inhibits the action of acetylcholine by binding to acetylcholine receptors.

Automatic Continuous Air Monitoring System (ACAMS): This system can detect G agents, VX, or mustard agents at very low levels. It is an automatic gas chromatograph that first collects agent on a solid sorbent and then thermally desorbs the agents into a separation column for analysis.

Automatic Liquid Agent Detector (ALAD): A liquid agent devise that can detect droplets of GD, VX, HD. and Lewisite as well as thickened agents. It transmits its alarm by field wire to a central alarm unit.

Bacteria: Small, free-living, microscopic organisms that reproduce by simple division; the diseases they produce often respond to treatment with antibiotics. Bacteria are single-celled, can exist independently, and vary in size from about 0.3 um to 10 um (microns). Bacteria can cause disease either by directly invading body tissue or by producing toxins once inside the body.

BAL: British Anti-Lewisite. Dimercaprol, a treatment for toxic inhalations.

Biochemical Oxygen Demand: A numerical estimate of contamination in water expressed in milligrams per liter of dissolved oxygen. A measure of the amount of oxygen consumed in biological processes that break down organic matter in water.

Biological Agent: A microorganism that causes disease in people, plants, or animals or causes the deterioration of material.

Biological Agents: BACTERIA (Anthrax, Brucellosis, Cholera, Plague, Tularemia). VIRUSES (Crimean-Congo Hemorrhagic Fever, Rift Valley Fever, Smallpox, Venezuelan Equine Encephalitis (VEE), Viral Hemorrhagic Fever (Ebola)). TOXINS (Botulinum, Ricin, Staphylococcal Enterotoxin B (SEB), Trichothecene Mycotoxins/T-2).

Biological Integrated Detection System (BIDS) (Military): BIDS, better known as XM31 in the military, consists of a lightweight, multipurpose, collective protection shelter mounted on a heavy high-mobility, multipurpose, wheeled military vehicle equipped with a biological detection suite. At the time of this writing, BIDS can detect the bacteria *Bacillus anthracis* and *Yersinia pestis,* and the toxins botulism toxic A and staphylococcal enterotoxin B. Since biological toxins are most likely to be dispersed as aerosols, ambient air is continuously sampled and the background distribution of aerosols particles determined.

Biological Sampling Kit (Military): The BSK is required to perform three types of biological sampling: surface, liquid, and solid. The kit contains the required equipment for monitor/survey teams in the field to collect and forward biological samples needed by medical facilities.

Biosafety Levels For Infectious Agents: The essential elements of four biosafety levels for activities involving infectious microorganisms are set out below. The levels are designated in ascending order, by degree of protection provided to personnel, the environment, and the community. Although the entire four levels are very involved, there is room here for only the most basic requirements for each of the four levels.

Biosafety Level 1 (BSL-1) Suitable for work involving well-characterized agents not known to consistently cause disease in healthy adults humans, and of minimal potential hazard to laboratory personnel and the environment requiring standard microbiological practices.

Biosafety Level 2 (BSL-2) Similar to Biosafety Level 1 and is suitable for work involving agents of moderate potential hazard to personnel and the environment. Required practices are BSL-1 practices plus limited access, biohazard warning signs, "sharps" precautions, and a biosafety manual defining any needed waste decontamination or medical surveillance policies.

Biosafety Level 3 (BSL-3) Applicable to clinical, diagnostic, teaching, research, or production facilities in which work is done with indigenous or exotic agents which may cause serious or potentially lethal disease as a result of exposure by the inhalation route. Laboratory personnel have specific training in handling pathogenic and potentially lethal agents, and are supervised by competent scientists who are experienced in working with these agents. Required practices are BSL-2 practices plus controlled access, decontamination of clothing before laundering, cages decontaminated before bedding removed, and disinfectant foot bath as needed.

Biosafety Level 4 (BSL-4) Required for work with dangerous and exotic agents that pose a high individual risk or aerosol-transmitted laboratory infections and life threatening disease. Agents with a close or identical antigenic relationship to biosafety level 4 agents are handled at this level until sufficient data are obtained either to confirm continued work at this level, or to work with them at lower levels. Members of the laboratory staff have specific and thorough training in handling extremely hazardous infectious agents and they understand the primary and secondary containment functions of the standard and special practices, the containment equipment, and the laboratory design characteristics. They are supervised by competent scientists who are trained and experienced in working with these agents. Access to the laboratory is strictly controlled by the laboratory director. The facility is either in a separate building or in a controlled area within the building, which is completely isolated from all other areas of the building. A specific facility operations manual is prepared or adopted. Required practices are BSL-3 practices plus entrance through a chance room where personal clothing is removed and laboratory clothing put on, and a shower on exiting; and all waste are decontaminated before removal from the facility (Interested readers desiring more complete information about this particular subject are directed to the book, *Biosafety In Microbiological And Biomedical Laboratories*, 4th Edition, CDC/NIH, U.S. Department of Health and Human Services, Public Health Service, May 1999. This publication is available through the U.S. Government Printing Office, Washington, D.C. The stock number is 017-040-00547-4, and the price is $12.00 per copy; paperback).

BLEVE: Boiling Liquid Expanding Vapor Explosion.

Blister Agents: Substances that cause blistering of the skin. Exposure is through liquid or vapor contact with any exposed tissue.

Blood Agents: Substances that injure a person by interfering with cell respiration by dealing with the exchange of oxygen and carbon dioxide between blood and tissues.

B-NICE: B-NICE is an acronym pertaining to "Biological-Nuclear-Incendiary-Chemical-Explosives."

Boiling Point: The temperature at which the vapor pressure of a liquid is equal to the surrounding atmospheric pressure so that the liquid becomes a vapor.

Boom: A floating physical barrier serving as a continuous obstruction to the spread of a contaminant.

Bootie: A sock-like over-boot protector worn to minimize contamination.

Botulism: Poisoning by toxin derived from *Clostridium botilinum.*

Boyles Law: The volume of gas is inversely proportional to its pressure at constant temperature.

Breakthrough Time: The elapsed time between initial contact of a hazardous chemical with the outside surface of a protective clothing material and the time at which the chemical can be detected at the inside surface of the material.

Bronchiolitis: The inflammation of the bronchioles often associated with bronchopnuemonia.

Brucella: A genus of encapsulated, nonmotile bacteria (family Brucellaceae) containing short, rod-shaped to coccoid, Gram-negative cells. These organisms are parasitic, invading all animal tissues and causing infection of the genital organs, the mammary gland, and the respiratory and intestinal tracts, and are pathogenic for man and various species of domestic animals. They do not produce gas from carbohydrates.

Bubo: Inflammatory swelling of one or more lymph nodes, usually in the groin; the confluent mass of nodes usually suppurates and drains pus.

Buddy System: The organizing of employees into work groups so that each employee in the work group is designated to be observed by at least one other employee in the work group.

Bulla (Plural = Bullae): A large blister appearing as a circumscribed area of separation of the epidermis from the subepidemal structure or as a circumscribed area of separation of epidermal cells caused by the presence of serum, or occasionally by and injected substance.

CAER: Community Awareness and Emergency Response: a program developed by the Chemical Manufacturers Association to provide guidance for chemical plant managers to assist them in cooperating with local communities to develop integrated hazardous materials response plans.

CAM: Chemical Agent Monitor used by the U.S. military; it detects chemical agent vapors and provides a readout of the relative concentration of the vapors present. It is a hand-held, battery-operated devise for the monitoring of decontamination procedures and effectiveness on personnel and equipment. It can detect, identify and provide relative vapor concentration readouts for G and V-type nerve agents and H-type blister agents. The ICAM (Improved Chemical Agent Monitor) is a hand-held, soldier-operated, post-attack devise for monitoring chemical agent decontamination on people and equipment. It detects vapors of chemical agents by sensing molecular ions of specific mobilities (time of flight) and uses timing and microprocessor techniques to reject interference.

CAMEO: Computer Aided Management of Emergency Operations: A computer data base for storage and retrieval of pre-planning data for on scene use at hazardous materials incidents,

CANA: Convulsant Antidote for Nerve Agent, also called diazepam.

CANUTEC: Canadian Transport Emergency Center: A twenty-four hour, government sponsored hot line for chemical emergencies.

Carbamates: Organic chemical compounds that can be neurotoxic by competitively inhibiting acetylcholinesterase binding to acetylcholine.

Carcinogen: A material that has been found to cause cancer in humans or in animals.

C.A.S. Registration Number: Chemical Abstracts Service. An assigned number used to identify a material. CAS numbers identify specific chemicals and are assigned sequentially; the number is a concise, unique means of material identification. A product of more than one component will have a specific number for each component (i.e., the CAS# for lethal nerve agent GA, or tabun, is 77-81-9; while the CAS# for the liquid nerve agent GB, or sarin, is 107-44-8, and 50.642-23-4).

Cascade System: Several air cylinders attached in series to fill SCBA (self-contained breathing apparatus) bottles.

Casual Contact: A person who has been in the proximity to an infected person as in sharing a bus, taxi, or airplane; but has not been associated with body fluids or excretions.

Catastrophic Incident: A event that significantly exceeds the resources of a jurisdiction.

cc: Cubic centimeter, a volumetric measurement which is also equal to one milliliter (ml).

CERCLA: Comprehensive Environmental Response, Compensation and Liability Act. Known as CERCLA, or the SUPERFUND amendment, this federal law deals with hazardous substances releases to the environment and the cleanup of hazardous waste sites.

CFR: Code of Federal Regulations. A collection of federal regulations established by law.

CG: Phosgene.

Charles Law: The volume of gas is directly proportional to its absolute temperature at constant pressure.

CK: Cyanogen Chloride.

Chemical Agent: A chemical substance that is intended for use in military operations to kill, seriously injure, or incapacitate people through its physiological effects (including blood, nerve, choking, blister, and incapacitating agents). Not included in this category are riot control agents, chemical herbicides, or smoke and flame materials.

Chemical Agent GA: The chemical Ethyl N, N-dimethylphosphoramidocyanidate (CAS# 77-81-6) also known as tabun, is a nerve agent.

Chemical Agent GB: The chemical Isopropyl methyl phosphonofluoridate (CAS# 107-44-8) also known as sarin, is a nerve agent.

Chemical Agent GD: The chemical Pinacolyl methyl phosphonofluoridate (CAS# 96-64-0) also known as soman, is a nerve agent.

Chemical Agent H: Levinstein mustard (CAS# 471-03-4) is a mixture of 70 percent bis (2-chloroethyl) sulfide and 30 percent sulfur impurities produced by the Levinstein process and is a blister agent.

Chemical Agent HD: Distilled mustard (HD), or bis (2-chloroethyl) sulfide, (CAS# 505-60-2) is mustard (H) that has been purified by washing and vacuum distillation to reduce sulfur impurities; Agent HD is a blister agent.

Chemical Agent HT: Agent T is bis [2-(2-chloroethylthio) ethyl] ether (CAS# 63918-89-8) and is a sulfur, oxygen and chlorine compound similar in structure to HD. It is 60 percent HD and 40 percent T plus a variety of sulfur contaminants and impurities, and is a blister agent.

Chemical Agent L: (Lewisite) Agent L is a blister agent, Dichloro 2-chlorovinyldichloro-arsine, (CAS# 541-25-3) with a chemical formula of C2H2AsCl3.

Chemical Contamination: The deposition of chemical agents of personnel, clothing, equipment, strictures, or areas. Chemical contamination mainly consists of liquid, solid particles, and vapor hazards. Vapor hazards are probably the most prevalent means of contaminating the environment, although they are not necessarily a contact hazard.

Chemical Name: The scientific designation of a chemical or a name that will clearly identify the chemical for hazard evaluation purposes.

Chemical Reaction: A change in the arrangement of atoms or molecules to yield substances of different composition and properties. (See REACTIVITY)

Chemical Substance: A substance usually associated with some description of its toxicity or exposure hazard, including solids, liquids, mists, vapors, fumes, gases, and particulate aerosols. Exposure, via inhalation, ingestion, or contacts with skin or eyes, may cause toxic effects, usually in a dose-dependent manner.

Chemical Surety: Controls, procedures, and actions which contribute to the safety, security, and reliability of chemical agents and their associated weapon systems throughout their life cycle without degrading operational performance.

Chemical Warfare: All aspects of military operations involving the use of lethal munitions/agents and warning and protective measures associated with such offensive operations.

Chemoprophylaxis: Prevention of disease by use of chemicals or drugs.

CHEMTREC: The Chemical Transportation Emergency Center located in Washington, D.C., a public service provided by the private Chemical Manufacturers Association, provides emergency response information and assistance twenty-four hours a day for responders to hazardous materials incidents.

Choking Agents: Substances that cause physical injury to the lungs by exposure through inhalation. Death results through lack of oxygen.

Cholinergic: Relating to nerve cells or fibers that employ acetylcholine as their neurotransmitter.

Cholinesterase: (ChE) An enzyme that catalyzes the hydrolysis of acetocholine to choline (a vitamin) and acetic acid.

Chemical Degradation: A chemical action involving the molecular breakdown of the material due to contact with a chemical. The action may cause the personal protective equipment to swell, shrink, blister, discolor, become brittle, sticky, soft or to deteriorate. These changes permit chemicals to get through the suit more rapidly or to increase the probability of permeation.

Chemical Hazards Response Information System/Hazard Assessment Computer System: (CHRIS/HACS) Developed by the Coast Guard, HACS is a computerized model of the CHRIS manuals. It is used by federal on scene coordinators during a chemical spill or response.

Chemical Penetration: The movement of material through a suit's closures, such as zippers, buttonholes, seams, flaps, or other design features. Abraded, torn, or ripped suits will also allow penetration.

Chemical Protective Clothing Material: Any material or combination of materials used in an item of clothing for the purpose of isolating parts of the wearer's body from contact with a hazardous material.

Chemical Protective Suit: A single or multi-piece garment constructed of chemical protective clothing materials designed and configured to protect the wearer's torso, head, arms, legs, hands, and feet.

Chemical Resistance: The ability to resist chemical attack. The attack is dependent on the method of test and its severity is measured by determining the changes in physical properties. Time, temperature, stress, and reagent, may all be factors that affect the chemical resistance of a material.

Chemical Resistant Materials: Materials that are specifically designed to inhibit or resist the passage of chemicals into and through the materials by the processes of penetration and permeation.

Chronic: Persistent, prolonged, or repeated conditions

Chronic Exposure: A prolonged exposure occurring over a period of days, weeks, or years.

Chronic Toxicity: Adverse health effects from repeated doses of a toxic chemical or other toxic substance over a relatively prolonged period of time, generally greater than one year.

Ciprofloxacin: An antibiotic drug useful in treating bacterial infections; the recommended antibiotic for treating anthrax infections as well as prophylaxis in a biological warfare setting.

CNS: Abbreviation for central nervous system.

Coccobacillus: A short, thick bacterial rod of the shape of an oval or slightly elongated coccus.

Code of Federal Regulations: The federal government's official publication of federal regulations. Volumes are divided into fifty titles according to subject matter. Titles are divided into chapters which are divided into parts and sections.

Combustible: According to the DOT and NFPA, combustible liquids are those having a flash point at or above 100 degrees F (37.8 degrees C), or liquids that will burn. They do not ignite as easily as flammable liquids. However, combustible liquids can be ignited under certain circumstances, and must be handled with caution. Substances such as wood, paper, etc., are termed "Ordinary Combustibles."

Communicable: Capable of being transmitted from human to human, animal to animal, or human to animal.

Compatibility Charts: Permeation and penetration data supplied by the manufacturers of protective clothing to indicate chemical resistance and breakthrough time of various garment materials as tested against a battery of chemicals.

Competence: Having skills, knowledge, and judgment necessary to perform certain objectives in a satisfactory manner.

Concentration: The amount of a chemical agent present in a unit volume of air, usually expressed in milligrams per cubic meter (mg/m3).

Conjunctiva (Plural = Conjunctivae): The mucous membrane investing the anterior surface of the eyeball and the posterior surface of the lids.

Containment: All activities necessary to bring the incident to a point of stabilization, and to establish a degree of safety for emergency personnel greater than existed upon arrival.

Contamination: A substance or process that poses a threat to life, health, or the environment; or the deposit and/or absorption of NBC contamination on and by structures, areas, personnel and objects.

Contamination Control Line: The established line around a contamination reduction zone that separates it from the support zone.

Contingency Plan: A pre-planned document presenting an organized and coordinated plan of action to limit potential pollution in case of fire, explosion or discharge of hazardous materials which defines specific responsibilities and tasks.

Control: The procedures, techniques, and methods used in the mitigation of a hazardous materials incident, including containment, confinement, and extinguishment.

Control Zones: The designation of areas at a hazardous materials incident based upon safety and the degree of hazard.

Coordination: To bring together in a uniformed and controlled manner the functions of all agencies on scene.

Corrosive: A substance that causes visible destruction or visible changes in human skin tissues at the site of contact.

Cost Recovery: The procedure that allows for the agency having jurisdiction to pursue reimbursement for all costs associated with a hazardous materials incident.

Cryogenic: Gases, usually liquefied, that induce freezing temperatures of −150 degrees F and below; such as liquid oxygen, liquid helium, liquid natural gas, and liquid hydrogen.

CSF: Abbreviation for cerebrospinal fluid.

Cubic Meter: (m3) A measure of volume in the metric system.

Cutaneous: Pertaining to or affecting the skin.

CX: Phosgene Oxime, a blister agent.

Cyanosis: A dark bluish or purplish coloration of the skin and mucous membrane due to deficient oxygenation of the blood, evident when reduced hemoglobin in the blood exceeds 5 g per 100 ml.

Cytotoxin: Toxin that directly damages and kills the cell with which it makes contact.

Decomposition: The breakdown of a chemical or substance into different parts or simpler compounds. Decomposition can occur due to heat, chemical reaction, decay, etc.

Decontamination: The physical and/or chemical process of reducing and preventing the spread of contamination from persons and equipment used at a hazardous materials incident. The process of making any person, object, or area safe by absorbing, destroying, neutralizing, making harmless, or removing the hazardous material.

Decontamination Corridor: A corridor that acts as a protective buffer and bridges between the hot zone and the cold zone and is located in the warm zone within which decontamination stations and personnel are located to apply decontamination procedures.

Degradation: The process of decomposition. A chemical action involving the molecular breakdown of protective clothing material due to contact with a chemical. Degradation is evidenced by visible signs such as charring, shrinking, or dissolving. Testing clothing for weight of thickness changes, or loss of tensile strength, will also reveal degradation.

Degree of Hazard: A relative measure of how much harm a substance can cause.

Dermal: Pertaining to or affecting the skin.

Dermal Exposure: Exposure to or absorption through the skin.

Detection: The determination of the presence of a chemical agent, biological agent, or nuclear substance.

Detection, Biological: The military has a limited number of *field* detectors that can identify biological agents, but if you happen to be a local or state responder the bulk of biological agent identification work will actually be done in a biological laboratory such as found in hospitals, research universities, or the Centers of Disease Control in Atlanta, Georgia.

Detoxification Rate: The rate at which the body's own action overcome or neutralize chemicals or toxins.

Dike: An embankment or ridge, natural or man made, used to control the movement of liquids, sludges, solids or other materials. An overflow dike would be constructed in a fashion to allow the uncontaminated water to flow unobstructed over the dike keeping the contaminant behind the dike. An underflow dike allows the uncontaminated water to flow unobstructed under the dike keeping the contaminant behind the dike.

Direct Reading Instrument: A portable device that measures, and displays, in a short period of time, the concentration of a contaminant in the environment.

Dispersion: To spread, scatter, or diffuse through air, soil, surface or ground water.

Disposal Drum: A specially constructed drum used to overpack damaged or leaking containers of hazardous materials for shipment.

Diversion: The intentional movement of a hazardous material in a controlled manner so as to relocate it in an area where it will pose less harm to the community and the environment.

DKIE: Decontamination Kit, Individual Equipment (Military).

Dose: The amount of substance ingested, absorbed, and/or inhaled per exposure period.

Dose Rate: How fast a dose is absorbed or taken into a body.

DOT Identification Numbers: Four-digit numbers proceeded by UN (United Nations) or NA (North American) that are used to identify the particular hazardous materials for regulation of transportation (e.g., the UN code number for chlorine trifluoride is UN1749). Haz Mat Response Teams (HMRTs) often refer to such numbers to identify specific chemicals listed in the Department of Transportation's *Emergency Response Guidebook* Double Gloving: An additional set of gloves worn in addition to the already in-place protection.

DP: Diphosgene, a choking agent.

Draeger Tubes: Draeger tubes are designed to detect specific compounds in the air. Users will draw a specific volume of air through the tube with a Draeger Pump, and then read a color change reaction against a series of quantifying markers on the tube. The glass tube are filled with one or more substances that undergo chemical reactions in the presence of specific chemicals or types of chemicals. The basis of any direct reading Draeger tube is the chemical reaction of the measured substance with the chemicals of the filling preparation. This reaction will result in a color change which the user can identify and therefore quantify the amount of the measured substance. Each tube has specific interference's that will indicate a positive reading for the compound of interest. The user must have a general knowledge of what the compound is to minimize the number of separate tests. The Draeger tubes for specific chemical warfare agents are as follows:

Blister Agents:

    Lewisite: Organic Arsenic Compounds and Arsine.

    Nitrogen Mustard: Organic Basic Nitrogen Compounds.

    Phenyldichloroarsine: Organic Arsenic Compounds and Arsine.

    Phosgene Oxime: Cyangen Chloride 0.25/a (but is limited to only a "Yes" or "No").

    Sulfur Mustard: Thioether.

Blood Agents:

    Arsine: Arsine 0.05/a.

    Cyanogen Chloride: Cyanogen Chloride 0.25/a

    Hydrogen Arsenide: Organic Arsenic Compounds and Arsine.

    Hydrogen Cyanide: Hydrocyanic Acid 2/a.

Pulmonary/Choking Agents:

    Chlorine: Chlorine 0.2/a.

    Choropicrin: Carbon Tetrachloride 1/a.

    Diphosgene: (None)

    Phosgene: Phosgene 0.25/b.

Nerve Agents:

    Dichlorvos (DDVP):   Phosphoric Acid Ester 0.05/a.

    DFP: Phosphoric Acid Ester 0.05/a.

    Metasystox: Phosphoric Acid Ester 0.05/a/

    Sarin (GB):   Phosphoric Acid Ester 0.05/a.

    Soman (GD):   Phosphoric Acid Ester 0.05/a.

    Tabun (GA):   Phosphoric Acid Ester 0.05/a.

    VX (VX):   Phosphoric Acid Ester 0.05/a.

Drugs (for paramedics and other qualified medical first responders): Atropine sulfate belongs to the following drug classes; anticholinergic, antimuscarinic, parasympatholytic, antiparkinson drug, antidote, diagnostic agent, and belladonna alkaloid. The drug competitively blocks the effects of acetylcholine at muscarinic cholinergic receptors that mediate the effects of parasympathetic postganglionic impulses, depressing salivary and bronchial secretions, dilating the bronchi, inhibiting vagal influences on the heart, relaxing the CI (gastrointesinal) and GU (genitourinary) tracts, inhibiting gastric acid secretion (high doses), relaxing the pupil of the eye (mydriatic effect), and prevent accommodation for near vision (cycloplegic effect); it also blocks the effects of acetylcholine in the CNS (central nervous system). There are a number of different uses for atropine sulfate, but two come to mind for first responders to chemical agent incidents: treatment of closed head injuries that cause acetylcholine release into the CSF (cerebrospinal fluid), EEG (electrencephalogram) abnormalities, stupor, and neurologic signs; and as an antidote (with external cardiac massage) for CV (cardiovascular) collapse from overdose of parasympathomimetic (cholinergic) drugs (choline esters, pilocarpine), or cholinesterase inhibitors (e.g., physostigmine, isoflurophate, organophosphorus insecticides). Dosage as an antidote for poisoning due to cholinesterase inhibitors would require doses of at least two to three mg (milligram). There are a number of contraindications and cautions to the use of this drug. This drug should be taken as prescribed, thirty minutes before meals; avoid excessive dosage. A person who takes this drug should avoid hot environments and will be heat intolerant, and dangerous reactions may occur. Empty your bladder taking this drug.

> **ciprofloxacin hydrochloride**   is an antibacterial drug that is bactericidal in that it interferes with DNA replication in susceptible gram-negative bacteria presenting cell reproduction. Among a number of uses, this drug provides prevention of anthrax following exposure to anthrax bacilla including prophylactic use in areas suspected of using biological germ warfare or in terrorism attacks. The required dose for anthrax exposure is 500 mg by mouth four times a day for three to six months. Ciprofloxacin is *not recommended* for pediatric use as it produced lesions of joint cartilage in immature experimental animals. This drug could have increased risk of severe photosensitivity reactions if combined with St. John's Wort therapy. Take oral drug on an empty stomach, one hour before or two hours after meals; victim should drink plenty of fluids while on ciprofloxacin, and use caution if driving or using dangerous equipment.

> **Diazepam**   is a C-IV controlled substance plus being a benzodiazepine, antianxiety agent, antiepileptic agent, and a skeletal muscle relaxant. This drug acts mainly at the limbic system and reticular formation; may act in spinal cord and at supraspinal sites to produce skeletal muscle relation; potentiates the effects of GABA, an inhibitory neurotransmitter; anxiolytic effects occur at doses well below those necessary to cause sedation, ataxia; has little effect on cortical function. It is basically for the management of anxiety disorders or for short-term relief of symptoms of anxiety, as a muscle relaxant, and for treatment of panic attacks as well as for uses. This drug should be used cautiously with elderly or debilitated patients, or those with impaired liver of kidney function. It can also cause birth defects and should not be used during pregnancy because it can cause cleft lip or palate, inguinal hernia, cardiac defects, microcephaly and pyloric stenosis when used in the first trimester; and neonatal withdrawal syndrome has been reported in babies. The adult dose when taken orally for anxiety disorders, skeletal muscle spasms, and convulsive disorders is two to ten milligrams twice-a-day or four-times a day. Caution: Do not administer intra-arterially, may produce arteriospasm, gangrene. Change from intravenous (IV) therapy to oral therapy as soon as possible. Do not use small veins (dorsum of hand or wrist) for IV injection. Reduce dose of narcotic analgesics with IV (intravenous) diazepam; dose should be reduced by at least one-third or eliminated. Carefully monitor pulse, blood pressure, respiration during IV administration. Main-

tain patients receiving parenteral benzodiazepines in bed for three hours; do not permit ambulatory patients to, operate a vehicle following an injection. Monitor electroencephalogram (EEG) for patients treated for staus epilepticus; seizure may recur after initial control, presumably because of short duration of drug effect. Monitor liver and kidney functions, complete blood count during long-term therapy. Taper dosage gradually after long-term therapy, especially in epileptic patients. Arrange for epileptic patients to wear medic alert ID indicating that they are epileptics taking diazepam.

Doxycycine: Doxycycline is both an antibiotic and a tetracycline antibiotic, and is bacteriostatic in that it inhibits protein synthesis of susceptible bacteria, causing cell death. This drug is effective against rickettsiae; M. pneumonia; agents of psittacosis, ornithosis, P.pestis; P.tularensis; Brucella, and other illnesses. There is decreased effectiveness of this drug if taken with food or dairy products, sensitivity to sunlight to the extent that patients are cautioned to wear protective clothing and use sunscreen.

Lorazepam: This drug is a C-IV controlled substance and a benzodiazepine, antianxiety agent, and a sedative/hypnotic. Exact mechanisms of lorazepam are not understood, but the drug acts mainly at subcortical levels of the central nervous system, leaving the cortex relatively unaffected. Main site of the action may be the limbic system and reticular formation; benzodiazepines potentiate the effects of GABA, an inhibitory neurotransmitter; anxiolytic effects occur at doses well below those necessary to cause sedation and ataxia. This drug is used for the management of anxiety disorders or for short-term relief of symptoms of anxiety associated with depression; and as a preanesthetic medication in adults to produce sedation, relieve anxiety, and decrease recall of events related to surgery. This medicine should be used cautiously with impaired liver or kidney function or debilitation. The usual oral dose for adults is two to six milligrams a day, with a range of one to ten milligrams a day given in divided doses with largest dose at bedtime. For insomnia due to transient stress, the standard dose would be two to four milligrams given at bedtime. This drug should *not* be given to children who are less than twelve years old. *The can be drug dependence with withdrawal syndrome when drug is discontinued, but this is much more common with abrupt discontinuation of higher does used for longer than four months.* Caution: Do not administer intra-arterially; arteriospasm, gangrene may result. Give intra-muscular injections of undiluted drug deep into muscle mass, and monitor injection sites. Do not use solutions that are discolored or contain a precipitate. Protect drug from light, and refrigerate solution. Keep equipment to maintain a patient airway on standby when the drug is given intravenously. Reduce dose of narcotic analgesics by at least half in patients who have received parenteral loazepam, and keep patients who have received parenteral under close observation, preferably in bed, up to three hours. Do not permit ambulatory patients to drive following an injection. Taper dosage gradually after long-term therapy, especially in epileptic patients.

Morphine sulfate: This drug is a C-II controlled substance and a narcotic agonist analgesic. This drug is a principal opium alkaloid; acts as agonist at specific opioid receptors in the central nervous system to produce analgesia, euphoria, sedation; the receptors mediating these effects are thought to be the same as those mediating the effects of endogenous opioids (enkephalins, endorphins). Morphine is used for relief of moderate to severe acute and chronic pain, preoperative medication to sedate and allay apprehension, facilitate induction of anesthesia, reduce anesthetic dosage, and to provide an analgesic adjunct during anethesia. Morphine is also used as a component of most preparations that are referred to as Bromptom's Cocktail, an oral alcoholic solution that is used for chronic severe pain, especially in terminal cancer patients; and for intraspinal use with microinfusion devices for the relief of intractable pain. This medicine should not be used with premature infants. Respiratory depression may occur in the elderly, the very ill, and those with respiratory problems. Reduced dosage may be necessary. Major hazards of using morphine include respiratory depression, apnea, circulatory depression, respiratory arrest, shock and cardiac arrest.

penicillin V (potassium): This drug is an antibiotic and a penicillin (acid stable) drug. It is also a bactericidal that inhibits cell wall synthesis of sensitive organisms, causing cell death. It is used in mild to moderately severe infections caused by sensitive organisms — streptococci, pneumococci, staphylococci and fusospirochetes; and is a prophylaxis against bacterial endorcarditis in a patient with valvular heart disease undergoing dental or upper respiratory tract surgery. Take drug on an empty stomach with a full glass of water. Avoid self-treating other infections with this antibiotic because it is specific for the infection being treated.

Caution: You would have to go to either penicillin G (aqueous) or penicillin G (sodium) for treatment of *severe* infections caused by sensitive organisms including bacillus anthracis.

Tetracycline hydrochloride: This drug by-class is an antibiotic and a tetracycline and also a bacteriostatic that inhibits protein synthesis of susceptible bacteria, preventing cell replication. It can deal with infections caused by rickettsiae; Mycoplasma pneumoniae; agents of psittacosis, ornithosis; Pasteurella tularesis, Brucella and Staphylococcus aureus. When penicillin is contraindicated, this medicine can handle infections caused by a number of diseases, including Bacillus anthracis. This medicine can be used in systemic administration and as a dermatologic solution. Caution: Pregnancy (toxic to the fetus) and lactation (causes damage to the teeth of an infant) are contraindications to the use of this drug. Administer oral medication on an empty stomach, one hour before or two to three hours after meals. Do not give with antacids; if antacids must be used, give them three hours after the dose of tetracycline. *Culture* infection before beginning drug therapy. Do not use outdated drugs; a degraded drug is highly nephrotoxic and should not be used. Do not give oral drug with meals, antacid, or food. Arrange for regular renal tests with long-term therapy. Use topical preparations of this drug only when clearly indicated. Sensitization from the topical use may preclude its later use in serious infections. Topical preparations containing antibiotics that are not ordinarily given systemically are preferable.

DS2: Decontaminating Solution No. 2. A military decon solution for the battlefield (skin contact with DS2 must be avoided at all times, do not breath fumes, contact with liquid or vapors can be fatal).

Dysarthria: A disturbance of speech and language due to emotional stress, to brain injury, or to paralysis, incoordination, or spasticity of the muscles used for speaking.

Dysphagia, dysphagy: Difficulty in swallowing.

Dysphonia: Altered voice production.

Dyspnea: Shortness of breath, difficult or labored breathing.

Edema: An accumulation of an excessive amount of watery fluid in cells, tissues, or serous cavities.

Emergency Decontamination versus Full Decontamination: Every emergency response first responder should know the difference between emergency decontamination and full decontamination. Emergency decontamination is the physical process of immediately reducing contamination of persons in potentially life- threatening situations without the formal establishment of full decontamination. Emergency decontamination is the method used to prevent to victims, site personnel or bystanders, and rescuers from exposure to dangerous chemical or biological agents or weapons. Emergency decontamination provides only for gross decontamination, so there may still be the potential for secondary contamination. If you set up full decontamination, or the physical or chemical process of reducing and preventing the spread of all contamination, the team doing decontamination should wear the appropriate level of protective gear.

Emergency Operations Plan: A document that identifies the available personnel, equipment, facilities, supplies, and other resources in the jurisdiction, and state the method or scheme for coordinated actions to be taken by individuals and government services in the event of natural, man-made, or attack related disasters.

Emergency Response: Response to any occurrence which has or could result in a release of a hazardous substance.

Emergency Response Guidebook: A manual for first responders during the initial phase of a hazardous materials/dangerous goods incident developed under the supervision of the Office Hazardous Materials Issues and Training, Research & Special Programs Administration, U.S. Department of Transportation.

Emergency Response Plan: A plan that establishes guidelines for handling hazardous materials incidents as required by 29 CFR 1910.120.

Emergency Response to Terrorism Job Aid: (for Fire/EMS/HazMat/Law Enforcement Personnel) A very well done, pocket-sized manual on plastic stock that actually tells first responders extensive basic information about what needs to be done at a terrorist incident. This handbook was designed, produced and distributed through a joint partnership of the Federal Emergency Management Agency, the United States Fire Administration, and the

National Fire Academy; as well as the United States Department of Justice, Office of Justice Programs. Edition 1.0 received wide distribution to first responders when it was published in May or 2000. It is a very handy resource for all levels of government responders as well as commercial responders. The *Job Aid* is divided into five primary sections that are tabbed and color coded for rapid access to information even with gloved hands, the plastic gives the user the ability to write with dry marker or to permanently inscribe contact information with indelible markers, and the text is written in simple language using recognizable terms. The *Job Aid* is available free of charge from the USFA Publications Center to response organizations ordering five or fewer copies. Other organizations and individuals may order one free copy. For more information, contact the U.S. Fire Administration online at: http://www.usfa.fema.gov/nfa/tr_ssadd.htm

Endemic: A disease process that is continuously present in a given community, population, or geographic location

Endotoxin: A toxin produced in an organism and liberated only when the organism disintegrates.

Endotracheal Intubation: Passage of a tube through the nose or the mouth into the trachea for maintenance of the airway during anesthesia or for maintenance of an imperiled airway.

Enterotoxin: Toxins of bacterial origin that affect the intestines, causing diarrhea (e.g., toxins from Vibrio cholera, Staphylococcus, Shigella, E. Coli, Clostridium perfringens, Pseudomonas).

Entry Point: A specified and controlled access into a hot zone at a hazardous materials incident.

Entry Team Leader: The entry leader is responsible for the overall entry operations of assigned personnel within the hot zone.

Enzyme: A protein formed by living cells which acts as a catalyst on physiological chemical processes.

Enzyme-Linked Immunosorbent Assay (ELISA):   An immunological technique used to quantify the amount of antigen or antibody in a sample such as blood plasma or serum.

EPA Number: The number assigned to chemicals regulated by the Environmental Protection Agency (EPA).

Epidemiology: The study of disease in human populations

Epizootic: 1. Denoting a temporal pattern of disease occurrence in an animal population in which the disease occurs with a frequency clearly in excess of the expected frequency in that population during a given time interval. 2. An outbreak (epidemic) of disease in an animal population; often with the implication that it may also affect human populations.

Erythema: A reddening of the skin.

Etiologic Agent: A viable microorganism or its toxin that causes, or may cause, human disease.

Evacuation: To quickly and calmly leave an area in order to avoid exposure to a potentially harmful situation.

Evaporation Rate: The rate at which a material is converted to vapor (evaporates) at a given temperature and pressure when compared to the evaporation rate of a given substance. Health and fire hazard evaluations of materials involve consideration of evaporation rates as one aspect of the evaluation.

Exotoxin: A toxin secreted by a microorganism into the surrounding medicine.

Explosive: A materials that releases pressure, gas, or heat suddenly when subjected to shock, heat, or high pressure.

Explosive Ordnance Disposal: The detection, identification, field evaluations, rendering safe, recovery, and final disposal of unexploded ordnance or munitions chemical agents.

Exposure: The subjection of a person to a toxic substance or harmful physical agent through any route of entry.

Exposure Routes: The major routes of exposure include ingestion, inhalation, and absorption though the skin.

Febrile: Denoting or relating to fever.

Filter: A High-Efficiency Particulate Air (HEPA) filter is at least 99.97 percent efficient in removing particles with a diameter of 0.3 microns.

First Responder, Awareness Level: Individuals who are likely to witness or discover a hazardous substance release and who have been trained to initiate an emergency response sequence by notifying proper authorities.

First Responder, Operations Level: Individuals who respond to releases or potential releases of hazardous substances as part of the initial response to the site for the purpose of protecting nearby persons, property, or the environment from the effects of the release. They are trained to respond in a defensive fashion without actually trying to stop the release. Their function is to control the release from a safe distance, keep it from spreading, and prevent exposures.

Flammable: A material that catches on fire easily or spontaneously under conditions of standard temperature and pressure.

Flammable (explosive) Range: The range of gas or vapor concentration (percentage by volume in air) that will burn or explode if an ignition source is present. Limiting concentrations are commonly called the lower explosive limit and the upper explosive limit. Below the lower explosive limit, the mixture is too lean to burn; above the upper explosive limit, the mixture is to rich to burn.

Flaring: A process that is used with high vapor pressure liquids or liquefied compressed gases for the safe disposal of the product. Flaring is the controlled burning of material in order to reduce or control pressure and/or to dispose of a product.

Flash Point: The lowest temperature at which a flammable liquid gives off sufficient vapor to form an ignitable mixture with air near its surface or within a vessel.

Fomite: Objects, such as clothing, towels, and utensils that possibly harbor a disease agent and are capable of transmitting it.

Formalin: A 37 percent aqueous solution of formaldehyde.

Freezing Point: Temperature at which crystals start to form as a liquid is slowly cooled; alternatively, the temperature at which a solid substance begins to melt as it is slowly heated.

Full Protective Clothing: Protective gear, to include SCBA, and designed to keep gases, vapor, liquids, and solids from any contact with the skin while preventing ingestion or inhalation.

Fully Encapsulating Suits: Chemical protective suits that are designed to offer full body protection, including SCBA, are gas tight, and meet the design criteria as outlined in NFPA Standard 1991.

Fumes: Tiny solid particles formed by the vaporization of a solid which then condense in air.

Fungus: A general term to denote a group of eukaryotic protist, including mushrooms, yeasts, rusts, molds, smuts, etc., which are characterized by the absence of a rigid cell wall composed of chitin, mannans, and sometimes cellulose.

G/Kg: See GRAMS PER KILOGRAM.

GA: Tabun, a nerve agent.

Gas: A state of matter in which the material is compressible, has a low density and viscosity, can expand or contract greatly in response to changes in temperature and pressure, and readily and uniformly distributes itself throughout any container.

GB: Sarin, a nerve agent, more toxic than tabun or soman.

GB2:   A binary nerve agent.

GC: Gas chromatography.

GD: Soman, a nerve agent.

Generalized vaccinia: Secondary lesions of the skin following vaccination which may occur in subjects with previously healthy skin but are more common in the case of traumatized skin, especially in the case of eczema(eczema vaccinatum). In the latter instance, generalized vaccinia may result from mere contact with a vaccinated person. Secondary vaccinial lesions may also occur following transfer of virus from the vaccination to another site by means of the fingers (autoinnoculation).

Glanders: A chronic debilitating disease of horses and other equids, as well as some members of the cat family, caused by *Pseudomonas mallei*; it is transmissible to humans. It attacks the mucous membranes of the nostrils of the horse, producing an increased and vitiated secretion and discharge of mucus, and enlargement and induration of the glands of the lower jaw.

GRAM (g):   A metric unit of weight. One ounce equals 28.4 grams.

Gram-Negative: Refers to the inability of many bacteria to retain crystal violet or similar stain through the standard Gram stain procedure. They show only the red counter-strain.

Gram-Positive: Refers to the ability of many bacteria to retain crystal violet or similar stain through the standard Gram stain procedure. They retain a purple color

Gram Stain: A staining procedure used in classifying bacteria. A bacterial smear on a slide is stained with a purple basic triphenyl methane dye, usually crystal violet, in the presence of iodine/potassium iodide. The cells are then rinsed with alcohol or other solvent, and then counter-stained, usually with safranin. The bacteria then appear purple or red according to their ability to keep the purple stain when rinsed with alcohol. This property is related to the composition of the bacterial cell wall.

Grams Per Kilogram (g/Kg):   This indicates the dose of a substance given to test animals in toxicity studies. For example, a dose may be 2 grams (of substance) per kilogram of body weight (of the experimental animal).

Grounding: A safety practice to conduct any electrical charge to the ground, preventing sparks that could ignite a flammable material.

G-Series Nerve Agents: A series of nerve agents developed by the Germans: tabun (GA), sarin (GB), and soman (GD).

Guardian BTA (Bio-Threat Alert) System: The Guardian BTA is a biological agent detecting device made by Alexeter Technologies (in U.S., Toll-Free, 877-591-5571), a commercial biological defense systems provider based in Wheeling, Illinois with BTA™ test strips developed by TETRACORE, Inc. of Gaithersburg, Maryland. The Anthrax BTA device sold by Guardian is a qualified purchase under the U.S. Department of Justice equipment grant program for first responders. The system delivers reliable fifteen-minute results under time-critical decisions. Radio Frequency ID chip technology ensures chain-of-custody documentation, and the Guardian BTA Reader promises optimal sensitivity and data management. At the present time, anthrax, ricin, staphylococcal enterotoxin B (SEB), botulinum, plague tests are available while pox, tularemia and Brucella are in development. The Guardian BTA device *is designed for Level A use (in addition to positive pressure breathing apparatus, Level A use by National Fire Protection Association standards requires a vapor protective suit).* A Guardian BTA™ test strip reader is delivered in a watertight, airtight protective case for portability and includes a built-in personal printer for on site results. The BTA™ test strip employs agent-specific antibodies to identify a suspect material, solid or liquid, when mixed with an aqueous solution. Five drops of liquid mixture are added to the sample port on the test strip, and the sample interacts with the reagents and moves along the test material inside the test strip's plastic case. Thin-layer chromatography then provides the results.

Guide for the Selection of Chemical Agent and Toxic Industrial Material Detection Equipment for Emergency First Responders: This two-volume report, done by the National Institute of Justice, is available online at http://www.ojp.usdoj.gov/. Once at the site, go to OJP Publications: A to Z and scan by exact title wording. This guide for emergency first responders provides information about detecting chemical agents and toxic industrial materials and selecting equipment for different applications. Because of the large numbers of items identified in this guide, Volume I features the guide, and Volume II contains the detection equipment data sheets. These commercially available products described in this report were those known to the authors as of May of 2000. The full text of the report is available through use of the Adobe Acrobat File, which can be provided without cost. The guide contains the following contents: Foreword, Executive Summary, Introduction, Introduction to Chemical Agents & Toxic Industrial Materials, Overview of Chemical Agents and Toxic Industrial Materials (point detection technologies, standoff detectors, analytical instruments), Selection Factors (chemical agents detected, toxic industrial materials detected, sensitivity, resistance to interferants, response time, start-up time, detection states, alarm capability, portability, power capabilities, battery needs, operational environment, durability, procurement costs, operator skill level, training requirements) Equipment Evaluation (equipment usage categories, evaluation results), Recommended Questions on Detectors, and References. There are also a number of tables and figures.

The U.S. Department of Justice also has two very necessary reports for first responder organizations and hazardous materials response teams. The first such report is entitled *"An Introduction to Biological Agent Detection Equipment for Emergency First Responders"*; the second such report is entitled *"Guide for the Selection of Chemical and Biological Detection Equipment for Emergency First Responders."* Both are available online at http://www.ojp. usdoj.gov.

H: Levinstein mustard, a blister agent.

HD: Sulfur Mustard, a blister agent.

Half-life: The time in which the concentration of a chemical in the environment is reduced by half.

Hazard Assessment: A process used to qualitatively or quantitatively assess risk factors to determine incident operations.

Hazard Class: A series of nine descriptive terms that have been established by the UN Committee of Experts to categorize the hazardous nature of chemical, physical, and biological materials. These categories are flammable liquids, flammable solids, explosives, gases, oxidizers, radioactive materials, corrosives, poisonous and infectious substances, and dangerous substances.

Hazardous Material: A substance which by its nature, containment and reactivity has the capability of inflicting harm during an accidental occurrence; characterized as being toxic, corrosive, flammable, reactive, an irritant or a strong sensitizer and thereby posing a threat to health and the environment when improperly managed.

Hazardous Materials Incident: The uncontrolled release, or potential release, of a hazardous material from its container into the environment.

Hazardous Materials Response Team: An organized group of trained response personnel operating under an emergency response plan and appropriate standard operating procedures, who are expected to perform work to handle and control actual or potential leaks or spills of hazardous materials requiring close approach to the material to control or stabilize an incident.

HCN: Hydrogen Cyanide.

Hemolysis: Alteration, dissolution, or destruction or red blood cells in such a manner that hemoglobin is liberated into a medium in which the cells are suspended (e.g., by specific complement-fixing antibodies, toxins, various chemical agents, tonicity, alteration of temperature.

High-Efficiency Particulate Air (HEPA) Filter: A filter that is at least 99.97 percent efficient in removing particles with a diameter of 0.3 u (microns) used to treat exhaust air from equipment that may generate aerosols.

Host: An organism that serves as a home to, and often as a food supply for, a parasite, such as a virus.

Hot Zone: The area adjacent to and surrounding a hazardous materials incident that extends far enough to prevent the effects of hazardous materials releases from endangering personnel outside the zone; also known as the restricted zone or the exclusion zone.

H-Series Agents: A series of persistent blister agents including distilled mustard (HD) and the nitrogen mustards (HN-1, HN-2, and HN-3).

HTH: Calcium Hypochlorite.

Hydrolysis: Process of an agent reacting with water. It does not materially affect the agent cloud in tactical use, because the rate of hydrolysis is too slow.

Hyperemia: The presence of an increased amount of blood in a part or organ.

Hypotension: Subnormal arterial blood pressure.

Hypovolemia: A decreased amount of blood in the body.

ICt50:   Inhalation dose of a chemical agent (vapor or aerosol) that produces a given, defined level of "incapacitation" in 50 percent of the exposed subjects.

ID50:   Dose of a liquid chemical agent needed to produce "incapacitation" in 50 percent of the exposed subjects.

Immediately Dangerous To Life or Health (IDLH):   This is a standard set by the National Institute of Occupational Safety and Health (NIOSH) which limits exposure to any toxic, corrosive, or asphyxiant substance that poses an immediate threat to life, or would cause irreversible or delayed adverse health effects, or world interfere with an individual's ability to escape from a dangerous atmosphere.

Immune Globulin (IG):   IG is a sterile solution containing antibodies from human blood. It is obtained by cold ethanol fractionation of large pools of blood plasma and contains 15–18 percent protein. Intended for intramuscular administration, IG is primarily indicated for routine maintenance of immunity of certain immunodeficient persons and for passive immunity against measles and hepatitis. IG does not transmit hepatitis B virus, human immunodeficiency virus (HIV), or other infectious diseases.

Immunity: a) Resistance usually associated with the presence of antibodies or cells in a body that effectively resist the effects of an infectious disease organism or toxin. b) A condition of being able to resist a particular disease, especially through preventing growth and development of a pathogenic microorganism or by counteracting the effects of its products.

Immunization: Administration either of a non-toxic antigen to confer active immunity or antibody to confer passive immunity to a person or animal in order to render them insusceptible to the toxic effects of a pathogen or toxin.

Immunoassay: Detection and assay of substances by serological (immunological) methods; in most applications the substance in question serves as antigen, both in antibody production and in measurement of antibody by the test substance.

Improved (Chemical Agent) Point Detection System (IPDS): This detection system is a new shipboard point detector and alarm that replaces the chemical Agent Point Detection System. This system can detect nerve and blister agent vapors at low levels and automatically provides an alarm to the ship.

Incident: An event involving a hazardous material or a release or potential release of a hazardous material.

Incident Action Plan: A plan which is initially prepared at the first meeting of emergency personnel who have responded to an incident. The plan contains general control objectives reflecting overall incident strategy, and specific action plans.

Incident Commander: The incident commander will assume control of the incident scene beyond the first responder incident level, and must demonstrate competency in the Incident Command System. The incident commander is responsible for developing an effective organizational structure, allocating resources, making appropriate assignments, managing information, and continually attempting to mitigate the incident. The employer shall certify that the incident commander meets the requirements of 29 CFR 1910.120.

Incident Command System (ICS): An organized system of responsibilities, roles, and standard operating procedures used to manage and direct emergency operations.

Incident Safety Officer: The incident safety officer is a position mandated by Occupational Safety and Health laws. He/she is attached to the incident commander, and should be the person with the most knowledge about the various safety aspects at a hazardous materials scene. As provided under OSHA law, the incident safety officer has the power and authority to alter, suspend, or terminate the operation when, in his/her opinion, the conditions are unsafe.

Incompatible: The term applied to two substances to indicate that one material cannot be mixed with the other without the possibility of a dangerous reaction.

Individual Chemical Agent Detector (ICAD): A miniature lightweight chemical agent detector that can be worn by an individual. It detects and alarms to nerve, blood, choking, and blister agents and is intended for a variety of applications. It may be used as a point detector.

Industrial Agents: Chemicals manufactured for industrial purposes rather than to specifically kill or maim human beings. Hydrogen cyanide, cyanogen chloride, phosgene, and chloropicrin are industrial chemicals that can be military agents as well; AC, CK, CG, and PS. Many herbicides and pesticides are industrial chemicals that also can be chemical agents.

Ingestion: Swallowing (such as eating and drinking). Chemicals can get into or onto food, drink, utensils, cigarettes, or hands where they can be ingested.

Inhalation: Breathing. Once inhaled, contaminants can be deposited in the lungs, taken into the blood, or both.

Inhibitor: A substance that is added to another to prevent or slow down an unwanted reaction or change.

Interim Biological Agent Detector (IBAD) — Rapid Prototype: A detector that provides a near-term solution to a deficiency in shipboard detection of biological warfare agents. This equipment is capable of detecting an increase in the particulate background, which may indicate a man-made biological attack is underway, and sampling the air for identification analysis. It can also detect a change in background within 15 minutes and can identify biological agents within an additional thirty minutes.

Intravenous Immune Globulin (IGIV):   IGIV is a product derived from blood plasma from a donor pool similar to the immune globulin (IG) pool, but prepared so it is suitable for intravenous use. IGIV does not transmit infectious diseases. It is primarily used for replacement therapy in primary antibody-deficiency disorders, for the treatment of Kawasaki disease, immune thrombocytopenic purpura, hypogammaglobulinemia in chronic lymphocytic leukemia, and in some cases of HIV infection.

In vitro: In an artificial environment, referring to a process or reaction occurring therein, as in a test tube or culture media.

In vivo: In the living body, referring to a process or reaction occurring therein.

Inoculation: Introduction into the body of the causative organism of a disease.

Isolation: Separation of infected persons or animals from others to prevent or limit direct or indirect transmissions of the infectious agent.

Joint Biological Point Detection System (JBPDS):   The Army, Navy, Air Force, and Marine Corps use this detection system. The developmental system will replace all existing biological detection systems (Biological Integrated Detection System, Interim Biological Agent Detector, and Air Base/Port Advanced Concept Technology Demonstration), and provide biological detection capabilities throughout the services and throughout the battle-space. The common biological detection suite will consist of four functionalities:

1. Trigger (detects a significant change in the ambient aerosol in real time).
2. Collector (collects samples of the suspect aerosol for analysis by the JBPDS, and for confirmatory analysis by supporting laboratories in the Communications Zone and the continental United States).
3. Detector (able to broadly categorize the contents of the aerosol and lend confidence to the detection process; e.g., biological material in the aerosol or not, bacteriological, spore, protein, etc.).
4. Identification (provide presumptive identification of the suspect biological warfare agent and increases confidence in the detection process). The JBPDS program consists of two phases (Block I and Block II) to allow the fastest possible fielding of a joint biological detection system, while at the same time preparing to take advantage of the

rapid advances taking place in the biological detection/identification, information processing and engineering services.

Joint Chemical Agent Detector (JCAD):   This detector will employ surface acoustic wave technology to detect nerve and blister agents. It will also allow detection of new forms of nerve agents.

Joint Service Lightweight Standoff Chemical Agent Detector (JSLSCAD):   This detector is a fully coordinated joint service Research, Development, Test, and Evaluation program, chartered to develop a lightweight standoff chemical detector for the quad-services. It will be capable of scanning 360 degrees × 60 degrees, and automatically detecting nerve or blister agents up to a distance of five kilometers. The system will be light, compact, and operate from a stationary position on the move. The JSLSCAD Michelson interferometer employs a passive infrared system that will detect presence of chemical agents by completing a spectral analysis of target vapor agent chemical clouds. This detector is envisioned for employment on various platforms and in various roles, including fixed site defense, unmanned aerial vehicles, tank and other vehicles, and onboard ships.

Joint Service Warning and Identification LIDAR Detector (JSWILD):   This detector is a joint effort chartered to develop a chemical warning and identification system for various military services. This JSWILD will be a lightweight, vehicle-mountable, contamination-monitoring system, which detects and quantifies all type of chemical agent contamination (including agent rain, vapors, and aerosols) in a standoff mode from a distance of 20 kilometers. In addition, it will provide similar short-range (one to five km) capabilities in biological standoff detection as the LR-BSDS. It will operate from fixed sites and ground vehicles. The system has distance-ranging and contamination-mapping capabilities and transmits the information to a battlefield information network.

L: Lewisite, a blister agent.

Labpack: Generally refers to any small containers of hazardous waste in an overpacked drum, but not restricted to laboratory wastes.

LC50:   The lethal concentration (LC) of a toxicant in air which is lethal to fifty percent of the exposed lab animal population.

LD50:   The lethal dose (LD) of a toxicant when taken orally or absorbed through the skin, which is lethal to fifty percent of the exposed lab animal population.

Leak Control Compounds: Substances used for the plugging and patching of leaks in non-pressure and some low-pressure containers, pipes, and tanks.

Leak Control Devices: Tools and equipment used for patching and plugging of leaks in non-pressure and some low-pressure containers, pipes, and tanks.

LEL: Lower Explosive Limit. The lowest concentration of gas or vapor by percent of volume in air that will burn or explode if an ignition source is present at ambient temperatures. The LEL is constant up to 250 degrees F.

Lethal Concentration 50:   The concentration of an air contaminant (LC50) that will kill 50 percent of the test animals in a group during a single exposure.

Lethal Dose 50:   The Dose of a substance or chemical that will (LD50) kill 50 percent of the test animals in a group within the first thirty days following exposure.

Leukopenia: The antithesis of leukocytosis; any situation in which the total number of leukocytes in the circulating blood is less than normal, the lower limit of which is generally regarded as 4000–5000 per cu mm.

Level of Protection: In addition to positive pressure breathing apparatus, designations of types of personal protective equipment to be worn based on NFPA standards.

> **Level A:**   Vapor protective suit for hazardous chemical emergencies.

> **Level B:**   Liquid splash protective suit for hazardous chemical emergencies.

> **Level C:**   Limited use protective suit for hazardous chemical emergencies.

Level One Incident: Hazardous materials incidents which can be contained, extinguished, and/or abated using equipment, supplies, and resources immediately available to first responders having jurisdiction, and whose qualifications are limited to and do not exceed the scope of the training explained in 29 CFR 1910.

Level Two Incident: Hazardous materials incidents which can only be identified, tested, sampled, contained, extinguished, and/or abated utilizing the resources of a HMRT (hazardous materials response team), which requires the use of specialized chemical protective clothing, and whose qualifications are explained in 29 CFR-1910.

Level Three Incident: A hazardous materials incident which is beyond the controlling capability of a HMRT (technician or specialist level) whose qualifications are explained in 29 CFR 1910; and/or which must be additionally assisted by qualified specialty teams or individuals.

Local Emergency Planning Committee (LEPC):   A committee appointed by the state emergency response commission, as required by Title III of SARA, to formulate a comprehensive emergency plan.

Long Range Biological Standoff Detection System (LR – BSDS) P31:   This detection system uses infrared light detection and ranging technology to detect, range, and track aerosol clouds that are indicative of a biological warfare attack; the LR-BSDS cannot discriminate biological from non-biological clouds. The system has three major components: 1) Aiode pulsed ionizing radiation laser transmitter operating at infrared wavelength. 2) A receiver and telescope. 3) An information processor and display.

Lower Explosive Limit (LEL):   (Also known as Lower Flammable Limit). The lowest concentration of a substance that will produce a fire or flash when an ignition source (flame, spark, etc.) is present. It is expressed in percent of vapor or gas in the air by volume. Below the LEL or LFL, the air/contaminant mixture is theoretically too "lean" to burn. (See also UEL)

M8 Chemical Agent Detection Paper: A chemically treated, dye-impregnated paper, issued in a book of twenty-five sheets. It is designed to detect liquid V, G, and H agents. M8 paper will change colors to identify non-persistent G-type nerve agent (yellow), V-type nerve agent (black or dark green), or blister agents (red). It is included in the M256A I Kit and in the M18A2 Chemical Agent Detection Kit.

M8A1 Automatic Chemical Agent Alarm (ACAA) System: The only remote continuous air-sampling alarm in the U.S. Army. This alarm will sample the air for the presence of nerve agent vapors (GA, GB, GD, or VX) *only*. It is capable of detecting nerve agent levels in two minutes or less. The system is an electrochemical, point-sampling, chemical sampling alarm that can be hand-carried, backpacked, or mounted on a tactical unit. It consists of a M43A1 detector, as many as five M42 alarm units, and various power supplies. The M8A1 will automatically signal the presence of the nerve agent in the air by providing troops with both an audible and visible warning. It requires a Nuclear Regulatory Commission license.

M9 Chemical Detection Paper: A self-adhesive paper that can be readily attached to the body or to vehicles, shelters, and other equipment. It cannot distinguish the identity of agents. The agent sensitive dye will turn red upon contact with liquid nerve agents (G and V) and blister agents (H and L). The paper produces colored spots when in contact with nerve and blister agents.

M11 Portable Decontamination Apparatus: A device containing DS-2 used to decontaminate small areas, such as the steering wheel of other equipment that military troops must touch. It is filled with one and one-third quarts of DS-2.

M13 Portable Decontamination Apparatus: The M13 is about the size of a five-gallon gasoline can and is used to decontaminate vehicles and military crew-served weapons larger than a .50-caliber piece.

M17 Lightweight Decontamination System (LDS): The M17 is a portable pump and water-heating unit for producing hot water and steam. The system incorporates a 1,580-gallon collapsible water tank, two wand assemblies, connecting hoses, and a shower rail. It is issued to Army battalion-size units and to chemical decontamination companies and battalions.

M18A2 Chemical Agent Detector Kit: A kit used by technical escort teams, located in storage depots, and consisting of portable tests capable of detecting and classifying selected choking agents and blood agents as well as nerve agents and blister (mustards, arsenicals, urticants) agents. The kit is used to identify dangerous concentrations of toxic chemical agents in the air and liquid chemical agent contamination on exposed surfaces.

M21Remote Sensing Chemical Agent Automatic Alarm (RSCAAL): A two-man portable tripod-mounted, automatic scanning, passive, infrared sensor that detects nerve and blister agent vapor cloud based on changes in the infrared energy emitted from remote objects, or from a cloud formed by the agent. The M21 is line-of-sight dependent with a detection range up to three miles and a field of view of 1.5 degrees vertical and 60 degrees horizontal. It is used for surveillance and reconnaissance missions and will search areas between enemy and friendly forces.

M22 Automatic Chemical Agent Alarm (ACADA): An advanced, point-sampling, chemical agent alarm system employing ion-mobility spectrometry. It is man-portable, operates independently after system start-up, and provides an audible and visual alarm. The system detects and identifies nerve and blister agents. It also provides communications interface for automatic battlefield warning and reporting. The M22 system replaces the M8A1 Alarm as an automatic point detector and augments the CAM as a survey instrument.

M90 Automatic Agent Detector (AMAD):   An automatic nerve and mustard agent detector that detects agents in vapor form. It transmits an alarm by radio to a central alarm unit. It is currently used in the Air Force.

M90 DIA Chemical Agent Detector (CAD):   A man-portable instrument designed to determine and indicate the hazard from nerve or blister (mustard) agent vapors present in the air. Hazard levels are indicated as High, Medium, and Low concentrations. The detector is programmable, with the capability to add new agents as they are developed. It is operable over a multitude of operational platforms including day and night conditions. It can be used to verify clean areas, perform area surveys, identify contamination, and verify the effectiveness of decontamination operations. This unit is currently fielded within the Air Force.

M93 and M93A1 FOX (NBCRS):   The FOX Nuclear Biological Chemical Reconnaissance System is a field detection and protection platform vehicle, equipped with a mobile mass spectrometer capable of detecting, identifying, and quantifying up to 60 chemical units simultaneously. This system is also capable of collecting biological sample of solids and liquids. FOX vehicles are equipped with an overpressure system, six-wheel drive, seats for a crew of four, and are able to operate on both land and water. The FOX is equipped with an industrial chip that allows detecting 115 industrial chemical agents. The M93FOX NBCRS is the best bet the military presently has for providing a comprehensive solution to the problem of NBC attacks in a terrorism and weapons of mass destruction (WMD) battle scenario. The six-tire vehicle has onboard an M21RSCAAL detector, a meteorological sensor, central data processing unit, M21 monitor/control panel, printer, monitor, mass spectrometer, air sample vacuum intake, surface sampler wheels, grappler, marker, extension glove, swim propeller, marker storage racks, chemical agent detector, AN/VDR-2 Radiation Detector, position location navigation system, commander's display and keyboard, and GPS (global positioning system).

M256-Series Chemical Agent Detector Kit: A kit used by military personnel to detect and identify field concentrations of nerve, blister or blood agent vapors. The kit consists of twelve samplers/detectors and a packet of M8 detector paper. It is used at the squad, crew or section level to detect and identify field concentrations of nerve, blister or blood agents vapors. It is usually used to determine when it is safe to unmask, to locate and identify chemical hazards, and to monitor decontamination effectiveness.

M258A1 Skin Decontamination Kit: A kit issued to each soldier containing wipes with solutions that will neutralize most nerve and blister agents.

M272 Water Testing Kit: A lightweight portable kit used to detect and identify dangerous levels of common chemical warfare agents in raw and treated water in about seven minutes. It is a test water sampler and is not a continuous monitor. Each kit includes twenty-five tests for each agent.

M291 Skin Decontamination Kit: This kit is used to decontaminate the soldier's hands, face, ears, and neck. Packets in the kit consist of a foil-laminated fiber material containing a reactive resin. It replaces the M258A1 Skin Decontamination Kit.

M295:   Decontamination Packet, Individual Equipment (DPIE) (Military).

Macula, plural: maculae: 1) A small spot, perceptibly different in color from the surrounding tissue. 2) A small, discolored patch or spot on the skin, neither elevated above nor depressed below the skin's surface.

Mark I & II: Nerve Agent Antidote Kit (Military).

Material Safety Data Sheets (MSDS):   An MSDS contains descriptive information on hazardous chemicals under OSHA's Hazard Communication Standard. The data sheets also provide precautionary information on safe handling, health effects, chemical and physical properties, emergency phone numbers, and first aid procedures.

Material Safety Data Sheets for Biological Agents: M.S.D.S. sheets have been available for years for general industry workers, but Health CANADA has produced 169 Material Safety Data Sheets for Infectious Diseases including bacillus anthracis, bacillus cereus, chlamydia psittaci, chamydia trachomatis, Colorado tick fever virus, Crimean-Congo hemorrhagic fever virus, dengue fever(1,2,3, 4), Eastern (Western) equine encephalitis virus, Ebola virus, hantavirus, hepatitis virus (A through E), human immunodeficiency virus, influenza virus, Marburg virus, rickettsia akari, rickettsia rickettsii, vaccinia virus, Venezuelan equine encephalitis, yellow fever virus, and yersinia pestis. Each Material Safety Data Sheet — Infectious Diseases contains the following references: Sexction I — Infectious Agent (Name, Synonym or Cross Reference, Characteristics); Section II — Health Hazar (Pathogenicity, Epidemiology, Host range, Mode of transmission, Incubation period, Communicability); Section III — Dissemination (Reservoir, Zoonosis, Vectors,); Section IV — Viability (Drug susceptibility, Susceptibility to disinfectants, Physical inactivation, Survival outside host); Section V — Medical (Surveillance, First Aid/Treatment, Immunization, Prophylaxis); Section VI — Laboratory Hazards (Laboratory-acquired infections, Source/Specimens, Primary hazards, Special hazards); Section VII — Recommended Precautions (Containment precautions, Protective clothing, Other precautions); Section VIII — Handling Information (Spills, Disposal, Storage); and Section IX — Miscellaneous Information. Such information is produced for personnel working in the life sciences as quick safety reference material relating to infectious microorganisms. The intent of such M.S.D.S.s is to provide a safety resource for laboratory workers working with these infectious substances. Because such workers are usually working in a scientific setting and are potentially exposed to much higher concentrations of these human pathogens than the general public, the terminology in these M.S.D.S. is technical and detailed and containing information that is relevant specifically to the laboratory setting. Contact: Health Canada, Health Protection Branch — Laboratory Center for Disease Control online at, http://www.hc-sc.gc.ca/hpb/lcdc/biosafty/msds/index.html, then click "MSDS for Infectious Diseases."

MD: Methyldichloroarsine, a type of blister/vesicant agent. .

mg: See MILLIGRAM.

mg/Kg: See MILLIGRAMS PER KILOGRAM.

mg/M3:   See MILLIGRAMS PER CUBIC METER

Melting Point: The temperature at which a solid changes to a liquid. A melting range may be given for mixtures.

Merck Index: Includes basic information on several thousand compounds that are important in general chemical and biochemical practice.

Micron: A unit of measurement equal to one-millionth of a meter

Milligrams Per Kilogram: This indicates the dose of a substance (mg/kg) given to test animals in toxicity studies. For example, a dose may be 2 milligrams (of substance) per kilogram of body weight (of the experimental animal).

Mini-Cam: Miniature chemical agent monitor (Military).

Miscible: Able to mix (but not chemically combine) in any ratio without separating into two phases.

Mists: Suspended liquid droplets generated by condensation from the gaseous to the liquid state or by breaking up a liquid into a dispersed state; such as by splashing, foaming, or atomizing. Mist is formed when a finely divided liquid is suspended in air.

Mitigation: An action employed to contain, reduce, or eliminate the harmful effects of a spill or release of a hazardous material.

Monitoring: To determine contamination levels and atmospheric conditions by observation and sampling using instruments and devices to identify and quantify contaminants and other factors.

MOPP: Mission-Oriented Protective Posture, the protective clothing used by members of the U.S. military who engage in nuclear, biological, and chemical warfare. MOPP gear provides a flexible system requiring personnel to wear only that protective clothing and equipment appropriate to the threat level, work rate imposed by the mission, temperature, and humidity.

Moribund: Dying; at the point of death.

Mutagen: Anything that can cause a change (mutation) in the genetic material of a living cell.

Mutual Aid: An agreement to supply specifically agreed upon aid or support in an emergency situation between two or more agencies, jurisdictions, or political sub-divisions.

Myalgia: Muscular pain.

Mycotoxin: A fungal toxin.

Mydriasis: Dilation of the pupil.

Narcosis: General and nonspecific reversible depression of neuronal excitability, produced by a number of physical and chemical agents, usually resulting in stupor rather than in anesthesia.

National Contingency Plan: Created by CERCLA to define the federal response authority and responsibility for oil and hazardous materials spills. The regulations are codified at 40 CFR 300.

National Fire Protection Association (NFPA):   A voluntary membership agency to promote fire safety and allied considerations. NFPA publishes standards of interest to hazardous materials responders; such as, NFPA-471 Recommended Practice for Responding to Hazardous Materials Incidents, NFPA-472 Standard for Professional Competence of Responders to Hazardous Materials Incidents, and NFPA-473, Competencies for EMS Personnel Responding to Hazardous Materials Incidents.

National Institute for Occupational Safety and Health (NIOSH):   A federal agency that performs research on occupational disease and injury, and recommends limits for substances and assists OSHA in investigations and research.

National Response Center (NRC):   The national response center in Washington, D.C. is operated by the U.S. Coast Guard. The center must be informed by the spiller within twenty-four hours of any spill of a reportable quantity of a hazardous substance.

NBC: Nuclear, biological, and chemical.

Necrosis: Pathologic death of one or more cells, or of a portion of tissue or organ, resulting from irreversible damage.

Nerve Agent: Substances that interfere with the central nervous system. Organic esters of phosphoric acid used as a chemical warfare agent because of their extreme toxicity (tabun-GA, sarin-GB, soman-GD, GF, and VX). All are potent inhibitors of the enzyme, acetylcholinesterase, which is responsible for the degradation of the neurotransmitter, acetylcholine in neuronal synapses or myoneural junctions. Nerve agents are readily absorbed by inhalation and/or through intact skin.

Nerve Agent Antidote Kit: (NAAK) Also called the MARK I, containing atropine and 2 PAM chloride (Military).

Nerve Agent Pyrdostigmine Pretreatment: (NAPP)

Neutralize: To render chemically harmless; to bring a solution of an acid or base to a pH of 7.0.

Nonpersistent Agent: An agent that upon release loses its ability to cause casualties after ten to fifteen minutes. It has a high evaporation rate, is lighter than air, and will disperse rapidly in open air.

North American Emergency Response Guidebook: This popular manual is in every fire response vehicle in the United States. It is primarily a guide to assist first responders in quickly identifying the specific or generic hazard(s) involved in an incident, and protecting themselves and the general public during the initial response phase of the incident. This book was developed jointly by Transport Canada, the U.S. Department of Transportation, and the Secretariat of Communication and Transportation of Mexico for use by firefighters, police, and other emergency service personnel. The various guides by chemical name or number deal with Potential Hazards, Public Safety, Protective Clothing, Evacuation and Emergency Response (Fire, Spill or Leak, and First Aid). If you have the name of a specific chemical, look it up in the blue pages of the guide to get the proper guide number for critical response information. If you do not know the exact name of the chemical you are searching for, use the agent class. As an example, "Infectious Substances" have only one guide number, Guide 158.

Nosocomial: Denoting a new disorder (not the patient's original condition) associated with being treated in a hospital, such as a hospital-acquired infection.

Occupational Safety and Health Administration (OSHA): The U.S. Department of Labor through OSHA has safety and health regulatory and enforcement control over worker health in most industries, businesses, and states in the nation.

Odor Threshold: The minimum concentration of a substance at which a majority of test subjects can detect and identify the substance's characteristic odor.

Off-gassing: Giving off a vapor or gas.

Organic Materials: Compounds composed of carbon, hydrogen, and other elements with chain or ring structures.

Organophosphorous Compound: Containing elements of phosphorous and carbon, the physiological effects of such a compound include inhibition of acetylcholinesterase. A number of pesticides including parathion and malathione, and virtually all nerve agents, are organophosphorous compounds.

Osteomyelitis: Inflammation of the bone marrow and adjacent bone.

Oxidation: The process of combining oxygen with some other substance to a chemical change in which an atom loses electrons.

Oxidizer: Is a substance that gives up oxygen easily to stimulate combustion of organic material.

Oxime: A chemical compound containing one or more oxime groups. Although some blister agents are bad, some oximes are beneficial. 2-PAM chloride is used in treatment of nerve agent poisoning. This drug increases the effectiveness in poisoning by some, but not all, cholinesterase inhibitors (nerve agents).

Oxygen Deficiency: An atmosphere having less than the normal percentage of oxygen found in normal air. Normal air contains 21 percent oxygen at sea level. OSHA defines any atmosphere containing 19.5 percent or less by volume in air as "oxygen deficient" and unsafe to enter without respiratory protection.

Oxygen sensors/Combustible gas indicators/Toxic substances: These are all component sensors of an average atmospheric monitor. The oxygen sensor detects the percentage of oxygen by volume in the air. Most monitors are set to alarm visually and audibly if oxygen levels fall below 19.5 percent or rise above 23.5 percent. Combustible gas indicators measure flammable gases as a percentage of the lower explosive limit (LEL). They are set to alarm visually and audibly if they detect gas levels at or above 10 percent of the LEL. Toxic substances are measured in parts per million (ppm). Most monitors are set to alarm if the selected toxic gas reaches or exceeds the permissible exposure limit (PEL) for that gas.

PD: Phenyldichoroarsine, a blister agent.

Pandemic: Denoting a disease affecting or attacking the population of an extensive region, country, continent; extensively epidemic.

Papule: A small, circumscribed, solid elevation on the skin.

Passive immunity: Providing temporary protection from disease through the administration of exogenously produced antibody (i.e., transplacental transmission of antibodies to the fetus or the injection of immune globulin for specific preventive purposes).

Pathogen: Biological agents that are disease-producing microorganisms, such as bacteria, mycoplasma, rickettsia, fungi, or viruses.

Percutaneous: Denoting the passage of substances through unbroken skin; for example, by needle puncture, including introduction of wire and catheters.

Permeation: The passage of chemicals, on a molecular level, through intact material such as protective clothing,

Permissible Exposure Limit (PEL):   The maximum time-weighted average concentration mandated by OSHA to which workers may be repeatedly exposed for eight hours a day, forty hours per week without adverse health effects.

Persistent Agent: Chemical agents that do not hydrolyze or volatilize readily, such as VX and HD. At the time of release, this agent can produce casualties for an extended period of time up to several days. Usually, it has a low evaporation rate. Since its vapor is heavier than air, its vapor cloud will hug the ground and accumulate in low areas. It is an inhalation hazard, but extreme care should be taken to avoid skin contact as well.

Personal Protective Equipment (PPE):   Equipment, provided to shield or isolate a person from the chemical, physical, and thermal hazards that may be encountered at a hazardous materials incident; and should include protection for the respiratory system, skin, eyes, face, hands, feet, head, body, and hearing.

pH: The value that represents the acidity or alkalinity of an aqueous solution. The number is a logarithm to the base 10 of the reciprocal of the hydrogen ion concentration of a solution. Pure water has a pH of 7. The pH scale is logarithmic and the intervals are exponential, so the progression of values represents a factor of ten.

Phosgene: Carbonyl Chloride. An extremely poisonous gas, but not immediately irritating even when fatal concentrations are inhaled.

Physical State: The (solid, liquid, or gas) of a chemical under specific conditions of temperature and pressure.

Plasma: The fluid portion of the blood, as opposed to the particulate bodies suspended in the blood.

Plug and Patch: Plugging and/or patching refers to the use of compatible plugs and/or patches to temporarily reduce or stop the flow of materials from small holes, rips, tears, or gashes in containers.

Plume: A vapor cloud formation which has shape and buoyancy.

Poison, Class A: A D.O.T. term for extremely dangerous poisons such as poisonous gases or liquids of such a nature that a very small amount of the gas or vapor of the liquid mixed with air is dangerous to life. Examples include phosgene, cyanogen, hydrocyanic acid, and nitrogen peroxide.

Poison, Class B: A D.O.T. term for liquid, solid, paste, or semisolid substances other than Class A poison of irritating materials that are known or presumed on the basis of animal tests to be so toxic to man as to afford a hazard to health during transportation.

Polymerase chain reaction: An in vitro method for enzymatically synthesizing and amplifying defined sequences of DNA in molecular biology. Can be used for improving DNA-based diagnostic procedures for identifying unknown BW agents.

Polymerization: A process in which a hazardous materials is reacted in the presence of a catalyst of heat or light or with itself or another material to form a polymeric system, which often times is violent.

ppb: Parts per billion.

ppm: Parts per million.

ppt: Parts per trillion.

Potentially Responsible Party: (PRP) An individual or company identified by EPA as potentially liable under CERCLA for cleanup costs at a hazardous waste site. PRPs may include generators of hazardous substances, present or former owners of hazardous substances that have been disposed, site property owners, and transporters of hazardous materials to the site.

Precursor: Any chemical reactant that takes part at any stage in the production by whatever method of a toxic chemical. This includes any key components of a binary or multi-component chemical system.

Protection Factor: The ratio of the concentration outside the personal protective equipment to the concentration inside the personal protective equipment. Measurement site are critical for proper determination (e.g., for a protective mask, the measurement inside the mask would be made at a subject's breathing zone, and the measurements outside the mask would be made in a corresponding zone).

Ptosis, plural, ptoses: In reference to the eyes, drooping of the eyelids.

Public Information Officer (PIO): Person who acts as a liaison between the incident commander and the news media.

Pulmonary edema: Edema of the lungs.

Pyridostigmine Bromide: An antidote enhancer that blocks acetylcholinesterase, protecting it from nerve agents. When taken in advance of nerve agent exposure, pyridostigmine bromide increases survival provided that atropine and oxime and other measures are taken.

Pyrogenic: Causing fever.

Quarantine: Quarantine is the isolation of patients with a communicable disease, or those exposed to a communicable disease, during the contagious period in order to control the spread of illness. Quarantine over the years has been a practice of holding travelers or ships, trucks, or airplanes coming from places of epidemic disease for the purpose of inspection or disinfection. In the age of weapons of mass destruction, quarantine is defined as the *restriction of activities or limitation of freedom of movement* of those presumed exposed to a communicable disease in such a manner as to prevent effective contact with those not so exposed. Although quarantine measures may be instituted and enforced for both individual persons and populations, the term is used more frequently to discuss methods undertaken at a population-wide level. The Centers of Disease Control and Prevention (CDC), Atlanta, Georgia, says the major source of legal authority for public health intervention is the police power, defined as the inherent authority of all sovereign governments to enact laws and promote regulation that safeguard the health, welfare, and the morals of its citizens. The tenth Amendment reserves to the states all powers not expressly granted to the federal government nor otherwise prohibited by the Constitution, including the police power. Courts have repeatedly held over the past years that state quarantine laws are a proper exercise of their police power. The laws may be used to detain persons within a dedicated area and to exclude healthy individuals from entering such an area. In current times, states should consider modern constitutional considerations like due process, freedom of movement, and bodily integrity when considering amendments or changes to current quarantine laws.

In October of 2001, the CDC put together a draft model law for the states, "Model State Emergency Health Powers Act." The proposed model law would give state officials broad powers to close buildings, take over hospitals, and order quarantines during a biological attack. The various state legislatures would have to decide to adopt, or not adopt, the model law. A CDC conference considered a summary of public health powers necessary for adequate response to a bioterrorism incident that is reproduced below.

Collection of Records and Data

> Reporting of diseases, unusual clusters, and suspicious events

Access to hospital and provider records

> Data sharing with law enforcement agencies
>
> Veterinary reporting
>
> Reporting of workplace absenteeism
>
> Reporting from pharmacies

Control of Property

Right of access to suspicious premises

Emergency closure of facilities

Temporary use of hospitals and the ability to transfer patients

> Temporary use of hotel rooms and drive-through facilities
>
> Procurement or confiscation of medicines and vaccines
>
> Seizure of cell phones and other "walkie-talkie" type equipment
>
> Decontamination of buildings
>
> Seizure and destruction of contaminated articles

Management of Persons

Identification of exposed persons

Mandatory medical examinations

Collect laboratory specimens and perform tests

Rationing of medicines

Tracking and follow-up of persons

Isolation and quarantine

Logistical authority for patient management

Enforcement authority through police or National Guard

Suspension of licensing authority for medical personnel from outside jurisdictions

Authorization of other doctors to perform functions of medical examiner

Access to Communications and Parking Relations

Identification of public health officers, e.g., badges

Dissemination of accurate information, rumor control, 1-800 numbers

Establishment of command center

> Access to elected officials
>
> Access to experts in human relations and post-traumatic stress syndrome
>
> Diversity in training, cultural differences, dissemination of information in multiple languages

Federal assistance may be provided to state and local authorities in enforcing their quarantine and other health regulations pursuant to section 311 of the Public Health Service Act. (42 U.S.C. 243(a)). In addition, CDC's quarantine regulations authorized Federal intervention "in the event of inadequate local control." (Please refer to 42 CFR 70.2, and 21 CFR 1240.30)

Reactivity: A substance's susceptibility to undergoing a chemical reaction or change that may result in dangerous side effects, such as explosion, burning, and corrosive or toxic emissions.

Reagent: A chemical substance used to produce a chemical reaction.

Rebreathers: The system provides a duration of use for up to two hours, weighs only 30 pounds, and is both NIOSH and MSHA approved. The Rebreather system is positive pressure which will keep gases away from the user's face.

Reference Library: Chemical textbooks, references, computer data programs, and similar materials carried by response personnel.

Release: Controlled or uncontrolled escape of chemical agent(s) into the environment. Any spilling, leaking, pumping, pouring, emitting, emptying, discharging, injecting, escaping, leaching, dumping, or disposing into the environment (including the abandonment or discarding of barrels, containers, and other closed receptacles containing any hazardous substance or pollutant or contaminant).

Rem: Radiation Equivalent Man; the unit of dose equivalence commonly used in the United States.

Reservoir: Any person, animal, anthropod, plant, soil, or substance (or combination of these) in which an infectious agent normally lives and multiplies, on which it depend for survival, and in which it reproduces itself in such a manner that it can be transmitted to a susceptible vector

Respirator: A device that is designed to protect the wearer from inhaling harmful contaminants.

Respiratory Hazard: A particular concentration of an airborne contaminant that, when it enters the body by way of the respiratory system or by being breathed into the lungs, results in some bodily function impairment.

Response Actions, Basic, In a Weapons of Mass Destruction Incident: The incident site and any downwind hazard area must be secured, and casualty rescue undertaken by personnel in proper protective gear who will not end up becoming victims themselves. Speed must be present in the identification of the agent(s) by detecting instruments available on scene and watching for signs and symptoms of chemical agent contamination of injured on scene (victims of biological agents may be unlikely to show any signs and symptom immediately other than possible flu-like symptoms). The number of victims who may have to be decontaminated could be very large, action in this regard will have to happen quickly, and decontamination staff will have to have access to proper personal protective equipment so they do not become contaminated by off-gassing or by other factors. Response personnel will have to preserve possible evidence of an attack while they also must neutralize potentially contaminated areas while being extremely cautious about finding and identifying possible secondary devices.

RETECS: Registry of Toxic Effects of Chemical Substances, is published by NIOSH, and presents basic toxicity data on thousands of materials.

Retinitis: Inflammation of the retina.

Rhinorrhea: A discharge from the nasal mucous membrane.

Ribavirin: An antiviral drug used in treatment of viral hemorrhagic fevers.

Rickettsia: A microorganism of the genus Rickettsia. made up of small rod-shaped coccoids occurring in fleas, lice, ticks, and mites by which they are transmitted to man and other animals causing diseases such as typhus, scrub typhus, and Rocky Mountain Spotted Fever in humans.

Risk Assessment: The scientific process of evaluating the toxic properties of a chemical and the conditions of human exposure to it, in order to ascertain the likelihood that exposed humans will be adversely affected, and to characterize the nature of the effects they may experience. It may contain some or all of the following four steps: hazard identification, dose-response assessment, exposure assessment, and risk characterization.

Route of Exposure: The avenue by which a chemical comes into contact with and organism (such as a person). Possible routes include inhalation, ingestion, and dermal contact.

RSCAAL: Remote Sensing Chemical Agent Alarm used by the military.

Rupture: The physical failure of a container or mechanical device, releasing or threatening to release a hazardous material.

SA: Arsine.

Sample: To take a representative portion of a material for evidence or analytical reasons.

Scarification: The making of a number of superficial incisions in the skin. It is the technique used to administer tularemia and smallpox vaccines.

Scenario: An outline of a natural or expected course of events.

Scene: The location impacted or potentially impacted by a hazard.

Secondary Contamination: Contamination that occurs due to contact with a contaminated person or object rather than direct contact with agent aerosols; cross contamination.

Self-Contained Breathing Apparatus (SCBA): SCBA is protective equipment consisting of an enclosed facepiece and an independent, individual supply (tank) of air; used for breathing in atmospheres containing toxic substances, including a positive pressure SCBA or a combination SCBA/supplied air breathing apparatus certified by NIOSH and the Mine Safety Health Administration or an appropriate approval agency. Positive pressure units maintain a slight positive pressure in the face-piece during both inhalation and exhalation. If a leak or failure of a seal within the mask occurs, air will leak out, rather than contaminants being drawn in. SCBAs normally come with air cylinders that are most commonly rated for thirty, forty-five, or sixty minutes of use. Depending on the degree of effort due to increased physical requirements in wearing such gear in a response, users should never plan on being able to use an air bottle for its fully rated capacity. One advantage of the SCBA is that there is no risk of air lines getting tangled. However, the air supply will run out in a limited amount of time, and the units are very heavy and bulky.

Sensitizer: A substance that may cause no reaction in a person during initial exposures, but afterwards, further exposures will cause an allergic response to the substance.

Septic shock: 1) shock associated with sepsis, usually associated with abdominal and pelvic infection complicating trauma or operations; 2) shock associated with septicemia caused by Gram-negative bacteria.

Shigellosis: Bacillary dysentery caused by bacteria of the genus *Shigella*, often occurring in epidemic patterns.

Shipboard Chemical Agent Point Detection System (CAPDS): A fixed system capable of detecting nerve agents in vapor form using a baffle tube ionization spectrometer. This CAPDS obtains a sample of external air, ionizes airborne vapor molecules, and collects them on a charged plate after eliminating lighter molecules via the baffle structure. The system is installed in an upper superstructure level and provides ships with the capability to detect nerve agents. It will be activated when ships enter high threat areas and during operation in littoral waterways. The system is installed on most surface combatant's ships.

Shipping Papers: Term used to refer to the shipping documents that must accompany all shipments of hazardous materials and waste.

Shock: An upset in the body caused by inadequate amounts of blood circulating in the blood stream. It can be caused by marked blood loss, overwhelming infection, severe injury to tissue, emotional factors, etc.

Short Term Exposure Limit (STEL): The time weighted average concentration to which workers can be exposed continuously for a short period of time, normally fifteen minutes, without suffering irritation, chronic or irreversible tissue damage, etc,..

SIC Code: Standard Industrial Classification (SIC) codes are a system of numerical codes that categorizes industrial facilities by the type of activity in which they are engaged. All companies conducting the same type of business, regardless of their size, have the same SIC code. As an example, SIC code 2911 refers to petroleum refineries.

Simulant: A chemical that appears and acts like an agent.

SKIN: This designation sometimes appears alongside a TLV or PEL. It refers to the possibility of absorption of the particular chemical through the skin and eyes. Thus, protection of large surface areas of skin should be considered to prevent skin absorption so that the TLV is not invalidated.

Smart Air Sampling System (SASS): The SASS is an air sampler designed to collect and concentrate biological aerosols into a liquid media for subsequent analysis. The collected liquid sample is then provided to a bio-detection field device to determine whether biological warfare agents are present. The SASS can take a 5–7cc sample in ten minutes, and is battery-operated for up to eight hours.

SMART Tickets: ETG, located in Baltimore, distributes and markets SMART tickets as well as other domestic preparedness products. ETG is a prime Department of Defense contractor and an international supplier of military products. They deal in chemical Environmental Technology Group, and biological detection systems, as well as other products. The SMART™ Biological Warfare Agent Detection Ticket employs patented immuno-chemistry tests for specific biological agents (including anthrax, plague, ricin, botulinum toxins, brucella and several others). They utilize antibody/antigen reactions. A reaction vial contains colloidal gold particles. If an agent is present in sample, a complex forms between gold labeled antibodies and agent, a dacron swab transfers this complex to a ticket, the complex is filtered and concentrated onto a membrane, and the complex become visible as a red spot. If an agent is *not* present in the sample, a complex does not form, gold particles diffuse through membrane and are not visually detectable. To operate the system, wipe suspected area with a swab, place six drops of buffer solution into vial, tap tube with finger to mix pellet, place swab into vial, squeeze swab against vial wall to mix, and place swab into upper portion of the ticket and wait five to fifteen minutes. For test results, observe test spot; if any distinct red color appears which may be in the shape of a dot or crescent, the test is *positive* if there is a stronger color in the test spot than in the negative control spot. The control spot must be free of any color other than a very faint pink. If a reaction occurs in the negative control spot, or, detection spot is difficult to read; place one drop of buffer on a clean swab, and wipe the reaction area. Positive results will not wash away.

ETG also distributes the APD 2000 (Advanced Portable Detector) which can simultaneously detect nerve and blister chemical agents and provides agent identification, recognizes pepper spray and mace, and identifies hazardous compounds. Sensitivity for V agents is four parts-per-billion (ppb) with a response time of 30 seconds, for G agents is 15 ppb/30 seconds, for H agents is 300 ppb/15 seconds, and for lewisite is 200 ppb/15 seconds. For high concentrations of these agents, detection time is ten seconds. Selectable settings allow the APD 2000 to be used as a detector which automatically clears down following an alarm, or as a continuously-sampling monitor. A fixed site remote detector featuring the APD 2000 system can also be supplied for force protection, fixed installation monitoring, building installation monitoring, perimeter security, remote detector networks, or decon hot/warm zone monitoring.

Another detection instrument, the miniature chemical agent detector (ICAD/TM), can simultaneously detect nerve, blister, blood, and choking agents and warn responders or military troops through both audible and visual alarms. This detector is currently in use by U.S. and NATO forces. It weighs only eight ounces, and incorporates a replaceable sensor module which allows it to operate continuously for up to four months. Contact: Environmental Technology Group, Inc. 1400 Taylor Avenue, P.O. Box 9840, Baltimore, MD 21284-9840, Tel. 419-321-5370. FAX: 410-321-5255.

Solubility: The ability of one material to dissolve in or blend uniformly with another.

Specific Gravity: The ratio of the mass of a unit of volume of a substance to the mass of the same volume of a standard substance (usually water) at a standard temperature.

Specific Immune Globulin: This solution is a special preparation obtained from blood plasma from donor pools pre-selected for a high antibody content against a specific antigen (e.g., hepatitus B immune globulin, varicella-zoster immune globulin, rabies immune globulin, tetanus immune globulin, vaccinia immune globulin, and cytomegalovirus immune globulin). Like IG and IGIV, these preparations do not transmit infectious diseases.

Spill: The release of a liquid, powder, or solid hazardous material in a manner that poses a threat to air, water, ground, or the environment.

Spores: Resistant, dormant cells of some bacteria.

Stabilization: The point in time where the dangerous effects or results of a hazardous material has been controlled.

Staging Area: The safe area established for the temporary location of available resources closer to the incident to reduce response time.

State Emergency Response Commission: State commissions required under the Superfund Amendments and Reauthorization Act (SARA) which designates emergency planning districts, and appoint local emergency planning committees and supervise and coordinate their activities.

STB: Supertropical Bleach (a decontamination agent), a mixture of calcium oxide and bleaching powder.

STEL: Short-Term Exposure Limit. The maximum concentration for a continuous fifteen minute exposure period, with four exposure periods a day with sixty minutes minimum between exposure periods.

Strict Liability: The responsible party is liable even though they have exercised reasonable care.

Stridor: A high-pitched, noisy respiration, like the blowing of the wind; a sign of respiratory obstruction, especially in the trachea or larynx.

Sump: A pit or tank that catches liquid runoff for drainage or disposal.

Superantigen: An antigen that interacts with the T cell receptor in a domain outside of the antigen recognition site. This type of interaction induces the activation of larger numbers of T cells compared to antigens that are presented in the antigen recognition site.

Supplied Air Breathing Apparatus (SABA): SABA, when combined with an emergency escape bottle, provides the safest combination for respiratory protection. It provides air from a remote supply connected via an extended air hose, and has an escape cylinder rated for five to fifteen minutes of use, in case the remote air supply in compromised. SABAs are not as bulky or cumbersome as SCBAs, and enable longer work periods. The disadvantages of SABAs are the air lines can get tangled, and damaged by cuts, abrasions, and chemical contamination. Also, the airlines will limit your range, and the user will have to exit in the same manner as he or she entered the area.

Symptoms: Functional evidence of disease. Information related by an individual that may indicate illness or injury.

Synapse: A site at which neutrons make functional contacts with other neurons or cells.

Synergistic Effect: Joint action of agents that when taken together increase each other's effectiveness.

Synonym: Another name by which the same chemical may be known.

Systemic: Spread throughout the body; affecting many or all body systems or organs; not localized in one spot or area.

TAP Apron: Toxicological Agent Protective Apron (Military).

TC50: Toxic Concentration 50 percent is the concentration in inhaled air needed to produce an observed toxic effect in 50 percent of the test animals in a given time period.

Tachycardia: Rapid beating of the heart, conventionally applied to rates over 100 per minute.

Technical Escort Unit: Individuals technically qualified and properly equipped to accompany designated material that requires a high degree of safety and security during shipment.

Teratogen: An agent or substance that may cause physical defects in the developing embryo or fetus when a pregnant female is exposed to that substance.

Thickened Agent: An agent to which a polymer or plastic has been added to retard evaporation and cause it to adhere to surfaces.

Threshold: A level of chemical exposure below which there is no adverse effects and above which there is significant toxicological effect.

Threshold Limit Value: Airborne concentrations of substances devised by the ACGIH that represent conditions under which it is believed that nearly all workers may be exposed day after day with no adverse effect. TLV's are advisory exposure guidelines, not legal standards, based on evidence from industrial experience, animal studies, or human studies when they exist. There are three different types of TLV's: Time Weighted Average (TLV-TWA):   Short Term Exposure Limit (TLV-STEL) and Ceiling (TLV-C). (See also PEL)

Time weighted average: The average time, over a given work period (e.g., eight-hour work day), of a person's exposure to a chemical or an agent. The average is determined by sampling for the contaminant throughout the time period. Represented as TLV-TWA.

TLV: Threshold Limit Value. An estimate of the average safe airborne concentration of a substance; conditions under which it is believed that nearly all workers may be repeatedly exposed day after day without adverse effect.

Toxic: A poison relating to or caused by a toxin, able to cause injury by contact or systemic action. The ability of a material to injure biological tissue; another word for poisonous.

Toxicity: The potential for a substance to exert a harmful effect on humans or animals and a description of the effect and the conditions or concentrations under which the effect takes place.

Toxicology: The study of nature, effects, and detection of poisons in living organisms. The basic assumption of toxicology is that there is a relationship among the dose, the concentration at the affected site, and resulting effects.

Toxin: A colloidal poisonous substance that is a specific product of the metabolic activities of a living organism and notably toxic when introduced into living tissue.

Toxoid: A modified bacterial toxin that has been rendered nontoxic (commonly with formaldehyde) but retains the ability to stimulate the formation of antitoxins (antibodies) and thus producing an active immunity. Examples include Botulinum, tetanus, and diphtheria toxoids.

TRACEM-P: An acronym meaning seven types of harm that could be encountered at a terrorist incident: Thermal-Radioactive-Asphyxiation-Chemical-Etiological-Mechanical-Psychological.

Trade name: The commercial name or trademark by which a chemical is known. One chemical may have a variety of trade names depending on the manufacturers or distributors involved.

Transfer: The process of moving a liquid, gas, or some forms of solids from a leaking or damaged container to a secure container. Care must be taken to ensure the pump, transfer hoses and fittings, and the container selected are compatible with the hazardous materials. When hazardous substances are transferred, proper concern to electrical continuity (bonding/grounding) must be observed.

Triage: The sorting of and allocation of treatment to patients, particularly in warfare or disasters, according to a system of priorities according to the urgency of their need for care designed to maximize the numbers of survivors.

TWA: Time Weighted Average. Usually, a personal, eight-hour exposure concentration to an airborne chemical hazard.

T-2:   Trichothecene, one type of microtoxin.

2-PAM Chloride: Trade names protopam chloride, or pralidoxime chloride. 2-PAM chloride can be used in the treatment of nerve agent poisoning.

United Nations/North American Identification Number(s):   UN/NA identification numbers are four-digit numbers assigned to identify and cross-reference a hazardous materials (i.e., nitric acid, fuming = 2032; butane = 1011; white phosphorus, dry = 1381).

Universal Precautions: Methods for healthcare workers to avoid infection from blood-borne diseases first developed by the Centers for Disease Control and Prevention (CDC) in 1987. Their guidelines include use of protective gloves, masks, and eyewear when in contact with blood or body fluids.

Unstable liquid: A liquid that, in its pure state or as commercially produced, will react vigorously in some hazardous way under shock conditions (i.e., dropping), certain temperatures, or pressures.

Upper Explosive Limit: Also known as Upper Flammable Limit. Is the highest concentration (expressed in percent of vapor or gas in the air by volume) of a substance that will burn or explode when an ignition source is present. Theoretically above this limit the mixture is said to be too "rich" to support combustion. The difference between the LEL and the UEL constitutes the flammable range or explosive range of a substance. That is, if the LEL is one ppm and the UEL is five ppm, then the explosive range of the chemical is one ppm to Five ppm. (See also LEL)

Upwind: In or toward the direction from which the wind blows.

V-Agents: Persistent, highly toxic nerve agents absorbed primarily through the skin.

Vaccine: A preparation of killed or weakened infective or toxic agent used as an inoculation to produce active artificial immunity; that is, a suspension of live (usually attenutated) or inactivated microorganisms (e.g., bacteria, virus, or rickettsiae) administered to induce immunity and prevent infectious disease.

Vaccinia: An infection, primarily local and limited to the site of inoculation, induced in man by inoculation with the vaccinia (coxpox) virus in order to confer resistance to smallpox (variola). On about the third day after vaccination, papules form at the site of inoculation which become transformed into umbilicated vesicles and later pustules; they then dry up, and the scab falls off on about the twenty-first day, leaving a pitted scar; in some cases there are more or less marked constitutional disturbances.

Vapor: The gaseous form of substances that are normally in a solid or liquid state that can be changed to this state by increasing the pressure or decreasing the temperature.

Vapor Density: The weight of a given volume of vapor of gas compared to the weight of an equal volume of dry air, both measured at the same temperature and pressure.

Vapor Dispersion: Vapors from certain materials can be dispersed or moved using water spray or air movement. Reducing the concentration of the material may bring the material into its flammable range.

Vapor Pressure: Vapor pressure is a function of the substance and the temperature and is often used as a measure of how rapidly a liquid will evaporate. The pressure exerted when a solid or liquid is in equilibrium with its own vapor.

Vapor Suppression: Vapor suppression refers to the reduction or elimination of vapors emanating from the spilled or released material through the application of specially designed agents, also called blanketing. Vapor suppression can also be accomplished by the use of solid activated material to treat hazardous materials. This process results in the formation of a solid that affords easier handing but results in a hazardous solid that must be disposed of properly.

Variola: Synonym for smallpox.

Vector: A carrier, or a host that carries a pathogen from one host to another.

Venting: Venting is the process that is used to deal with liquids or liquefied compressed gases where a danger, such as an explosion or mechanical rupture of the container or vessel, is considered likely. The method of venting will depend on the nature of the hazardous material. In general, it involves the controlled release of material to reduce and contain the pressure and diminish the probability of an explosion.

Vesicant: An agent that operates on the eyes and lungs, and is capable of producing blisters.

Viremia: The presence of virus in the bloodstream.

Virus: Any of various submicroscopic pathogens consisting essentially of a core of a single nucleic acid surrounded by a protein coat, having the ability to replicate only inside a living cell.

Volatility: A measure of how readily a substance will vaporize. Volatility is directly related to vapor pressure.

Warm Zone: The area where personnel and equipment decontamination takes place.

Weapons of Mass Destruction (WMD):   Any explosive, incendiary, or poison gas, bomb, grenade, rocket having a propellant charge of more than four ounces; a missile having an explosive or incendiary charge of more than one quarter ounce; a mine or device similar to the above; poison gas; any weapon involving a disease organism; or any weapon that is designed to release radiation or radioactivity at a level dangerous to human life.

Yellow Rain: A lethal yellow substance thought to have been dispersed aerially as a warfare agent in Southeast Asia and Afghanistan; the lethal component is though to have been a trichothecene mycotoxin that was reported to produce severe nausea and vomiting, disturbances in the central nervous system. Fever, chills, and abnormally low blood pressure with a case mortality of approximately 50 percent.

Zoonosis: An infection or infestation shared in nature by humans and other animals that are the normal or usual host; a disease of humans acquired from an animal source.

APPENDIX 1:   Lethal Nerve Agent (VX)

Date: 14 September 1988
Revised: 13 August 2001
In the event of an emergency:
Telephone the SBCCOM Operations
Center's 24-hour emergency
Number: 410-436-2148

Section I: General Information

Manufacturer's Address:
U.S. Army Soldier and Biological Chemical Command (SBCCOM)
Edgewood Chemical Biological Center (ECBC)
ATTN: AMSSB-RCB-RS
Aberdeen Proving Ground, MD 21010-5424

CAS Registry Numbers:
50782-69-9, 51848-47-6, 53800-40-1, 70938-84-0

Chemical Name:
O-ethyl S-[2-(diisopropylamino)ethyl] methylphosphonothiolate

Trade Name And Synonyms:
Phosphonothioic acid, methyl-, S-(2-bis(1-methylethylamino)ethyl) 0-
ethyl ester O-ethyl S-(2-diisopropylaminoethyl) methylphosphonothiolate
S-2-Diisopropylaminoethyl O-ethyl methylphosphonothioate
S-2((2-Diisopropylamino)ethyl) O-ethyl methylphosphonothiolate
O-ethyl S-(2-diisopropylaminoethyl) methylphosphonothioate
O-ethyl S-(2-diisopropylaminoethyl) methylthiolphosphonoate
S-(2-diisopropylaminoethyl) o-ethyl methyl phosphonothiolate
Ethyl-S-dimethylaminoethyl methylphosphonothiolate VX
EA 1701
TX60

Chemical Family: Sulfonated organophosphorous compound

Formula/Chemical Structure:
$C_{11} H_{26} N O_2 P S$

```
CH3 0           CH(CH3)2

   \||        /

   P - S - CH2CH2 - N

   /          \

CH3CH2O         CH(CH3)2
```

NFPA 704 Signal:
Health - 4
Flammability - 1
Reactivity - 1
Special - 0

Section II: Ingredients

Ingredients/Name: VX

Percentage by Weight: 100%

Threshold Limit Value (TLV):   0.00001mg/m^3

Section III: Physical Data

Boiling Point @ 760 mm Hg: 568 ^{o}F (298 ^{o}C)

Vapor Pressure: 0.00063 mm Hg @ 25 ^{o}C

Vapor Density (Air = 1 STP):   9.2 @ 25 °C

Solubility (g/100g solvent):   5.0 @ 21.5 °C and 3.0 @ 25 ^{o}C in water. Soluble in organic solvents.

Specific Gravity (H$_2$0=1g/mL@25°C):   1.0113

Freezing/Melting Point (^{o}C):   -50°C

Liquid Density: 1.0083 g/mL@25°C

Volatility: 8.9 mg/m^3 @ 25 ^{o}C

Viscosity (CENTISTOKES):   9.958 @ 25 ^{o}C

Appearance and Odor: Colorless to straw colored liquid and oderless, similar in appearance to motor oil.

Section IV: Fire and Explosion Data

Flashpoint: 159 ^{o}C (McCutchan - Young)

Flammability Limits (% By Volume):   Not Available

Lower Explosive Limit: Not Applicable.

Upper Explosive Limit: Not Applicable

Extinguishing Media: Water mist, fog, foam, CO_2. Avoid using extinguishing methods that will cause splashing or spreading of the VX.

Special Fire Fighting Procedures: All persons not engaged in extinguishing the fire should be immediately evacuated from the area. Fires involving VX should be contained to prevent contamination to uncontrolled areas. When responding to a fire alarm in buildings or areas containing VX, fire fighting personnel should wear full firefighter protective clothing during chemical agent firefighting and fire rescue operations. Respiratory protection is required. Positive pressure, full face piece, NIOSH-approved self-contained breathing apparatus (SCBA) will be worn where there is danger of oxygen deficiency and when directed by the fire chief or chemical accident/incident (CAI) operations officer. In cases where firefighters are responding to a chemical accident/incident for rescue/reconnaissance purposes they will wear appropriate levels of protective clothing (See Section VIII). Do not breathe fumes. Skin contact with nerve agents must be avoided at all times. Although the fire may destroy most of the agent, care must still be taken to assure the agent or contaminated liquids do not further contaminate other areas or sewers. Contact with liquid VX or vapors can be fatal.

Unusual Fire And Explosion Hazards: None known.

Section V: Health Hazard Data

Airborne Exposure Limits (AEL): The permissible airborne exposure concentration for VX for an 8-hour workday of a 40-hour work week is an 8-hour time weighted average (TWA) of 0.00001 mg/m³. This value can be found in "DA Pam 40-8, Occupational Health Guidelines for the Evaluation and Control of Occupational Exposure to Nerve Agents GA, GB, GD, and VX." To date, however, the Occupational Safety and Health Administration (OSHA) has not promulgated a permissible exposure concentration for VX.

VX is not listed by the International Agency for Research on Cancer (IARC), American Conference of Governmental Industrial Hygienists (ACGIH), Occupational Safety and Health Administration (OSHA), or National Toxicology Program (NTP) as a carcinogen.

Effects of Overexposure: VX is a lethal cholinesterase inhibitor. Doses which are potentially life-threatening may be only slightly larger than those producing least effects. Death usually occurs within 15 minutes after absorption of a fatal dosage.

Route	Form	Effect	Type	Dosage
ocular	vapor	miosis	ECt50	<0.09 mg-min/m³
Inhalation	vapor	runny nose	ECt50	<0.09 mg-min/m³
Inhalation (15 l/min)	vapor	severe incapacitation	ICt50	<25 mg-min/m³
Inhalation (15 l/min)	vapor	death	LCt50	<30 mg-min/m³
Percutaneous	liquid	death	LD50	<10 mg/70 kg man minutes

Effective dosages for vapor are estimated for exposure durations of 2-10 minutes.

Symptoms of overexposure may occur within minutes or hours, depending upon the dose. They include: miosis (constriction of pupils) and visual effects, headaches and pressure sensation, runny nose and nasal congestion, salivation, tightness in the chest, nausea, vomiting, giddiness, anxiety, difficulty in thinking, difficulty sleeping, nightmares, muscle twitches, tremors, weakness, abdominal cramps, diarrhea, involuntary urination and defecation. With severe exposure symptoms progress to convulsions and respiratory failure.

Emergency and First Aid Procedures:

Inhalation: Hold breath until respiratory protective mask is donned. If severe signs of agent exposure appear (chest tightens, pupil constriction, incoordination, etc.), immediately administer, in rapid succession, all three Nerve Agent Antidote Kit(s), Mark I injectors (or atropine if directed by a physician). Injections using the Mark I kit injectors may be repeated at 5 to 20 minute intervals if signs and symptoms are progressing until three series of injections have been administered. No more injections will be given unless directed by medical personnel. In addition, a record will be maintained of all injections given. If breathing has stopped, give artificial respiration. Mouth-to-mouth resuscitation should be used when mask-bag or oxygen delivery'systems are not available. Do not use mouth-to-mouth resuscitation when facial contamination exists. If breathing is difficult, administer oxygen. Seek medical attention Immediately.

Eye Contact: **Immediately** flush eyes with water for 10-15 minutes, then don respiratory protective mask. Although miosis (pinpointing of the pupils) may be an early sign of agent exposure, an injection will not be administered when miosis is the only sign present. Instead, the individual will be taken **Immediately** to a medical treatment facility for observation.

Skin Contact: Don respiratory protective mask and remove contaminated clothing. Immediately wash contaminated skin with copious amounts of soap and water, 10% sodium carbonate solution, or 5% liquid household bleach. Rinse well with water to remove excess decontaminant. Administer nerve agent antidote kit, Mark I, only if local sweating and muscular twitching symptoms are observed. Seek medical attention **Immediately**.

Ingestion: Do not induce vomiting. First symptoms are likely to be gastrointestinal. **Immediately** administer Nerve Agent Antidote Kit, Mark I. Seek medical attention **Immediately**.

Section VI: Reactivity Data

Stability: Relatively stable at room temperature. Unstabilized VX of 95% purity decomposes at a rate of 5% a month at 71 °C.

Incompatibility: Negligible on brass, steel, and aluminum.

Hazardous Decomposition Products: During a basic hydrolysis of VX up to 10% of the agent is converted to diisopropylaminoethyl methylphosphonothioic acid (EA2192). Based on the concentration of EA2192 expected to be formed during hydrolysis and its toxicity (1.4 mg/kg dermal in rabbit at 24 hours in a 10/90 wt.% ethanol/water solution), a Class B poison would result. The large scale decon procedure, which uses both HTH and NaOH, destroys VX by oxidation and hydrolysis. Typically the large scale product contains 0.2 - 0.4 wt.% EA2192 at 24 hours. At pH 12, the EA2192 in the large scale product has a half-life of about 14 days. Thus, the 90-day holding period at pH 12 results in about a 64-fold reduction of EA2192 (six half-lives). This holding period is sufficient to reduce the toxicity of the product below that of a Class B poison. Other less toxic products are ethyl methylphosphonic acid, methylphosphinic acid, diisopropyaminoethyl mercaptan, diethyl methylphosphonate, and ethanol. The small scale decontamination procedure uses sufficient HTH to oxidize all VX thus no EA2192 is formed.

Hazardous Polymerization: Does not occur.

Section VII: Spill, Leak, And Disposal Procedures

Steps To Be Taken In Case Material Is Released Or Spilled: If leaks or spills of VX occur, only personnel in full protective clothing (See Section VIII) will remain in the area. In case of personnel contamination see Section V for emergency and first aid instructions.

Recommended Field Procedures (For Quantities Greater Than 50 Grams): *Note:* These procedures can only be used with the approval of the Risk Manager or qualified safety professionals). Spills must be contained by covering with vermiculite, diatomaceous earth, clay or fine sand. An alcoholic HTH mixture is prepared by adding 100 milliliters of denatured ethanol to a 900-milliliter slurry of 10% HTH in water. This mixture should be made just before use since the HTH can react with the ethanol. Fourteen grams of alcoholic HTH solution are used for each gram of VX. Agitate the decontamination mixture as the VX is added. Continue the agitation for a minimum of one hour. This reaction is reasonably exothermic and evolves substantial off gassing. The evolved reaction gases should be routed through a decontaminate filled scrubber before release through filtration systems. After completion of the one hour minimum agitation, 10% sodium hydroxide is added in a quantity equal to that necessary to assure that a pH of 12.5 is maintained for a period not less than 24 hours. Hold the material at a pH between 10 and 12 for a period not less than 90 days to ensure that a hazardous intermediate material is not formed (See Section VI). Scoop up all material and place in a DOT approved container. Cover the contents of the with decontaminating solution as above. After sealing the exterior, decontaminate and label according to EPA and DOT regulations. All leaking containers will be over packed with sorbent (e.g., vermiculite) placed between the interior and exterior containers. Decontaminate and label according to EPA and DOT regulations. Dispose of decontaminant according to Federal, state, and local laws. Conduct general area monitoring to confirm that the atmospheric concentrations do not exceed the airborne exposure limits (See Sections II and VIII).

If the alcoholic HTH mixture is not available, then the following decontaminants may be used instead and are listed in the order of preference: Decontaminating Agent (DS2), Supertropical Bleach Slurry (STB), and Sodium Hypochlorite.

Recommended Laboratory Procedures (For Quantities Less Than 50 Grams): If the active chlorine of the Calcium Hypochlorite (HTH) is at least 55%, then 80 grams of a 10% slurry are required for each gram of VX. Proportionally more HTH is required if the chlorine activity of the HTH is lower than 55%. The mixture is agitated as the VX is added and the agitation is maintained for a minimum of one hour. If phasing of the VX/decon solution continues after 5 minutes, an amount of denatured ethanol equal to a 10 wt.% of the total agent/decon will be added to help miscibility. Place all material in a DOT approved container. Cover the contents with decontaminating solution as above. After sealing, decontaminate the exterior of the container and label according to EPA and DOT regulations. All leaking containers will be over packed with sorbent placed between the interior and exterior containers. Decontaminate and label according to EPA and DOT regulations. Dispose of according to Federal, State, and local laws. Conduct general area monitoring to confirm that the atmospheric concentrations do not exceed the airborne exposure limits (See Sections II and VIII).

Note: Ethanol should be reduced to prevent the formation of a hazardous waste.

Upon completion of the one hour agitation the decon mixture will be adjusted to a pH between 10 and 11. Conduct general area monitoring to confirm that the atmospheric concentrations do not exceed the airborne exposure limits (See Sections II and VIII).

Waste Disposal Method: Open pit burning or burying of VX or items containing or contaminated with VX in any quantity is prohibited. The detoxified VX(using procedures above) can be thermally destroyed by in a EPA approved incinerator according to appropriate provisions of Federal, State, or local Resource Conservation and Recovery Act (RCRA) regulations.

Note: Some decontaminate solutions are hazardous waste according to RCRA regulations and must be disposed of according to those regulations.

Section VIII: Special Protection Information

Respiratory Protection

Concentration	Respiratory Protective Equipment
<0.00001 mg/m³	A full face piece, chemical canister air-purifying protective mask will be on hand for escape. M40-series masks are acceptable for this purpose. Other masks certified as equivalent may be used.
>0.00001 or = 0.02 mg/m³	A NIOSH/MSHA approved pressure demand full face piece SCBA or supplied air respirators with escape air cylinder may be used. Alternatively, a full face piece, chemical canister air-purifying protective mask is acceptable for this purpose (See DA Pam 385-61 for determination of appropriate level.
>0.02 or unknown	NIOSH/MSHA approved pressure demand full face piece SCBA suitable for use in high agent concentrations with protective ensemble. (See DA Pam 385-61 for examples)

Ventilation: Local exhaust: Mandatory. Must be filtered or scrubbed to limit exit concentrations to <0.00001 mg/m³. Air emissions will meet local, state, and federal regulations.

Special: Chemical laboratory hoods will have an average inward face velocity of 100 linear feet per minute (lfpm) ±20% with the velocity at any point not deviating from the average face velocity by more than 20%. Existing laboratory hoods will have an inward face velocity of 150 lfpm ±20%. Laboratory hoods will be located such that cross-drafts do not exceed 20% of the inward face velocity. A visual performance test using smoke-producing devices will be performed in assessing the ability of the hood to contain agent VX.

Other: Recirculation or exhaust air from chemical areas is prohibited. No connection between chemical areas and other areas through ventilation system is permitted. Emergency backup power is necessary. Hoods should be tested at least semiannually or after modification or maintenance operations. Operations should be performed 20 centimeters inside hood face.

Protective Gloves: Butyl Rubber Glove M3 and M4 Norton, Chemical Protective Glove Set.

Eye Protection: At a minimum chemical goggles will be worn. For splash hazards use goggles and face shield.

Other Protective Equipment: For laboratory operations, wear lab coats, gloves and have mask readily accessible. In addition, daily clean smocks, foot covers, and head covers will be required when handling contaminated lab animals.

Monitoring: Available monitoring equipment for agent VX is the M8/M9 detector paper, detector ticket, M256/M256A1 kits, bubbler, Depot Area Air Monitoring System (DAAMS), Automated Continuous Air Monitoring System (ACAMS), Real-Time Monitor (RTM), Demilitarization Chemical Agent Concentrator (DCAC), M8/M43, M8A1/M43A1, CAM-M1, Hydrogen Flame Photometric Emission Detector (HYFED), the Miniature Chemical Agent Monitor (MINICAM), and the Real Time Analytical Platform (RTAP). Real-time, low-level monitors (with alarm) are required for VX operations. In their absence, an Immediately Dangerous to Life and Health (IDLH) atmosphere must be presumed. Laboratory operations conducted in appropriately maintained and alarmed engineering controls require only periodic low-level monitoring.

Section IX: Special Precautions

Precautions To Be Taken In Handling And Storing: When handling agents, the buddy system will be incorporated. No smoking, eating, or drinking in areas containing agents is permitted. Containers should be periodically inspected for leaks, (either visually or using a detector kit). Stringent control over all personnel practices must be exercised. Decontaminating equipment will be conveniently located. Exits must be designed to permit rapid evacuation. Chemical showers, eyewash stations, and personal cleanliness facilities must be provided. Wash hands before meals and shower thoroughly with special attention given to hair, face, neck, and hands using plenty of soap and water before leaving at the end of the work day.

Other Precautions: Agent containers will be stored in a single containment system within a laboratory hood or in double containment system.

For additional information see "AR 385-61, The Army Toxic Chemical Agent Safety Program," "DA Pam 385-61, Toxic Chemical Agent Safety Standards," and "DA Pam 40-8, Occupational Health Guidelines for the Evaluation and Control of Occupational Exposure to Nerve Agents GA, GB, GD, and VX."

Section X: Transportation Data

Note: Forbidden for transport other than via military (Technical Escort Unit) transport according to 49 CFR 172.

Proper Shipping Name: Toxic liquids, organic, n.o.s.

DOT Hazard Class: 6.1, Packing Group I, Hazard Zone A.

DOT Hazard Class: 6.1, Packing Group I, Hazard Zone A.

DOT Label: Poison.

DOT Marking: Toxic liquids, organic, n.o.s. (O-ethel S-(2-diisopropylaminoethyl)meth ylphosphonothiolate) UN 2810, Inhalation Hazard.

DOT Placard: Poison.

Emergency Accident Precautions And Procedures: See Sections IV, VII, and VIII.

Precautions to be taken in transportation: Motor vehicles will be placarded regardless of quantity. Drivers will be given full information regarding shipment and conditions in case of an emergency. AR 50-6 deals specifically with the shipment of chemical agents. Shipment of agents will be escorted in accordance with AR 740-32.

The Edgewood Chemical Biological Center (ECBC), Department of the Army believes that the data contained herein are actual and are the results of the tests conducted by ECBC experts. The data are not to be taken as a warranty or representation for which the Department of the Army or ECBC assumes legal responsibility. They are offered solely for consideration. Any use of this data and information contained in this MSDS must be determined by the user to be in accordance with applicable Federal, State, and local laws and regulations.

HYDROGEN CYANIDE, LIQUEFIED ICSC: 0492
HYDROGEN CYANIDE, LIQUEFIED
Hydrocyanic acid
Prussic acid
(liquefied)
HCN
Molecular mass: 27.03

CAS # 74-90-8
RTECS # MW6825000
ICSC # 0492
UN # 1051
EC # 006-006-00-X

TYPES OF
HAZARD/
EXPOSURE ACUTE HAZARDS/
SYMPTOMS PREVENTION FIRST AID/
FIRE FIGHTING
FIRE Extremely flammable. Gives off irritating or toxic fumes (or gases) in a fire.
NO open flames, NO sparks, and NO smoking.
Shut off supply; if not possible and no risk to surroundings, let the fire burn itself out; in other cases extinguish with powder, water spray, foam, carbon dioxide.

EXPLOSION Gas/air mixtures are explosive.
Closed system, ventilation, explosion-proof electrical equipment and lighting.
In case of fire: keep cylinder cool by spraying with water. Combat fire from a sheltered position.

EXPOSURE
AVOID ALL CONTACT!
IN ALL CASES CONSULT A DOCTOR!

INHALATION
Confusion. Drowsiness. Headache. Nausea. Shortness of breath. Unconsciousness. Collapse. Death.
Ventilation, local exhaust, or breathing protection.
Fresh air, rest. Half-upright position. Avoid mouth to mouth resuscitation, administer oxygen by trained personnel. Refer for medical attention. See Notes.

SKIN
MAY BE ABSORBED! (Further see Inhalation).
Protective gloves. Protective clothing.

Rinse skin with plenty of water or shower. Refer for medical attention. Wear protective gloves when administering first aid.

EYES
VAPOUR WILL BE ABSORBED! Redness (see Inhalation).
Safety goggles, face shield, or eye protection in combination with breathing protection.
First rinse with plenty of water for several minutes (remove contact lenses if easily possible), then take to a doctor.

INGESTION
Burning sensation (further see Inhalation).
Do not eat, drink, or smoke during work.
Rinse mouth. See inhalation. Do NOT induce vomiting. Refer for medical attention. See Notes.

SPILLAGE DISPOSAL STORAGE PACKAGING & LABELLING
Evacuate danger area immediately! Consult an expert! Ventilation. Absorb remaining liquid in sand or inert absorbent and remove to safe place. Do NOT wash away into sewer. NEVER direct water jet on liquid. Prevent from entering confined spaces. Do NOT let this chemical enter the environment (extra personal protection: gas-tight chemical protection suit including self-contained breathing apparatus).
Fireproof. Separated from food and feedstuffs. Cool. Store only if stabilized.

F+ symbol
T+ symbol
R: 12-26
S: (1/2-)7/9-16-36/37-38-45
UN Hazard Class: 6.1
UN Subsidiary Risks: 3
UN Packing Group: I
Marine pollutant.

ICSC: 0492 Prepared in the context of cooperation between the International Programme on Chemical Safety & the Commission of the European Communities © IPCS CEC 1993

International Chemical Safety Cards
HYDROGEN CYANIDE, LIQUEFIED ICSC: 0492

PHYSICAL STATE; APPEARANCE:
COLOURLESS GAS OR LIQUID , WITH CHARACTERISTIC ODOUR.

PHYSICAL DANGERS:
The gas mixes well with air, explosive mixtures are easily formed.

CHEMICAL DANGERS:
The substance may polymerize due to warming, under the influence of base(s), over 2% water, or temperatures above 184°C, or if not chemically stabilized, with fire or explosion

hazard. On combustion, forms toxic and corrosive gases, including nitrogen oxides. The solution in water is a weak acid. Reacts violently with oxidants, hydrogen chloride in alcoholic mixtures, causing fire and explosion hazard.

OCCUPATIONAL EXPOSURE LIMITS (OELs):
TLV: 4.7 ppm; 5 mg/m3 (ceiling values) (skin) (ACGIH 1997).

ROUTES OF EXPOSURE:
The substance can be absorbed into the body by inhalation, through the skin and by ingestion.

INHALATION RISK:
A harmful contamination of the air can be reached very quickly on evaporation of this substance 20°C.

EFFECTS OF SHORT-TERM EXPOSURE:
The substance irritates the eyes and the respiratory tract. The substance may cause effects on the central nervous system , resulting in impaired respiratory and circulatory functions. Exposure may result in death. See Notes.

EFFECTS OF LONG-TERM OR REPEATED EXPOSURE:

PHYSICAL PROPERTIES
Boiling point: 26°C
Melting point: -13°C
Relative density (water = 1): 0.69
Solubility in water: miscible
Vapour pressure, kPa at 20°C: 82.6
Relative vapour density (air = 1): 0.94
Flash point: -18°C c.c.
Auto-ignition temperature: 538°C
Explosive limits, vol% in air: 5.6-40.0
Octanol/water partition coefficient as log Pow: 0.35

ENVIRONMENTAL DATA
The substance is very toxic to aquatic organisms.

NOTES
Mineral acids are commonly used as stabilizers. The occupational exposure limit value should not be exceeded during any part of the working exposure. Specific treatment is necessary in case of poisoning with this substance; the appropriate means with instructions must be available. The odour warning when the exposure limit value is exceeded is insufficient. The recommendations on this Card also apply to hydrogen cyanide, stabilized, absorbed in a porous inert material. Another UN number: 1614, hydrogen cyanide stabilized, absorbed in a porous inert material. Transport Emergency Card: TEC (R)-61G60
NFPA Code: H4; F4; R2;

ADDITIONAL INFORMATION

ICSC: 0492 HYDROGEN CYANIDE, LIQUEFIED
© IPCS, CEC, 1993

IMPORTANT LEGAL NOTICE: Neither the CEC or the IPCS nor any person acting on behalf of the CEC or the IPCS is responsible for the use which might be made of this information. This card contains the collective views of the IPCS Peer Review Committee and may not reflect in all cases all the detailed requirements included in national legislation on the subject. The user should verify compliance of the cards with the relevant legislation in the country of use.

In general all information presented in these pages and all items available for download are for public use. However, you may encounter some pages that require a login password and id. If this is the case you may assume that information presented and items available for download therein are for your authorized access only and not for redistribution by you unless you are otherwise informed.

This page last reviewed: September 25, 2001
Public Inquiries
English (888) 246-2675
Español (888) 246-2857
Mon-Fri 8am-11pm EST
Sat-Sun 10am-8pm EST

Centers for Disease Control and Prevention
1600 Clifton Rd.
Atlanta, GA 30333
U.S.A
(404) 639-3311

- ☐ What should I know about smallpox?

- ☐ Are we expecting a smallpox attack?

- ☐ Is there an immediate smallpox threat?

- ☐ If I am concerned about a smallpox attack, can I go to my doctor and request the smallpox vaccine?

- ☐ Are there plans to manufacture more vaccine in case of a bioterrorism attack using smallpox?

- ☐ If someone comes in contact with smallpox, how long does it take to show symptoms?

- ☐ Is smallpox fatal?

- ☐ How is smallpox spread?

- ☐ If someone is exposed to smallpox, is it too late to get a vaccination?

- ☐ If people got the vaccination in the past when it was used routinely, will they be immune?

- ☐ How many people have not had the vaccination?

- ☐ Is it possible for people to get smallpox from the vaccination?

- ☐ How safe is the smallpox vaccine?

- ☐ Is there any treatment for smallpox?

- ☐ Is there a test to indicate if smallpox is in the environment like there is for anthrax?

- ☐ If smallpox is discovered or released in a building, or if a person develops symptoms in a building, how can that area be decontaminated?

- ☐ What should people do if they suspect a patient has smallpox or suspect that smallpox has been released in their area?

- ☐ How can we stop the spread of smallpox after someone comes down with it?

- ☐ What is the smallpox vaccine, and is it still required?

- ☐ Should I get vaccinated against smallpox?

- ☐ Many vaccinations are required, why don't people get the smallpox vaccine?

- ☐ Is the smallpox virus, variola, that causes smallpox still around?

- ☐ Are some people still receiving the smallpox vaccination today?

- ☐ What are the risks of the smallpox vaccines? Are there side effects?

- ☐ Do other countries have smallpox vaccine stores?

☐ What are the symptoms of smallpox?

☐ What is the HHS/CDC smallpox plan? When will it be released? Does the plan address mass vaccination?

☐ Does the death rate differ for those who have been vaccinated from those who have not?

☐ Should you get the smallpox vaccine if you're immuno-compromised?

☐ How long does a smallpox vaccination last?

☐ Is every American going to be vaccinated for smallpox?

☐ How many people would have to get smallpox before it is considered an outbreak?

☐ What should be done if there is a smallpox outbreak?

☐ What should be done to isolate someone with smallpox?

☐ Will ciprofloxacin protect me against smallpox?

☐ If the decision is made that everyone needs to be vaccinated, how will this occur and who will pay for it?

☐ Is there a test to indicate whether smallpox is in the environment like there is for anthrax?

☐ When will additional smallpox vaccine be ready?

☐ Is there a test to determine whether or not you have any protection from smallpox from vaccinations received years ago?

☐ Is smallpox contagious before the smallpox symptoms show?

☐ What is the difference between a "live vaccine" and a "killed vaccine"?

☐ If people got the vaccination when it was available in the past, will they be immune?

☐ If someone had smallpox once, are they immune? Would they need the vaccine?

☐ Is it possible for someone to receive the smallpox vaccine and have it not "take," i.e., work? How does someone who has been vaccinated for smallpox know that he or she is immune or that the vaccine has "taken"?

☐ Are diluted doses of smallpox vaccine as effective?

☐ Would a more diluted smallpox vaccine be an effective booster shot?

☐ How soon will results be available for dilution studies on the smallpox vaccine?

☐ What is the smallpox vaccine made of?

☐ Is there a risk of accidental exposure to persons involved in the production of smallpox vaccine?

☐ Why are health responders being vaccinated against smallpox, but the general public is not?

☐ Who is asking for volunteers to receive the smallpox vaccine? Who should people call if interested in volunteering?

What Should I Know about Smallpox?

Vaccination is not recommended, and the vaccine is not available to health providers or the public. In the absence of a confirmed case of smallpox anywhere in the world, there is no need to be vaccinated against smallpox. There also can be severe side effects to the smallpox vaccine, which is another reason we do not recommend vaccination. In the event of an outbreak, the CDC has clear guidelines to swiftly provide vaccine to people exposed to this disease. The vaccine is securely stored for use in the case of an outbreak. In addition, Secretary of Health and Human Services Tommy Thompson recently announced plans to accelerate production of a new smallpox vaccine.

☐ Are we expecting a smallpox attack?
We are not expecting a smallpox attack, but the recent events that include the use of biological agents as weapons have heightened our awareness of the possibility of such an attack.

☐ Is there an immediate smallpox threat?
At this time we have no information that suggests an imminent smallpox threat.

☐ If I am concerned about a smallpox attack, can I go to my doctor and request the smallpox vaccine?
The last naturally acquired case of smallpox occurred in 1977. The last cases of smallpox, from laboratory exposure, occurred in 1978. In the United States, routine vaccination against smallpox ended in 1972. Since the vaccine is no longer recommended, the vaccine is not available. The CDC maintains an emergency supply of vaccine that can be released if necessary, since post-exposure vaccination is effective.

☐ Are there plans to manufacture more vaccine in case of a bioterrorism attack using smallpox?
Yes. In 2000, CDC awarded a contract to a vaccine manufacturer to produce additional doses of smallpox vaccine.

☐ If someone comes in contact with smallpox, how long does it take to show symptoms?
The incubation period is about 12 days (range: 7 to 17 days) following exposure. Initial symptoms include high fever, fatigue, and head and back aches. A characteristic rash, most prominent on the face, arms, and legs, follows in 2-3 days. The rash starts with flat red lesions that evolve at the same rate. Lesions become pus-filled after a few days and then begin to crust early in the second week. Scabs develop and then separate and fall off after about 3-4 weeks.

☐ Is smallpox fatal?
The majority of patients with smallpox recover, but death may occur in up to 30% of cases.

☐ How is smallpox spread?

In the majority of cases, smallpox is spread from one person to another by infected saliva droplets that expose a susceptible person having face-to-face contact with the ill person. People with smallpox are most infectious during the first week of illness, because that is when the largest amount of virus is present in saliva. However, some risk of transmission lasts until all scabs have fallen off.

Contaminated clothing or bed linen could also spread the virus. Special precautions need to be taken to ensure that all bedding and clothing of patients are cleaned appropriately with bleach and hot water. Disinfectants such as bleach and quaternary ammonia can be used for cleaning contaminated surfaces.

☐ If someone is exposed to smallpox, is it too late to get a vaccination?

If the vaccine is given within 4 days after exposure to smallpox, it can lessen the severity of illness or even prevent it.

☐ If people got the vaccination in the past when it was used routinely, will they be immune?

Not necessarily. Routine vaccination against smallpox ended in 1972. The level of immunity, if any, among persons who were vaccinated before 1972 is uncertain; therefore, these persons are assumed to be susceptible. For those who were vaccinated, it is not known how long immunity lasts. Most estimates suggest immunity from the vaccination lasts 3 to 5 years. This means that nearly the entire U.S. population has partial immunity at best. Immunity can be boosted effectively with a single revaccination. Prior infection with the disease grants lifelong immunity.

☐ How many people have not had the vaccination?

Approximately half of the U.S. population has never been vaccinated.

☐ Is it possible for people to get smallpox from the vaccination?

No, smallpox vaccine does not contain smallpox virus but another live virus called vaccinia virus. Since this virus is related to smallpox virus, vaccination with vaccina provides immunity against infection from smallpox virus.

☐ How safe is the smallpox vaccine?

Smallpox vaccine is considered very safe. However, some people with pre-existing conditions such as eczema or immune system disorders have a higher risk for having complications from the vaccine. Adverse reactions have been known to occur that range from mild rashes to rare fatal encephalitis and disseminated vaccina. Smallpox vaccine should not be administered to persons with a history or presence of eczema or other skin conditions, pregnant women, or persons with immunodeficiency diseases and among those with suppressed immune systems as occurs with leukemia, lymphoma, generalized malignancy, or solid organ transplantation.

☐ Is there any treatment for smallpox?

There is no proven treatment for smallpox, but research to evaluate new antiviral agents is ongoing. Patients with smallpox can benefit from supportive therapy (e.g., intravenous fluids, medicine to control fever or pain) and antibiotics for any secondary bacterial infections that may occur.

☐ Is there a test to indicate if smallpox is in the environment like there is for anthrax?

Various agencies are currently validating tests designed to test for the smallpox virus in the environment.

☐ If smallpox is discovered or released in a building, or if a person develops symptoms in a building, how can that area be decontaminated?
The smallpox virus is fragile and in the event of an aerosol release of smallpox, all viruses will be inactivated or dissipated within 1-2 days. Buildings exposed to the initial aerosol release of the virus do not need to be decontaminated. By the time the first cases are identified, typically 2 weeks after the release, the virus in the building will be gone. Infected patients, however, will be capable of spreading the virus and possibly contaminating surfaces while they are sick. Therefore, standard hospital grade disinfectants such as quaternary ammonias are effective in killing the virus on surfaces should be used for disinfecting hospitalized patients' rooms or other contaminated surfaces. Although less desirable because it can damage equipment and furniture, hypochlorite (bleach) is an acceptable alternative. In the hospital setting, patients' linens should be autoclaved or washed in hot water with bleach added. Infectious waste should be placed in biohazard bags and autoclaved before incineration.

☐ What should people do if they suspect a patient has smallpox or suspect that smallpox has been released in their area?
Report suspected cases of smallpox or suspected intentional release of smallpox to your local health department. The local health department is responsible for notifying the state health department, the FBI, and local law enforcement. The state health department will notify the CDC.

☐ How can we stop the spread of smallpox after someone comes down with it?
Symptomatic patients with suspected or confirmed smallpox are capable of spreading the virus. Patients should be placed in medical isolation so that they will not continue to spread the virus. In addition, people who have come into close contact with smallpox patients should be vaccinated immediately and closely watched for symptoms of smallpox. Vaccine and isolation are the strategies for stopping the spread of smallpox.

☐ What is the smallpox vaccine, and is it still required?
The vaccine against smallpox is made with a virus related to smallpox virus called vaccinia virus. It is not made with smallpox virus called variola. The vaccine is a highly effective immunizing agent against smallpox infection. It was successfully used to eradicate smallpox from the human population.

☐ Should I get vaccinated against smallpox?
Vaccination is not recommended at this time, and the vaccine is not available to healthcare providers or to the public. In the absence of a confirmed case of smallpox anywhere in the world, there is no need to be vaccinated against smallpox. CDC has clear guidelines for providing vaccine to people exposed to smallpox if a case did occur. Healthcare workers and close contacts of the person or persons with confirmed smallpox disease would receive the vaccine. Through CDC, healthcare workers would have access to the vaccine if it were needed to prevent the disease.

☐ Many vaccinations are required, why don't people get the smallpox vaccine?

The last known naturally occurring case of smallpox occurred in Somalia in 1977. In May 1980, the World Health Assembly certified that the world was free of naturally occurring smallpox. By the 1960s, because of vaccination programs and quarantine regulations, the risk for importation of smallpox into the United States had been reduced. As a result, recommendations for routine smallpox vaccination were rescinded in 1971. In 1976, the recommendation for routine smallpox vaccination of health-care workers was also discontinued. In 1982, the only active licensed producer of vaccinia vaccine in the United States discontinued production for general use, and in 1983, distribution to the civilian population was discontinued. All military personnel continued to be vaccinated, but that practice ceased in 1990. Since January 1982, smallpox vaccination has not been required for international travelers, and International Certificates of Vaccination forms no longer include a space to record smallpox vaccination.

☐ Is the smallpox virus, variola, that causes smallpox still around?
Although smallpox disease has been eradicated, two countries still keep smallpox virus (variola) stocks. Two laboratories hold stocks of smallpox virus (variola). These are the WHO Collaborating Centres in Atlanta, USA and Koltsovo, Russian Federation.

☐ Are some people still receiving the smallpox vaccination today?
Yes. Vaccinia vaccine is recommended for laboratory workers who directly handle cultures, animals contaminated or infected with, nonhighly attenuated vaccinia virus, recombinant vaccinia viruses derived from nonhighly attenuated vaccinia strains, or other orthopoxviruses that infect humans. These would include monkeypox, cowpox, vaccinia, and variola. Other health-care workers, such as physicians and nurses whose contact with nonhighly attenuated vaccinia viruses is limited to contaminated materials such as medical dressings but who adhere to appropriate infection control measures, are at lower risk for accidental infection than laboratory workers. However, because a theoretical risk for infection exists, vaccination can be offered to this group. Vaccination is not recommended for people who do not directly handle nonhighly attenuated virus cultures or materials or who do not work with animals contaminated or infected with these viruses.

☐ What are the risks of the smallpox vaccines? Are there side effects?
Side effects from successful vaccination, particularly in those receiving their first dose of vaccine, include tenderness, redness, swelling, and a lesion at the vaccination site. In addition, the vaccination may cause fever for a few days and the lymph nodes in the vaccinated arm may become enlarged and tender. These symptoms are more common in those receiving their first dose of vaccine (15%–20% of those vaccinated) than in those being re-vaccinated (5%–10% of those vaccinated). The overall risks of serious complications of smallpox vaccination are low, and occur more frequently in those receiving their first dose of vaccine, and among young children. The most frequent serious complications are encephalitis (brain inflammation), progressive destruction of skin and other tissues at the vaccination site, and severe and destructive infection of skin affected already by eczema or other chronic skin disorder. The complication of encephalitis occurs in about one in 300,000 doses in children and one in 200,000 doses in adults. The vaccine is not recommended for those who have abnormalities

of their immune system because the complication of progressive destruction of skin and other tissues at the vaccination site has occurred only among recipients in this group. The vaccine is also not recommended for recipients who have eczema or other chronic skin disorders because the complication of severe and destructive infection of skin has occurred only among recipients in this group.

☐ Do other countries have smallpox vaccine stores?
In addition to the stock of smallpox vaccine in the US, an additional 50–100 million doses are estimated to exist worldwide. Many countries still hold smallpox vaccine (vaccinia) stocks. WHO recommends that countries that still have stocks of smallpox vaccine (vaccinia) maintain these stocks. This recommendation has been made for two reasons. Firstly, small amounts of vaccine are still needed to vaccinate laboratory personnel handling vaccinia virus and other members of this virus family. Some of these viruses are found in nature and cause illness among animals, and some are used in research to make new, safer vaccines against a variety of infectious diseases. Secondly, smallpox vaccine, vaccinia, will also be needed in case of a deliberate or accidental release of smallpox virus, variola.

☐ What are the symptoms of smallpox?
Variola virus causes smallpox. The incubation period is about 12 days with a range of 7 to 17 days following exposure. Initial symptoms include high fever, fatigue, and head and back aches. A characteristic rash, most prominent on the face, arms, and legs, follows in 2–3 days. The rash starts with flat red lesions that evolve at the same rate. Lesions become pus-filled and begin to crust early in the second week. Scabs develop and then separate and fall off after about 3–4 weeks. Most patients with smallpox recover, but death occurs in up to 30% of cases.

☐ What is the HHS/CDC smallpox plan? When will it be released? Does the plan address mass vaccination?
CDC has been preparing for some time for the remote possibility of an outbreak of smallpox as an act of terror. That process has intensified since September 11, 2001. Although we are planning for this possibility to protect public health, we have no indication that there is an imminent threat. As part of the ongoing effort to increase awareness, CDC has distributed a draft of a smallpox preparedness plan to reviewers for comment. It will then be reviewed by state health departments, which will participate in its implementation. However, if needed, it could be put in operation immediately.

☐ Does the death rate differ for those who have been vaccinated from those who have not?
For people exposed to smallpox, the vaccine can lessen the severity or even prevent illness if it is given within 4 days after exposure. Vaccine administered after exposure has been shown to provide significant protection against death from smallpox

☐ Should you get the smallpox vaccine if you're immuno-compromised?
No, not unless there is a smallpox outbreak. Vaccinations could cause deaths in people with weakened immune systems: those undergoing chemotherapy, organ transplant patients, and those with AIDS. There is no need to take that risk until there is

evidence of an outbreak. But the U.S. should have the vaccine ready if needed as an "insurance policy."

☐ How long does a smallpox vaccination last?
It is not known exactly how long the immunity from the smallpox vaccination will last. Most estimates suggest that immunity lasts from three to five years.

☐ Is every American going to be vaccinated for smallpox?
If there is an outbreak of smallpox, vaccinations of people may only be needed in the area around the cases of smallpox to contain the spread. If health officials are not able to contain the outbreak, vaccination of a wider group of people may be required. U.S. health officials are increasing the stock of smallpox vaccine to be ready to vaccinate as needed.

☐ How many people would have to get smallpox before it is considered an outbreak?
One suspected case of smallpox is considered a public health emergency. Smallpox surveillance in the United States includes detecting a suspected case or cases, making a definitive diagnosis with rapid laboratory confirmation at CDC, and preventing further smallpox transmission. A suspected smallpox case should be reported immediately by telephone to state or local health officials. They should immediately obtain advice regarding isolation of the patient or patients, and on laboratory specimen collection. State or local health officials should notify CDC immediately at (404) 639-2184 or (404) 639-0385 if a suspected case of smallpox is reported.

☐ What should be done if there is a smallpox outbreak?
If an outbreak occurs, the first step would be to properly isolate those with the disease. Health officials should be diligent regarding use of adequate isolation facilities and precautions. If they are at all uncertain about correct procedures for isolating patients, they should contact the state or local health department or CDC. All the contacts of the patients should be vaccinated as soon as possible. In the event that there are many cases in a city vaccinations may be given to the entire population of that city.

☐ What should be done to isolate someone with smallpox?
Isolation of confirmed or suspected smallpox patients will be necessary to limit the potential exposure of nonvaccinated and, therefore, nonimmune persons.

Airborne precautions using correct ventilation including negative air-pressure rooms with high-efficiency particulate air filtration should be initiated for hospitalized confirmed or suspected smallpox patients, unless the entire facility has been restricted to smallpox patients and recently vaccinated persons.

Although personnel who have been vaccinated recently and who have a demonstrated immune response should be fully protected against infection with smallpox virus, they should continue to observe standard contact precautions including using protective clothing and shoe covers when in contact with smallpox patients or contaminated materials to prevent inadvertent spread of variola virus to susceptible persons and potential self-contact with other infectious agents.

Personnel should remove and correctly dispose of all protective clothing before contact with nonvaccinated people.

Reusable bedding and clothing can be autoclaved or laundered in hot water with bleach to inactivate the virus.

Laundry handlers should be vaccinated before handling contaminated materials.

Nonhospital isolation of confirmed or suspected smallpox patients should be of a sufficient degree to prevent the spread of disease to nonimmune persons during the time the patient is considered potentially infectious, which includes from the onset of symptoms until all scabs have separated.

Private residences or other nonhospital facilities that are used to isolate confirmed or suspected smallpox patients should have nonshared ventilation, heating, and air-conditioning systems. Access to those facilities should be limited to recently vaccinated persons with a demonstrated immune response. If suspected smallpox patients are placed in the same isolation facility, they should be vaccinated to guard against accidental exposure caused by misclassification as someone with smallpox.

In addition to isolation of infectious smallpox patients, careful surveillance of contacts during their potential incubation period is required.

Transmission of smallpox virus rarely occurs before the appearance of the rash that develops 2–4 days after the initial fever.

If a vaccinated or unvaccinated contact experiences a fever >101° F (38° C) during the 17-day period after his or her last exposure to a smallpox patient, the contact should be isolated immediately to prevent contact with nonvaccinated or nonimmune persons until smallpox can be ruled out by clinical or laboratory examination.

☐ Will ciprofloxacin protect me against smallpox?
No. Because smallpox is a virus, antibiotics such as ciprofloxacin will not fight the smallpox infection. The only cure is to get the vaccine within a few days of exposure to the virus.

☐ If the decision is made that everyone needs to be vaccinated, how will this occur and who will pay for it?
There will be a systematic administration of the vaccine that will be paid for by the United States government.

☐ Is there a test to indicate whether smallpox is in the environment like there is for anthrax?
Scientists believe that if smallpox virus is released as an aerosol and not exposed to UV light, it may persist for as long as 24 hours or somewhat longer under favorable conditions. However, by the time patients become ill, which takes about 10 days to 12 days after infection with the virus, and it has been determined that an aerosol release of smallpox virus had occurred, there would be no viable smallpox virus left in the environment to detect. Trying to detect the virus everywhere at all times without any indications of any illness in people would not be feasible.

The occurrence of smallpox infection among people who handled laundry from infected patients is well documented, and it is believed that virus in such material remains viable for extended periods. In this situation, the virus could be detected in the environment, but investigators would already know it was there because of the presence of the associated illness.

In studies conducted during the smallpox eradication program and by surveillance for cases in newly smallpox-free areas it was reasoned that if the virus were able to persist in nature and infect humans, there would be cases occurring for which no source could be identified. Cases of this type were not observed. When cases were found, there were human cases in people who had direct contact with another infected person.

☐ When will additional smallpox vaccine be ready?
President Bush's recent budget request (October 17, 2001) proposed spending $509 million to speed the development and acquisition of smallpox vaccine in order to reach any American potentially exposed to the virus in a potential bioterrorist attack. Currently, more than 15 million doses of smallpox are available. The additional funds will allow the department to stockpile as much vaccine as needed to protect the nation in the event of an outbreak of smallpox.

☐ Is there a test to determine whether or not you have any protection from smallpox from vaccinations received years ago?
No. Routine vaccination against smallpox ended in 1972. The level of immunity, if any among persons who were vaccinated before 1972 is uncertain; therefore, these persons are assumed to be susceptible. For those were vaccinated, it is not known how long immunity lasts. Most estimates suggest that vaccination protection lasts from 3 to 5 years. Immunity can be boosted effectively with a single revaccination. Prior infection with the disease grants lifelong immunity.

☐ Is smallpox contagious before the smallpox symptoms show?
Smallpox patients are most infectious during the first week of the rash. At this time, patients have sores in their mouths. These sores release smallpox virus into the patient's saliva. The virus may spread through the air when the infected person breathes, talks, laughs, or coughs. A patient is no longer infectious after all scabs have fallen off, usually about 3 or 4 weeks after the start of the rash.
Symptoms of smallpox begin 12-14 days (range 7-17 days) after exposure. The disease starts with 2-3 days of high fever and extreme tiredness with severe headache and backache. The rash usually begins about 2-4 days after the fever and, at first, is a few red spots on the face and forearms and in the mouth. It then spreads to the trunk and legs. Sores might form on the palms and soles as well. By the fourth day of rash, the spots have turned to blisters (vesicles), and by the seventh day the blisters turn to pustules (blisters filled with pus). Smallpox skin sores are deeply embedded in the skin (dermis) and feel like firm round objects in the skin. The pustules form scabs by the fourteenth day. As the sores heal, the scabs separate and pitted scarring gradually develops.

☐ What is the difference between a "live vaccine" and a "killed vaccine"?
There are two basic types of vaccines: live (live-attenuated) and killed (inactivated).
Live vaccines are made from viruses or bacteria, sometimes called "wild," that cause disease. These wild viruses or bacteria are weakened in a laboratory.
Live vaccine works when the virus replicates in the body of a vaccinated person. This turns on the immune system and prepares the body to fight the disease when exposed to it. The immune response to a live vaccine is almost the same as from natural infection. Sometimes a person getting a live vaccine has mild symptoms of the disease.

Live vaccines rarely may cause severe or fatal reactions as a result of uncontrolled replication (growth) of the vaccine virus. This may occur in persons with weak immune systems, including persons with leukemia or human immunodeficiency virus (HIV) infection or persons undergoing treatment with certain drugs. This is why it is so important to know a person's health status before giving a live vaccine.

Currently available live vaccines include measles, mumps, polio, rubella, vaccinia (smallpox), varicella (chickenpox), and yellow fever. All of these are made from viruses. There are two live bacterial vaccines: 1) Bacillus of Calmette and Guérin (BCG) vaccine for tuberculosis and 2) oral typhoid.

Killed vaccines are made by growing bacteria or virus and then treating it with heat and/or chemicals (usually formalin). These vaccines cannot cause disease from infection, even in someone with a weakened immune system.

Killed vaccines always require multiple doses. The first dose does not produce protective immunity. It "primes" the immune system, getting it ready to react. A protective immune response develops after the second or third dose.

Available killed vaccines include acellular pertussis, anthrax, botulism, cholera, diptheria, hepatitis A, hepatitis B, Haemophilus influenzae type b (Hib), influenza, Lyme disease, meningococcus, pertussis, plague, pneumococcus, polio, rabies, tetanus, typhoid, and typhoid VI.

☐ If people got the vaccination when it was available in the past, will they be immune?
Not necessarily. It is not clear how long protection from smallpox vaccine lasts. Most experts believe that protection from vaccination lasts 3 to 5 years. Persons who were vaccinated before 1972 may have some protection against smallpox, but it is uncertain. This means that the U.S. population has partial immunity at best. Immunity can be boosted with a single revaccination. Routine vaccination against smallpox ended in 1972.

☐ If someone had smallpox once, are they immune? Would they need the vaccine?
Most people who have had smallpox disease are protected from the disease for life and do not need to be vaccinated. However, few people living in the United States have had smallpox.

☐ Is it possible for someone to receive the smallpox vaccine and have it not "take," i.e., work? How does someone who has been vaccinated for smallpox know that he or she is immune or that the vaccine has "taken"?
Evaluating a person's immunity against smallpox is difficult. In the past, a process called "re-challenging" was the only way.

Centers for Disease Control and Prevention
Public Inquiries: (888) 246-2675
1600 Clifton Road
Atlanta, GA 30333
U.S.A.
(404) 639-3311

APPENDIX 4:   Lewisite

Date: 16 April 1988
Revised: 4 October 1999

In the event of an emergency
Telephone the SBCCOM Operations
Center's 24-hour emergency
Number: 410-436-2148

Section I: General Information

Manufacturer's Address:
U.S. Army Soldier and Biological Chemical Command (SBCCOM)
Edgewood Chemical Biological Center (ECBC)
ATTN: AMSSB-RCB-RS
Aberdeen Proving Ground, MD 21010-5424

CAS Registry Number: 541-25-3

Chemical Name: Dichloro- (2-chlorovinyl) arsine

Trade name and synonyms:
Arsine, (2-chlorovinyl) dichloro-
Arsonous dichloride, (2-chloroethenyl)
Chlorovinylarsine dichloride
2-Chlorovinyldichloroarsine
Beta-Chlorovinyldichloroarsine
Lewisite
L
EA 1034

Chemical Family: Arsenical (vesicant)

Formula/Chemical Structure:
$C_2 H_2 As Cl_3$
<FONT=+2

```
                    Cl

                    /

    Cl CH = CH - As

                    \

                    Cl
```

363

NFPA 704 Signal:
Health - 4
Flammability - 1
Reactivity - 1
Special - 0

Section II: Ingredients

Ingredients/Name: Lewisite

Percentage by Weight: 100%

Threshold Limit Value (TLV):   0.003 mg/m³ (This is a ceiling value)

Section III: Physical Data

Boiling Point °F (°C):   Calculated 374 °F (190 °C)

Vapor Pressure (mm Hg):   0.22 @ 20 °C 0.35 @ 25 °C

Vapor Density (Air=1):   7.1

Solubility (g/100g solvent):   Insoluble in water and dilute mineral acids. Soluble in organic solvents, oils and alcohol.

Specific Gravity (H₂0=1):   1.891 @ 20 °C

Freezing/Melting Point (°C):   -18.2 to 0.1 (Depending on purity)

Liquid Density (g/mL):   1.888 @ 20 °C

Volatility (mg/m³):   2,500 @ 20 °C

Viscosity (Centipoise):   2.257 @ 20 °C

Molecular Weight (g/mol):   207.32

Appearance and Odor: Pure Lewisite is a colorless oily liquid. "War gas" is amber to dark brown liquid. A characteristic odor is usually geranium-like; very little odor when pure.

Section IV: Fire and Explosion Data

Flashpoint: Does not flash

Flammability Limits (% by volume):   Not Applicable

Extinguishing Media: Water, fog, foam, CO_2. Avoid use of extinguishing methods that will cause splashing or spreading of L.

Special Fire Fighting Procedures: All persons not engaged in extinguishing the fire should be immediately evacuated from the area. Fires involving L should be contained to prevent contamination to uncontrolled areas. When responding to a fire alarm in buildings or areas containing agents, fire-fighting personnel should wear full firefighter protective clothing (flame resistant) during chemical agent fire-fighting and fire rescue operations. Respiratory protection is required. Positive pressure, full facepiece, NIOSH-approved self-contained breathing apparatus (SCBA) will be worn where there is danger of oxygen deficiency and when directed by the fire chief or chemical accident/incident (CAI) operations officer. In cases where firefighters are responding to a chemical accident/incident for rescue/reconnaissance purposes they will wear appropriate levels of protective clothing (See Section VIII).

Do not breathe fumes. Skin contact with agents must be avoided at all times. Although the fire may destroy most of the agent, care must still be taken to assure the agent or contaminated liquids do not further contaminate other areas or sewers. Contact with the agent liquid or vapor can be fatal.

Unusual Fire and Explosion Hazards: None known

Section V: Health Hazard Data

Airborne Exposure Limit (AEL):   The permissible airborne exposure concentration of L for an 8-hour workday or a 40-hour workweek is an 8-hour time weighted average (TWA) of 0.003 mg/m^3 as a ceiling value. A ceiling value may not be exceeded at any time. The ceiling value for Lewisite is based upon the present technologically feasible detection limits of 0.003 mg/m^3. This value can be found in "DA Pam 40-173, Occupational Health Guidelines for the Evaluation and Control of Occupational Exposure to Mustard H, HD, and HT." To date, however, the Occupational Safety and Health Administration (OSHA) has not promulgated permissible exposure concentration for L.

Effects Of Overexposure: L is a vesicant (blister agent), also, it acts as a systemic poison, causing pulmonary edema, diarrhea, restlessness, weakness, subnormal temperature, and low blood pressure. In order of severity and appearance of symptoms, it is a blister agent, a toxic lung irritant, absorbed in tissues, and a systemic poison. When inhaled in high concentrations, L may be fatal in as short a time as 10 minutes. L is not detoxified by the body. Common routes of entry into the body are ocular, percutaneous, and inhalation.

Lewisite is generally considered a suspect carcinogen because of its arsenic content.

Toxicological Data:
Man:
LCt50 (inhalation, man) = 1200 - 1500 mg min/m^3
LCt50 (skin vapor exposure, man) = 100,000 mg min/m^3
LDLO (skin, human) = 20 mg/kg
LCt50 (skin, man): >1500 mg/min^3. L irritates eyes and skin and gives warning of its presence. Minimum effective dose (ED min) = 200 mg/m^3 (30 min).
ICt50 (eyes, man): <300 mg min/m^3.

Animal:
LD50 (oral, rat) = 50 mg/kg
LD50 (subcutaneous, rat) = 1 mg/kg
LCtLO (inhalation, mouse) = 150 mg/m^3 10m
LD50 (skin, dog = 15 mg/kg)
LD50 (skin, rabbit) = 6 mg/kg
LD50 (subcutaneous, rabbit) = 2 mg/kg

LD50 (intravenous, rabbit) = 2 mg/kg
LD50 (skin, guinea pig) = 12 mg/kg
LD50 (subcutaneous, guinea pig) = 1 mg/kg
LCt50 (inhalation, rat) = 1500 mg min/m^3 (9 min)
LD50 (vapor skin, rat) = 20,000 mg min m 25 min)
LD50 (skin, rat) = 15 - 24 mg/kg
LD50 (ip, dog) = 2 mg/kg

Acute Exposure:

Eyes: Severe damage. Instant pain, conjunctivitis and blepharospasm leading to closure of eyelids, followed by corneal scarring and iritis. Mild exposure produces reversible eye damage if decontaminated instantly. More permanent injury or blindness is possible within one minute of exposure.

Skin: Immediate stinging pain increasing in severity with time. Erythema (skin reddening) appears within 30 minutes after exposure accompanied by pain with itching and irritation for 24 hours. Blisters appear within 12 hours after exposure with more pain that diminishes after 2-3 days. Skin burns are much deeper than with HD. Tender skin, mucous membrane, and perspiration-covered skin are more sensitive to the effects of L. This, however, is counteracted by L's hydrolysis by moisture, producing less vesicant and higher vapor pressure product.

Respiratory Tract: Irritating to nasal passages and produces a burning sensation followed by profuse nasal secretions and violent sneezing. Prolonged exposure causes coughing and production of large quantities of froth mucus. In experimental animals, injury to respiratory tracts, due to vapor exposure is similar to mustards; however, edema of the lung is more marked and frequently accompanied by pleural fluid.

Systemic Effects: L on the skin, and inhaled vapor, cause systemic poisoning. A manifestation of this is a change in capillary permeability, which permit's loss of sufficient fluid from the bloodstream to cause hemoconcentration, shock and death. In nonfatal cases, hemolysis of erythrocytes has occurred with a resultant hemolytic anemia. The excretion of oxidized products into the bile by the liver produces focal necrosis of that organ, necrosis of the mucosa of the biliary passages with periobiliary hemorrhages, and some injury to the intestinal mucosa. Acute systematic poisoning from large skin burns cause's pulmonary edema, diarrhea restlessness, weakness, subnormal temperature, and low blood pressure in animals.

Chronic Exposure: Lewisite can cause sensitization and chronic lung impairment. Also, by comparison to agent mustard and arsenical compounds, it can be considered as a suspected human carcinogen.

Emergency and First Aid Procedures:

Inhalation: Hold breath until respiratory protective mask is donned. Remove from the source **Immediately**. If breathing is difficult, administer oxygen. If breathing has stopped, give artificial respiration. Mouth-to-mouth resuscitation should be used when approved mask-bag or oxygen delivery systems are not available. Do not use mouth-to-mouth resuscitation when facial contamination is present. Seek medical attention **Immediately**.

Eye Contact: Speed in decontaminating the eyes is absolutely essential. Remove the person from the liquid source, flush the eyes **Immediately** with water for at least 15 minutes by tilting the head to the side, pulling the eyelids apart with the fingers and pouring water slowly into the eyes. Do not cover eyes with bandages but, if necessary, protect eyes by means of dark or opaque goggles. Transfer the patient to a medical facility **Immediately.**

Skin Contact: Don respiratory protective mask. Remove the victim from agent sources immediately. **Immediately** wash skin and clothes with 5% solution of sodium hypochlorite or liquid household bleach within one minute. Cut and remove contaminated clothing, flush contaminated skin area again with 5% sodium hypochlorite solution, then wash contaminated skin area with soap and water. Seek medical attention **Immediately.**

Ingestion: Do not induce vomiting. Give victim milk to drink. Seek medical attention **Immediately.**

Section VI: Reactivity Data

Stability: Stable in steel or glass containers at temperatures below 50 °C

Incompatibility: Corrosive to steel at a rate of 1×10^{-5} to 5×10^{-5} in/month at 65 °C

Hazardous Decomposition Products: Reasonably stable; however, in presence of moisture, it hydrolyses rapidly, losing its vesicant property. It also hydrolyses in acidic medium to form HC1 and non-volatile (solid) chlorovinylarsenious oxide, which is less vesicant than Lewisite. Hydrolysis in alkaline medium, as in decontamination with alcoholic caustic or carbonate solution or DS2, produces acetylene and trisodium arsenate ($Na_3 As O_4$). Therefore, decontaminated solution would contain toxic arsenic.

Hazardous Polymerization: Does not occur.

Section VII: Spill, Leak, And Disposal Procedures

Steps To Be Taken In Case Material Is Released Or Spilled: If leaks or spills of L occur only personnel in full protective clothing will be allowed in the area (See Section VIII). See Section V for emergency and first aid instructions.

Recommended Field Procedures: Lewisite should be contained using vermiculite, diatomaceous earth, clay, or fine sand and neutralized as soon as possible using copious amounts of alcoholic caustic, carbonate, or Decontaminating Agent (DS2). Caution must be exercised when using these decontaminates since acetylene will be given off. Household bleach can also be used if accompanied by stirring to allow contact. Scoop up all material and place in a DOT approved container. Cover the contents with decontaminating solution as above. After sealing, the exterior decontaminated and labeled according to EPA and DOT regulations. All leaking containers will be over packed with sorbent (e.g. vermiculite) placed between the interior and exterior containers. Decontaminate and label according to EPA and DOT regulations. Dispose of decontaminate according to Federal, state, and local laws. Conduct general area monitoring to confirm that the atmospheric concentrations do not exceed the airborne exposure limits (See Sections II and VIII).

Recommended Laboratory Procedures: A 10 wt.% alcoholic sodium hydroxide solution is prepared by adding 100 grams of denatured ethanol to 900 grams of 10 wt.% NaOH in water. A minimum of 200 grams of decon is required for each gram of L. The decon and agent solution is agitated for a minimum of one hour. At the end of the hour the resulting pH should be checked and adjusted to above 11.5 using additional NaOH, if required. It is permitted to substitute 10 wt.% alcoholic sodium carbonate made and used in the same ratio as the NaOH listed above. Reaction time should be increased to 3 hours with agitation for the first hour. Final pH should be adjusted to above 10. Scoop up all material and place in an approved DOT container. Cover the contents with decontaminating solution as above. The exterior of the container will be decontaminated and labeled according to EPA and DOT regulations. All leaking containers will be over packed with sorbent (e.g., vermiculite) placed between the interior and exterior containers. Decontaminate and label according to EPA and DOT regulations. Dispose of the material in accordance with waste disposal methods provided below. Conduct general area monitoring with an approved monitor to confirm that the atmospheric concentrations do not exceed the airborne exposure limits (See Sections II and VIII).

It is permitted to substitute 5.25% sodium hypochlorite for the 10% alcoholic sodium hydroxide solution above. Allow one hour with agitation for the reaction. Adjustment of the pH is not required. Conduct general area monitoring to confirm that the atmospheric concentrations do not exceed the airborne exposure limit (See Section VIII).

Waste Disposal Method: All neutralized material should be collected and contained for disposal according to land ban RCRA regulations or thermally decomposed in an EPA permitted incinerator equipped with a scrubber that will scrub out the chlorides and equipped with an electrostatic precipitator or other filter device and containerize and label according to DOT and EPA regulations. The arsenic will be disposed of according to land ban RCRA regulations. Any contaminated materials or protective clothing should be decontaminated using alcoholic caustic, carbonates, or bleach analyzed to assure it is free of detectable contamination (3X) level. The clothing should then be sealed in plastic bags inside properly labeled drums and held for shipment back to the DA issue point.

Note: Some decontaminate solutions are hazardous waste according to RCRA regulations and must be disposed of IAW those regulations.

Section VIII: Special Protection Information

Concentration	Respiratory Protective Equipment
< 0.003 mg/m^3	A full face piece, chemical canister, air-purifying protective mask will be on hand for escape. M40-series masks are acceptable for this purpose. Other masks certified as equivalent may be used.
> or = 0.003 mg/m3 or unknown	NIOSH/MSHA approved pressure demand full face piece SCBA suitable for use in high Lewisite concentrations with protective ensemble (See DA Pam 385-61 for examples).

Ventilation

Local Exhaust: Mandatory. Must be filtered or scrubbed. Air emissions shall meet local, state and federal regulations.

Special: Chemical laboratory hoods will have an average inward face velocity of 100 linear feet per minute (lfpm) ±20% with the velocity at any point not deviating from the average face velocity by more than 20%. Existing laboratory hoods will have an inward face velocity of 150 lfpm ±20%. Laboratory hoods will be located such that cross drafts do not exceed 20% of the inward face velocity. A visual performance test using smoke producing devices will be performed in assessing the ability of the hood to contain Lewisite.

Other: Recirculation of exhaust air from agent areas is prohibited. No connection between agent area and other areas through the ventilation system is permitted. Emergency backup power is necessary. Hoods should be tested semiannually or after modification or maintenance operations. Operations should be performed 20 centimeters inside hoods.

Protective Gloves: Butyl Rubber gloves M3 and M4, Norton, Chemical Protective Glove Set

Eye Protection: As a minimum, chemical goggles will be worn. For splash hazards use goggles and face shield.

Other Protective Equipment: For laboratory operations, wear lab coats, gloves and have mask readily accessible. In addition, daily clean smocks, foot covers, and head covers will be required when handling contaminated lab animals.

Monitoring: Available monitoring equipment for agent Lewiste is the M18A2 (yellow band), bubblers (arsenic and GC method), and M256 and A1 kits.

Real-time, low-level monitors (with alarm) are required for Lewisite operations. In their absence, an Immediately Dangerous to Life and Health (IDLH) atmosphere must be presumed. Laboratory operations conducted in appropriately maintained and alarmed engineering controls require only periodic low-level monitoring.

Section IX: Special Precautions

Precautions To Be Taken In Handling And Storing: When handling agents, the buddy system will be incorporated. No smoking, eating, or drinking in areas containing agents is permitted. Containers should be periodically inspected for leaks, (either visually or using a detector kit). Stringent control over all personnel practices must be exercised. Decontaminating equipment will be conveniently located. Exits must be designed to permit rapid evacuation. Chemical showers, eyewash stations, and personal cleanliness facilities must be provided. Wash hands before meals and shower thoroughly with special attention given to hair, face, neck, and hands using plenty of soap and water before leaving at the end of the workday.

Other Precautions: L should be stored in containers made of glass for Research, Development, Test and Evaluation (RDTE) quantities or one-ton steel containers for large quantities. Agent will be stored in a single containment system within a laboratory hood or in a double containment system.

For additional information see "AR 385-61, The Army Toxic Chemical Agent Safety Program," "DA Pam 385-61, Toxic Chemical Agent Safety Standards," and "DA Pam 40-173, Occupational Health Guidelines for the Evaluation and Control of Occupational Exposure to Mustard Agents H, HD, and HT."

Section X: Transportation Data

Note: Forbidden for transport other than via military (Technical Escort Unit) transport according to 49 CFR 172

Proper Shipping Name: Toxic liquids, n.o.s.

Dot Hazard Class: 6.1, Packing Group I

Dot Label: Poison

Dot Marking: Toxic liquids, n.o.s.; Dichloro-(2-chlorovinyl)arsine UN 2810

Dot Placard   Poison

Emergency Accident Precautions and Procedures: See Sections IV, VII, and VIII.

Precautions To Be Taken In Transportation: Motor vehicles will be placarded regardless of quantity. Drivers will be given full information regarding shipment and conditions in case of an emergency. AR 50-6 deals specifically with the shipment of chemical agents. Shipment of agents will be escorted in accordance with AR 740-32.

The Edgewood Chemical Biological Center (ECBC), Department of the Army believes that the data contained herein are actual and are the results of the tests conducted by ECBC experts. The data are not to be taken as a warranty or representation for which the Department of the Army or ECBC assumes legal responsibility. They are offered solely for consideration. Any use of this data and information contained in this MSDS must be determined by the user to be in accordance with applicable Federal, State, and local laws and regulations.

APPENDIX 5: Recognition of Illness Associated with the Intentional Release of a Biologic Agent

On September 11, 2001, following the terrorist incidents in New York City and Washington, D.C., CDC recommended heightened surveillance for any unusual disease occurrence or increased numbers of illnesses that might be associated with the terrorist attacks. Subsequently, cases of anthrax in Florida and New York City have demonstrated the risks associated with intentional release of biologic agents.[1] This report provides guidance for health-care providers and public health personnel about recognizing illnesses or patterns of illness that might be associated with intentional release of biologic agents.

Health-Care Providers

Health-care providers should be alert to illness patterns and diagnostic clues that might indicate an unusual infectious disease outbreak associated with intentional release of a biologic agent and should report any clusters or findings to their local or state health department. The covert release of a biologic agent may not have an immediate impact because of the delay between exposure and illness onset, and outbreaks associated with intentional releases might closely resemble naturally occurring outbreaks. Indications of intentional release of a biologic agent include 1) an unusual temporal or geographic clustering of illness (e.g., persons who attended the same public event or gathering) or patients presenting with clinical signs and symptoms that suggest an infectious disease outbreak (e.g., ≥2 patients presenting with an unexplained febrile illness associated with sepsis, pneumonia, respiratory failure, or rash or a botulism-like syndrome with flaccid muscle paralysis, especially if occurring in otherwise healthy persons); 2) an unusual age distribution for common diseases (e.g., an increase in what appears to be a chickenpox-like illness among adult patients, but which might be smallpox); and 3) a large number of cases of acute flaccid paralysis with prominent bulbar palsies, suggestive of a release of *botulinum* toxin.

CDC defines three categories of biologic agents with potential to be used as weapons, based on ease of dissemination or transmission, potential for major public health impact (e.g., high mortality), potential for public panic and social disruption, and requirements for public health preparedness.[2] Agents of highest concern are *Bacillus anthracis* (anthrax), *Yersinia pestis* (plague), variola major (smallpox), *Clostridium botulinum* toxin (botulism), *Francisella tularensis* (tularemia), filoviruses (Ebola hemorrhagic fever, Marburg hemorrhagic fever); and arenaviruses (Lassa [Lassa fever], Junin [Argentine hemorrhagic fever], and related viruses). The following summarizes the clinical features of these agents.[3-6]

Anthrax.   A nonspecific prodrome (i.e., fever, dyspnea, cough, and chest discomfort) follows inhalation of infectious spores. Approximately 2–4 days after initial symptoms, sometimes after a brief period of improvement, respiratory failure and hemodynamic collapse ensue. Inhalational anthrax also might include thoracic edema and a widened mediastinum on chest radiograph. Gram-positive bacilli can grow on blood culture, usually 2–3 days after onset of illness. Cutaneous anthrax follows deposition of the organism onto the skin, occurring particularly on exposed areas of the hands, arms, or face. An area of local edema becomes a pruritic macule or papule, which enlarges and ulcerates after 1–2 days. Small, 1–3 mm vesicles may surround the ulcer. A painless, depressed, black eschar usually with surrounding local edema subsequently develops. The syndrome also may include lymphangitis and painful lymphadenopathy.

Plague. Clinical features of pneumonic plague include fever, cough with muco-purulent sputum (gram-negative rods may be seen on gram stain), hemoptysis, and chest pain. A chest radiograph will show evidence of bronchopneumonia.

Botulism. Clinical features include symmetric cranial neuropathies (i.e., drooping eyelids, weakened jaw clench, and difficulty swallowing or speaking), blurred vision or diplopia, symmetric descending weakness in a proximal to distal pattern, and respiratory dysfunction from respiratory muscle paralysis or upper airway obstruction without sensory deficits. Inhalational botulism would have a similar clinical presentation as foodborne botulism; however, the gastrointestinal symptoms that accompany foodborne botulism may be absent.

Smallpox (variola). The acute clinical symptoms of smallpox resemble other acute viral illnesses, such as influenza, beginning with a 2–4 day nonspecific prodrome of fever and myalgias before rash onset. Several clinical features can help clinicians differentiate varicella (chickenpox) from smallpox. The rash of varicella is most prominent on the trunk and develops in successive groups of lesions over several days, resulting in lesions in various stages of development and resolution. In comparison, the vesicular/pustular rash of smallpox is typically most prominent on the face and extremities, and lesions develop at the same time.

Inhalational tularemia. Inhalation of *F. tularensis* causes an abrupt onset of an acute, nonspecific febrile illness beginning 3–5 days after exposure, with pleuropneumonitis developing in a substantial proportion of cases during subsequent days.[7]

Hemorrhagic fever (such as would be caused by Ebola or Marburg viruses). After an incubation period of usually 5–10 days (range: 2–19 days), illness is characterized by abrupt onset of fever, myalgia, and headache. Other signs and symptoms include nausea and vomiting, abdominal pain, diarrhea, chest pain, cough, and pharyngitis. A maculopapular rash, prominent on the trunk, develops in most patients approximately 5 days after onset of illness. Bleeding manifestations, such as petechiae, ecchymoses, and hemorrhages, occur as the disease progresses.[8]

Clinical Laboratory Personnel

Although unidentified gram-positive bacilli growing on agar may be considered as contaminants and discarded, CDC recommends that these bacilli be treated as a "finding" when they occur in a suspicious clinical setting (e.g., febrile illness in a previously healthy person). The laboratory should attempt to characterize the organism, such as motility testing, inhibition by penicillin, absence of hemolysis on sheep blood agar, and further biochemical testing or species determination.

An unusually high number of samples, particularly from the same biologic medium (e.g., blood and stool cultures), may alert laboratory personnel to an outbreak. In addition, central laboratories that receive clinical specimens from several sources should be alert to increases in demand or unusual requests for culturing (e.g., uncommon biologic specimens such as cerebrospinal fluid or pulmonary aspirates).

When collecting or handling clinical specimens, laboratory personnel should 1) use Biological Safety Level II (BSL-2) or Level III (BSL-3) facilities and practices when working with clinical samples considered potentially infectious; 2) handle all specimens in a BSL-2 laminar flow hood with protective eyewear (e.g., safety glasses or eye shields), use closed-front laboratory coats with cuffed sleeves, and stretch the gloves over the cuffed sleeves; 3) avoid any activity that places persons at risk for infectious exposure, especially activities that might create aerosols or droplet dispersal; 4) decontaminate laboratory benches after each use and dispose of supplies and equipment in proper receptacles; 5) avoid touching mucosal surfaces with their hands (gloved or ungloved), and never eat or drink in the laboratory; and 6) remove and reverse their gloves before leaving the laboratory and dispose of them in a biohazard container, and wash their hands and remove their laboratory coat.

When a laboratory is unable to identify an organism in a clinical specimen, it should be sent to a laboratory where the agent can be characterized, such as the state public health laboratory or, in some large metropolitan areas, the local health department laboratory. Any clinical specimens suspected to contain variola (smallpox) should be reported to local and state health authorities and then transported to CDC. All variola diagnostics should be conducted at CDC laboratories. Clinical laboratories should report any clusters or findings that could indicate intentional release of a biologic agent to their state and local health departments.

Infection-Control Professionals

Heightened awareness by infection-control professionals (ICPs) facilitates recognition of the release of a biologic agent. ICPs are involved with many aspects of hospital operations and several departments and with counterparts in other hospitals. As a result, ICPs may recognize changing patterns or clusters in a hospital or in a community that might otherwise go unrecognized.

ICPs should ensure that hospitals have current telephone numbers for notification of both internal (ICPs, epidemiologists, infectious diseases specialists, administrators, and public affairs officials) and external (state and local health departments, Federal Bureau of Investigation field office, and CDC Emergency Response office) contacts and that they are distributed to the appropriate personnel.[9] ICPs should work with clinical microbiology laboratories, on- or off-site, that receive specimens for testing from their facility to ensure that cultures from suspicious cases are evaluated appropriately.

State Health Departments

State health departments should implement plans for educating and reminding health-care providers about how to recognize unusual illnesses that might indicate intentional release of a biologic agent. Strategies for responding to potential bioterrorism include 1) providing information or reminders to health-care providers and clinical laboratories about how to report events to the appropriate public health authorities; 2) implementing a 24-hour-a-day, 7-day-a-week capacity to receive and act on any positive report of events that suggest intentional release of a biologic agent; 3) investigating immediately any report of a cluster of illnesses or other event that suggests an intentional release of a biologic agent and requesting CDC's assistance when necessary; 4) implementing a plan, including accessing the Laboratory Response Network for Bioterrorism, to collect and transport specimens and to store them appropriately before laboratory analysis; and 5) reporting immediately to CDC if the results of an investigation suggest release of a biologic agent.

Reported by: National Center for Infectious Diseases; Epidemiology Program Office;
Public Health Practice Program Office; Office of the Director, CDC.

Editorial Note

Health-care providers, clinical laboratory personnel, infection control professionals, and health departments play critical and complementary roles in recognizing and responding to illnesses caused by intentional release of biologic agents. The syndrome descriptions, epidemiologic clues, and laboratory recommendations in this report provide basic guidance that can be implemented immediately to improve recognition of these events.

After the terrorist attacks of September 11, state and local health departments initiated various activities to improve surveillance and response, ranging from enhancing communications (between state and local health departments and between public health agencies and health-care providers) to conducting special surveillance projects. These special projects have included active surveillance for changes in the number of hospital admissions, emergency department visits, and occurrence of specific syndromes. Activities in bioterrorism preparedness and emerging infections over the past few years have better positioned public health agencies to detect and respond to the intentional release of a biologic agent. Immediate review of these activities to identify the most useful and practical approaches will help refine syndrome surveillance efforts in various clinical situations.

Information about clinical diagnosis and management can be found elsewhere.[1-9] Additional information about responding to bioterrorism is available from CDC at http:// www.bt.cdc.gov; the U.S. Army Medical Research Institute of Infectious Diseases at http:// www.usamriid.army.mil/education/bluebook.html; the Association for Infection Control Practitioners at http://www.apic.org; and the Johns Hopkins Center for Civilian Biodefense at http://www.hopkins-biodefense.org.

Notes

1. CDC. Update: investigation of anthrax associated with intentional exposure and interim public health guidelines, October 2001. *MMWR* 2001, 50, 889–93.
2. CDC. Biological and chemical terrorism: strategic plan for preparedness and response. *MMWR* 2000, 49(no. RR-4).
3. Arnon, S.S., Schechter, R., Inglesby, T.V., et al. Botulinum toxin as a biological weapon: medical and public health management. *JAMA* 2001, 285, 1059–70.
4. Inglesby, T.V., Dennis, D.T., Henderson, D.A., et al. Plague as a biological weapon: medical and public health management. *JAMA* 2000, 283, 2281–90.
5. Henderson, D.A., Inglesby, T.V., Bartlett, J.G., et al. Smallpox as a biological weapon: medical and public health management. *JAMA* 1999, 281, 2127–37.

6. Inglesby, T.V., Henderson, D.A., Bartlett, J.G., et al. Anthrax as a biological weapon: medical and public health management. *JAMA* 1999, 281, 1735–963.
7. Dennis, D.T., Inglesby, T.V., Henderson, D.A., et al. Tularemia as a biological weapon: medical and public health management. *JAMA* 2001, 285, 2763–73.
8. Peters, C.J. Marburg and Ebola virus hemorrhagic fevers. In: Mandell, G.L., Bennett, J.E., Dolin, R., eds. Principles and practice of infectious diseases. 5th ed. New York, New York: Churchill Livingstone 2000, 2, 1821–3.
9. APIC Bioterrorism Task Force and CDC Hospital Infections Program Bioterrorism Working Group. Bioterrorism readiness plan: a template for healthcare facilities. Available at http:// www.cdc.gov/ncidod/hip/Bio/bio.htm. Accessed October 2001.
10. Use of trade names and commercial sources is for identification only and does not imply endorsement by the U.S. Department of Health and Human Services.
11. References to non-CDC sites on the Internet are provided as a service to MMWR readers and do not constitute or imply endorsement of these organizations or their programs by CDC or the U.S. Department of Health and Human Services. CDC is not responsible for the content of pages found at these sites.

APPENDIX 6: Antimicrobial Prophylaxis to Prevent Anthrax Among Decontamination/Cleanup Workers Responding to an Intentional Distribution of *Bacillus anthracis*

Decontamination/cleanup workers working in environments known to be contaminated with *Bacillus anthracis* spores may be at risk for inhalational anthrax. These workers should wear appropriate personal protective equipment (PPE) and follow appropriate procedures, as outlined in other CDC Guidance documents.

Despite appropriate PPE and procedures, however, there will remain a potential for breaches of protection and contamination of the workers. Furthermore, there is potential that such a breach or contamination will not be recognized at the time of occurrence. Finally, while it may be appropriate to conduct medical surveillance of cleanup workers for epidemiologic monitoring of the effectiveness of the protective measures, monitoring may not be reliable enough or timely enough to rely on for clinical decisions regarding the need for antimicrobial prophylaxis on an individual basis.

Recommendations

CDC recommends that decontamination/cleanup workers receive antimicrobial prophylaxis, using standard regimens starting in conjunction with or prior to the time of first entry into a contaminated location and continuing for 60 days after final opportunity for exposure.

The current recommended regimens (for adults) are as follows:

> ciprofloxacin, 500 mg by mouth every 12 hours or doxycycline, 100 mg by mouth every 12 hours

These recommendations may be modified as additional information becomes available.

A medical protocol should be developed to implement prophylaxis, and this program should be under the supervision of an experienced physician. At a minimum, the protocol should include the following components: there should be a pre-deployment assessment, including ascertainment of history of drug allergies, current medication that might interact adversely with the selected prophylactic antimicrobial, presence of any medical conditions that might contraindicate use of the selected antimicrobial, and education regarding potential side effects and how to report symptoms or problems. There should also be education regarding recognition of potential breaches in protection and regarding anthrax and it's symptoms, emphasizing the need for prompt reporting of both breaches and symptoms. Provision should also be made for periodic re-assessment of workers receiving prophylaxis; this assessment should include both monitoring for evidence of side effects of medications and epidemiologic surveillance for evidence of exposures. There are no available data to guide selection of an appropriate interval for re-assessments, so as an interim guidance, this should be left to the professional judgment of the supervising physician. If workers develop adverse side effects during prophylaxis, alternative prophylactic antimicrobial therapies may be available and warranted.

It is important to emphasize that this is an interim guidance and that a number of important issues remain unresolved. This guidance is subject to revision as these issues become better clarified.

Centers for Disease Control and Prevention
1600 Clifton Road
Atlanta, GA 30333
(404) 639-3311

APPENDIX 7:   Melioidosis

Program Contents

, Topic Home
, Programs
, Offices
, Disease Listing

Quick Links

Get Smart on the Farm
11 Aug 2005
CDC Foundation interview with
Tom Chiller, epidemiologist and
medical director of CDC's Get
Smart on the Farm program.

Contact Info

1600 Clifton Road NE, MS-D63
Atlanta, GA 30033
Phone: + 1-800-311-3435

Email Us

Division of Bacterial and Mycotic Diseases

Home > Home > Disease Listing > Melioidosis

Melioidosis
(Burkholderia pseudomallei)

Disease Listing | General Information | Technical Information | Additional Information

Frequently Asked Questions

▶ What is melioidosis?
▶ Why has melioidosis become a current issue?
▶ How common is melioidosis and where is it found?
▶ How is melioidosis transmitted and who can get it?
▶ What are the symptoms of melioidosis?
▶ How is melioidosis diagnosed?
▶ Can melioidosis be spread from person and person?
▶ Is there a way to prevent infection?
▶ Is there a treatment for melioidosis?

What is melioidosis?

Melioidosis, also called Whitmore's disease, is an infectious disease caused by the bacterium *Burkholderia pseudomallei*. Melioidosis is clinically and pathologically similar to glanders disease, but the ecology and epidemiology of melioidosis are different from glanders. Melioidosis is predominately a disease of tropical climates, especially in Southeast Asia where it is endemic. The bacteria causing melioidosis are found in contaminated water and soil and are spread to humans and animals through direct contact with the contaminated source. Glanders is contracted by humans from infected domestic animals.

Why has melioidosis become a current issue?

Burkholderia pseudomallei is an organism that has been considered as a potential agent for biological warfare and biological terrorism.

How common is melioidosis and where is it found?

Melioidosis is endemic in Southeast Asia, with the greatest concentration of cases reported in Vietnam, Cambodia, Laos, Thailand, Malaysia, Myanmar (Burma), and northern Australia. Additionally, it is seen in the South Pacific, Africa, India, and the Middle East. In many of these countries, *Burkholderia pseudomallei* is so prevalent that it is a common contaminate found on laboratory cultures. Moreover, it has been a common pathogen isolated from troops of all nationalities that have served in areas with endemic disease. A few isolated cases of melioidosis have occurred in the Western Hemisphere in Mexico, Panama, Ecuador, Haiti, Brazil, Peru, Guyana, and in the states of Hawaii and Georgia. In the United States, confirmed cases range from none to five each year and occur among travelers and immigrants.

How is melioidosis transmitted and who can get it?

Besides humans, many animal species are susceptible to melioidosis. These include sheep, goats, horses, swine, cattle, dogs, and cats. Transmission occurs by direct contact with contaminated soil and surface waters. In Southeast Asia, the organism has been repeatedly isolated from agriculture fields, with infection occurring primarily during the rainy season. Humans and animals are believed to acquire the infection by inhalation of dust, ingestion of contaminated water, and contact with contaminated soil especially through skin abrasions, and for military troops, by contamination of war wounds. Person-to-person transmission can occur. There is one report of transmission to a sister with diabetes who was the caretaker for her brother who had chronic melioidosis. Two cases of sexual transmission have been reported. Transmission in both cases was preceded by a clinical history of chronic prostatitis in the source patient.

What are the symptoms of melioidosis?

Illness from melioidosis can be categorized as acute or localized infection, acute pulmonary infection, acute bloodstream infection, and chronic suppurative infection. Inapparent infections are also possible. The incubation period (time between exposure and appearance of clinical symptoms) is not clearly defined, but may range from 2 days to many years.

Acute, localized infection: This form of infection is generally localized as a nodule and results from inoculation through a break in the skin. The acute form of melioidosis can produce fever and general muscle aches, and may progress rapidly to infect the bloodstream.

Pulmonary infection: This form of the disease can produce a clinical picture of mild bronchitis to severe pneumonia. The onset of pulmonary melioidosis is typically accompanied by a high fever, headache, anorexia, and general muscle soreness. Chest pain is common, but a nonproductive or productive cough with normal sputum is the hallmark of this form of melioidosis.

Acute bloodstream infection: Patients with underlying illness such as HIV, renal failure, and diabetes are affected by this type of the disease, which usually results in septic shock. The symptoms of the bloodstream infection vary depending on the site of original infection, but they generally include respiratory distress, severe headache, fever, diarrhea, development of pus-filled lesions on the skin, muscle tenderness, and disorientation. This is typically an infection of short duration, and abscesses will be found throughout the body.

Chronic suppurative infection: Chronic melioidosis is an infection that involves the organs of the body. These typically include the joints, viscera, lymph nodes, skin, brain, liver, lung, bones, and spleen.

How is melioidosis diagnosed?

Melioidosis is diagnosed by isolating *Burkholderia pseudomallei* from the blood, urine, sputum, or skin lesions. Detecting and measuring antibodies to the bacteria in the blood is another means of diagnosis..

Can melioidosis be spread from person to person?

Melioidosis can spread from person to person by contact with the blood and body fluids of an infected person. Two documented cases of male-to-female sexual transmission involved males with chronic prostatic infection due to melioidosis.

Is there a way to prevent infection?

There is no vaccine for melioidosis. Prevention of the infection in endemic-disease areas can be difficult since contact with contaminated soil is so common. Persons with diabetes and skin lesions should avoid contact with soil and standing water in these areas. Wearing boots during agricultural work can prevent infection through the feet and lower legs. In health care settings, using common blood and body fluid precautions can prevent transmission.

Is there a treatment for melioidosis?

Most cases of melioidosis can be treated with appropriate antibiotics. *Burkholderia psuedomallei*, the organism that causes melioidosis, is usually sensitive to imipenem, penicillin, doxycycline, amoxicillin-clavulanic acid, azlocillin, ceftazidime, ticarcillin-vulanic acid, ceftriaxone, and aztreonam. Treatment should be initiated early in the course of the disease. Although bloodstream infection with melioidosis can be fatal, the other types of the disease are nonfatal. The type of infection and the course of treatment can predict any long-term sequelae.

Date: October 12, 2005
Content source: Coordinating Center for Infectious Diseases / Division of Bacterial and Mycotic Diseases

SAFER · HEALTHIER · PEOPLE™
Centers for Disease Control and Prevention, 1600 Clifton Rd, Atlanta, GA 30333, U.S.A
Tel: (404) 639-3311 / Public Inquiries: (404) 639-3534 / (800) 311-3435

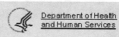

Department of Health and Human Services

APPENDIX 8:   Glanders

Department of Health and Human Services
Centers for Disease Control and Prevention

⊠ Health & Safety Topics | ⊠ Publications & Products | ⊠ Data & Statistics | ⊠ Conferences & Events

Division of Bacterial and Mycotic Diseases

Home > Home > Disease Listing > Glanders

Glanders
(Burkholderia mallei)

For comprehensive CDC information about bioterrorism and related issues, please visit **http://www.bt.cdc.gov**.

Disease Listing | General Information | Technical Information | Additional Information

Frequently Asked Questions

- ▸ What is glanders?
- ▸ Why has glanders become a current issue?
- ▸ How common is glanders?
- ▸ How is glanders transmitted and who can get it?
- ▸ What are the symptoms of glanders?
- ▸ Where is glanders usually found?
- ▸ How is glanders diagnosed?
- ▸ Can glanders spread from person and person?
- ▸ Is there a way to prevent infection?
- ▸ Is there a treatment for glanders?

What is glanders?

Glanders is an infectious disease that is caused by the bacterium *Burkholderia mallei*. Glanders is primarily a disease affecting horses, but it also affects donkeys and mules and can be naturally contracted by goats, dogs, and cats. Human infection, although not seen in the United States since 1945, has occurred rarely and sporadically among laboratory workers and those in direct and prolonged contact with infected, domestic animals.

Why has glanders become a current issue?

Burkholderia mallei is an organism that is associated with infections in laboratory workers because so very few organisms are required to cause disease. The organism has been considered as a potential agent for biological warfare and of biological terrorism.

Program Contents

- ▸ Topic Home
- ▸ Programs
- ▸ Offices
- ▸ Disease Listing

Quick Links

Get Smart on the Farm
11 Aug 2005
CDC Foundation interview with Tom Chiller, epidemiologist and medical director of CDC's Get Smart on the Farm program.

....

Contact Info

1600 Clifton Road NE, MS-D63
Atlanta, GA 30033
Phone: + 1-800-311-3435

....

How common is glanders?

The United States has not seen any naturally occurring cases since the 1940s. However, it is still commonly seen among domestic animals in Africa, Asia, the Middle East, and Central and South America.

How is glanders transmitted and who can get it?

Glanders is transmitted to humans by direct contact with infected animals. The bacteria enter the body through the skin and through mucosal surfaces of the eyes and nose. The sporadic cases have been documented in veterinarians, horse caretakers, and laboratorians.

What are the symptoms of glanders?

The symptoms of glanders depend upon the route of infection with the organism. The types of infection include localized, pus-forming cutaneous infections, pulmonary infections, bloodstream infections, and chronic suppurative infections of the skin. Generalized symptoms of glanders include fever, muscle aches, chest pain, muscle tightness, and headache. Additional symptoms have included excessive tearing of the eyes, light sensitivity, and diarrhea.

Localized infections: If there is a cut or scratch in the skin, a localized infection with ulceration will develop within 1 to 5 days at the site where the bacteria entered the body. Swollen lymph nodes may also be apparent. Infections involving the mucous membranes in the eyes, nose, and respiratory tract will cause increased mucus production from the affected sites.

Pulmonary infections: In pulmonary infections, pneumonia, pulmonary abscesses, and pleural effusion can occur. Chest X-rays will show localized infection in the lobes of the lungs.

Bloodstream infections: Glanders bloodstream infections are usually fatal within 7 to 10 days.

Chronic infections: The chronic form of glanders involves multiple abscesses within the muscles of the arms and legs or in the spleen or liver.

Where is glanders usually found?

Geographically, the disease is endemic in Africa, Asia, the Middle East, and Central and South America.

How is glanders diagnosed?

The disease is diagnosed in the laboratory by isolating *Burkholderia mallei* from blood, sputum, urine, or skin lesions. Serologic assays are not available.

Can glanders spread from person to person?

In addition to animal exposure, cases of human-to-human transmission have been reported. These cases included two suggested cases of sexual transmission and several cases in family members who cared for the patients.

Is there a way to prevent infection?

There is no vaccine available for glanders. In countries where glanders is endemic in animals, prevention of the disease in humans involves identification and elimination of the infection in the animal population. Within the health care setting, transmission can be prevented by using common blood and body fluid precautions.

Is there a treatment for glanders?

Because human cases of glanders are rare, there is limited information about
antibiotic treatment of the organism in humans. Sulfadiazine has been found to be an
effective in experimental animals and in humans. *Burkholderia mallei* is usually
sensitive to tetracyclines, ciprofloxacin, streptomycin, novobiocin, gentamicin,
imipenem, ceftrazidime, and the sulfonamides. Resistance to chloramphenicol has
been reported.

Date: October 11 , 2005
Content source: Coordinating Center for Infectious Diseases / Division of Bacterial and Mycotic Diseases

SAFER・HEALTHIER・PEOPLE®

Centers for Disease Control and Prevention,1600 Clifton Rd, Atlanta, GA 30333, U.S.A
Tel: (404) 639-3311 / Public Inquiries: (404) 639-3534 / (800) 311-3435

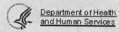

Department of Health
and Human Services

Department of Health and Human Services
Centers for Disease Control and Prevention

| ▣ Health & Safety Topics | ▣ Publications & Products | ▣ Data & Statistics | ▣ Conferences & Events |

Program Contents

- , Topic Home
- , Programs
- , Offices
- , Disease Listing

Quick Links

Get Smart on the Farm
11 Aug 2005
CDC Foundation interview with Tom Chiller, epidemiologist and medical director of CDC's Get Smart on the Farm program.

Contact Info

1600 Clifton Road NE, MS-D63
Atlanta, GA 30033
Phone: + 1-800-311-3435

Division of Bacterial and Mycotic Diseases

Home > Home > Disease Listing > Brucellosis

Brucellosis
(Brucella melitensis, abortus, suis, and canis)

For comprehensive CDC information about bioterrorism and related issues, please visit **http://www.bt.cdc.gov.**

Disease Listing | General Information | Technical Information | Additional Information

Frequently Asked Questions

- ▸ What is brucellosis?
- ▸ How common is brucellosis?
- ▸ Where is brucellosis usually found?
- ▸ How is brucellosis transmitted to humans, and who is likely to become infected?
- ▸ Can brucellosis be spread from person and person?
- ▸ Is there a way to prevent infection?
- ▸ My dog has been diagnosed with brucellosis. Is that a risk for me?
- ▸ How is brucellosis diagnosed?
- ▸ Is there a treatment for brucellosis?
- ▸ I am a veterinarian and I recently accidentally jabbed myself with the animal vaccine (RB-51 or strain 19, or REV-1) while I was vaccinating cows (or sheep, goats). What do I need to do?

What is brucellosis?

Brucellosis is an infectious disease caused by the bacteria of the genus *Brucella*. These bacteria are primarily passed among animals, and they cause disease in many different vertebrates. Various *Brucella* species affect sheep, goats, cattle, deer, elk, pigs, dogs, and several other animals. Humans become infected by coming in contact with animals or animal products that are contaminated with these bacteria. In humans brucellosis can cause a range of symptoms that are similar to the flu and may include fever, sweats, headaches, back pains, and physical weakness. Severe infections of the central nervous systems or lining of the heart may occur. Brucellosis can also cause long-lasting or chronic symptoms that include recurrent fevers, joint pain, and fatigue.

How common is brucellosis?

Brucellosis is not very common in the United States, where 100 to 200 cases occur each year. But brucellosis can be very common in countries where animal disease control programs have not reduced the amount of disease among animals.

Where is brucellosis usually found?

Although brucellosis can be found worldwide, it is more common in countries that do not have good standardized and effective public health and domestic animal health programs. Areas currently listed as high risk are the Mediterranean Basin (Portugal, Spain, Southern France, Italy, Greece, Turkey, North Africa), South and Central America, Eastern Europe, Asia, Africa, the Caribbean, and the Middle East. Unpasteurized cheeses, sometimes called "village cheeses," from these areas may represent a particular risk for tourists.

How is brucellosis transmitted to humans, and who is likely to become infected?

Humans are generally infected in one of three ways: eating or drinking something that is contaminated with *Brucella*, breathing in the organism (inhalation), or having the bacteria enter the body through skin wounds. The most common way to be infected is by eating or drinking contaminated milk products. When sheep, goats, cows, or camels are infected, their milk is contaminated with the bacteria. If the milk is not pasteurized, these bacteria can be transmitted to persons who drink the milk or eat cheeses made it. Inhalation of *Brucella* organisms is not a common route of infection, but it can be a significant hazard for people in certain occupations, such as those working in laboratories where the organism is cultured. Inhalation is often responsible for a significant percentage of cases in abattoir employees. Contamination of skin wounds may be a problem for persons working in slaughterhouses or meat packing plants or for veterinarians. Hunters may be infected through skin wounds or by accidentally ingesting the bacteria after cleaning deer, elk, moose, or wild pigs that they have killed.

Can brucellosis be spread from person to person?

Direct person-to-person spread of brucellosis is extremely rare. Mothers who are breast-feeding may transmit the infection to their infants. Sexual transmission has also been reported. For both sexual and breast-feeding transmission, if the infant or person at risk is treated for brucellosis, their risk of becoming infected will probably be eliminated within 3 days. Although uncommon, transmission may also occur via contaminated tissue transplantation.

Is there a way to prevent infection?

Yes. Do not consume unpasteurized milk, cheese, or ice cream while traveling. If you are not sure that the dairy product is pasteurized, don't eat it. Hunters and animal herdsman should use rubber gloves when handling viscera of animals. There is no vaccine available for humans.

My dog has been diagnosed with brucellosis. Is that a risk for me?

B. canis is the species of *Brucella* species that can infect dogs. This species has occasionally been transmitted to humans, but the vast majority of dog infections do not result in human illness. Although veterinarians exposed to blood of infected animals are at risk, pet owners are not considered to be at risk for infection. This is partly because it is unlikely that they will come in contact with blood, semen, or placenta of the dog. The bacteria may be cleared from the animal within a few days of treatment; however re-infection is common and some animal body fluids may be infectious for weeks. Immunocompromised persons (cancer patients, HIV-infected individuals, or transplantation patients) should not handle dogs known to be infected with *B. canis*.

How is brucellosis diagnosed?

Brucellosis is diagnosed in a laboratory by finding *Brucella* organisms in samples of blood or bone marrow. Also, blood tests can be done to detect antibodies against the bacteria. If this method is used, two blood samples should be collected 2 weeks apart.

Is there a treatment for brucellosis?

Yes, but treatment can be difficult. Doctors can prescribe effective antibiotics. Usually, doxycycline and rifampin are used in combination for 6 weeks to prevent reoccuring infection. Depending on the timing of treatment and severity of illness, recovery may take a few weeks to several months. Mortality is low (<2%), and is usually associated with endocarditis.

I am a veterinarian, and I recently accidentally jabbed myself with the animal vaccine (RB-51 or strain 19, or REV-1) while I was vaccinating cows (or sheep, goats). What do I need to do?

These are live vaccines, and strain 19 is known to cause disease in humans. Although we know less about the other vaccines, the recommendations are the same. You should see a health care provider. A baseline blood sample should be collected for testing for antibodies. We recommend that you take antibiotics (doxycycline and rifampin for strain 19 and REV-1, or doxycycline alone for RB-51) for 3 weeks. At the end of that time you should be rechecked and a second blood sample should be collected. (The sample can also be collected at 2 weeks.) The same recommendations hold true for spraying vaccine in the eyes (6 weeks of treatment in this case) or spraying onto open wounds on the skin.

Date: October 6, 2005
Content source: Coordinating Center for Infectious Diseases / Division of Bacterial and Mycotic Diseases

SAFER·HEALTHIER·PEOPLE™
Centers for Disease Control and Prevention,1600 Clifton Rd, Atlanta, GA 30333, U.S.A
Tel: (404) 639-3311 / Public Inquiries: (404) 639-3534 / (800) 311-3435

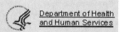

Department of Health
and Human Services

APPENDIX 10:   Key Facts about Tularemia

This fact sheet provides important information that can help you recognize and get treated for tularemia. For more detailed information, please visit the Centers for Disease Control and Prevention (CDC) Tularemia Web site (www.bt.cdc.gov/agent/tularemia).

What Is Tularemia?

Tularemia is a potentially serious illness that occurs naturally in the United States. It is caused by the bacterium *Francisella tularensis* found in animals (especially rodents, rabbits, and hares).

What Are the Symptoms of Tularemia?

Symptoms of tularemia could include:

- sudden fever
- chills
- headaches
- diarrhea
- muscle aches
- joint pain
- dry cough
- progressive weakness

People can also catch pneumonia and develop chest pain, bloody sputum and can have trouble breathing and even sometimes stop breathing.

Other symptoms of tularemia depend on how a person was exposed to the tularemia bacteria. These symptoms can include ulcers on the skin or mouth, swollen and painful lymph glands, swollen and painful eyes, and a sore throat.

How Does Tularemia Spread?

People can get tularemia many different ways:

- being bitten by an infected tick, deerfly or other insect
- handling infected animal carcasses
- eating or drinking contaminated food or water
- breathing in the bacteria, *F. tularensis*

Tularemia is not known to be spread from person to person. People who have tularemia do not need to be isolated. People who have been exposed to the tularemia bacteria should be treated as soon as possible. The disease can be fatal if it is not treated with the right antibiotics.

How Soon Do Infected People Get Sick?

Symptoms usually appear 3 to 5 days after exposure to the bacteria, but can take as long as 14 days.

What Should I Do if I Think I Have Tularemia?

Consult your doctor at the first sign of illness. Be sure to let the doctor know if you are pregnant or have a weakened immune system.

How Is Tularemia Treated?

Your doctor will most likely prescribe antibiotics, which must be taken according to the directions supplied with your prescription to ensure the best possible result. Let your doctor know if you have any allergy to antibiotics.

A vaccine for tularemia is under review by the Food and Drug Administration and is not currently available in the United States.

What Can I Do to Prevent Becoming Infected with Tularemia?

Tularemia occurs naturally in many parts of the United States. Use insect repellent containing DEET on your skin, or treat clothing with repellent containing permethrin, to prevent insect bites. Wash your hands often, using soap and warm water, especially after handling animal carcasses. Be sure to cook your food thoroughly and that your water is from a safe source.

Note any change in the behavior of your pets (especially rodents, rabbits, and hares) or livestock, and consult a veterinarian if they develop unusual symptoms.

Can Tularemia Be Used as a Weapon?

Francisella tularensis is very infectious. A small number (10-50 or so organisms) can cause disease. If F. tularensis were used as a weapon, the bacteria would likely be made airborne for exposure by inhalation. People who inhale an infectious aerosol would generally experience severe respiratory illness, including life-threatening pneumonia and systemic infection, if they are not treated. The bacteria that cause tularemia occur widely in nature and could be isolated and grown in quantity in a laboratory, although manufacturing an effective aerosol weapon would require considerable sophistication.

What is CDC Doing about Tularemia?

The CDC operates a national program for bioterrorism preparedness and response that incorporates a broad range of public health partnerships. Other things CDC is doing include:

- Stockpiling antibiotics to treat infected people
- Coordinating a nation-wide program where states share information about tularemia

- Creating new education tools and programs for health professionals, the public, and the media.

For more information, visit www.bt.cdc.gov/agent/tularemia, or call the CDC public response hotline at (888) 246-2675 (English), (888) 246-2857 (Español), or (866) 874-2646 (TTY)

Plague is an infectious disease that affects animals and humans. It is caused by the bacterium Yersinia pestis. This bacterium is found in rodents and their fleas and occurs in many areas of the world, including the United States.

Y. pestis is easily destroyed by sunlight and drying. Even so, when released into air, the bacterium will survive for up to one hour, although this could vary depending on conditions.

Pneumonic plague is one of several forms of plague. Depending on circumstances, these forms may occur separately or in combination:

- **Pneumonic plague** occurs when *Y. pestis* infects the lungs. This type of plague can spread from person to person through the air. Transmission can take place if someone breathes in aerosolized bacteria, which could happen in a bioterrorist attack. Pneumonic plague is also spread by breathing in *Y. pestis* suspended in respiratory droplets from a person (or animal) with pneumonic plague. Becoming infected in this way usually requires direct and close contact with the ill person or animal. Pneumonic plague may also occur if a person with bubonic or septicemic plague is untreated and the bacteria spread to the lungs.
- **Bubonic plague** is the most common form of plague. This occurs when an infected flea bites a person or when materials contaminated with Y. pestis enter through a break in a person's skin. Patients develop swollen, tender lymph glands (called buboes) and fever, headache, chills, and weakness. Bubonic plague does not spread from person to person.
- **Septicemic plague** occurs when plague bacteria multiply in the blood. It can be a complication of pneumonic or bubonic plague or it can occur by itself. When it occurs alone, it is caused in the same ways as bubonic plague; however, buboes do not develop. Patients have fever, chills, prostration, abdominal pain, shock, and bleeding into skin and other organs. Septicemic plague does not spread from person to person.

Symptoms and Treatment

With pneumonic plague, the first signs of illness are fever, headache, weakness, and rapidly developing pneumonia with shortness of breath, chest pain, cough, and sometimes bloody or watery sputum. The pneumonia progresses for 2 to 4 days and may cause respiratory failure and shock. Without early treatment, patients may die.

Early treatment of pneumonic plague is essential.   To reduce the chance of death, antibiotics must be given within 24 hours of first symptoms. Streptomycin, gentamicin, the tetracyclines, and chloramphenicol are all effective against pneumonic plague.

Antibiotic treatment for 7 days will protect people who have had direct, close contact with infected patients. Wearing a close-fitting surgical mask also protects against infection.

A plague vaccine is not currently available for use in the United States.

For more information, visit www.bt.cdc.gov or
call the CDC public response hotline at
(888) 246-2675 (English),
(888) 246-2857 (Español), or
(866) 874-2646 (TTY)

APPENDIX 12:   Botulism Facts for Health Care Providers

Information and guidance for clinicians can be found on the **Botulism: Clinical Guidance** site.

Agent	Toxin produced by *Clostridium botulinum*, an encapsulated, anaerobe, gram-positive, spore-forming, rod-shaped (bacillus) bacterium
Disease	Botulism is a neuroparalytic (muscle-paralyzing) disease. There are three forms of naturally occurring botulism:

Disease (continued):
- **Foodborne botulism**
  Caused by ingestion of pre-formed toxin
- **Infant botulism**
  Caused by ingestion of *C. botulinum* which produces toxin in the intestinal tract
- **Wound Botulism**
  Caused by wound infection with *C. botulinum* that secretes the toxin

Botulinum Toxin as a Biological Weapon
- Aerosolized botulinum toxin is a possible mechanism for a bioterrorism attack
- **Inhalational botulism does not occur naturally**
- Inhalational botulism cannot be clinically differentiated from the 3 naturally occurring forms
- Indications of intentional release of a biologic agent may include:
  o An unusual geographic clustering of illness (e.g., persons who attended the same public event or gathering)
  o A large number of cases of acute flaccid paralysis with prominent bulbar palsies, especially if occurring in otherwise healthy persons

Transmission
Botulism is not transmissible from person-to-person

Incubation
Symptoms begin within 6 hours to 2 weeks after exposure (often within 12-36 hours)

Symptoms/Signs
- Symmetrical cranial neuropathies
  o Difficulty swallowing or speaking, dry mouth
  o Diplopia (double vision), blurred vision, dilated or non-reactive pupils, ptosis (drooping eyelids)
- Symmetric descending weakness respiratory dysfunction (requiring mechanical ventilation)
- Descending flaccid paralysis
- Intact mental state
- No sensory dysfunction
- No fever

Diagnosis/Lab/Reporting
- **Clinicians should contact their state health departments to report suspected cases**
- Diagnosis: history and clinical exam
- Laboratory confirmation:
  o Demonstrating the presence of toxin in serum, stool, or food
  o Culturing *C. botulinum* from stool, wound or food

Differential Diagnoses	Differential Diagnoses for Adults	Differential Diagnoses for Infants
	<ul><li>Guillain-Barre syndrome</li><li>Myasthenia gravis</li><li>Cerebrovascular accident (CVA)</li><li>Bacterial and/or chemical food poisoning</li><li>Tick paralysis</li><li>Chemical intoxication (e.g., carbon monoxide)</li><li>Mushroom poisoning</li><li>Poliomyelitis</li></ul>	<ul><li>Sepsis</li><li>Meningitis</li><li>Electrolyte-mineral imbalance</li><li>Reye's syndrome</li><li>Congenital myopathy</li><li>Werdnig-Hoffman disease</li><li>Leigh disease</li></ul>
Treatment	<ul><li>Prompt diagnosis is essential</li><li>Antitoxin is effective in reducing the severity of symptoms, if administered early</li><li>A supply of antitoxin against botulism is maintained by the CDC</li><li>State health departments should contact CDC to arrange for a clinical consultation by phone, and (if indicated) the release of the antitoxin</li><li>Supportive care as needed, including mechanical ventilation</li></ul>	
Prophylaxis	<ul><li>Botulism can be prevented by the administration of neutralizing antibody in the bloodstream</li><li>Passive immunity can be provided by equine botulinum antitoxin or by specific human hyperimmune globulin, while endogenous immunity can be induced by immunization with botulinum toxoid</li></ul>	
Control Measures	<ul><li>Medical personnel caring for patients with suspected botulism should use standard precautions</li><li>Patients with suspected botulism do not need to be isolated</li><li>If meningitis is suspected in a patient with flaccid paralysis, medical personnel should use droplet precautions</li><li>Heating to an internal temperature of 85°C for at least 5 minutes will detoxify contaminated food or drink</li><li>When exposure is anticipated, some protection may be conferred by covering the mouth and nose with clothing such as an undershirt, shirt, scarf, or handkerchief</li><li>In contrast with mucosal surfaces, intact skin is impermeable to botulinum toxin</li><li>After exposure to botulinum toxin, clothing and skin should be washed with soap and water</li><li>Contaminated objects or surfaces should be cleaned with 0.1% hypochlorite bleach solution if they cannot be avoided for the hours to days required for natural degradation</li></ul>	
For more information	For more information, please visit the Botulism Emergency Preparedness and Response page. You may also contact 1-800-CDC-INFO, or e-mail coca@cdc.gov.	

Page last modified April 19, 2006

Page Located on the Web at http://www.bt.cdc.gov/agent/botulism/hcpfacts.asp

APPENDIX 13: Fact Sheet: Anthrax Information for Health Care Providers

Inhalation Anthrax

Incubation Period
- Usually <1 week; may be prolonged for weeks (up to 2 months)

Typical Signs/Symptoms (often biphasic, but symptoms may progress rapidly) Initial Phase
- Non-specific symptoms such as low-grade fever, nonproductive cough, malaise, fatigue, myalgias, profound sweats, chest discomfort (upper respiratory tract symptoms are rare)
- Maybe rhonchi on exam, otherwise normal
- Chest X-ray:
  - o mediastinal widening
  - o pleural effusion (often)
  - o infiltrates (rare)

Subsequent phase
- 1–5 days after onset of initial symptoms
- May be preceded by 1–3 days of improvement
- Abrupt onset of high fever and severe respiratory distress (dyspnea, stridor, cyanosis)
- Shock, death within 24–36 hours

Laboratory
- Coordinate all aspects of testing, packaging, and transporting with public health laboratory/Laboratory Response Network (LRN).
- Obtain specimens appropriate to system affected:
  - o blood (essential)
  - o pleural fluid
  - o cerebral spinal fluid (CSF)
  - o skin lesion

Clues to diagnosis
- Gram-positive bacilli on unspun peripheral blood smear or CSF
- Aerobic blood culture growth of large, gram-positive bacilli provides preliminary identification of *Bacillus* species.

Treatment (See "Inhalational Anthrax Treatment Protocol" at * http://www.cdc.gov/mmwr/preview/mmwrhtml/mm5042a1.htm for specific therapy)
- Obtain specimens for culture BEFORE initiating antimicrobial therapy.
- Initiate antimicrobial therapy immediately upon suspicion.
- Do NOT use extended-spectrum cephalosporins or trimethoprim/sulfamethoxazole because anthrax may be resistant to these drugs.
- Supportive care including controlling pleural effusions

Precautions
- Standard contact precautions

Gastrointestinal Anthrax

Incubation Period
- Usually 1–7 days

Typical Signs/Symptoms

Initial phase
- Nausea, anorexia, vomiting, and fever progressing to severe abdominal pain, hematemesis, and diarrhea that is almost always bloody
- Acute abdomen picture with rebound tenderness may develop.
- Mesenteric adenopathy on computed tomography (CT) scan likely. Mediastinal widening on chest X-ray has been reported.

Subsequent phase
- 2–4 days after onset of symptoms, ascites develops as abdominal pain decreases.
- Shock, death within 2–5 days of onset

Laboratory
- Coordinate all aspects of testing, packaging, and transporting with public health laboratory/LRN.
- Obtain specimens appropriate to system affected:
  o blood (essential)
  o ascitic fluid

Clues to diagnosis
- Gram-positive bacilli on unspun peripheral blood smear or ascitic fluid
- Pharyngeal swab for pharyngeal form
- Aerobic blood culture growth of large, gram-positive bacilli provides preliminary identification of *Bacillus* species.

Treatment (See "Inhalational Anthrax Treatment Protocol" at http://www.cdc.gov/ mmwr/preview/mmwrhtml/mm5042a1.htm for specific therapy)
- Obtain specimens for culture BEFORE initiating antimicrobial therapy.
- Early (during initial phase) antimicrobial therapy is critical.
- Do **NOT** use extended-spectrum cephalosporins or trimethoprim/sulfamethoxazole because anthrax may be resistant to these drugs.

Precautions
- Standard precautions

Oropharyngeal Anthrax

Incubation Period
- Usually 1–7 days

Typical Signs/Symptoms

Initial phase

- Fever and marked unilateral or bilateral neck swelling caused by regional lymphadenopathy
- Severe throat pain and dysphagia
- Ulcers at the base of the tongue, initially edematous and hyperemic

Subsequent phase

- Ulcers may progress to necrosis
- Swelling can be severe enough to compromise the airway

Laboratory

- Coordinate all aspects of testing, packaging, and transporting with public health laboratory/LRN.
- Obtain specimens appropriate to system affected:
  - o blood (essential)
  - o throat

Clues to diagnosis

- Aerobic blood culture growth of large, gram-positive bacilli provides preliminary identification of Bacillus species.

Treatment (See "Inhalational Anthrax Treatment Protocol" at http://www.cdc.gov/mmwr/preview/mmwrhtml/mm5042a1.htm for specific therapy)

- Obtain specimens for culture BEFORE initiating antimicrobial therapy.
- Do NOT use extended-spectrum cephalosporins or trimethoprim/sulfamethoxazole because anthrax may be resistant to these drugs.
- Supportive care including controlling ascites

Precautions

- Standard contact precautions

For more information,
visit www.bt.cdc.gov/agent/anthrax, or
call CDC at 800-CDC-INFO (English and Spanish) or
888-232-6348 (TTY).

The Medical Management of Chemical Casualties

Search | E-Mail | Credits

Medical Supplies Database

Home
About this CD
Agent Scenarios
Minicourses
Books & Manuals
Related Sites
In-House Courses
Satellite Broadcast
FTX
Video Segments
CCCD Information
Glossary

This database is a list of everything that the military medical system thinks it will need to care for chemical and biological casualties. It was produced by the Joint Readiness Clinical Advisory Board (JRCAB), COL Cornelius Maher, commander, based upon the recommendations of the chemical and biological panels of 1999 and 2001. The chemical panels included members of the US Army Medical Research Institute of Chemical Defense staff. The biological panels included members of the staff at the US Army Medical Research Institute of Infectious Diseases.

This list of items derives from JRCAB "patient conditions", hypothetical patients with syndromes caused by chemical and biological warfare agents. The supplies are listed alphabetically, not by patient, since there is enormous overlap between patients. You may contact JRCAB directly, if you would like, for example, the specific list of items required to care for a severe nerve agent casualty. In the database, as presented here, all of the items are listed together, along with National Stock Numbers (NSN's) and associated costs. You may use this list as a reference for supplies you wish to order, or to identify or print out selected items with their costs. Products listed in the database may be purchased through the approved military purchasing system. If you are outside of the military, you may purchase them through commercial vendors.

This list is supplied to us, for distribution in this format, freely by JRCAB, who maintains it and updates it every two years. JRCAB may be contacted directly at Fort Detrick, Maryland, at 301-619-2001 or DSN 343-2001, website: http://www.armymedicine.army.mil/jrcab.

Database Usage Instructions

You must have Microsoft Access 2000 installed on your computer in order to open this database. Two versions of this database are available:

Limited - can be run from the CD or your hard drive
Full - can only be run from your hard drive (requires 1MB Hard Drive Space)

To run either version of the database, print and follow the Instructions for Using the Medical Supplies Database.

- Medical Supplies Database (Limited Version)
- Medical Supplies Database (Full Version)
- Medical Supplies List (Adobe Acrobat)

**For more information contact
the Chemical Casualty Care Division of the USAMRICD at (410) 436-2230**

NBC Medical Supplies List

QTY	ITEM	NSN	UI	COST	ITEM TOTAL
	ACETAMINOPHEN SUPPOSITORIES 650MG ADULT RECTAL I.S. 100/PACKAGE	6505011677283	PG	$37.98	
	ACETAMINOPHEN TABLETS USP 0.325GM 1000S	6505009857301	BT	$7.30	
	ACETAZOLAMIDE TABLETS USP 250MG 100 TABLETS PER BOTTLE	6505006640857	BT	$3.22	
	ACETIC ACID GLACIAL USP 1 LB	6505001002470	BT	$13.73	
	ACETONE ACS LIQUID 1 PT (473 ML)	6810007534780	BT	$3.30	
	ACYCLOVIR CAPSULES 200MG 100 CAPSULES PER BOTTLE	6505012069246	BT	$136.17	
	ADAPTER INJECTION-ASPIRATION SITE MALE LUER REUSABLE STERILE200S	6515010701497	PG	$79.96	
	ADAPTER MEDICINAL 22MM ID ODORLESS PLASTIC DISPOSABLE 50S	6515011732039	PG	$14.54	
	ADAPTER OXYGEN FLOW METER TO TUBING 1.50"LG NON-STER GREEN 50S	6515012988018	PG	$29.92	
	ADENOSINE INJECTION USP 2ML SINGLE DOSE VIAL 10 PER PACKAGE	6505013809548	PG	$13.06	
	ADHESIVE MOLESKIN 4 YDS LONG 9" WIDE SOFT 12 PER PACKAGE	6510014562000	PG	$279.09	
	ADHESIVE TAPE SURGICAL POROUS WOVEN 1"X 10YD 12S	6510009268882	PG	$12.81	
	ADHESIVE TAPE SURGICAL POROUS WOVEN 3 INCHES BY 10 YARDS 4S	6510009268884	PG	$13.05	
	ADHESIVE TIES SURGICAL 11.125 BY 7.250 INCHES 24S	6510000033058	PG	$59.25	
	ADHESIVE TISSUE OCTLYCANOACRYLATE ADH USE TO CLOSE SURG WOUND12S	6515014590249	PG	$260.52	
	ADMINISTRATION SET	6515014819766	PG	$81.39	
	AIRWAY NASOPHARYNGEAL 6MM ID 8MM OD SMOOTH RD EDGES PLAS DISP10S	6515012331920	PG	$38.43	
	AIRWAY NASOPHARYNGEAL ARGYLE DESIGN 28FR 6.9X9.3MM CRVD PLAS 10S	6515011295437	PG	$43.78	
	AIRWAY PHARYNGEAL 4" AIRWAY/CUT AWAY FLANGE KINK RES 30FR 30S	6515011649637	PG	$31.37	
	AIRWAY PHARYNGEAL BERMAN DSGN MEDIUM 90MM LG OPEN SIDE DISP 12S	6515013343346	PG	$13.43	
	AIRWAY PHARYNGEAL GUEDEL DSGN SZ 4 MED/LGE ADL 90MM LG PLAS 30S	6515012331913	PG	$32.65	
	ALBUMIN HUMAN USP 25% 100ML CAN	6505002998179	CN	$147.93	
	ALBUTEROL INHALATION AEROSOL 17GM CONTAINER 200 METERED SPRAYS	6505011169245	PG	$3.18	
	ALBUTEROL SULFATE INHALATION SOLUTION 20ML BOTTLE WITH DROPPER	6505012579953	BT	$10.41	
	AMIODARONE HYDROCHLORIDE INJECTION 50MG/ML 3ML AMPUL 10/PACKAGE	6505014232851	PG	$855.15	
	AMOXICILLIN AND POTASSIUM CLAVULANATE TABLETS 30 TABLETS/BOTTLE	6505012036259	BT	$67.43	
	AMPICILLIN SODIUM STERILE USP POWDER FORM 1GM BOTTLE	6505009933518	BT	$0.42	
	ANESTHESIA BREATHING CIRCUIT	6515013968089	PG	$114.30	
	ANESTHESIA SET EPIDURAL 10S 2 COMPONENTS	6515011486997	PG	$91.93	
	ANTIMICROBIAL INOCULATION & PREPARATION KIT 60 PROCEDURES	6550013100613	EA	$169.20	
	APPLICATOR DISP POV-IOD 7.2ML IMPREG TIP .125DIA 3.75 TO 4.5"L	6510010087917	PG	$23.05	
	APPLICATOR PLASTIC/WOOD ROD 6" LG 0.083" DIA STERILE DISP 2000S	6515009051473	PG	$19.40	
	APPLICATOR WOOD ROD 6"LG .083" DIA PLAIN TIP CYL SHAPE DISP 864S	6515003038100	PG	$4.90	
	ARTIFICIAL TEARS SOLUTION 15ML BOTTLE WITH OPHTHALMIC TIP	6505010156456	BT	$1.86	
	ASPIRIN TABLETS USP 0.324GM 100S	6505001009985	BT	$1.49	
	ATROPINE SULFATE INJECTION USP 0.1MG/CC 10ML BOTTLE 10 PER BOX	6505010946196	BX	$30.58	
	ATROPINE SULFATE INJECTION USP 0.4MG/ML 20ML VIAL	6505007542547	VI	$4.28	
	AZITHROMYCIN FOR INJECTION 500MG/VIAL 10 PER PACKAGE	6505014590414	PG	$146.45	
	AZITHROMYCIN TABLETS 250MG 18 TABLETS PER PACKAGE	6505014491618	PG	$70.70	
	BACITRACIN OINTMENT USP 7100 UNITS 0.5OZ TUBE 12 TUBES/PACKAGE	6505001596625	PG	$0.88	
	BAG AND TAPES 6IN X 3.5IN BLACK NYLON CODURA PACK BAG	6530014726786	EA	$50.00	
	BAG BIOHAZARD DISPOSAL A4FLAT 45X38IN PLAS RED HEAT SEAL TOP100S	8105013806463	PG	$101.35	
	BAG OSTOMY FLAT POUCH CUT TO FIT CAN BE CUT TO 4IN STER 5S	6515014679307	PG	$25.66	
	BAG PLASTIC 39 X 33 IN DARK BROWN & GREEN 33 GAL 50 LB 125S	8105011839769	BX	$25.44	
	BAG PLASTIC 39"HT 36"W SING WALL HEAT SEAL 44GAL CAPACITY 200S	8105012708749	PG	$84.74	
	BAG PLASTIC A4 FLAT 12IN X 12IN HEAT SEAL CLEAR	8105008377757	BX	$34.53	

NBC Medical Supplies List

QTY	ITEM	NSN	UI	COST	ITEM TOTAL
	BAG PLASTIC A4 FLAT 8 X 8"	8105008377755	MX	$36.89	
	BAG WASTE RECEPTACLE 3 BUNDLES OF 100 300S	8105014683177	PG	$97.86	
	BAG WASTE RECEPTACLE 48X40"1.5MM THK RED NO SEAMS PLASTIC 250S	6530011554062	PG	$72.78	
	BAG,SAND	8105002854744	HD	$42.94	
	BAND PERSONAL IDENTIFICATION ADULT PATIENT PLASTIC RED 250S	6530011663494	PG	$57.65	
	BAND PERSONNEL IDENTIFICATION ADULT PATIENT PLASTIC 400S	6530001047631	PG	$41.61	
	BANDAGE ADH.75X3" FLESH/CLEAR STER DRESS AFFIXED TO PLAS ADH100S	6510005977469	PG	$2.08	
	BANDAGE ADHESIVE .75 BY 3 INCHES FLESH 300S	6510009137909	BX	$10.00	
	BANDAGE ADHESIVE RAYON WOVEN FABRIC 100S	6510011460988	PG	$9.44	
	BANDAGE CAST FIBERGLASS RESIN IMPREGNATED 4YDX4IN WOVEN 10S	6510012445742	PG	$94.73	
	BANDAGE CAST IMPREGNATED 4YD X 5IN MOISTURE RESIST LT-WEIGHT100S	6510014649109	PG	$50.64	
	BANDAGE CAST IMPREGNATED4YD X3IN MOISTURE RESIST LIGHT-WEIGHT100	6510014649117	PG	$33.31	
	BANDAGE ELAS 10YDX1.75"RBBR W/NYLON & PLYSTR WHITE OPEN NET	6510010888470	EA	$18.83	
	BANDAGE ELAS 4"X5YDS STER NONWVN W/ELAS FIBERS COHESIVE MATL 18S	6510013549509	PG	$51.31	
	BANDAGE ELASTIC CLOTH FLANNEL COTTON TAN 4 WING DRESSING 3X3"50S	6510011621548	PG	$11.09	
	BANDAGE ELASTIC COTTON ROLLED 54X6" W/ADHESIVE COMPOUND STER 10S	6510010972199	PG	$41.83	
	BANDAGE ELASTIC COTTON RUBBER WARP THREADS 4 IN X 1.50 YARDS 10S	6510001039749	PG	$23.68	
	BANDAGE ELASTIC FLESH ROLLED 4.5YDX4"WASHABLE PRESSURE 12S	6510009355822	PG	$8.50	
	BANDAGE ELASTIC FLESH ROLLED NONSTER 4.5YDX3" PRESSURE 12S	6510009355821	PG	$8.71	
	BANDAGE ELASTIC FLESH ROLLED NONSTERILE 6"X 4.5 YDS 12S	6510009355823	PG	$21.15	
	BANDAGE FELT ORTHOPEDIC WHITE NONSTERILE ROLLED 6IN X 4YD 36S	6510008172632	PG	$79.60	
	BANDAGE FELT ORTHOPEDIC WHITE ROLLED NONSTERILE 4IN X 4 YD 72S	6510008172634	PG	$52.47	
	BANDAGE GAUZE 6-PLY ROLLED WHITE STERILE 4YD X 4.5IN I.S. 100S	6510000583047	PG	$98.55	
	BANDAGE GAUZE COMPRESSED CAMOUFLAGED OLIVE GREEN 216"X3" STERILE	6510002003185	EA	$2.54	
	BANDAGE GAUZE TUBULAR 50YD X .875 W/APPLICATOR WHITE NONSTERILE	6510002007015	RO	$23.99	
	BANDAGE KIT ELASTIC	6510014600849	EA	$3.50	
	BANDAGE MUSLIN COMPRESSED OLIVE DRAB37X37X52" TRIANG W/SFTY PINS	6510002011755	EA	$2.86	
	BASIN SITZ BATH PATIENT-OPERATED PLASTIC POLYETHYLENE 10S	6530010306861	PG	$19.80	
	BASIN WASH 7QT STERILIZABLE PLASTIC FITS RING STAND 12S	6530014588301	PG	$1.00	
	BATTERY NONRECHARGEABLE 9V 2 TERM .656"W X 1.031"L X 1.906"H DRY	6135010631978	PG	$5.52	
	BATTERY NONRECHARGEABLE ALKALINE 1.5 VOLT C CELL 12 PER PACKAGE	6135009857846	PG	$6.51	
	BATTERY NONRECHARGEABLE ALKALINE 1.5V 1 YEAR SHELF LIFE	6135009857845	PG	$4.74	
	BECLOMETHASONE DIPROPIONATE INHALATION AEROSOL 16.8GM OR 5.88OZ	6505012400587	PG	$38.52	
	BECLOMETHASONE DIPROPIONATE NASAL SUSPENSION 25GM BOTTLE	6505012754811	EA	$43.02	
	BEDPAN FRACTURE-TYPE PLAS AUTOCLAVABLE 12S	6530014617951	PG	$46.67	
	BEDPAN PONTOON-TYPE STACKABLE AUTOCLAVABLE 12S	6530014617882	PG	$32.20	
	BLADE ORTHOPEDIC CAST CUTTER 2IN LG 2.5IN WIDTH 2.5IN DIA	6515014410907	EA	$21.50	
	BLADE ORTHOPEDIC CAST CUTTER 2IN LG 2IN WIDTH 2IN DIA BLADE	6515014410905	EA	$21.50	
	BLADE SURG KNIFE DET NO.10 SMALL TANG U/W 3 3L 7 9 HDL CS 150S	6515010095295	PG	$15.16	
	BLADE SURG KNIFE DET NO.15 SMALL TANG U/W 3 3L 7 9 HDL STEEL150S	6515010095293	PG	$32.04	
	BLADE SURG KNIFE DETACHABLE NO.64 USE WITH 3H 3HL 3K HANDLE 24S	6515007822619	PG	$38.40	
	BLANKET CASUALTY PLASTIC FILM ALUM COATED 96X56INCHES GREEN	7210009356666	EA	$6.98	
	BLANKET THERMAL 4X7 FT DISP F/USE TO PRESERVE BODY TEMP 25S	6532014586602	PG	$131.00	
	BLOOD & INTRAVENOUS WARMING SET 8'LG 18MM PRI VOL STER PATH 30S	6515013719617	PG	$444.69	
	BLOOD GP SERUM ANTI-A LIQ 10ML VI F/SLIDE/TUBE/MICROPLATE 15'S	6550013170288	PG	$57.65	
	BLOOD GROUPING SERUM ANTI-B 10ML VI F/SLIDE/TUBE/MICROPLATE 15'S	6550013170289	PG	$69.17	

NBC Medical Supplies List

QTY	ITEM	NSN	UI	COST	ITEM TOTAL
	BLOOD GROUPING SERUM BT W/DROPPER 10 ML RH POS/NEG DET 15'S	6550013299842	PG	$145.05	
	BLOOD RECIPIENT SET INDIRECT TRANSFUSION	6515014885514	PG	$106.06	
	BOOT COVER KNEE HI SURE GRIP AQUA TRACK BLACK ELAS ANKLE&TOP200S	6532014702782	PG	$145.00	
	BRIMONIDINE TARTRATE 0.2% 5ML USED IN TREATMENT OF GLAUCOMA	6505014590429	BT	$14.26	
	BRUSH SANITARY STRAIGHT BRUSH HEAD SPIRAL WOUND WOOD HANDLE 17"	7920002349317	EA	$5.51	
	BRUSH SET INSTRUMENT CLEANING NYLON VARIOUS SIZES AND LENGTHS	7920011721118	SE	$56.63	
	BRUSH-SPONGE SURG SCRUB4%CHLORLLHEXIDINE GLUCONATE DISP 200S	6530012114810	PG	$87.84	
	BRUSH-SPONGE SURGICAL SCRUB DIPS NONSTER 3% PCMX I.P. 300S	6530013643353	PG	$153.72	
	BRUSH-SPONGE SURGICAL SCRUB STER NON-IRRITATING IND SEALED 300S	6530013344379	PG	$128.47	
	BUCKET CAST DISP WAX COVERED DOUBLE WRAPPED PAPER 5QT	6530014883954	PG	$28.50	
	BUMETANIDE INJECTION 0.25MG/ML 2ML AMPUL 10 AMPULES PER PACKAGE	6505011742389	PG	$10.49	
	BUPIVACAINE HYDROCHLORIDE INJECTION USP .50% 30ML VIAL 10S	6505011277946	PG	$13.40	
	CABLE ASSEMBLY ELECTROSURGICAL APPARATUS 12' LONG DISP 10/PG	6515013488746	PG	$68.00	
	CABLE MONITORING SET ARTERIAL PRESSURE REUSABLE 15FT ABBOTT	6515014346106	EA	$118.58	
	CALCIUM CHLORIDE INJECTION USP 10% 10CC NEEDLE W/SYRINGE 10S	6505001394548	PG	$28.95	
	CALCIUM CHLORIDE SOLUTION 15ML CONTAINER 6 PER PACKAGE	6550011197686	PG	$31.82	
	CANE WALKING 36.5" LG WITHOUT TIP	6515007740000	EA	$7.46	
	CANISTER ASSY SUCTION SURG PLASTIC 200CC CAP LID HYDROP FILTER	6515014862077	PG	$205.00	
	CANNUAL NASAL OXYGEN PLASTIC 252" DISPOSABLE NON-FLARED TIP	6515014737526	PG	$29.21	
	CANNULA LARYNGEAL YANKAUER DSGN W/O CONT VENT STR DISTAL END 50S	6515010747812	PG	$34.99	
	CAP OPERATING SURGICAL WOMEN BLUE MAIN BODY 24IN DISPOSABLE 600S	6532013085341	PG	$66.56	
	CAP THERMAL ADULT BOUFFANT STYLE 100S	6532014586597	PG	$76.00	
	CARD INDICATING STERILZ DBL INFO/DATA OF GENESIS TIME/DATE500S	6530013702241	PG	$23.73	
	CATHETER & CONN SUCT TRACH E14 18FR WHISTLE TIP 2 EYES MAS 50S	6515004588416	PG	$25.31	
	CATHETER INTRAVENOUS D9 14 GA PLAST POLY RD 2" STR NEEDLE 200S	6515014884911	PG	$209.00	
	CATHETER INTRAVENOUS D9 16GA PLAST POLY RD 200S	6515014884971	PG	$153.59	
	CATHETER INTRAVENOUS D9 18GA PLAST POLY RD 1 1/4" STR NEEDLE 200	6515014884994	PG	$216.07	
	CATHETER INTRAVENOUS D9 20GA PLAST POLY RD 1" STR NDL 200S	6515014885454	PG	$186.00	
	CATHETER KIT PACING TRANSVENOUS W/BIPOLAR SWAN-GANZ 5FR CATHETER	6515011669008	EA	$108.33	
	CATHETER URETHRAL FOLEY 24FR ROUND TIP SILICONIZED RBBR DISP 12S	6515001049006	BX	$25.65	
	CATHETERIZATION KIT CARDIO 20GAX1.75" RADPQ TEFLON CATH STER 50S	6515011660154	PG	$403.63	
	CATHETERIZATION KIT SUPRAPUBIC PUNCTURE 5CC BALLOON CAP 4 COMP	6515013405434	EA	$88.94	
	CATHETERIZATION KT URETHRAL PRECISION400URINE METER FOLEY16FR5CC	6515014784788	PG	$180.00	
	CEFAZOLIN SODIUM INJECTION USP 1GM/GM CEFAZOLIN 10ML VIAL 25S	6505014802501	PG	$43.64	
	CEFTAZIDIME FOR INJECTION 2GM VIAL 10 VIALS PER PACKAGE	6505012314807	PG	$185.44	
	CEFTRIAXONE SODIUM STERILE USP 250MG VIAL 10 VIALS PER PACKAGE	6505012277028	PG	$68.71	
	CEPHALEXIN CAPSULES USP EQUIVALENT TO 250MG 100 PER BOTTLE	6505001656545	BT	$5.69	
	CEPHALOPLASTIN REAGENT 2ML 10S USE W/6550-01-119-7686	6550011159182	PG	$16.31	
	CHLORAMPHENICOL SODIUM SUCCINATE STERILE USP 1 GRAM VIAL 10/BOX	6505007540280	BX	$37.33	
	CHLORHEXIDINE GLUCONATE CLEANSING SOLUTION 118ML BOTTLE	6505011534431	BT	$2.22	
	CHLORPROMAZINE HYDROCHLORIDE INJECTION USP 25MG/ML 2ML AMPUL 10S	6505001296709	PG	$48.37	
	CIPROFLOXACIN CONCENTRATE FOR INJECTION 40ML VIAL 60 VIALS/PG	6505013370320	PG	$273.40	
	CIPROFLOXACIN TABLETS USP 500MG I.S. 100 TABLETS PER PACKAGE	6505012738650	PG	$153.50	
	CLEANER TOBACCO PIPE COTTON OR COTTON/SYNTHETIC BLEND 6"LG 32S	9920002929946	BX	$12.22	
	CLEANER&LUBRICANT SURGICAL INSTRUMENTS 1GALLON CONCENTRATE	6840013791945	BT	$28.17	
	CLINDAMYCIN INJECTION USP STERILE 150MG/ML 60ML VIAL	6505012468718	VI	$25.92	

NBC Medical Supplies List

QTY	ITEM	NSN	UI	COST	ITEM TOTAL
	CLIP HEMOSTATIC LIGATION DSGN 6X4.4MM MED TITANIUM ALLOY DISP120	6515013719646	PG	$434.11	
	CLIP HEMOSTATIC LIGATION MEDIUM 6X4.4MM RELOADING TITANIUM 240S	6515013719669	PG	$434.11	
	CLOTH CLEANING WHITE 15 X12" 900S	7920011263250	PG	$82.50	
	COLLAGEN HEMOSTATIC 2 X 2" WAFERS 3% CONCENTRATION 100/PACKAGE	6510014479959	PG	$1,266.00	
	CONNECTOR TUBING STRAIGHT A18 TAPER SERRATIONS 3.5"LG .188"ID	6515009269201	PG	$14.99	
	CONTROL BLD CHEMISTRY ABNORMAL LYOPHILIZED HUMAN SERA = TO 5ML10	6550010348336	BX	$129.56	
	CONTROL COAGULATION ABNORMAL LYOPHILIZED 1ML/VIAL 10S	6550010380792	PG	$60.09	
	CONTROL COAGULATION NORMAL LYOPHILIZED 1ML VIAL 10'S	6550010380793	PG	$20.90	
	CONTROL STRIPS VISUAL URINALYSIS 25S	6550010934010	BT	$35.54	
	COVER ELECTRONIC THERMOMETER PROBE PLASTIC DISPOSABLE 100S	6515013738659	PG	$9.35	
	COVER XRAY CASSETTE LARGE SIZE PILLOWCASE TYPE DISPOSABLE 20S	6525012538295	PG	$116.04	
	CRUTCH AUXILLARY 48-59" LG WOOD VARNISHED ADJUSTABLE	6515007777325	PR	$14.17	
	CULTURE MEDIUM GC AGAR BASE DEHYDRATED 0.25 POUNDS	6550009355870	BT	$21.41	
	CULTURE MEDIUM MACCONKEY AGAR POWDER	6550001258750	BT	$48.43	
	CULTURE MEDIUM MUELLER HINTON 500GM BOTTLE	6550008805608	BT	$53.01	
	CULTURE MEDIUM SOYBEAN-CASEIN DIGEST DEHYDRATED 500 GRAMS	6550011561629	BT	$23.63	
	CULTURE MEDIUM THIOGLYCOLLATE POWDER	6550010339865	BT	$14.78	
	CUP DISPOSABLE ROUND TWO-PIECE ROLLED HOT LIQUID 8 OZ WHITE2000S	7350001623006	BX	$64.42	
	CUP MEDICINE PLAS POLYPROPYLENE 1OZ CAPACITY GRAD100/SLVE/50SLVE	6530012701529	PG	$40.08	
	CURETTE EAR BILLEAU MEDIUM SZ WAX CURETTE 3IN DISP	6515014607918	PG	$64.20	
	CUSHION CRUTCH SPONGE RBBR 8X1.50" 6" OPENING LG 0.688" MEDIAL W	6515007777340	PR	$2.32	
	CUVETTE BLOOD SAMPLE PLAS DISP K31 U/W 6630-01-126-9960 1000'S	6640011414800	PG	$187.87	
	CYCLOPENTOLATE HYDROCHLORIDE OPHTHALMIC SOLUTION USP 1% 15ML	6505002999666	BT	$1.10	
	DEPRESSOR TONGUE 6X0.75X0.0675" PARALLEL EDGES ROUNDED ENDS 100S	6515013055120	PG	$2.75	
	DETERGENT ENZYMATIC	6850014715613	EA	$36.00	
	DETERGENT HOSPITAL GLASSWARE AND INSTRUMENT 1GALLON DISP PLAS 4S	6640007642245	PG	$77.92	
	DETERGENT SURGICAL 4% CHLORHEXIDINE GLUCONATE 32FL OZ	6505010453255	BT	$4.80	
	DEXTROSE INJECTION USP 50% 50ML CARTRIDGE 10 PER BOX	6505001394460	PG	$18.45	
	DIAZEPAM INJECTION USP 5MG/ML 10ML VIAL	6505012406894	VI	$1.88	
	DIAZEPAM INJECTION USP 5MG/ML 2ML SYRINGE WITH NEEDLE 10/PACKAGE	6505001375891	PG	$12.57	
	DIAZEPAM TABLETS USP 5MG 500 TABLETS PER BOTTLE	6505007837218	BT	$2.48	
	DICLOXACILLIN SODIUM CAPSULES USP 250MG 100 CAPSULES PER BOTTLE	6505003697289	BT	$7.07	
	DIFFERENTIATION DISCS MICROORGANISM FACTOR X 300S	6550010316437	PG	$115.13	
	DIFFERENTIATION DISKS MICROORGANISM FACTOR V 300S	6550010324239	PG	$146.42	
	DIFFERENTIATION DISKS MICROORGANISM PNEUMOCOCCUS 300S	6550010316436	PG	$45.02	
	DIFFERENTIATION DISKS MICROORGANISMS FACTORS XV 10S	6550010371058	PG	$97.24	
	DIGOXIN INJECTION USP 0.25MG/ML 2ML AMPUL 10 AMPULS PER PACKAGE	6505005317761	PG	$38.48	
	DILTIAZEM HYDROCHLORIDE INJECTION 5MG/ML 10ML VIAL 6 VIALS/PG	6505013539827	PG	$157.92	
	DIPHENHYDRAMINE HYDROCHLORIDE INJ USP 50MG/ML 1ML SYRINGE 10/BX	6505001487177	BX	$28.61	
	DISC ADHESIVE 1.25" DIAMETER 11/32" APERATURE DOUBLE FACED 612S	6515012478232	PG	$85.26	
	DISH CULTURE PETRI TOP & BOTTOM COMPLETE DISP 15 X 100 MM 500S	6640002400035	PG	$57.97	
	DISH CULTURE PETRI TOP & BOTTOM COMPLETE DISPOSABLE 100S	6640010313139	PG	$89.86	
	DISINFECTANT GENERAL	6840014763011	BT	$48.80	
	DISINFECTANT-DETERGENT GENERAL PURP POWDER NINETY 1/2OZ PACKETS	6840011357409	PG	$30.74	
	DISPOSAL CONTAINER HYPO NEEDLE&SYRINGE PLAS AUTOCLAV2GL W/LID20S	6530011732269	PG	$112.41	
	DISSECTOR KITTNER 200 PER PACKAGE	6515011562354	PG	$101.43	

NBC Medical Supplies List

QTY	ITEM	NSN	UI	COST	ITEM TOTAL
	DOBUTAMINE HYDROCHLORIDE INJECTION 250MG 20ML SINGLE DOSE VIAL	6505012394660	VI	$4.94	
	DOPAMINE HYDROCHLORIDE AND DEXTROSE INJECTION USP 250ML BAG 12S	6505013146681	PG	$223.93	
	DOPAMINE HYDROCHLORIDE INJECTION USP 40MG/ML 10ML UNIT 10/PG	6505011231060	PG	$89.96	
	DOXYCYCLINE HYCLATE FOR INJECTION USP 100MG BOTTLE 5 PER PACKAGE	6505011084828	PG	$10.54	
	DOXYCYCLINE HYCLATE TABLETS USP 100MG 500 TABLETS PER BOTTLE	6505011534335	BT	$25.45	
	DRAINAGE KIT CLO SUCT JACKSON-PRATT RADPQ 10MM W DRAIN DISP 5S	6515013414523	PG	$85.51	
	DRAPE SURGICAL 75X57" I.P. DISP STER FAN-FOLDED 3/4 SH BLUE 20S	6530011732311	PG	$42.11	
	DRAPE X-RAY DISPOSABLE STERILE FITS IMAGE INTENSIFIERS 44X57"10S	6530012114791	PG	$142.56	
	DRESSING BURN 24X36 INCHES STERILE ABSORBENT COTTON GAUZE 15S	6510011532857	PG	$104.57	
	DRESSING BURN 4X16IN SATURATED W/WATER GEL INDIVUALLY WRAPPED28S	6510012435894	PG	$159.32	
	DRESSING BURN FIRST AID WATER-GEL TYPE POLYESTER 8X18"STERILE20S	6510014575844	PG	$103.00	
	DRESSING CHEST WOUND ONE WAY VALVE SELF ADHESIVE KRATON 10/PG	6510014081920	PG	$74.67	
	DRESSING FIRST AID FIELD CAMOUFLAGED 11.5-12"W 11.5-12"LG ABS	6510002017425	EA	$7.24	
	DRESSING FIRST AID FIELD CAMOUFLAGED 7.75-8.25"LG 7.25-7.75"W	6510002017430	EA	$5.04	
	DRESSING FIRST AID FIELD WHITE 4"W X 6.250-7.250" LG ABSORBENT	6510000835573	EA	$2.86	
	DRESSING FIRST AID FLD CAMOUFLAGED 6.250-7.250X4" COMPRESSSTER	6510001594883	EA	$2.90	
	DRESSING OCCLUSIVE ADHESIVE 3IN LG 2IN WIDE TRANSPARENT 100S	6510013122963	PG	$44.25	
	DRESSING OCCLUSIVE ADHESIVE 4.25 OZ AEROSOL SPRAY 12S	6510011562319	PG	$67.67	
	DRUG DELIVERY SYSTEM INHALER METERED DOSE KIT WITH CARRYING CASE	6515013200225	EA	$7.82	
	EARPIECE STETHOSCOPE NOISE PROTECTION DISPOSABLE 50S	6515013690937	PG	$223.45	
	ELECTRODE COAGULATION D8 6" LG SHARP PT EXTENDED BLADE 50S	6515012708828	PG	$167.74	
	ELECTRODE ELECTROCARDIOGRAPH ADULT SIZE PREJELLED DISPOSABLE	6515012787003	PG	$23.06	
	ELECTRODE ELECTRONIC NERVE STIMULATION F/USE W/6515-01-463-0896	6515014630896	EA	$10.00	
	ELECTRODE GEL ELECTROMEDICAL 4 OZ CAP TU CONDUCTIVE DERMA GEL12S	6515013930404	PG	$26.35	
	ELECTRODE GROUNDING ELECTROSURG DISP PREJELLED SELF ADHERING 50S	6515011563051	PG	$198.00	
	ELECTRODE MONITORING ADULT DISP SIL/SIL CHL FOAM ADH PAD 200S	6515011535627	PG	$75.22	
	EMOLLIENT LOTION 2 FL OZ BOTTLE 100 BOTTLES PER PACKAGE	6508014341725	PG	$1.67	
	ENVELOPE STERILIZATION B7FLAT3.5BY23.5IN W/18MO SHL LIFE 500S	6530012247421	PG	$175.68	
	ENVELOPE STERILIZATION PAPER AND PLAS SELF-SEALING 10X5.25" 400S	6530011511807	PG	$44.19	
	ENVELOPE STERILIZATION SELF-SEALING STEAM-GAS 15X12IN 400S	6530011690244	PG	$119.82	
	EPHEDRINE SULFATE INJECTION USP 50MG/ML 1ML AMPUL 100/PACKAGE	6505013859409	PG	$10.97	
	EPINEPHRINE INJECTION USP AQUEOUS 1ML AMPUL 25 AMPULS/PG	6505002998760	PG	$9.88	
	EPINEPHRINE INJECTION USP0.1MG PER ML SYRINGE-NEEDLE UNIT10ML10S	6505010932384	PG	$19.83	
	ERYTHROMYCIN OPHTHALMIC OINTMENT USP 5MG/GM 3.5GM TUBE	6505009820288	TU	$1.86	
	ETOMIDATE INJECTION 20ML SYRINGE-NEEDLE UNIT 10 PER PACKAGE	6505012040681	PG	$250.39	
	EVACUATOR SURGICAL WOUND 48"TUBE LG 4"WOUND EVACUATOR NEEDLE	6515001490671	EA	$12.33	
	EXERCISE BANDS RESI 6"X50YD STRETCHABLE LATEX RUBBER GREEN/HEAVY	6510014658816	RO	$65.75	
	EXERCISER HAND REGULAR SZ FOAM RUBBER PLAS COATED 12S	6530014656777	EA	$4.99	
	EXERCISER LUNG INHAL PLAS 3 CHAM 600CC TO 1200CC PER SEC 12S	6530013598546	PG	$109.76	
	FASTENER TAPE HOOK NYLON WHITE 2" WIDE 25YD ROLL	8315013583555	RO	$69.17	
	FASTENER TAPE PILE NYLON WHITE 2" WIDE 25YD ROLL	8315013583556	RO	$49.00	
	FELT PADDING ORTHO 36X21X0.25" HIGH GRADE WHITE W/DISPEN BOX 1YD	6515013652042	RO	$21.77	
	FENTANYL CITRATE INJECTION USP 5ML AMPUL 10 AMPULES PER BOX	6505010731316	PG	$5.98	
	FILM RADIOGRAPHIC NON-INTERLEAVED 43X35CM T-MAT G/RA 100S	6525013548695	PG	$185.39	
	FILTER STERILIZATION KIMGARD CO FILTERS 7-1/2IN 1000S	6530014585587	PG	$76.00	
	FILTER URLOGICAL DISP W/CLOTH FILTER AND TAB GRIPS 1000S	6515012106172	PG	$101.91	

NBC Medical Supplies List

QTY	ITEM	NSN	UI	COST	ITEM TOTAL
	FLASHLIGHT PENLIGHT	6230006354998	EA	$1.19	
	FLUCONAZOLE TABLETS 150MG 12S	6505013953039	PG	$152.66	
	FLUORESCEIN SODIUM OPHTHALMIC STRIPS MODIFIED 300S	6505011591493	PG	$125.24	
	FOOTWEAR COVERS OPERATING ROOM NON-CONDUCTIVE EXTRA LARGE 100S	6532011522796	PG	$24.63	
	FUROSEMIDE INJECTION USP 10MG/ML 2ML AMPUL 25 PER PACKAGE	6505011575117	PG	$6.35	
	FUROSEMIDE INJECTION USP 10MG/ML STERILE 10ML VIAL 50/PACKAGE	6505014806901	PG	$40.60	
	GAUZE ABSORBENT IODOFORM IMPREGNATED .25 INCH BY 5 YARDS 12S	6510010037697	PG	$34.45	
	GAUZE ABSORBENT IODOFORM IMPREGNATED .500" BY 5YD STERILE 12S	6510011534534	PG	$26.59	
	GAUZE PETROLATUM STEP BACK ACCORDION WHITE 18X3" 12S	6510002020800	PG	$8.54	
	GEL TRANSMISSION ULTRASOUND 250ML SQUEEZE BT 6.50-6.95 PH 12S	6515005293346	PG	$36.94	
	GENTAMICIN SULFATE INJECTION USP 40MG EQUIV/ML 2ML VIAL 25/PG	6505012139514	PG	$7.09	
	GLOVE PATIENT EXAMINING AND TREATMENT LARGE POWDER FREE 1000/PG	6515013244676	PG	$290.70	
	GLOVE PATIENT EXAMINING AND TREATMENT MED RBBR POWDER FREE 1000S	6515013252370	PG	$73.57	
	GLOVE PATIENT EXAMINING AND TREATMENT X-LARGE POWDER FREE 1000S	6515013940006	PG	$82.35	
	GLOVES SURG GEN SURG SZ8 POWDER-FREE LIGHTWEIGHT 50S	6515012522484	PG	$69.17	
	GLOVES SURGEONS' GEN SURG RUBBER NATURAL SZ 7 DISP STER 50S	6515012611137	PG	$107.60	
	GLOVES SURGEONS GEN SURG SZ 6.5 PWDR-FREE LTWT RBBR STER DISP50S	6515012547700	PG	$73.57	
	GLOVES SURGEONS GENERAL SURGERY SZ 8.5 POWDER FREE LIGHTWEIGHT2S	6515012513759	PG	$60.39	
	GLOVES SURGEONS LATEX CREAM COLOR SIZE 6	6515014736347	PG	$195.00	
	GLOVES SURGEONS LATEX CREAM COLOR SIZE 6 1/2	6515014736344	PG	$195.00	
	GLOVES SURGEONS LATEX CREAM COLOR SIZE 7	6515014736341	PG	$195.00	
	GLOVES SURGEONS LATEX CREAM COLOR SIZE 8	6515014736331	PG	$195.00	
	GLOVES SURGEONS LATEX CREAM COLOR SIZE 9	6515014736329	PG	$195.00	
	GLOVES SURGEONS RUBBER LATEX CREAM COLOR SIZE 8 1/2	6515014736323	PG	$195.00	
	GLOVES SURGEONS RUBBER POWDER-FREE SZ 7.5 STERILE DISPOSABLE 50S	6515012534260	PG	$65.88	
	GLOVES SURGRONS LATEX CREAM COLOR SIZE 7 1/2	6515014736340	PG	$195.00	
	GLUCAGON FOR INJECTION USP 1 MG UNIT FOR EMERGENCY USE	6505014667505	EA	$44.13	
	GLYCERIN OPHTHALMIC SOLUTION USP 7.50 ML	6505011636333	BT	$43.78	
	GLYCOPYRROLATE INJECTION USP 0.2 MG PER ML 20 ML	6505010197627	BT	$1.10	
	GOGGLES PROTECTIVE INFECTION DISTORTION-FREE LENSES ADJ 100S	6540012901157	PG	$153.72	
	GOWN ISOLATION DISPOSABLE YELLOW LARGE ELASTIC CUFF 100S	6532011536517	PG	$77.90	
	GRAM STAINING KIT	6550002619053	EA	$37.39	
	HALOPERIDOL INJECTION USP 5MG/ML 1ML AMPUL 10 AMPULES/PACKAGE	6505002688530	PG	$4.94	
	HAND GRIP CRUTCH CLOSED TYPE RUBBER CREAM FITS ALL STD CRUTCHES	6515012538141	PR	$2.75	
	HANDLE ELECTRODE ELECTROSURGERY ACCOM 2.38MM SHANK DIA DISP 50S	6515011535386	PG	$196.82	
	HEMORRHOIDAL SUPPOSITORIES ADULT RECTAL 24 PER PACKAGE	6505013750132	PG	$8.67	
	HEPARIN SODIUM IN SODIUM CHLORIDE INJECTION 500ML BAG 24/PACKAGE	6505013770444	PG	$60.00	
	HEPARIN SODIUM INJECTION USP 1000 UNITS PER ML 10 ML	6505001539740	VI	$0.54	
	HETASTARCH IN SODIUM CHLORIDE INJECTION 500ML BAG 12 BAGS/PG	6505012811247	PG	$295.14	
	HOLDER BLOOD COLLECTING TUBE PLAS POLYPROP 2.438"LG .750"ID 10'S	6630012309964	PG	$3.48	
	HOMATROPINE HYDROBROMIDE OPHTHALMIC SOLUTION USP 5% 15ML BOTTLE	6505006895532	BT	$4.39	
	HUMIDIFIER HYGROSCOPIC HUMID-VENT 2 PORTS	6515014779007	PG	$101.70	
	HUMIDIFIER INHALATION THERAPY APPAR 50CC HYGROSCOPIC PLASTIC 25S	6515012750093	PG	$120.49	
	HUMIDIFIER INHALATION THERAPY APPARATUS 300MM RESERVOIR CAP 50S	6515010988264	PG	$81.86	
	HYDROCODONE BITARTRATE AND ACETAMINOPHEN TABLETS 100 TABS/BOTTLE	6505011899903	BT	$4.61	
	HYDROXYZINE HYDROCHLORIDE TABLETS USP 25MG 500 TABLETS/BOTTLE	6505005799717	BT	$12.55	

NBC Medical Supplies List

QTY	ITEM	NSN	UI	COST	ITEM TOTAL
	IBUPROFEN TABLETS USP 800 MG 500 TABLETS PER BOTTLE	6505012149062	BT	$14.12	
	IMMERSION OIL MICROSCOPY 1 OZ BOTTLE	6640002999807	BT	$3.01	
	IMPLANT MESH SURGICAL 12X12" SIZE WOVEN MESH VICRYL ABSORBABLE3S	6515013254985	PG	$712.11	
	INDICATOR STERILZ STRIP RAPID READOUT U/W GRAVITY STM STERILIZER	6530013663253	PG	$393.40	
	INDICATOR STERILZ STRIP STEAM TAMPERPROOF U/W STERILX SYS 1000S	6530013511698	PG	$130.66	
	INFUSOR PRESSURE BLOOD INTRAVENOUS BAG 14X6" 1000ML COLOR CODED	6515012808163	EA	$17.37	
	INJECTOR TUBE PLAS ACCOM 1ML & 2ML MEDICATION NEEDLE CART UNIT	6515013448487	EA	$5.00	
	INSECT STING TREATMENT KIT	6505010436795	EA	$11.01	
	INTRAVENOUS INJECTION SET 5 COMPONENTS	6515014625967	PG	$56.55	
	INTRODUCER SET CATHETER PERCUTANEOUS 28 COMPONENTS 10S	6515013110361	PG	$187.82	
	IODOQUINOL TABLETS USP 650MG 100 TABLETS PER BOTTLE	6505012300578	BT	$43.67	
	IRRIGATION KIT CATHETER-NASOGASTRIC TUBE	6515014844182	PG	$110.00	
	ISOFLURANE USP 100ML BOTTLE 6 BOTTLES PER PACKAGE	6505014437083	PG	$121.80	
	ISOPROTERENOL HYDROCHLORIDE INJECTION USP 0.200MG/ML 1ML AMPUL25	6505011171996	PG	$30.59	
	ISOSORBIDE SOLUTION 220ML 12S	6505011562292	BX	$317.21	
	KETAMINE HYDROCHLORIDE INJECTION 100MG/ML 5ML VIAL 10 VIALS/PG	6505011533733	PG	$175.45	
	KETOROLAC TROMETHAMINE INJECTION USP 30MG/ML 1ML UNIT 10/PACKAGE	6505014821064	PG	$37.78	
	KIT CIRCUIT VENTILATOR C/O DISP HOSES 25 CIRCUITS	6530014465763	EA	$130.05	
	KNIFE GENERAL SURG 1.50" BLADE LG SIZE 15 BLADES SS/CS DISP 100S	6515009786133	PG	$11.74	
	LABEL PSA WHITE A2 RECT 1.75IN W X 2.75IN L 1000/ROLL 5 ROLLS/PG	7530011728862	PG	$27.38	
	LABEL X-RAY FILM ID PRESSURE SENSITIVE ADHESIVE RADIOPAQUE 300S	6525008807257	PG	$83.58	
	LEAD SET TRANSCUTAN STIMULATOR F/USE W/6515-01-463-0901	6515014630898	EA	$24.00	
	LEGGINGS SURGICAL ADJUSTABLE DISP 100S	6532014586595	PG	$264.00	
	LIDOCAINE AND EPINEPHRINE INJECTION USP 1:100,000 20 ML VIAL	6505014554200	VI	$1.39	
	LIDOCAINE HYDROCHLORIDE AND DEXTROSE INJ 500ML 18S	6505011947265	PG	$139.13	
	LIDOCAINE HYDROCHLORIDE AND DEXTROSE INJECTION USP 2ML AMPUL 10S	6505011264915	PG	$47.71	
	LIDOCAINE HYDROCHLORIDE INJECTION USP 10ML BOTTLE 5 PER PACKAGE	6505011065499	PG	$15.78	
	LIDOCAINE HYDROCHLORIDE INJECTION USP 50ML VIAL 25 VIALS/PACKAGE	6505014478094	PG	$12.16	
	LIDOCAINE HYDROCHLORIDE INJECTION USP 5ML SYRINGE-NEEDLE UNIT10S	6505011561797	PG	$13.02	
	LIDOCAINE HYDROCHLORIDE JELLY USP 2% 30 ML	6505005843131	PG	$7.60	
	LIDOCAINE HYDROCHLORIDE&EPINEPHRINE INJ USP 1.8ML CARTRIDGE 100S	6505011461139	PG	$29.87	
	LIDOCAINE OINTMENT USP 5% 35 GM	6505007854357	TU	$1.71	
	LIGHT BULB LARYNGOSCOPE 2.5V GAS FILLED HIGH INTENSITY	6515012867592	EA	$10.25	
	LINER WET WEATHER PONCHO: POLYESTER NYLON QUILTED CAMOUFLAGE	8405008893683	EA	$29.20	
	LINERS KICK BUCKET PLAS 12X8X22IN DISP ANTI-STATIC 500S	7240014685066	PG	$30.00	
	LITTER FOLDING RIGID ALUM POLE PLAS NYLON DUCK CVR 91.60"O/A L	6530013807309	EA	$297.21	
	LOG BOOK RAPID READOUT F/USE W/6530-01-336-3253	6530014587631	PG	$7.38	
	LORATADINE TABLETS 10MG 100 TABLETS PER BOTTLE	6505013771432	BT	$230.52	
	LORAZEPAM INJECTION USP 2MG/ML 1ML CARTRIDGE/NEEDLE UNIT 10/PG	6505011789760	PG	$40.63	
	LORAZEPAM INJECTION USP 4MG/ML 1ML CARTRIDGE UNIT 10/PACKAGE	6505013932145	PG	$86.10	
	LUBRICANT OPHTHALMIC TOPICAL 1/8OZ OR 3.5GM W/.5% CHLRBTNL	6505001507622	TU	$2.64	
	LUBRICANT SURGICAL 4 OZ (113.4 GM)	6505001538809	TU	$3.57	
	LUMBAR PUNCTURE KIT 20 COMP ADULT W/2.50ML SYRINGE STER DISP 20S	6515000828264	PG	$178.47	
	MAGNESIUM SULFATE INJECTION USP 20 ML VIAL 25 PER PACKAGE	6505012650056	PG	$12.38	
	MAGNESIUM SULFATE INJECTION USP 2ML AMPUL 25 AMPULS PER PACKAGE	6505013018175	PG	$5.76	
	MANNITOL INJECTION USP 25% 50ML SINGLE DOSE VIALS 25 VIALS/PG	6505011253253	PG	$33.73	

NBC Medical Supplies List

QTY	ITEM	NSN	UI	COST	ITEM TOTAL
	MARKER TUBE TYPE BLACK PERMANENT ULTRA FINE POINT SHARPIE 12S	7520013964722	PG	$14.27	
	MASK & REBREATHING BAG ORONASAL W/ADAPTER CHIN STYLE FLEX DISP50	6515013180463	PG	$106.19	
	MASK AIRWAY LARYNGEAL REUSABLE SILICONE PEDIATRIC SZ3 19CM	6515014196395	EA	$241.56	
	MASK AIRWAY LARYNGEAL REUSABLE SILICONE SIZE .4 19CM	6515014196393	EA	$241.56	
	MASK AIRWAY LARYNGEAL REUSABLE SILICONE SIZE 5 20CM	6515014196394	EA	$241.56	
	MASK FACE CARDIOPULMONARY RESUSCITATION LARGE DISP FILTERED 10S	6515014871109	PG	$62.50	
	MASK OXYGEN PLASTIC DISPOSABLE 50S	6515011535216	PG	$121.11	
	MASK SURG ANTI-FOG NON-GLASS PLEATED LINT-FREE GREEN DISP 300S	6515011535988	PG	$46.51	
	MASK TRACHEOSTOMY AIRLIFE ADULT COLLAR SHAPED PLAS CLR FLEX 50S	6515009154141	PG	$33.27	
	MASK-SHIELD SURGICAL FEMALE OPENING CONN HALF RD SHAPE CLR 100S	6515013277257	PG	$175.00	
	MEDICAL EQUIPMENT SET EAR-NOSE-THROAT TEAM FIELD	6545014586057	EA	$89.00	
	MEPERIDINE HYDROCHLORIDE INJECTION USP 100MG/ML 1ML VIAL 25S	6505014830103	PG	$22.12	
	MEPIVACAINE HYDROCHLORIDE INJECTION USP 15MG PER ML 30ML	6505009141742	VI	$8.43	
	METER PEAK FLOW RATE INDICATING EXPIRATORY FULL RANGE I.W. 12S	6515014311204	PG	$150.00	
	METHYLPREDNISOLONE SODIUM SUCCINATE FOR INJECTION USP 125MG	6505011080809	CO	$0.98	
	METOCLOPRAMIDE INJECTION USP 5MG/ML 2ML VIAL 25 PER PACKAGE	6505012683738	PG	$6.70	
	METOPROLOL TARTRATE INJECTION USP 1MG/ML 5ML AMPUL 12/PACKAGE	6505013092742	PG	$46.98	
	METRONIDAZOLE TABLETS USP 250MG 250 TABLETS PER BOTTLE	6505008901840	BT	$3.36	
	MIDAZOLAM HYDROCHLORIDE INJECTION 5MG/ML 1ML VIAL 10/PACKAGE	6505012444736	PG	$63.01	
	MIDAZOLAM HYDROCHLORIDE INJECTION 5ML MULTI DOSE VIAL 10VIALS/PG	6505012721975	PG	$66.54	
	MONITORING SET ARTERIAL PRESSURE STERILE 20S	6515014341421	PG	$871.81	
	MORPHINE SULFATE INJECTION USP 10MG AUTOMATIC INJECTOR	6505013025530	EA	$5.88	
	MORPHINE SULFATE INJECTION USP 10MG/ML 1ML VIAL 25 PER PACKAGE	6505014830274	PG	$21.60	
	NALOXONE HYDROCHLORIDE INJECTION USP 0.4MG/ML 1ML AMPUL 10/BX	6505000797867	BX	$4.85	
	NALOXONE HYDROCHLORIDE INJECTION USP 1MG/ML 2ML AMPUL 10/PACKAGE	6505012405812	PG	$27.14	
	NAPHAZOLINE HCL&ANTAZOLINE PHOSPHATE OPHTHALMIC SOLUTION 15 ML	6505009746353	BT	$5.93	
	NEBULIZER MEDICINAL C/O MOUTHPIECE TEE 7FT TUBING 6IN FLEX TU50S	6515014660371	PG	$68.25	
	NEEDLE HYPO C10A DENTAL 25GA SELF-THREADING/THREADED STER 100S	6515001817404	PG	$8.20	
	NEEDLE HYPO C12 GP 22GA 1.50" LG LUER LOCK PLASTIC HUB STER 100S	6515011727650	PG	$4.36	
	NEEDLE HYPO C13A GP 18GA 1.438-1.562" LG LUER LOCK REG STER 100S	6515007542834	PG	$6.63	
	NEEDLE HYPO C13A GP 20GA 1.438-1.562" LG LUER LOCK REG STER 100S	6515007542836	PG	$6.17	
	NEEDLE HYPO C13A GP 25GA .578-.672" LG LUER LOCK STER DISP 100S	6515006555751	PG	$3.75	
	NEEDLE HYPO C20A SPINAL 22GA 3.375-3.625" LG LUER LOCK REG 25S	6515011039996	PG	$30.36	
	NEEDLE HYPO C20A SPINAL 25GAX3.375-3.625"LG LUER LOCK HUB REG25S	6515011039995	PG	$28.69	
	NEEDLE HYPO C5A BLD COLL 20GAX1.50" REG THD MULTIPLE SMPL 1000S	6515010032368	PG	$114.45	
	NEEDLE HYPO SPINAL ANES DISP 25 GA X 2 1/2 IN 100S	6515011562729	BX	$170.15	
	NEEDLE HYPODERMIC REGIONAL BLOCK 22GA C12A GP CRS STER DISP 50S	6515012564979	PG	$107.55	
	NEEDLE SPINAL ANEST	6515014817119	PG	$140.00	
	NEEDLE SUTURE FRENCH SPRING EYE SZ 2 B3 1/2 CIR STERILE DISP 72S	6515012931880	PG	$115.37	
	NEISSERIA ENRICHMENT MEDIUM DEHYDRATED 10ML 5S	6550010309068	PG	$87.99	
	NEOSTIGMINE METHYLSULFATE INJECTION USP 10ML MULTIPLE DOSE VIAL	6505009586325	VI	$1.63	
	NIMODIPINE CAPSULES 30MG I.S. 100 CAPSULES PER PACKAGE	6505013069502	PG	$347.42	
	NITROGLYCERIN IN DEXTROSE INJECTION 250ML BAG 12 BAGS/PACKAGE	6505013432489	PG	$58.53	
	NITROGLYCERIN INJECTION USP 5MG/ML 10ML VIAL 10 PER PACKAGE	6505014703206	PG	$122.00	
	NOREPINEPHRINE BITARTRATE INJECTION USP 4ML AMPUL 10 PER PACKAGE	6505002999496	BX	$50.44	
	OLOPATADINE HYDROCHLORIDE OPHTHALMIC SOLUTION STERILE 0.1% 5ML	6505014590420	BT	$38.19	

NBC Medical Supplies List

QTY	ITEM	NSN	UI	COST	ITEM TOTAL
	OPHTHALMIC IRRIGATING SOLUTION 15-18ML BOTTLE 36 BOTTLES/PACKAGE	6505011197693	PG	$63.24	
	OPHTHALMIC IRRIGATING SOLUTION 500ML BOTTLE W/HANGING DEVICE 6S	6505011197694	PG	$5.49	
	OPHTHALMOSCOPE 2.5 VOLT COMPACT W/HALOGEN LIGHT&POLARIZED FILTER	6540014587337	EA	$133.50	
	OXYCODONE AND ACETAMINOPHEN TABLETS USP 100 TABLETS PER PACKAGE	6505013625340	PG	$8.40	
	OXYGEN USP 99% CYLINDER TYPE D 95GL	6505001325181	EA	$119.74	
	OXYGEN USP 99% CYLINDER TYPE H 1650 GALLON	6505001325199	EA	$278.68	
	OXYMETAZOLINE HYDROCHLORIDE NASAL SOLUTION 15ML SPRAY BOTTLE	6505008694177	BT	$4.16	
	PAD ABDOMINAL 7.5X8" WHITE 3LAYERS STER SEAL PKG POST OP 240S	6510007755706	PG	$44.48	
	PAD ABDOMINAL WHITE 16X12" COVERED W/GAUZE/NONWOVEN FABRIC 144S	6510012717671	PG	$67.46	
	PAD BED LINEN NONWOVEN PLASTIC POLYPROPYLENE DISP 23X36"	6530013667671	PG	$74.00	
	PAD COOLING CHEMICAL PLASTIC WHITE DISPOSABLE 8X4" 16S	6530012761366	PG	$15.33	
	PAD COTTON GAUZE COVERED 2.5 X 2-1/8 INCHES CREAM OR WHITE 50S	6510011077575	PG	$9.16	
	PAD ELECTRODE DEFIBRILLATOR 6X4.50" PREGELLED SELF-ADH DISP 200S	6515013838387	PG	$401.10	
	PAD HEAT TREATMENT INSTANT HEAT PACK KWIK HEAT BECOMES 110 DEG 4	6530013813848	PG	$14.22	
	PAD ISOPROPYL ALCOHOL IMPREGNATED NONWVN COTTON/RAYON WHITE 200S	6510007863736	PG	$2.15	
	PAD NONADHERENT 8X3" NONWOVEN COTTEN I.P. STERILE WHITE 50S	6510009862942	PG	$8.57	
	PAD POST-SURGICAL-OBSTETRICAL BELTLESS 13"LG 3"W EXTRA-ABS 180S	6510012935596	PG	$47.21	
	PAD SCOURING ALUMINUM OXIDE ABRASIVE BACKED PLASTICS SPONGE	7920006555290	DZ	$2.30	
	PADDING CAST-SPLINT ORTHOPEDIC FELT WHITE 1/2IN X 21IN X 36IN W	6515014670904	EA	$70.98	
	PAPER ELECTROCARDIOGRAPH 150YD L X 60MM W 100 ROLLS	6515013960711	PG	$271.35	
	PAPER FILTER F/USE WITH RIGID STERILIZATION CONTAINER SYS 1000S	6640013785773	PG	$44.60	
	PAPER FILTER QUANTITATIVE 110MM DIA DOUBLE ACID WASH 100S	6640004357800	PG	$12.35	
	PAPER LENS PAD WHITE BIBULOUS PAPER 6IN LONG 4IN WIDE PERFORATED	6640009370760	PG	$22.40	
	PAPER SHEETING EXAMINATION-TREATMENT TABLES 125'LG 17.75"W 12S	6530010923914	PG	$22.63	
	PAPER TOILET 11LBS/500 SHEETS 24X36" ROLL NONCOMPRESSED 1-PLY96S	8540005303770	BX	$45.61	
	PATIENT UTILITY KIT ADMISSION	6530014766157	EA	$121.00	
	PEN BALL-POINT BLACK FINE POINT POCKET RETRACTABLE	7520009357135	DZ	$4.65	
	PENICILLIN G BENZATHINE SUSPENSION STERILE USP 4ML UNIT 10/PG	6505011561722	PG	$280.52	
	PENICILLIN G POTASSIUM FOR INJECTION USP 20,000,000 UNITS 10S	6505002689593	PG	$20.86	
	PENICILLIN V POTASSIUM TABLETS USP 400000 UNITS 40 TABS/BOTTLE	6505001178579	BT	$1.59	
	PHENOBARBITAL SODIUM INJ USP 130MG/ML 1ML VIAL 25 VIALS/PACKAGE	6505012052393	PG	$40.68	
	PHENYLEPHRINE HYDROCHLORIDE INJECTION USP 1% 1 ML 25S	6505001049320	BX	$11.80	
	PHENYLEPHRINE HYDROCHLORIDE OPHTHALMIC SOLUTION USP 2.5% 15ML BT	6505002719220	BT	$15.57	
	PHENYTOIN SODIUM INJECTION USP 50MG/ML VIAL 5ML 25S	6505013329024	PG	$4.36	
	PHYSOSTIGMINE SALICYLATE INJECTION USP 1MG/ML 2ML 10 AMPULES/PG	6505014667522	PG	$19.83	
	PILLOWCASE NONWOV FAB DISP AQUA 20X29 IN 200S	7210011563617	BX	$136.06	
	PILOCARPINE HYDROCHLORIDE OPHTHALMIC SOLUTION USP 2% 15 ML	6505005824679	BT	$1.78	
	PIN SAFETY BRASS NICKEL 2" LONG SZ 3	8315007877000	BX	$1.84	
	PINWHEEL SENSORY NEUROLOGICAL DISP STER 14S	6515014342346	PG	$197.64	
	PIPERACILLIN SODIUM AND TAZOBACTAM FOR INJECTION 10 VIALS/PG	6505013845767	PG	$147.09	
	PIPET SEROLOGICAL PLASTIC DISPOSABLE 1 ML 100S	6640008294023	PG	$16.69	
	PIPET TIP PLAS POLYPROP M30 LARGE 1.88IN LG DISP NONSTERILE	6640005028231	PG	$18.00	
	POLYVINYL ALCOHOL OPHTHALMIC SOLUTION 1.4% 15ML	6505009617486	BT	$0.71	
	POTASSIUM CHLORIDE CONCENTRATE FOR INJECTION USP 20ML VIAL 25/PG	6505010801988	PG	$9.99	
	POVIDONE-IODINE CLEANSING SOLUTION USP 7.5% 1GL OR 3.780LI	6505009947224	BT	$15.45	
	POVIDONE-IODINE CLEANSING SOLUTION USP 7.5% 4 FL OUNCES OR 118ML	6505004917557	BT	$1.00	

NBC Medical Supplies List

QTY	ITEM	NSN	UI	COST	ITEM TOTAL
	POVIDONE-IODINE TOPICAL SOLUTION USP 1GL (3.780 LITER)	6505007540374	BT	$8.18	
	PREDNISOLONE ACETATE OPHTHALMIC SUSPENSION 1% 5 ML	6505001335843	BT	$4.19	
	PREPARATION KIT BLOOD DONOR STERILE 20S	6510011139208	PG	$14.05	
	PROBE BILIARY DUCT FOGARTY DESIGN 5FR 40CM LG 8MM BALLOON DISP	6515012229953	EA	$54.42	
	PROBE BILIARY DUCT FOGARTY DESIGN 6FR 23CM LG 13MM BALLOON DISP	6515011562469	EA	$57.59	
	PROBENECID TABLETS USP 0.5 GRAM 100S	6505005276885	BT	$6.05	
	PROCAINAMIDE HYDROCHLORIDE INJECTION USP 100 MG PER ML 10 ML	6505002998614	BT	$0.47	
	PROMETHAZINE HYDROCHLORIDE INJECTION USP 25MG/ML 1ML AMPUL 25/BX	6505006807352	BX	$9.70	
	PROMETHAZINE HYDROCHLORIDE SUPPOSITORIES 25 MG 12S	6505000654214	PG	$34.64	
	PROMETHAZINE HYDROCHLORIDE TABLETS USP 25 MG 1000 TABLETS/BOTTLE	6505005843277	BT	$10.31	
	PROPOFOL INJECTION 10MG/ML 20ML AMPUL 25 PER PACKAGE	6505014340430	PG	$480.25	
	PROTAMINE SULFATE INJECTION USP 10MG/ML 5ML AMPUL 25/PACKAGE	6505013707519	PG	$27.63	
	PSEUDOEPHEDRINE HYDROCHLORIDE TABLETS USP 60MG 100S	6505014737775	BT	$2.50	
	RANITIDINE INJECTION USP 25MG/ML 40ML VIAL	6505012859011	VI	$30.18	
	RAZOR SURG PREPARATION PLAS STATIONARY DSGN SOLID HDL DISP 100S	6515013225898	PG	$27.54	
	REGULATOR FLOW RATE IV INJECTION SET ONE-WAY DSGN 18" TU LG 48S	6515013648148	PG	$246.00	
	REMOVER SURGICAL STAPLE STERILE DISPOSABLE 12S	6515012968419	PG	$22.25	
	RESUSCITATOR MOUTH TO MASK ADULT SINGLE USE 10S	6515012801841	PG	$90.42	
	RETAINER TUBE TRACH THOMAS PLAS FOAM-PADDING ADULT 6-9 DISP 25S	6515014311214	PG	$98.27	
	RIFAMPIN CAPSULES USP 300MG 100 CAPSULES PER BOTTLE	6505001656575	BT	$48.52	
	RINGER'S INJECTION LACTATED USP 1000ML BAG 12 BAGS PER PACKAGE	6505013306267	PG	$6.30	
	RINGER'S INJECTION LACTATED USP 500ML BAG 24 BAGS PER PACKAGE	6505013306266	PG	$15.00	
	ROCURONIUM BROMIDE INJECTION 10MG/ML 5ML VIAL 10 PER PACKAGE	6505013932144	PG	$115.22	
	RUBBER BAND 1-4LB NO.19	7510002051438	BG	$0.39	
	RUBBER BANDS EXERCISE RESISTIVE 6" X 50YD ROLL LATEX RUBBER	6510014659368	RO	$57.50	
	SCOPOLAMINE HYDROBROMIDE INJECTION USP 0.4MG/ML 1ML VIAL 25/BOX	6505010880499	BX	$14.00	
	SENSITIVITY DISC DIAGNOSTIC PENICILLIN G 10 UNITS 500S	6550010316170	PG	$41.69	
	SENSITIVITY DISCS DIAGNOSTIC 50 DISKS PER TUBE 10 TUBES/PACKAGE	6550011397512	PG	$42.70	
	SENSITIVITY DISCS DIAGNOSTIC AMPICILLIN 10 MICROGRAMS 500S	6550010316158	PG	$32.51	
	SENSITIVITY DISCS DIAGNOSTIC CEPHALOTHIN 30 MICROGRAMS 500S	6550010316161	PG	$43.37	
	SENSITIVITY DISCS DIAGNOSTIC CHLORAMPHENICOL 30 MICROGRAMS 500S	6550010316162	PG	$43.87	
	SENSITIVITY DISCS DIAGNOSTIC CLINDAMYCIN 50 DISKS/TUBE 10 TU/PG	6550011397515	PG	$43.37	
	SENSITIVITY DISCS DIAGNOSTIC ERYTHROMYCIN 15 MICROGRAMS 500S	6550010316163	PG	$41.69	
	SENSITIVITY DISCS DIAGNOSTIC GENTAMICIN 10 MICROGRAMS 500S	6550010316166	PG	$41.67	
	SENSITIVITY DISCS DIAGNOSTIC METHICILLIN 5 MICROGRAMS 500S	6550010316169	PG	$48.31	
	SENSITIVITY DISCS DIAGNOSTIC TETRACYCLINE 30 MICROGRAMS 500S	6550010316173	PG	$61.93	
	SENSITIVITY DISCS DIAGNOSTIC VANCOMYCIN 30 MCG 500S	6550011397516	PG	$38.38	
	SHAMPOO BABY 7 FL OZ 24S	8520011494129	PG	$51.91	
	SHAVE KIT PATIENT PERSONAL LIGHTWEIGHT 100S	6530011669517	PG	$58.57	
	SHAVER POWDER MIX W/WATER REMOVES BEARD W/NO NEED F/RAZOR 5OZ12S	8520014765113	PG	$26.66	
	SHEET BED NONWOVEN FABRIC WHITE DISPOSABLE 96X60INCHES 100S	7210001446082	PG	$189.73	
	SHIELD EYE SURG GOFFMAN SGL CONSTRUCTION CATARACT OPERATION 50S	6515012538165	PG	$53.50	
	SHOES SHOWER BLACK RBBR W/CROSS STRAP NON-SLIP ADULT X-LARGE 36S	6532014697522	PG	$57.60	
	SHOES SHOWER RBBR BLACK ADULT SZ SHOE W/CROSS STRAP SZ MEDIUM72S	6532014697512	PG	$57.60	
	SHOES SHOWER RBBR BLACK W/CROSS STRAP NON-SLIP ADULT SZ SMALL72S	6532014697517	PG	$57.60	
	SHOES SHOWER RUBBER BLACK W/CROSS STRAP ADULT SZ LARGE 72S	6532014697520	PG	$57.60	

NBC Medical Supplies List

QTY	ITEM	NSN	UI	COST	ITEM TOTAL
	SHORTS MEN'S PJ ORHTO MID-THIGH ELAS WAIST DISP BLUE LGE 50S	6532014695587	PG	$29.90	
	SHORTS MEN'S PJ ORTHO MID THIGH ELA WAIST DISP BLUE MEDIUM 50S	6532014695676	PG	$29.90	
	SKIN CLEANSER 4OZ FL BT WATERLESS ALCOHOL BASE 24S	8520013469200	PG	$36.38	
	SKIN CLOSURE ADHESIVE SURGICAL POROUS .25 BY 4 INCHES 500S	6510000547255	PG	$65.71	
	SLIDE MICROSCOPE PLAIN FROSTED END 75.4 X 25 MM 72S	6640000744191	PG	$4.37	
	SLIPPERS CONVALESCENT PATIENTS LGE SZ 8-10 POLYUR AVOCADO DISP	6532000797902	PR	$0.54	
	SLIPPERS CONVALESCENT PATIENTS MEDIUM SZ 6-8 POLYUR BLUE DISP	6532000797899	PR	$0.64	
	SOAP TOILET NONMEDICATED CAKE I.W. 0.75OZ 1000S	8520005510375	MX	$47.85	
	SODA LIME NF 37LB (16.79 KG)	6505006873564	CN	$56.23	
	SODIUM BICARBONATE INJ USP 8.4% SYRINGE-NEEDLE UNIT 50ML 10S	6505002165370	PG	$18.24	
	SODIUM CHLORIDE INJECTION USP 0.9% 100ML BAG 64 BAGS PER PACKAGE	6505013308924	PG	$43.93	
	SODIUM CHLORIDE INJECTION USP 0.9% 250ML PLASTIC BAG 24 BAGS/PG	6505011828013	PG	$22.53	
	SODIUM CHLORIDE INJECTION USP 0.900% 10ML VIAL 25 VIALS/PACKAGE	6505012870626	PG	$15.75	
	SODIUM CHLORIDE INJECTION USP 1000ML BAG 12 BAGS PER PACKAGE	6505013306269	PG	$5.41	
	SODIUM CHLORIDE INJECTION USP 500 ML PLASTIC BAG 24 PER PACKAGE	6505012910333	PG	$31.30	
	SODIUM CHLORIDE IRRIGATION USP 0.9% 1000ML BOTTLE 12 PER BOX	6505010750678	BX	$9.17	
	SODIUM CHLORIDE IRRIGATION USP 3000ML BAG 4 BAGS PER PACKAGE	6505004434582	PG	$25.92	
	SODIUM CITRATE AND CITRIC ACID ORAL SOLUTION USP 473ML BOTTLE	6505010974766	BT	$4.61	
	SODIUM HYPOCHLORITE SOL 5% 6GAL	6810005987316	BX	$12.52	
	SODIUM NITROPRUSSIDE STERILE USP 50 MG	6505010095019	BT	$5.49	
	SODIUM PHOSPHATES ENEMA USP DISP ENEMA UNIT 4-1/2 FL OZ (133 ML)	6505006198215	BT	$0.69	
	SODIUM POLYSTYRENE SULFONATE SUSPENSION USP 500ML BOTTLE	6505011932830	BT	$16.47	
	SPECIMEN KIT URINE STERILE DISP 50S	6530011732453	PG	$40.15	
	SPECIMEN TRAP MUCOUS 40 ML STER DISP GRADUATED INCR 2 ML 50S	6530013950398	PG	$164.70	
	SPECULUM OTOSCOPE AURAL PLASTIC 4MM DISP 1000 PER PG	6515014850294	PG	$22.35	
	SPLINT FINGER ALUM MALLEABLE 18"LG X 3/4"W FOAM PADDED	6515014854041	PG	$36.30	
	SPLINT UNIVERSAL 36X4.50" MALLEABLE ALUM RADIOLUCENT LTWT GRAY12	6515012254681	PG	$62.63	
	SPLINTING MATERIAL 45IN X 6IN OCL CONTOUR SINGLE SPLINT 5S	6515014618338	PG	$117.34	
	SPLINTING MATERIAL ORTHO DESIGN 72X4" ROLL FORM ALUM/RUBBER FOAM	6515013589488	EA	$19.76	
	SPLINTING MATERIAL ORTHOPEDIC 30IN X4IN OCL CONTOUR SGL SPLINT5S	6515014618357	PG	$79.55	
	SPLINTING MATERIAL ORTHOPEDIC OCL CONTOUR SINGLE SPLINT 4X15IN5S	6515014618369	PG	$60.85	
	SPONGE LAPAROTOMY DISPOSABLE RADIOPAQUE 18"X18" PREWASHED 100S	6510011603261	PG	$18.35	
	SPONGE SURG GAUZE 6.75X6" STERILE 32 PLY QUANTITY WHITE 600S	6510002940009	PG	$161.92	
	SPONGE SURG GAUZE ABS 4X8" WHITE RADPQ STERILE U/I O.R. 80S	6510001161285	PG	$171.83	
	SPONGE SURGICAL GAUZE 2X2" STERILE WHITE 20-12 MESH 3000S	6510000584421	PG	$89.60	
	SPONGE SURGICAL GAUZE 4PLY WHITE 36IN LONG 8IN WIDE 100S	6510012219064	PG	$126.02	
	SPONGE SURGICAL GAUZE 8X4" 800 SPONGES PER PACKAGE	6510014617854	PG	$148.19	
	SPONGE SURGICAL GAUZE ABS WHITE 4X4" WATERPROOF PEEL LABEL 1280S	6510001161311	PG	$139.74	
	STAPLE UNIT SURG GASTROINTESTINAL 3MM ID 52 CRS STAPLES DISP 12S	6515012078226	PG	$863.52	
	STAPLE UNIT SURG MULTIFIRE TA 3.5MM W/55 TITANIUM STAPLES DISP3S	6515013354620	PG	$329.40	
	STAPLER SKIN SURGICAL DISP 35 SZ 6.9X3.9MM STAPLES 6S	6515010715561	PG	$108.49	
	STAPLER SURG 55 UNITS CAP MULTIFIRE TA W/55 STAPLES BLUE DISP 3S	6515013354625	PG	$308.77	
	STETHOSCOPE ESOPHAGEAL 24FR PLASTIC STERILE DISPOSABLE 20S	6515010985770	PG	$143.31	
	STOCKINET SURG 25YDX6" CTN YARN WHITE/NAT TREAT WOUNDS DRESSINGS	6510002040000	RO	$17.11	
	STOCKINET SURGICAL SYNTHETIC 25YD X 3" PLASTIC POLYESTER	6510011978825	RO	$14.00	
	STOCKINET SURGICAL SYNTHETIC 25YDX4" WATER REPELLENT	6510011978826	RO	$16.36	

NBC Medical Supplies List

QTY	ITEM	NSN	UI	COST	ITEM TOTAL
	STOPCOCK IV THERAPY 3-WAY 20-30"LG W/MALE LUER CONNECTOR STER50S	6515008648864	PG	$93.95	
	STRAP WEBBING 72X1.5" OLIVE DRAB W/B10 BUCKLE F/SPLINT LEG	6515003745920	EA	$1.77	
	STRAP WEBBING PATIENT SECURING BUCKLE W/LOCKING DEVICE & SPRING	6530007844205	EA	$20.01	
	STYLET TRACHEAL TUBE 7.5MM TO 10MM TRACHEAL TUBES DISP 10S	6515013948327	PG	$43.62	
	SUCCINYLCHOLINE CHLORIDE INJECTION USP 100MG/ML 10ML VIAL 25S	6505014716410	PG	$69.42	
	SUCCINYLCHOLINE CHLORIDE INJECTION USP 20MG/ML 10ML VIAL 12/PG	6505014667167	PG	$15.83	
	SUCTION AND IRRIGATION ASSEMBLY SURGICAL	6515014441638	PG	$335.00	
	SUCTION SET TRACHEAL 14FR 22" LENGTH PLASTIC STERILE DISP 100S	6515011676666	PG	$56.00	
	SULFADIAZINE SILVER CREAM 1% TOPICAL 400GM JAR	6505005607331	JR	$14.54	
	SULFADOXINE AND PYRIMETHAMINE TABLETS I.S. 25 TABLETS/PACKAGE	6505011320257	PG	$57.71	
	SULFAMETHOXAZOLE AND TRIMETHOPRIM TABLETS USP 250 TABLETS/BOTTLE	6505011485041	BT	$159.13	
	SULFOSALICYLIC ACID DIHYDRATE ANALYZED REAGENT 4 OZ (113.4GM)	6550001464875	BT	$5.71	
	SUMATRIPTAN TABLETS 25 MG 9 TABLETS PER PACKAGE	6505014629906	PG	$94.78	
	SUPPORT KNEE 24" LG THIGH SIZE 30" CALF SIZE 20" W/STRAPS FOAM	6515012408712	EA	$15.00	
	SUPPORT KNEE CRS EITHER KNEE21IN MIN23IN MAX 2SLEEVE MULTI-DIREC	6515014669137	EA	$35.71	
	SUPPORT KNEE EITHER KNEE 18.5IN MIN 21IN MAX 2SLEEVE MULTI DIREC	6515014669053	EA	$35.71	
	SUPPORT KNEE NYLON EITHER KNEE 23IN MIN 26IN MAX O/A LG	6515014669148	EA	$35.71	
	SUPPORT KNEE NYLON O/A 15.25IN MIN 18IN MAX LG 2SLEEVE DESIGN	6515014669153	EA	$35.71	
	SUPPORT SHOULDER IMMOBILIZER MUSLIN LARGE	6515011534930	EA	$7.92	
	SUPPORT SHOULDER IMMOBILIZER MUSLIN MEDIUM	6515011534929	EA	$4.52	
	SUPPORT SHOULDER IMMOBILIZER MUSLIN SMALL	6515011534928	EA	$4.52	
	GOWN3S	6530014884417	PG	$118.00	
	SUTURE ABS GEN SURG SZ 0 12 18" STRANDS COATED DYED VIOLET 12S	6515011673781	PG	$75.08	
	SUTURE ABS SIZE 000 2.25'LG B3 1/2 CIR C7 RD TAPER ORDER BY 3S	6515002259585	DZ	$13.55	
	SUTURE ABS SURG GEN SURGERY SIZE 2-0 18" LG UNARMED STERILE 24S	6515012513744	PG	$150.16	
	SUTURE ABS SUTUPAK SZ3-0 12 18" STRANDS COATED VICRYL ORDER BY2S	6515012615870	DZ	$58.01	
	SUTURE ABS SZ 4-0 54"LG LIGAPAK LIGATING REEL VIOLET ORDER BY 1S	6515013807432	DZ	$17.00	
	SUTURE ABSORBABLE SURGICAL VIOLET 8"MIN -18" MAX LG	6515014825679	DZ	$132.14	
	SUTURE KIT SURGICAL INCL ADSON TISSUE FORCEPS WITH TEETH DISP20S	6515013093551	PG	$145.27	
	SUTURE NONABS GI SZ 3-0 8 18" STRANDS NDL SH UNCOATED ORDER BY2S	6515011384742	DZ	$180.53	
	SUTURE NONABS GI/NEUROSURGERY SZ 2-0 8 18" STRANDS BR ORDER BY2S	6515011477050	DZ	$180.53	
	SUTURE NONABS OB/GYN SZ 1-0 36" LG ATRALOC NDL CTX ORDER BY 2S	6515012025455	DZ	$20.00	
	SUTURE NONABS SKIN CLOSURE SZ 3-0 1 2.5' STRAND MONO BLACK 36S	6515001916527	PG	$15.68	
	SUTURE NONABS SURG CARDIO SZ 3-0 DBL ARMED MONOFILAMENT STER 36S	6515011535730	PG	$71.31	
	SUTURE NONABS SURG CARDIO SZ 4-0 DBL ARMED 36" LG ORDER BY 3S	6515013110338	DZ	$39.02	
	SUTURE NONABS SURG SZ 2-0 DBL ARMED MONOFILAMENT 2.5' STER 24S	6515012591734	PG	$106.43	
	SUTURE NONABS SURG SZ 3-0 UNARMED SILK BLACK BR 2.5'LG STER 36S	6515007638483	PG	$58.08	
	SUTURE NONABS SURG SZ 4-0 UNARMED SILK BR BLACK 2.5' LG STER 36S	6515007639605	PG	$61.92	
	SUTURE NONABS SZ 2-0 2.50'LG SILK BLACK UNARMED BRAIDED STER 36S	6515010758288	PG	$101.85	
	SUTURE REMOVAL KIT STER DISP F/REMOVING SILK/NYLON SKIN SUT 50S	6515004361881	PG	$47.48	
	SYRINGE AND NEEDLE HYPODERMIC SAFETY 3ML 22GAGE STER DISP 100S	6515014520470	PG	$50.00	
	SYRINGE CARTRIDGE DENTAL ASPIRATING SIDE LOADING SCREW CAP	6515010108761	EA	$20.25	
	SYRINGE HYPO 10/12ML CAPACITY LUER LOCK CONCENTRIC TIP DISP 100S	6515011688108	PG	$114.75	
	SYRINGE HYPODERMIC GENERAL PURPOSE 20CU CM CAPACITY DISP STER25S	6515014123101	PG	$7.86	
	SYRINGE HYPODERMIC GP 30ML CAPACITY LUER LOCK TIP STER DISP 25S	6515011656742	PG	$14.47	
	SYRINGE HYPODERMIC GP 3ML CAP LUER LOCK TIP PLAS STER DISP 100S	6515004627348	PG	$6.56	

NBC Medical Supplies List

QTY	ITEM	NSN	UI	COST	ITEM TOTAL
	SYRINGE IRRIGATING BULB TYPE 50ML CAP RBBR PROTECTIVE TIP CAP50S	6515008282462	PG	$46.18	
	SYRINGE IRRIGATING PISTON 60ML 0.5 OUNCES STER DISP 120S	6515014570288	PG	$101.00	
	TAG MASS CASUALTY INDICENT TRIAGE DESIGN COLOR CODED 50S	6530013983968	PG	$10.00	
	TAPE SEALING STERILIZATION INDICATOR OPAQUE 2160"LG .75"W 60 YD	6530009175821	RO	$1.06	
	TAPE TEXTILE BINDING BRAID TAPE WEBB 1/2IN	8315009037160	PG	$38.88	
	TEST KIT HUMAN CHORIONIC GONADOTROPIN DETERMINATION 50 TESTS	6550013022653	EA	$197.64	
	TEST KIT OCCULT BLOOD DETERMINATION 100 TESTS	6550001656538	EA	$21.91	
	TEST STRIPS&COLOR CHART BILIRUBIN BLD GLU KET NITRITE PH PROT100	6550011225540	BT	$35.11	
	TEST TUBE 10 BY 75 MM 4ML CAPACITY W/O LIP CULTURED DISP 1000S	6640011203552	PG	$18.59	
	TEST TUBE 12 BY 75 MM DISPOSABLE GLASS BOROSILICATE 1000S	6640011190013	PG	$21.46	
	TEST TUBE 16 BY 125MM SCREW CAP GLASS BOROSILICATE 15ML CAP 250S	6640006041102	PG	$34.84	
	TETRACAINE HYDROCHLORIDE OPHTHALMIC SOLUTION 0.5% 15 ML	6505005824737	BT	$3.39	
	TETRACAINE HYDROCHLORIDE STERILE USP 20MG AMPUL 100 AMPULS/PG	6505011533015	PG	$194.49	
	THERMOMETER CLINICAL HUMAN ORAL DIGTAL LTWT 300 HOUR USE PLASTIC	6515013737292	EA	$9.50	
	THIABENDAZOLE TABLETS USP 500MG INDIVIDUALLY SEALED 36 TABS/PG	6505012269909	PG	$22.10	
	THIAMINE HYDROCHLORIDE INJECTION USP 100MG/ML 1ML VIAL 25/PG	6505012380362	PG	$28.99	
	THIOPENTAL SODIUM FOR INJECTION USP 500MG PER PACKAGE	6505012226566	PG	$110.52	
	THIOPENTAL SODIUM FOR INJECTION USP 500MG VIAL 25 VIALS/PACKAGE	6505010410558	PG	$178.43	
	THROMBOPLASTIN LYOPHILIZED RABBIT BRAIN 10S	6550001613050	BX	$23.83	
	TIMOLOL MALEATE OPHTHALMIC SOLUTION USP 5ML BOTTLE WITH TIP	6505010696519	BT	$3.20	
	TIP CANE AND CRUTCH 0.75" DIAMETER 3" HIGH CORRUGATED RUBBER 12S	6515010133911	PG	$11.55	
	TISSUE FACIAL WHITE 2-PLY 24X36" BOX DISPENSER 100S	8540007935425	PG	$6.23	
	TOWEL PACK SURG BLUE/ GRAY OR GREEN STER DISP 19X26" 96S	6530001101854	PG	$55.58	
	TOWEL PACK SURGICAL 16X29" STER DISP NONWVN FABRIC I.S. 280S	6530009982739	PG	$51.85	
	TRACHEOSTOMY CARE SET DISP 24S STER ITEMS PACKED IN ORDER OF USE	6515011534716	PG	$51.66	
	TRAP FINGER ORTHOPEDIC 8IN BLUE STER DISP RONCI DESIGN LARGE 8S	6515013968953	PG	$108.70	
	TRAP FINGER ORTHOPEDIC STER DISP RNCI DESIGN X-LGE GREEN 8S	6515013968954	PG	$108.70	
	TRAP FINGER ORTHOPEDIC STER DISP RONCI DESIGN MED YELLOW 8IN 8S	6515013968955	PG	$108.70	
	TRAP FINGER ORTHOPEDIC STER DISP RONCI DESIGN SMALL 8IN RED 8S	6515013968952	PG	$108.70	
	TRAY SKIN PREP WET SKIN SCRUB PACKS STERILE 20TRAYS	6530013080984	PG	$62.40	
	TREATMENT SYSTEM LIQ4OZ BT UNIT DOSE PWDR DOES NOT DECONTAMINATE	6850014862035	PG	$200.49	
	TRIAMCINOLONE ACETONIDE INJECTABLE SUSP USP 40MG/ML 5ML VIAL	6505012104472	VI	$8.38	
	TRIFLURIDINE OPHTHALMIC SOLUTION 1% 7.5ML BOTTLE	6505011428314	BT	$93.41	
	TRIMETHOBENZAMIDE HCL & BENZOCAINE SUPPOSITORIES NF 50S	6505008901819	BX	$11.85	
	TUBE ASSY GAS ANES APPARATUS NONCONDUCTIVE 22MM DIA 40" LG 20S	6515013281901	PG	$147.27	
	TUBE BLOOD COLLECTING VACUUM 7ML W/O ANTICOAGULANT NONSTERILE100	6630001451137	PG	$7.50	
	TUBE CENTRIFUGE PLASTIC POLYETHYLENE 118MM LG 18MM OD 15ML CAP12	6640009266986	PG	$17.60	
	TUBE DRAINAGE SURG E3 0.625"DIA 15"LG LGE SUCTION LUMEN 5S	6515012240129	PG	$252.54	
	TUBE DRAINAGE SURG PENROSE E3 12" LG 0.625" DIA 0.011" THK 200S	6515013849023	PG	$112.23	
	TUBE DRAINAGE SURGICAL PENROSE 12X3/8X0.011" RBBR RADIOPAQUE200S	6515013852013	PG	$85.64	
	TUBE DRAINAGE SURGICAL PENROSE 1X18" RUBBER RADIOPAQUE STER 200S	6515011885316	PG	$69.44	
	TUBE DRINKING PLASTIC STRAIGHT INDIVIDUALLY WRAPPED 10000	7350014674327	PG	$48.20	
	TUBE ENDOBRONCHIAL LEFT MAINSTEM BRONCHUS E5 DB LUMEN 37FR STER	6515012259715	EA	$75.78	
	TUBE ENDOBRONCHIAL LEFT MAINSTEM BRONCHUS E5 DBL LUMEN 39FR STER	6515012259716	EA	$58.55	
	TUBE ENDOTRACHEAL MURPHY UNCUFFED 3.0MM 10S	6515011665076	PG	$28.49	
	TUBE NASOPHARYNGEAL 26FR CRVD 8.7MM OD 6.4MM ID STER DISP 10S	6515011295436	PG	$26.78	

NBC Medical Supplies List

QTY	ITEM	NSN	UI	COST	ITEM TOTAL
	TUBE STOMACH SURG SALEM E19 W/FUNNEL 16FR DBL LUMEN 48" LG 50S	6515001490316	PG	$93.91	
	TUBE SUCT SURG SUCTION DSGN 144" LG PLAS 9/32"ID TRANSPARENT 20S	6515011742354	PG	$44.80	
	TUBE TRACH NASAL/RAE E12 W/CUFF 7.5MM DIA LP CUFF MURPHY EYE 10S	6515013608941	PG	$57.65	
	TUBE TRACHEAL 37 FR DOUBLE LUMEN PILOT BALOON STERILE DISP 4S	6515014205264	PG	$206.41	
	TUBE TRACHEAL 4AFR PLASTIC NOT WOVEN STERILE DISP DOUBLE LUMEN4S	6515014211388	PG	$204.27	
	TUBE TRACHEAL MURPHY E12 6.0MM ID 8.1MM OD 28CM LG PLAS DISP 10S	6515001050720	PG	$21.19	
	TUBE TRACHEAL MURPHY E12 W/CUFF 7.0MM ID 30CM LG W/15MM CONN 10S	6515001050744	PG	$24.39	
	TUBE TRACHEAL MURPHY E12 W/CUFF 8.0MM ID 10.7MM DIA 32CM LG 10S	6515001050759	PG	$27.99	
	TUBING BREATHING AEROSOL PLASTIC CORRUGATED 100FT 22MM CUFF DISP	6515011405344	EA	$15.99	
	TUBING SURGICAL PLASTIC BULBOUS 5MM ID NONSTERILE FLEXIBLE 100FT	6515012774772	RO	$11.00	
	VALVE SURG DRAIN HEIMLICH 4.50" LG 1" DIA FLUTTER RBBR STER 10S	6515009269150	PG	$189.91	
	VANCOMYCIN HYDROCHLORIDE STERILE USP 1GM VIAL 10 VIALS/PACKAGE	6505012478801	PG	$61.03	
	VASOPRESSIN INJECTION USP 1ML VIAL 10 VIALS PER PACKAGE	6505006848625	PG	$74.66	
	VECURONIUM BROMIDE FOR INJECTION 10 MG 10ML VIALS 10/PKG	6505012580983	PG	$142.73	
	VENOUS CENTRIFUGAL HEMATOLOGY TUBE 1000S	6630012337593	PG	$325.40	
	VERAPAMIL HYDROCHLORIDE INJECTION 2.5MG/ML 2ML AMPUL	6505011313855	CO	$2.32	
	VERAPAMIL HYDROCHLORIDE INJECTION 2.5MG/ML 4ML SYRINGE 10/PG	6505012663771	PG	$22.62	
	WATER FOR IRRIGATION STERILE USP 1000 ML 12S	6505011533796	PG	$29.63	
	WATER FOR IRRIGATION STERILE USP 1000 ML 12S	6505006018965	PG	$12.96	
	WATER FOR IRRIGATION STERILE USP 1000ML CONTAINER 12 PER BOX	6505010750679	PG	$11.34	
	WOUND CLOSURE KIT FACIAL 18 COMPONENTS STERILE DISPOSABLE 20S	6515011534888	PG	$73.14	
	WRAPPER STERILIZATION 36X36"HEAVY-DUTY ONE STEP 72S	6530014588321	PG	$102.14	
	WRAPPER STERILIZATION 45X45IN HEAVY DUTY ONE-STEP 40S	6530014588312	PG	$95.64	
	WRAPPER STERILIZATION HEAVY DUTY 18X18IN 288S	6530014587733	PG	$102.13	
	WRAPPER STERILIZATION HEAVY DUTY24X24"1-STEP F/SURG SUPPLIES120S	6530014588305	PG	$74.24	
	WRIGHT GIEMSA STAIN 16 FL OZ (473 ML)	6550009269084	BT	$9.00	
	TOTAL COST				

UNIT OF ISSUE (UI) DESCRIPTIONS

BG = BAG
BT = BOTTLE
BX = BOX
CN = CAN
CO = CONTAINER
DZ = DOZEN
EA = EACH
HD = HUNDRED
JR = JAR
MX = THOUSAND
PG = PACKAGE
PR = PAIR
RO = ROLL
SE = SET
TU = TUBE
VI = VIAL

APPENDIX 15: Material Safety Data Sheet: Lethal Nerve Agent Sarin (GB)

Section I: General Information
Section II: Composition
Section III: Physical Data
Section IV: Fire and Explosion Data
Section V: Health Hazard Data
Section VI: Reactivity Data
Section VII: Spill, Leak, and Disposal Procedures
Section VIII: Special Protection Information
Section IX: Special Precautions
Section X: Transportation Data

Section I: General Information

MANUFACTURER'S NAME: Department of the Army

MANUFACTURER'S ADDRESS:
U.S. Army Chemical and Biological Defense Agency
Edgewood Research, Development and Engineering Center
ATTN: SCBRD-ODR-S
Aberdeen Proving Ground, MD 21010-5423

CAS REGISTRY NUMBER: 107-44-8 or 50642-23-4

CHEMICAL NAME AND SYMONYMS:
- Phosphonofluoridic acid, methyl-, isopropyl ester
- Phosphonofluoridic acid, methyl-, 1- methylethyl ester

ALTERNATE CHEMICAL NAMES:
- Isopropyl methylphosphonofluoridate
- Isopropyl ester of methylphosphonofluoridic acid
- Methylisoproposfluorophosphine oxide
- Isopropyl Methylfluorophosphonate
- O-Isopropyl Methylisopropoxfluorophosphine oxide
- O-Isopropyl Methylphosphonofluoridate
- Methylfluorophosphonic acid, isopropyl ester
- Isoproposymethylphosphonyl flouride

TRADE NAME AND SYNONYMS:
- GB
- Sarin
- Zarin

CHEMICAL FAMILY: Fluorinated organophosphorous compound

419

FORMULA   C4 H10 F02 P

NFPA 704 SIGNAL:
- Health - 4
- Flammability - 1
- Reactivity - 1

Section II: Composition

INGREDIENTS NAME: GB

FORMULA: C4 H10 FO2

PERCENTAGE: 100

AIRBORNE EXPOSURE LIMIT (AEL):   0.0001 mg/m3

Section III: Physical Data

BOILING POINT DEG F (DEG C):   316 (158)

VAPOR PRESSURE (mm hg):   2.9 @ 25 DEG C

VAPOR DENSITY (AIR=1):   4.86

SOLUBILITY IN WATER: Complete

SPECIFIC GRAVITY (H20=1):   1.0887 @ 25 DEG C

FREEZING/MELTING POINT: -56 DEG C

LIQUID DENSITY (g/cc):   1.0887 @ 25 DEG C/1.102 @ 20 DEG C

PERCENTAGE VOLATILE BY VOLUME: 22,000 m/m3 @ 25 DEG C/ 16,090 m/m3 @ 20 DEG C

APPEARANCE AND ODOR: Colorless liquid. Odorless in pure form.

Section IV: Fire and Explosion Data

FLASH POINT (METHOD USED):   Did not flash to 280 DEG F

FLAMMABLE LIMIT: Not applicable

LOWER EMPLOSIVE LIMIT: Not available

UPPER EXPLOSIVE LIMIT: Not available

EXTINGUISING MEDIA: Water mist, fog, foam, CO2. Avoid using extinguishing methods that will cause splashing or spreading of the GB

SPECIAL FIRE FIGHTING PROCEDURES: GB will react with steam or water to produce toxic & corrosive vapors. All persons not engaged in extinguishing the fire should be evacuated. Fires involving GB should be contained to prevent contamination to uncontrolled areas. When responding to a fire alarm in buildings or areas containing agents, firefighting personnel clothing (without TAP clothing) during chemical agent firefighting and fire rescue operations.

Respiratory protection is required. Positive pressure, full facepiece, NIOSH-approved self-contained breathing apparatus (SCBA) will be worn where there is danger of oxygen deficiency and when directed by the fire chief or chemical accident/incident (CAI) operations officer. In cases where firefighters are responding to a chemical accident/incident for rescue/reconnaissance purposes vice firefighting, they will wear appropriate levels of protective clothing (see Section 8).

UNUSUAL FIRE AND EXPLOSION HAZARDS: Hydrogen may be present

Section V: Health Hazard Data

AIRBORNE EXPOSURE LIMIT (AEL):   The permissible airborne exposure concentration for GB for an 6 hour workday or a 40 hour work week is an 8 hour time weight average (TWA) of 0.0001 mg/m3. This value is based on the TWA or GB which can be found in "AR 40-8, Occupational Health Guidelines for the Evaluation and Control of Occupational Exposure to Nerve Agents GA, GB, GD, and VX." To date, however, the Occupational Safety and Health Administration (OSHA) has not promulgated a permissible exposure concentration for GB.

EFFECTS OF OVEREXPOSURE: It is a lethal anticholinergic agent. Doses which are potentially life threatening may be only slightly larger than those producing minimal effects.

Route	Form	Effect	Type	Dosage
ocular	vapor	miosis	ECt50	less than 2 mg-min/m3
inhalation	vapor	runny nose	ECt50	less than 2 mg-min/m3
inhalation		severe incapacitation	ICt50	35 mg-min/m3
inhalation	vapor	death	LCt50	70 mg-min/m3
percutaneous	liquid	death	LD50	1700 mg/70 kg man

Effective dosages for vapor are estimated for exposure durations of 2-10 minutes.

Symptoms of overexposure may occur within minutes or hours—depending upon dose. They include: miosis (constriction of pupils) and visual effects, headache and pressure sensation, runny nose and nasal congestion, salivation, tightness in the chest, nausea, vomiting, giddiness, anxiety, difficulty in thinking, difficulty sleeping, nightmares, muscle twitches, tremors, weakness, abdominal cramps, diarrhea, involuntary urination and defecation.

With severe exposure, symptoms progress to convulsions and respiratory failure. GB is not listed by the International Agency for Research on Cancer (IARC), American Conference of Governmental Industrial Hygienists (ACGIH), Occupational Safety and Health Administration (OSHA), or National Toxicology Program (NTP) as a carcinogen.

EMERGENCY AND FIRST AID PROCEDURES:

- INHALATION: Hold breath until respiratory protective mask is donned. If severe signs of agent exposure appear (chest tightens, pupil constriction, incoordination, etc.), immediately administer, in rapid succession, all three Nerve Agent Antidote Kit(s), Mark I injectors (or atropine if directed by the local physician). Injections using the Mark I kit injectors may be repeated at 5 to 20 minute intervals if signs and symptoms are progressing until three series of injections have been administered. No more injections will be given unless directed by medical personnel. In addition, a record will be maintained of all injections given. If breathing has stopped, give artificial respiration. Mouth-to-mouth resuscitation should be used when approved mask-bag or oxygen delivery systems are not available. Do not use mouth-to-mouth resuscitation when facial contamination exists. If breathing is difficult, administer oxygen. Seek medical attention IMMEDIATELY.
- EYE CONTACT: Immediately flush eyes with water for 10-15 minutes, then don respiratory protective mask. Although miosis (pinpointing of the pupils) may be an early sign of agent exposure, an injection will not be administered when miosis is the only sign present. Instead, the individual will be taken IMMEDIATELY to the medical treatment facility for observation.
- SKIN CONTACT: Don respiratory protective mask and remove contaminated clothing. Immediately wash contaminated skin with copious amounts of soap and water, 10% sodium carbonate solution, or 5% liquid household bleach. Rinse well with water to remove decontaminant. Administer an intramuscular injection with the MARK I Kit injectors only if local sweating and muscular twitching symptoms are observed. SEEK MEDICAL ATTENTION IMMEDIATELY.
- INGESTION: Do not induce vomiting. First symptoms are likely to be gastrointestinal. Immediately administer an intramuscular injection of the MARK I kit autoinjectors. SEEK MEDICAL ATTENTION IMMEDIATELY.

Section VI: Reactivity Data

STABILITY: Stable when pure.

INCOMPATIBILITY: Attacks tin, magnesium, cadmium plated steel, some aluminums. Slight attack on copper, brass, lead, practically no attack on 1020 steel, Inconel & K-monel.

Hydrolyzes to form HF under acid conditions and isopropyl alcohol & polymers under basic conditions.

Section VII: Spill, Leak, and Disposal Procedures

STEPS TO BE TAKEN IN CASE MATERIAL IS RELEASED OR SPILLED: If leak or spills occur, only personnel in full protective clothing (see Section VIII) will remain in area. In case of personnel contamination see Section V.

RECOMMENDED FIELD PROCEDURES: Spills must be contained by covering with vermiculite, diatomaceous earth clay, fine sand, sponges, and paper or cloth towels. Decontaminate with copious amounts of aqueous Sodium Hydroxide solution (a minimum 10 wt percent). Scoop up all material and place in a fully removable head drum with a high density polyethylene liner. Cover the contents of the drum with decontaminating solution as above before affixing the drum head.

After sealing the head, the exterior of the drum shall be decontaminated and then labeled IAW EPA and DOT regulations. All leaking containers shall be overpacked with vermiculite placed between the interior and exterior containers. Decontaminate and label IAW EPA and DOT regulations. Dispose of the material IAW waste disposal methods provided below. Dispose of material used to decontaminate exterior of drum IAW Federal, state and local regulations. Conduct general area monitoring with an approved monitor (see Section VIII) to confirm that the atmospheric concentrations do not exceed the airborne exposure limit (see Sections II and VIII).

If 10 wt. percent aqueous Sodium Hydroxide solution is not available then the following decontaminants may be used instead and are listed in the order of preference: Decontamination Solution No. 2 (DS2), Sodium Carbonate, and Supertropical Bleach Slurry (STB).

RECOMMENDED LABORATORY PROCEDURES: A minimum of 56 grams of decon solution is required for each gram of GB. Decontaminant/agent solution is allowed to agitate for a minimum of one hour. Agitation is not necessary following the first hour. At the end of the one hour, the resulting solution should be adjusted to a pH greater than 11.5. If the pH is below 11.5, NaOH should be added until a pH above 11.5 can be maintained for 60 minutes.

An alternate solution for the decontamination of GB is 10 wt percent Sodium Carbonate in place of the 10 percent Sodium Hydroxide solution above. Continue with 56 grams of decon to 1 gram of agent. Agitate for one hour but allow three (3) hours for the reaction. The final pH should be adjusted to above 10. It is also permitted to substitute 5.25% Sodium Hypochlorite or 25 wt percent Monoethylamine (MEA) for the 10% Sodium Hydroxide solution above. MEA must be completely dissolved in water prior to addition of the agent. Continue with 56 grams of decon for each gram of GB and provide agitation for one hour. Continue with same ratios and time stipulations.

Scoop up all material and place in a fully removable head drum with a high density polyethylene liner. Cover the contents of the drum with decontaminating solution as above before affixing the drum head. After sealing the head, the exterior of the drum shall be decontaminated and then labeled IAW EPA and DOT regulations. All leaking containers shall be overpacked with vermiculite placed between the interior and exterior containers. Decontaminate and label IAW EPA and DOT regulations. Dispose of the material IAW waste disposal methods provided below. Dispose of material used to decontaminate exterior of drum IAW Federal, state and local regulations. Conduct general area monitoring with an approved monitor (see Section VIII) to confirm that the atmospheric concentrations do not exceed the airborne exposure limit (see Sections II and VIII).

WASTE DISPOSAL METHOD: Open pit burning or burying of GB or items containing or contaminated with GB in any quantity is prohibited. The detoxified GB using procedures above can be thermally destroyed by incineration in an EPA approved incinerator in accordance with appropriate provisions of Federal, state and local RCRA regulations.

Section VIII: Special Protection Information

RESPIRATORY PROTECTION:

Concentration	Respiratory Protective Equipment
less than 0.0001 mg/m3	A full face piece, chemical canister, air purifying protective mask will be onhand for escape. (The M9-, or M40-series masks are acceptable for this purpose.)
0.0001 to 0.2 mg/m3	A NIOSH/MSHA approved pressure demand full facepiece SCBA or supplied air respirator with escape air cylinder may be used.
	Alternatively, a full facepiece, chemical canister air purifying protective mask is acceptable for this purpose (for example, M9-, M17-, or M40-series mask or other mask certified as equivalent) is acceptable (see DA PAM 385-61 for determination of appropriate level).
greater than 0.2 mg/m3 or unknown	NIOSH/MSHA approved pressure demand full facepiece SCBA suitable for use in high agent concentrations with protective ensemble (see DA PAM 385-61 for examples).

VENTILATION: Local exhaust: Mandatory must be filtered or scrubbed to limit exit concentration to less than 0.0001 mg/m3 averaged over 8 hr/day indefinitely. Air emissions shall meet local, state and federal regulations.

- SPECIAL: Chemical laboratory hoods shall have an average inward face velocity of 100 linear feet per minute (1fpm) plus or minus 10 percent with the velocity at any point not deviating from the average face velocity by more than 20 percent. Existing laboratory hoods shall have an inward face velocity of 150 1fpm plus or minus 20 percent. Laboratory hoods shall be located such that cross drafts do not exceed 20 percent of the inward face velocity. A visual performance test utilizing smoke producing devices shall be performed in the assessment of the hood's ability to contain agent GB. Emergency backup power necessary. Hoods should be tested semi-annually or after modification or maintenance operations. Operations should be performed 20 cm inside hood face.
- OTHER: Recirculation of exhaust air from agent areas is prohibited. No connection is allowed between agent areas and other areas through ventilation system.

PROTECTIVE GLOVES: Butyl Glove M3 and M4, Norton, Chemical Protective Glove Set.

EYE PROTECTION: Chemical goggles. For splash hazards use goggles and faceshield.

OTHER PROTECTIVE EQUIPMENT: For general lab work, gloves and lab coat shall be worn with M9, M17 or M40 mask readily available.

MONITORING: Available monitoring equipment for agent GB is the M8/M9 Detector paper, detector ticket, blue band tube, M256/M256A1 kits, bubbler, Depot Area Air Monitoring System (DAAMS), Automatic Continuous Air Monitoring System (ACAMS), real time monitoring (RTM), Demilitarization Chemical Agent Concentrator (DCAC), M8/M43, M8A1/M43A2, Hydrogen Flame Photometric Emission Detector (HYPED), CAM-M1, Miniature Chemical Agent Monitor (MINICAM) and the Real Time Analytical Platform (RTAP).

Real-time, low-level monitors (with alarm) are required for GB operations. In their absence, an IDLH atmosphere must be presumed. Laboratory operations conducted in appropriately maintained and alarmed engineering controls require only periodic low-level monitoring.

Section IX: Special Precautions

PRECAUTIONS TO BE TAKEN IN HANDLING AND STORING: In handling, the buddy system will be incorporated. No smoking, eating and drinking in areas containing agent is permitted. Containers should be periodically inspected for leaks (either visually or by a detector kit). Stringent control over all personnel practices must be exercised. Decontamination equip shall be conveniently located. Exits must be designed to permit rapid evacuation. Chemical showers, eye-wash stations, and personal cleanliness facilities must be provided. Wash hands before meals and each worker will shower thoroughly with special attention given to hair, face, neck, and hands, using plenty of soap before leaving at the end of the work day.

OTHER PRECAUTIONS: Agents must be double contained in liquid and vapor tight containers when in storage or when outside of ventilation hood.

For additional information see "*AR 385-61, The Army Toxic Chemical Agent Safety Program*" "*DA PAM 385-61, Toxic Chemical Agent Safety Standards,*" and "*AR 40-8, Occupational Health Guidelines for the Evaluation and Control of Occupational Exposure to Nerve Agents GA, GB, GD, and VX.*"

Section X: Transportation Data

PROPER SHIPPING NAME: Poisonous liquids, n.o.s.

DOT HAZARD CLASSIFICATION: 6.1 Packing Group I Hazard Zone A

DOT LABEL: Poison

DOT MARKING: Poisonous liquid, n.o.s. (Isopropyl methylphosphonofluoridate) UN2810

DOT PLACARD   Poison

PRECAUTIONS TO BE TAKEN IN TRANSPORTATION   Motor vehicles will be placarded regardless of quantity. Driver shall be given full and complete information regarding shipment and conditions in case of emergency.

AR 50-6 deals specifically with the shipment of chemical agents. Shipments of agent will be escorted in accordance with AR 740-32.

EMERGENCY ACCIDENT PRECAUTIONS AND PROCEDURES: See Sections IV, VII, and VIII.

Material Safety Data Sheet: Lethal Nerve Agents Sulfur Mustards (HD and THD)

Section I: General Information

MANUFACTURER'S NAME: Department of the Army

MANUFACTURER'S ADDRESS:
U.S. Army Armament, Munitions and Chemical Command
Chemical Research, Development and Engineering Center
ATTN: SMCCR-CMS-E
Aberdeen Proving Ground, MD 21010-5423

CAS REGISTRY NUMBER: 505-60-2, 39472-40-7, 68157-62-0

CHEMICAL NAME AND SYNONYMS:
- Sulfide, bis (20chloroethyl)
- Bis (beta-chloroethyl) sulfide
- Bis (2-chloroethyl) sulfide
- (beta-chloroethylthio) ethane
- beta, beta'-dichlorodiethyl sulfide
- 2,2' dichlorodiethyl sulfide
- Di-2-chloroethyl sulfide
- beta, beta'-dichloroethyl sulfide
- 2,2'-dichloroethyl sulfide

TRADE NAME AND SYNONYMS:
- HD
- Senfgas
- H
- Sulfur mustard
- S-lost
- HS

- Iprit
- Suphur mustard gas
- Kampstoff "Lost"
- S-yperite
- Lost
- Yellow Cross Liquid
- Mustard Gas
- Yperite

CHEMICAL FAMILY: chlorinated sulfur compound

FORMULA: C4(H8)C12(S)

NFPA 704 SIGNAL:
- Health - 4
- Flammability - 1
- Reactivity - 1

Section II: Composition

Ingredients Name	Formula by Weight	Percentage	Airborne Exposure Limit (AEL)
Sulfur Mustard	C4(H8)C12(S)	100	0.003 mg/m3 (8 hr-TWA)

Section III: Physical Data

BOILING POINT DEG F (DEG C):   422 DEG F. (217 DEG C)

VAPOR PRESSURE (mm Hg):   0.072 mm Hg @ 20 DEG C (0.11 mm Hg @ 25 DEG C)

VAPOR DENSITY (AIR=1):   5.5

SOLUBILITY IN WATER: Negligible. Soluble in acetone, CH3(C1), tetrachloroethane, ethylbenzoate, and ether.

SPECIFIC GRAVITY (H20=1):   1.27 @ 20 DEG C

VOLATILITY: 610 mg/m3 @ 20 DEG C; 920 mg/m3 @ 25 DEG C

APPEARANCE AND ODOR: Water clear if pure. Normally pale yellow to black. Slight garlic type odor. The odor threshold for HD is 0.0006 mg/m3

Section IV: Fire and Explosion Data

FLASHPOINT (METHOD USED):   105 DEG C (ignited by large explosive charges)

FLAMMABILITY LIMITS (% by volume):   Unknown

EXTINGUISING MEDIA: Water, fog, foam, CO2. Avoid using extinguishing methods that will splash or spread mustard.

SPECIAL FIRE FIGHTING PROCEDURES: All persons not engaged in extinguishing the fire should be immediately evacuated from the area. Fires involving HD should be contained to prevent contamination to uncontrolled areas. When responding to a fire alarm in buildings or areas containing agents, firefighting personnel should wear full firefighter protective clothing (without TAP clothing) during chemical agent firefighting and fire rescue operations.

Respiratory protection is required. Positive pressure, full facepiece, NIOSH-approved self contained breathing apparatus (SCBA) will be worn where there is danger of oxygen deficiency and when directed by the fire chief of chemical accident/incident (CAI) operations officer. The M9 or M17 series mask may be worn in lieu of SCBA when there is no danger of oxygen deficiency. In cases where firefighters are responding to a chemical accident/incident for rescue/reconnaissance purposes vice firefighting, they will wear appropriate levels of protective clothing (see Section 8).

Section V: Health Hazard Data

AIRBORNE EXPOSURE LIMIT (REL)   The AEL for HD is 0.003 mg/m3 as proposed in the USAEHA Technical Guide No. 173, "Occupational Health Guidelines for the Evaluation and Control of Occupational Exposure to Mustard Agents H, HD and HT." No individual should be intentionally exposed to any direct skin or eye contact.

EFFECTS OF OVEREXPOSURE: HD is a vesicant (causing blisters) and alkylating agent producing cytotoxic action on the hematopoietic (blood-forming) tissues which are especially sensitive. The rate of detoxification of HD in the body is very slow and repeated exposures produce a cumulative effect. HD has been found to be a human carcinogen by the International Agency for Research on Cancer (IARC).

Median doses of HD in man are:

- LD50 (skin) = 100 mg/kg
- ICt50 (skin) = 2000 mg-min/m3 at 70-80 DEG F (humid environment); = 1000 mg-min/m3 at 90 DEG F (dry environment)
- ICt50 (eyes) = 200 mg-min/m3
- ICt50 (inhalation)=1500 mg-min/m3 (Ct unchanged with time)
- LD50 (oral) = 0.7 mg/kg

Maimum safe Ct for skin and eyes are 5 and 2 mg-min/m3, respectively.

ACUTE PHYSIOLOGICAL ACTION OF HD IS CLASSIFIED AS LOCAL AND SYSTEMIC:

- LOCALLY, HD affects both the eyes and the skin. SKIN damage occurs after percutaneous resorption. Being lipid soluble, HD can be resorbed into all organs. Skin penetration is rapid without skin irritation. Swelling (blisters) and reddening (erythema) of the skin occurs after a latency period of 4-24 hours following the exposure depending on degree of exposure and individual sensitivity. The skin healing process is very slow. Tender skin, mucous membrane and perspiration covered skin are more sensitive to the effects of HD. HD's effect on the skin, however, is less than on the eyes. Local action on the eyes produces severe necrotic damage and loss of eyesight. Exposure of eyes to HD vapor or aerosol produces lacrimation, photophobia, and inflammation of the conjunctiva and cornea.

- SYSTEMIC ACTIONS occur primarily through inhalation and ingestion. The HD vapor or aerosol is less toxic to the skin or eyes than the liquid form. When inhaled, the upper respiratory tract (nose, throat, trachea) is inflamed after a few hours latency period, accompanied by sneezing, coughing, and bronchitis, loss of appetite, diarrhea, fever, and apathy. Exposure to nearly lethal dose of HD can produce injury to bone marrow, lymph nodes, and spleen as indicated by a drop in WBC count and, therefore, results in increased susceptibilty to local and systemic infectons. Ingestion of HD will produce severe stomach pains, vomiting, and bloody stools after a 15-20 minute latency period.

- CHRONIC EXPOSURE to HD can cause sensitization, chronic lung impairment, (cough, shortness of breath, chest pain), and cancer of the mouth, throat, respiratory tract, skin, and leukemia. It may also cause birth defects.

EMERGENCY AND FIRST AID PROCEDURES:

- INHALATION: Remove from the source IMMEDIATELY. If breathing has stopped, give artificial respiration. If breathing is difficult, administer oxygen. Seek medical attention IMMEDIATELY.

- EYE CONTACT: Speed in decontaminating the eyes is absolutely essential. Remove person from the liquid source, flush the eyes immediately with water by tilting the head to the side, pulling the eyelids apart with the fingers and pouring water slowly into the eyes. Do not cover eyes with bandages but, if necessary, protect eyes by means of dark or opaque goggles. Transfer the patient to a medical facility IMMEDIATELY.

- SKIN CONTACT: Don respsiratory protective mask and gloves; remove victim from agent source immediately. Flush skin and clothes with 5 percent solution of sodium hypochlorite or liquid house hold bleach within one minute. Cut and remove contaminated clothing, flush contaminated skin area again with 5 percent sodium hypochlorite solution, then wash contaminated skin area with soap and water. If shower facilities are available, wash thoroughly and transfer to medical facility. If the skin becomes contaminated with a thickened agent, blot/wipe the material off immediately with an absorbent pad/paper towel prior to using decontaminating solution.

- INGESTION: Do not induce vomiting. Give victim milk to drink. Seek medical attention immediately.

Section VI: Reactivity Data

STABILITY: Stable at ambient temperatures. Decomposition temperature is 149 DEG C to 177 DEG C. Mustard is a persistent agent depending on pH and moisture, and has been known to remain active for up to three years in soil.

INCOMPATIBILITY: Conditions to avoid. Rapidly corrosive to brass @ 65 DEG C. Will corrode steel at a rate of .0001 in. of steel per month @ 65 DEG C.

HAZARDOUS DECOMPOSITION PRODUCTS: Mustard will hydrolize to form HCI and thiodiglycol.

HAZARDOUS POLYMERIZATION: Will not occur.

Section VII: Spill, Leak, and Disposal Procedures

STEPS TO BE TAKEN IN CASE MATERIAL IS RELEASED OR SPILLED: Only personnel in full protective clothing (see Section VIII) will remain in area where mustard is spilled.

RECOMMENDED FIELD PROCEDURES: The mustard should be contained using vermiculite, diatomaceious earth, clay or fine sand and neutralized as soon as possible using copious amounts of 5.25 percent Sodium Hypochlorite solution.

Scoop all material and place in an approved DOT container. Cover the contents of the drum with decontaminating solution as above. The exterior of the drum shall be decontaminated and then labeled IAW EPA and DOT regulations. All leaking containers shall be overpacked with vermiculite placed between the interior and exterior containers. Decontaminate and label IAW EPA and DOT regulations. Dispose of the material IAW waste disposal methods provided below. Dispose of the material used to decontaminate exterior of drum IAW Federal, state and local regulations. Conduct general area monitoring with an approved monitor (see Section 8) to confirm that the atmospheric concentrations do not exceed the airborne exposure limit (see Sections II and VIII).

If 5.25 percent Sodium Hypochlorite solution is not available then the following decontaminants may be used instead and are listed in the order of preference: Calcium Hypochlorite Decontamination Solution No. 2 (DS2), and Super Tropical Bleach Slurry (STB). WARNING: Pure, undiluted Calcium Hypochlorite (HTH) will burn on contact with liquid blister agent.

RECOMMENDED LABORATORY PROCEDURES: A minimum of 65 grams of decon per gram of HD is allowed to agitate for a minimum of one hour. Agitation is not necessary following the first hour if a single phase is obtained. At the end of 24 hours, the resulting solution shall be adjusted to a pH between 10 and 11. Test for presence of active chlorine by use of acidic potassium iodide solution to give free iodine color. Place 3 ml of the decontaminate in a test tube. Add several crystals of Potassium Iodine and swirl to dissolve. Add 3 ml of 50 wt percent Sulfuric Acid water and swirl. IMMEDIATE Iodine color indicates the presence of active chlorine. If negative, add additional 5.25 percent Sodium Hypochlorite solution to the decontamination solution, wait two hours, then test again for active chlorine. Continue procedure until positive chlorine is given by solution.

A 10 wt percent Calcium hypochlorite (HTH) mixture may be substituted for Sodium Hypochlorite. Use 65 grams of decon per gram of HD and continue the test as described for Sodium Hypochlorite.

Scoop up all material and place in approved DOT containers. Cover the contents of the drum with decontaminating solution as above. The exterior of the drum shall be decontaminated and then labeled IAW EPA and DOT regulations. All leaking containers shall be overpacked with vermiculite placed between the interior and exterior containers. Decontaminate and label IAW EPA and DOT regulations. Dispose of the material IAW waste disposal methods provided below. Dispose of the material used to decontaminate exterior of drum IAW federal, state and local regulations. Conduct general area monitoring with an approved monitor (see Section 8) to confirm that the atmospheric concentrations do not exceed the airborne exposure limits (see Section 8).

Note: Surfaces contaminated with HD and then rinse decontaminated may evolve sufficient mustard vapor to produce a physiological response.

WASTE DISPOSAL METHOD: All decontaminated material should be collected, contained and chemically decontaminated or thermally decomposed in an EPA approved incinerator, which will filter or scrub toxic by-products from effluent air before discharge to the atmosphere. Any contaminated protective clothing should be decontaminated using HTH or bleach and analyzed to assure it is free of detectable contamination (3X) level. The clothing should then be sealed in plastic bags inside properly labeled drums and held for shipment back to the DA issue point. Decontamination of waste or excess material shall be accomplished in accordance with the procedures outlined above with the following exception:

HD on laboratory glassware may be oxidized by its vigorous reaction with concentrated nitric acid.

Open pit burning or burying of HD or items containing or contaminated with HD in any quantity is prohibited.

Note: Some states define decontaminated surety material as a RCRA Hazardous Waste.

Section VIII: Special Protection Informaton

RESPIRATORY PROTECTION:

Concentration mg/m3	Respiratory Protection/Emsemble Required
Less than or equal to 0.003 as an 8-hr TWA	Protective mask not required provided that: (a) Continuous real-time monitoring (with alarm capability is conducted in the work area at the 0.003 mg/m3 level of detection. (b) M9, M17 or M40 mask is available and donned if ceiling concentrations exceed 0.003 mg/m3.

	(c) Exposure has been limited to the extent practicable by engineering controls (remote operations, ventilation, and process isolation) or work practices. If these conditions are not met then the following applies: Full facepiece, chemical canister, air purifying respirators. (The M9, M17 or M40 series or other certified equivalent masks are acceptable for this purpose in conjunction with the M3 toxicological agent protective (TAP) suit for dermal protection.)
Greater than 0.003 as an 8 hr TWA	The Demilitarization Protective Ensemble (DPE), 30 mil, may be used with prior approval from the AMC Field Safety Activity. Use time for the 30 mil DPE must be restricted to two hours or less.

Note: When 30 mil DPE is not available the M9 or M40 series mask with Level A protective ensemble including impregnated innerwear can be used. However, use time shall be restricted to the extent operationally feasible, and may not exceed one hour.

As an additional precaution, the cuffs of the sleeves and the legs of the M3 suit shall be taped to the gloves and boots respectively to reduce aspiration.

VENTILATION: Local Exhaust: Mandatory. Must be filtered or scrubbed.

Special: Chemical laboratory hoods shall have an average inward face velocity of 100 linear feet per minute (lfpm) plus or minus 10 percent with the velocity at any point not deviating from the average face velocity by more than 20 percent. Laboratory hoods shall be located such that cross- drafts do not exceed 20 percent of the inward face velocity. A visual performance test utilizing smoke-producing devices shall be performed in assessing the ability of the hood to contain agent HD.

Other: Recirculation of exhaust air from agent areas is prohibited. No connection between agent areas and other areas through ventilation system is permitted. Emergency backup power is necessary. Hoods should be tested semi-annually or after modificaton or maintenance operations. Operations should be performed 20 cm inside hoods.

PROTECTIVE GLOVES: MANDATORY. Butyl toxicological agent protective gloves (M3, M4, gloveset).

EYE PROTECTION: As a minimum, chemical goggles will be worn. For splash hazards use goggles and faceshield.

OTHER PROTECTIVE EQUIPMENT: Full protective clothing will consist of the m3 butyl rubber suit with hood, M2A1 butyl boots, M3 gloves, impregnated underwear, M9 series mask and coveralls (if desired), or the Demilitarization Protective Ensemble (DPE). For general lab work. gloves and lab coat shall be worn with M9 or M17 mask readily available.

   In addition, when handling contaminated lab animals, a daily clean smock, foot covers, and head covers are required.

MONITORING: Available monitoring equipment for agent HD is the M8/M9 Detector paper, blue bank tube, M256/M256A1 kits, bubbler. Depot Area Air Monitoring System (DAMMS), Automated Continuous Air Monitoring System (ACMS), CAM-M1, Hydrogen Flame Photometric Emission Detector (HYFED), and the Minature Chemical Agent Monitor (MINICAM).

Section IX: Special Precautions

PRECAUTIONS TO BE TAKEN IN HANDLING AND STORING: During handling, the "buddy" (two-man) system will be used. Containers should be periodically inspected for leaks either visually or using a detector kit, and prior to transfering the containers from storage to work areas. Stringent control over all personnel handling HD must be exercised. Chemical showers, eyewash stations, and personal cleanliness facilities must be provided. Each worker will wash their hands before meals and shower thoroughly with special attention given to hair, face, neck, and hands, using plenty of soap before leaving at the end of the workday. No smoking, eating, or drinking is permitted at the work site.

Decontaminating equipment shall be conveniently located. Exits must be designed to permit rapid evacuation. HD should be stored in containers made of glass for Research, Development, Test and Evaluation (RDTE) quantities or one-ton steel containers for large quantities. Agent shall be double-contained in liquid-tight containers when in storage.

OTHER PRECAUTIONS: For additional information, see AMC-R 385-131, "Safety Regulations for Chemical Agents H, HD, HT, GB, and VX" and USAEHA Technical Guide No. 173, "Occupational Health Guidelines for the Evaluation and Control of Occupational Exposure to Mustard Agents H, HD, and HT."

Section X: Transportation Data

PROPER SHIPPING NAME: Poisonous liquid, n.o.s.

DOT HAZARD CLASSIFICATION: Poison A

DOT LABEL: Poison gas

DOT MARKING: Poisonous liquid, n.o.s. (Sulfide, bis 2-chloroethyl) NA 1955

DOT PLACARD: POISON GAS

EMERGENCY ACCIDENT PRECAUTIONS AND PROCEDURES: See Section IV, and VIII.

PRECAUTIONS TO BE TAKEN IN TRANSPORTATION: Motor vehicles will be placarded regardless of quantity. Driver shall be given full and complete information regarding shipment and conditions in case of emergency. AR 50-6 deals specifically with the shipment of chemical agents. Shipments of agent will be escorted in accordance with AR 740-32.

While the Chemical Research Development and Engineering Center, Department of the Army believes that the data contained herein are factual and the opinions expressed are those of qualified experts regarding the results of the tests conducted, the data are not to be taken as a warranty or representation for which the Department of the Army or Chemical Research Development Engineering Center assumes legal responsibility. They are offered solely for your consideration, investigation, and verification. Any use of these data and information must be determined by the user to be in accordance with applicable Federal, State, and local laws and regulations.

Addendum A: Additional Information for Thickened HD

TRADE NAME AND SYNONYMS: Thickened HD, THD

HAZARDOUS INGREDIENTS: K125 (acryloid copolymer, 5%) is used to thicken HD, K125 is not known to be hazardous except in a finely-divided, powder form.

PHYSICAL DATA: Essentially the same as HD except for viscosity. The viscosity of HV is between 1000 and 1200 centistokes @ 25 DEG C.

FIRE AND EXPLOSION DATA: Same as HD.

HEALTH HAZARD DATA: Same as HD except for skin contact. For skin contact, don respiratory protective mask and remove contaminated clothing IMMEDIATELY. IMMEDIATELY scrape the HV from the skin surface, then wash the contaminated surface with acetone. Seek medical attention IMMEDIATELY.

SPILL, LEAK, AND DISPOSAL PROCEDURES: If spills or leaks of HV occur, follow the same procedures as those for HD, but dissolve the THD in acetone prior to introducing any decontaminating solution. Containment of THD is generally not necessary. Spilled THD can be carefully scraped off the contaminined surface and placed in a fully removable head drum with a high density, polyethylene lining. The THD can then be decontaminated, after it has been dissolved in acetone, using the same procedures used for HD. Contaminated surfaces should be treated with acetone, then decontaminated using the same procedures as those used for HD.

Note: Surfaces contaminated with THD or HD and then rinse-decontaminated may evolve sufficient mustard vapor to produce a physiological response.

SPECIAL PROTECTION INFORMATION: Same as HD.

SPECIAL PRECAUTIONS: Same as HD with the following addition. Handling the THD requires careful observation of the "stringers" (elastic, thread-like attachments) formed when the agents are transferred or dispensed. These stringers must be broken cleanly before moving the contaminating device or dispensing device to another location, or unwanted contamination of a working surface will result.

TRANSPORTATION DATA: Same as HD.

APPENDIX 17: Material Safety Data Sheet: Lethal Nerve Agents Soman (GD and Thickened GD)

Section I: General Information

MANUFACTURER'S NAME Department of the Army

MANUFACTURER'S ADDRESS:
U.S. Army Armament, Munitions and Chemical Command
Chemical Research, Development and Engineering Center
ATTN: SMCCR-CMS-E
Aberdeen Proving Ground, MD 21010-5423

CAS REGISTRY NUMBER: 96-64-0 or 50642-24-5

CHEMICAL NAME: Phosphonofluoridic acid, methyl-, 1, 2, 2-trimethylpropyl ester

ALTERNATE CHEMICAL NAMES:
- Pinacolyl methylphosphonofluoridate
- 1,2,2-Trimethylpropyl methylphosphonofluoridate
- Methylpinacolyloxyfluorophosphine oxide
- Pinacolyloxymethylphosphonyl flouride
- Pinacolyl methanefluorophosphonate
- Methylfluoropinacolylphosphonate
- Fluoromethylpinacolyloxyphosphine Oxide
- Methylpinacolyloxyphosphonyl flouride
- Pinacolyl methylfluorophosphonate
- 1,2,2,-Trimethylpropoxyfluoromethylphosphine oxide

TRADE NAME AND SYNONYMS:
- GD
- EA 1210

- Soman, Zoman
- PFMP

CHEMICAL FAMILY: Fluorinated organophosphorus compound

FORMULA: C7 H16 F02 P

NFPA 704 SIGNAL:
- Health - 4
- Flammability - 1
- Reactivity - 1

Section II: Hazardous Ingredients

Ingredients	Formula	Percentage by Weight	Airborne Exposure Limit
GD	C7 H16 FOP	100	0.00003 mg/m3

Section III: Physical Data

BOILING POINT DEG F (DEG C):   (198 DEC C) 388 DEG F

VAPOR PRESSURE: 0.40 mm Hg @ 25 DEG C

VAPOR DENSITY (AIR=1):   6.3

SOLUBILITY IN WATER: Moderate

SPECIFIC GRAVITY (H20=1):   1.022 @ 25 DEG C

VOLATILITY: 3900 mg/m3 @ 25 DEC C

MELTING POINT: -42 DEG C

APPEARANCE AND ODOR: When pure, colorless liquid with fruity odor. With impurities, amber or dark brown, with oil of camphor odor

Section IV: Fire and Explosion Data

FLASHPOINT: 121 DEG C (Open Cup)

FLAMMABILITY LIMITS: Unknown

LOWER EXPLOSIVE LIMIT: Not applicable

UPPER EXPLOSIVE LIMIT: Not applicable

EXTINGUISING MEDIA: Water, fog, foam, CO2 - Avoid using extinguishing methods that will cause splashing or spreading of the GD.

SPECIAL FIRE FIGHTING PROCEDURES: Fires involving GD should be contained to prevent contamination of uncontrolled areas. All persons not engaged in extinguishing the fire should be evacuated immediately. Contact with GD or its vapors can be fatal. When responding to a fire alarm in buildings or areas containing agents, firefighting personnel should wear full firefighter protective clothing (without TAP clothing) during chemical agent firefighting and fire rescue operations.

Respiratory protection is required. Positive pressure, full facepiece, NIOSH approved self contained breathing apparatus (SCBA) will be worn where there is danger of oxygen deficiency and when directed by the fire chief or chemical accident/incident (CAI) operations officer. The M9 or M17 series mask may be worn in lieu of SCBA when there is no danger of oxygen deficiency. In cases where firefighter are responding to a chemical accident/incident for rescue/reconnaissance purposes vice firefighting, they will wear appropriate levels of protective clothing (see Section 8).

UNUSUAL FIRE AND EXPLOSION HAZARDS: Hydrogen produced by the corrosive vapors reacting with metals, concrete, etc., may be present.

Section V: Health Hazard Data

AIRBORNE EXPOSURE LIMIT (AEL): The suggested permissible airborne exposure concentration of GD for an 8 hour workday or a 40 hour work week is an 8 hour time weighted average (TWA) of 0.00003 mg/m3 (2 x 10-5 ppm). This value is based on the TWA of GB as proposed in the USaEHA Technical Guide No. 169, "Occupational Health Guidelines for the Evaluation and Control of Occupational Exposure to Nerve Agents GA, GB, GD, and VX." To date, however, the Occupational Safety and Health Administration (OSHA) has not promulgated permissible exposure concentration for GD.

EFFECTS OF OVEREXPOSURE: GD is a lethal anticholinesterase agent with the median lethal dose in man being: LCt50 (inhalation) = 70 mg min/m3 (t = 10 min); LD50 (PC, bare skin) = 0.35 g/man (70 kg).

1. One to several minutes after overexposure to airborne GD the following acute symptoms appear:
    1. LOCAL EFFECTS (lasting 1-15 days, increase with dose)
        1. On eyes: Miosis (constriction of pupils); redness, pressure sensation on eyes.
        2. By inhalation: Rhinorrhea (runny nose), nasal congestion, tightness in chest, wheezing, salivation, nausea, vomiting
    2. SYSTEMIC EFFECTS (increases with dose): When inhaled GD will cause excessive secretion causing coughing/breathing difficulty: salivation and sweating: vomiting, diarrhea; stomach cramps; involuntary urination/defecation; generalized muscle twitching/muscle cramps; CNS depression including anxiety, restlessness, giddiness, insomnia, excessive dreaming and nightmares. With more severe exposure, also headache, tremor, drowsiness, concentration difficulty,

memory impairment, confusion, unsteadiness on standing or walking, and progressing to death.

2. After exposure to liquid GD, the following acute symptoms appear:
   1. LOCAL EFFECTS:
      1. On eyes: Miosis (constriction of pupils); redness, pressure sensation on eyes.
      2. By ingestion: salivation, anorexia, nausea, vomiting, abdominal cramps, diarrhea, involuntary defecation, heartburn.
      3. On skin: Sweating, muscle twitching
   2. Chronic exposure to GD causes forgetfulness, thinking difficulty, vision disturbances, muscular aches/pains. Although certain organophosphate pesticides have been shown to be teratogenic in animals, these effects have not been documented in carefully controlled toxicological evaluations for GD.

GD presently is not listed by the International Agency for Research on Cancer (IARC), National Toxicology Program (NTP), Occupational Safety and Health Administration (OSHA), or American Conference of Governmental Industrial Hygienists (ACGIH) as a carcinogen.

EMERGENCY AND FIRST AID PROCEDURES:
- INHALATION: Hold breath until respiratory protective mask is donned. If severe signs of agent exposure appear (chest tightens, pupil constriction, incoordination, etc.), immediately administer, in rapid succession, all three Nerve Agent Antidote Kit(s), Mark I injectors (or atropine if directed by the local physician). Injections using the Mark I kit injectors may be repeated at 5 to 20 minute intervals if signs and symptoms are progressing until three series of injections have been administered. No more injections will be given unless directed by medical personnel. In addition, a record will be maintained of all injections given. If breathing has stopped, give artificial respiration. Mouth-to-mouth resuscitation should be used when approved mask-bag of oxygen delivery systems are not available. Do not use mouth-to-mouth resuscitation when facial contamination exists. If breathing is difficult, administer oxygen. Seek medical attention IMMEDIATELY.
- EYE CONTACT: IMMEDIATELY flush eyes with water for 10-15 minutes, then don respiratory protective mask. Although miosis (pinpointing of the pupils) may be an early sign of agent exposure, an injection will not be administered when miosis is the only sign present. Instead, the individual will be taken IMMEDIATELY to the medical treatment facility for observation.
- SKIN CONTACT: Don respsiratory protective mask and remove contaminated clothing. Immediately wash contaminated skin with copious amounts of soap and water, 10 percent sodium carbonate solution, or 5 percent liquid household bleach. Rinse well with water to remove decontaminant. Administer nerve agent antidote kit, Mark I, only if local sweating and muscular twitching symptoms are present. Seek medical attention IMMEDIATELY.
- INGESTION: Do not induce vomiting. First symptoms are likely to be gastronintestinal. IMMEDIATELY administer Nerve Agent Antidote kit, Mark I. Seek medical attention immediately.

Section VI: Reactivity Data

STABILITY: Stable after storage in steel for 3 months at 65 Deg. C. GD corrodes steel at the rate of 1 x 10-5 inch/month. GD will hydrolyze to form HF--H-H-O-CH3 and (CH3) 3-C-C-O-P-OH

HAZARDOUS POLYMERIZATION: Will not occur.

Section VII: Spill, Leak, and Disposal Procedures

STEPS TO BE TAKEN IN CASE MATERIAL IS RELEASED OR SPILLED: If leak or spills occur, only personnel in full protective clothing (see Section VIII) will remain in area. In case of personnel contamination, see Section V "Emergency and First Aid Procedures."

RECOMMENDED FIELD PROCEDURES: Spills must be contained by covering with vermiculite, diatomaceous earth, clay, fine sand, sponges and paper or cloth towels. This containment is followed by treatment with copious amounts of aqueous Sodium Hydroxide solution (a minimum of 10 percent). Scoop up all material and place in a fully removable head drum with a high density polyethylene liner. Cover the contents of the drum with decontaminating solution as above before affixing the drum head. After sealing the head, the exterior of the drum shall be decontaminated and then labeled IAW EPA and DOT regulations.

All leaking containers shall be overpacked with vermiculite placed between the interior and exterior containers. Decontaminate and label IAW EPA and DOT regulations. Dispose of the material IAW waste disposal methods provided below. Dispose of material used to decontaminate exterior of drum IAW Federal, state and local regulations. Contaminated clothing will be placed in a fully removable head drum with a high density polyethylene liner and the contents shall be covered with decontaminating solution as above before affixing the drum head. Conduct general area monitoring to confirm that the atmospheric concentrations do not exceed the exposure limits (see Section VIII).

If 10 wt percent aqueous Sodium Hydroxide solution is not available then the following decontaminants may be used instead and are listed in the order of preference: Decontaminating Solution No. 2 (DS2), Sodium Carbonate, and Supertropical Tropical Bleach Slurry (STB).

RECOMMENDED LABORATORY PROCEDURES: A minimum of 55 grams of decon solution is required per gram of GD. Decontaminant/agent solution is allowed to agitate for a minimum of one hour. Agitation is not necessary following the first hour provided a single phase is obtained. At the end of the first hour the pH should be checked and adjusted up to 11.5 with additional NaOH as required.

An alternate solution for the decontamination of GD is 10 percent Sodium Carbonate in place of the 10 percent Sodium Hydroxide solution above. Continue with 55 grams of decon per gram of GD. Agitate for one hour and allow to react for 3 hours. At the end of the third hour adjust the pH to above 10. It is also permitted to substitute 5.25% Sodium

Hypochlorite for the 10% Sodium Hydroxide solution above. Continue with 55 grams of decon per gram of GD. Agitate for one hour and allow to react for 3 hours then adjust the pH to above 10.

Scoop up all material and place in a fully removable head and a high density polyethylene liner. Cover the contents with additional decontaminating solution before affixing the drum head. After sealing the head, the exterior of the drum shall be decontaminated and then labeled IAW EPA and DOT regulations. All contaminated clothing will be placed in a fully removable head drum with a high density polyethylene liner.

Cover the contents of the drum with decontaminating solution as above before affixing the drum head. After sealing the head, the exterior of the drum shall be decontaminated and then labeled IAW EPA and DOT regulations. All leaking containers shall be overpacked with vermiculite placed between the interior and exterior containers. Decontaminate and label IAW EPA and DOT regulations. Dispose of the material IAW waste disposal methods provided below. Conduct general area monitoring to confirm that the atmospheric concentrations do not exceed the exposure limits (see Section VIII).

WASTE DISPOSAL METHOD: Open pit burning or burying of GD or items containing or contaminated with GD in any quantity is prohibited. The detoxified GD (using procedures above) can be thermally destroyed by incineration in an EPA approved incinerator in accordance with appropriate provisions of Federal, state and local RCRA regulations.

Note: Some states define decontaminated surety material as a RCRA Hazardous Waste.

Section VIII: Special Protection Informaton

RESPIRATORY PROTECTION:

GD Concentration	Respiratory Protective Equipment
Less than 0.00003 mg/m3	M9, M17, or M40 series mask shall be available for escape as necessary
0.00003 mg/m3 to 0.06 mg/m3	M9, or M40 series mask with Level A or Level B ensemble (see AMCR 385-131 for determination of appropriate level). Demilitarization Protective Emsemble (DPE), or Toxicological Agent Protective Ensemble Self-Contained (TAPES), used with prior approval from AMC Field Safety Activity.
Greater than 0.06 mg/ m3 or unknown	DPE or TAPES used with prior approval from AMC Field Field Safety Activity *Note:* When DPE or TAPES is not available the M9 or M40 series mask with Level A protective ensemble can be used. However, use time shall be restricted to the extent operationally feasible, and may not exceed one hour. As an additional precaution, the cuffs of the sleeves and the legs of the M3 suit shall be taped to the gloves and boots respectively to reduce aspiration.

- Local Exhaust: Mandatory. Must be filtered or scrubbed to limit exit conc. to <.00001 mg/m3 (averaged over 8 hr/day, indefinitely).
- Special: Chemical laboratory hoods shall have an average inward face velocity of 100 linear feet per minute (lfpm) + 10 percent with the velocity at any point not deviating

from the average face velocity by more than 20 percent. Laboratory hoods shall be located such that cross-drafts do not exceed 20 percent of the inward face velocity. A visual performance test utilizing smoke- producing devices shall be performed in assessing the ability of the hood to contain agent GD.

- Emergency back-up power necessary: Hoods should be tested semi-annually or after modification or maintenance operations. Operations should be performed 20 cm inside hood face.
- Other: Recirculation of exhaust air from agent areas is prohibited. No connection between agent areas and other areas through ventilation system is permitted.

PROTECTIVE GLOVES: Butyl Glove M3 and M4; Northon, Chemical Protective Glove Set

EYE PROTECTION: Chemical Goggles. For splash hazards use goggles and faceshield.

OTHER PROTECTIVE EQUIPMENT: Full protective clothing will consist of M9 mask and hood, butyl rubber suit (M3), M2A1 butyl boots, M3 and M4 gloves, unimpregnated underwear, or demilitarization protective ensemble (DPE). For laboratory operations, wear lab coats and have a protective mask readily available.

MONITORING: Available monitoring equipment for agent GD is the Automatic Chemical Agent Detector Alarm (ACADA), bubblers (GC method), and Chemical Agent Monitor (CAM).

Section IX: Special Precautions

PRECAUTIONS TO BE TAKEN IN HANDLING AND STORING: In handling GD, the buddy system will be incorporated. No smoking, eating or drinking is permitted in areas containing agent GD. Containers should be periodically inspected for leaks (either visually or by a detector kit) and prior to transferring the containers from storage to work areas. Stringent control over all personnel practices must be exercised. Decontamination equipment shall be conveniently located. Exits must be designed to permit rapid evacuation. Chemical showers, eyewash stations, and personal cleanliness facilities shall be provided. Wash hands before meals and each worker will shower thoroughly with special attention given to hair, face, neck, and hands, using plenty of soap before leaving at the end of the workday.

OTHER PRECAUTIONS: Agent must be double-contained in liquid and vapor-tight containers when in storage or when outside of the ventilation hood.

For additional information, see AMC-R 385-131, "Safety Regulations for Chemical Agents H, HD, GB, and VX" and USaEHA Technical Guide No. 169, "Occupational Health Guidelines for the Evaluation and Control of Occupational Exposure to Nerve Agents GA, GB, GD, and VX."

Section X: Transportation Data

PROPER SHIPPING NAME: Poisonous liquid, n.o.s.

DOT HAZARD CLASSIFICATION: Poison A

DOT LABEL: Poison gas

DOT MARKING: Poisonous liquid, n.o.s. (Pinacolyl methylphosphonofluoridate) NA 1955

DOT PLACARD: POISON GA

EMERGENCY ACCIDENT PRECAUTIONS AND PROCEDURES: See Section IV, VII, and VIII.

PRECAUTIONS TO BE TAKEN IN TRANSPORTATION: Motor vehicles will be placarded regardless of quantity. Driver shall be given full and complete information regarding shipment and conditions in case of emergency.

AR 50-6 deals specifically with the shipment of chemical agents. Shipments of agent will be escorted in accordance with AR 740-32.

While the Chemical Research Development and Engineering Center, Department of the Army believes that the data contained herein are factual and the opinions expressed are those of qualified experts regarding the results of the tests conducted, the data are not to be taken as a warranty or representation for which the Department of the Army or Chemical Research Development Engineering Center assumes legal responsibility. They are offered solely for your consideration, investigation, and verification. Any use of these data and information must be determined by the user to be in accordance with applicable Federal, State, and local laws and regulations.

Addendum A: Physiological Effects

ACUTE PHYSIOLOGICAL EFFECTS:

Site of Action	Signs and Symptoms
Muscarine-like-	
Pupils	Miosis, marked, usually maximal (pinpoint), ometimes unequal.
Ciliary body	Frontal headache, eye pain on focusing, slight dimness of vision, occasional nausea and vomiting.
Conjunctivae	Hyperemia
Nasal mucous membranes	Rhinorrhea, hyperemia
Bronchial tree	Tightness in chest, sometimes with prolonged wheezing, expiration suggestive of broncho-constriction or increased secretion, cough.
	Following Systemic Absorption

Bronchial tree	Tightness in chest, with prolonged wheezing, expiration suggestive of broncho-constriction or increased secretion, dyspnea, slight pain in chest, increased, bronchial secretion, cough, pulmonary edema, cyanosis.
Gastrointestinal	Anorexia, nausea, vomiting, abdominal cramps, epigastric and substernal tightness (cardiospasm) with "heartburn" and eructation, diarrhea, tenesmus, involuntary defecation.
Sweat glands	Increased sweating
Salivary glands	Increased salivation
Lacrimal glands	Increased lacrimation
Heart	Slight bradycardia
Pupils	Slight miosis, occasionally unequal, later maximal miosis (pinpoint).
Ciliary body	Blurring of vision
Bladder	Frequent, involuntary micturition
Nicotine-like	
Striated muscle	Easy fatigue, mild weakness, muscular twitching, fasciculations, cramps, generalized weakness, including muscles of respiration, with dyspnea and cyanosis. Sympathetic ganglia Pallor, occasional elevation of blood pressure.
Central Nervous System	Giddiness, tension, anxiety, jitteriness, restlessness, emotional lability, excessive dreaming, insomnia, nightmares, headaches, tremor, withdrawal and depression, bursts of slow waves of elevated voltage in EEG, especially on over-ventilation, drowsiness, difficult concentration, slowness on recall, confusion, slurred speech, ataxia, generalized weakness, coma, with absence of reflexes, Cheyne-Stokes respirations, convulsions, depression of respiratory and circulatory centers, with dyspnea, cyanosis, and fall in blood pressure.

CHRONIC PHYSIOLOGICAL EFFECTS:

- Acute Exposure: If recovery from nerve agent poisoning occurs, it will be complete unless anoxia or convulsions have gone unchecked so long that irreversible central nervous system changes due to anoxemia have occurred.
- Chronic Exposure
   o The inhibition of cholinesterase enzymes throughout the body by nerve agents is more or less irreversible so that their effects are prolonged. Until the tissue cholinesterase enzymes are restored to normal activity, probably by very slow regeneration over a period of weeks or 2 to 3 months if damage is severe there is a period of increased susceptibility to the effects of another exposure to any nerve agent. During this period the effects of repeated exposures are cumulative; after a single exposure, daily exposure to concentrations of a nerve agent insufficient to produce symptoms may result in the onset of symptoms after several days. Continued daily exposure may be followed by increasingly severe effects. After symptoms subside, increased susceptibility persists for one to several days. The degree of exposure required to produce recurrence of symptoms, and the severity of these symptoms, depend on duration of exposure and time intervals between exposures. Increased susceptibility is not limited to the particular nerve agent initially absorbed.
   o Estimates have been made for the times as which 50 percent of exposed subjects would be affected (Et50's) at median incapacitating doses. These are presented below:

Et50	Degree of Effectiveness	ICt50	Exposure Time
min		mg min/m3	min
1.5	Moderate	27	0.5
3.0	Incap.	27	2.0
6.0		40	10.0
1.0	Severe	37	0.5
3.8	Incap.	37	2.0
7.8		56	10.0
2.0	Very	47	0.5
4.5	Severe	47	2.0
9.5	Incap.	72	10.0
6.5	Death	70	5.0
9.0		70	2.0
13.5		103	10.0

- Exposure to high concentrations of nerve agent may bring on incoordination, mental confusion and collapse so rapidly that the casualty cannot perform self-aid. If this happens, the man nearest to him will give first aid.
- Onset Time of Symptoms

Types of Effects	Route Absorption	Description of Effects	When Effects Appear after Exposure
Vapor Local	Lungs	Rhinorrhea, nasal Hyperemia tightness in chest, wheezing	One to several minutes
Vapor Local	Eyes	Miosis, conjectival hyperemia eye pain, frontal headache	One to several minutes
Vapor Systemic	Lungs or and central nervous system effects. Exposure: about 30 min. after mild exposure	Muscarine-like, nicotine-like	Less than 1 min. to a few min. after moderate or marked
Liquid Local	Eyes	Same as vapor effects	Instantly
Liquid Local	Ingestion gastrointestinal	About 30 min. after ingestion	
Liquid Local	Skin	Local sweating and muscular twitching	3 min. to 2 hours
Liquid Sy temic	Lungs		Several minutes
Liquid Systemic	Eyes	Same as for vapor	Several minutes
Liquid Systemic	Skin	Generalized sweating	15 min. to 2 hours
Liquid Systemic	Ingestion gastrointestinal		15 min. to 2 hours

Types of Effects	Route of Absorption	Duration of Effects After Mild Exposure	Severe Exposure
Vapor Local	Lungs	A few hours	1 to 2 days
Vapor Local	Eyes	Miosis - 24 hours	3 to 14 days 2 to 5 days
Vapor Systemic	Lungs or eyes	Several hours	8 days
Liquid Local	Eyes	Similar to effects of vapor	
Liquid Local	Ingestion	3 days	5 days
Liquid Local	Skin	3 days	5 days
Liquid Systemic	Lungs		1 to 5 days
Liquid Systemic	Eyes		2 to 4 days
Liquid Systemic	Skin		2 to 5 days
Liquid Systemic	Ingestion		3 to 5 days

Addendum B: First Aid Procedures

1. Exposed personnel will be removed immediately to an uncontaminated atmosphere. Personnel handling casualty cases will give consideration to their own safety and will take precautions and employ the prerequisite protective equipment to avoid becoming exposed themselves.

   CAUTION: Due to the rapid effects of nerve agents, it is extremely important that decontamination of personnel not be delayed by attempting to blot off excessive agent prior to decontamination with sodium hypochlorite.

2. The causalty will then be decontaminated by washing the contaminated areas with commercial liquid household bleach (nominal 5 percent solution hyprchlorite or 10 percent sodium carbonate solution) and flushing with clean water. Mask will be left on the victim until decontamination has been completed unless it has been determined that areas of the face were contaminated and the mask must be removed to facilitate decontamination. After decontamination, the contaminated clothing will be removed and skin contamination washed away. If possible, decontamination will be completed before the casualty is taken to the aid station of medical facility.

   CAUTION: Care must be taken when decontaminating facial areas to avoid getting the hypochlorite into the eye or mouth. Only clean water shall be used when flushing the eyes or mouth. Skin surfaces decontaminated with bleach should be thoroughly flushed with water to prevent skin irritation from the bleach.

3. If there is no apparent breathing, artifical resuscitation will be started immediately (mouth-to-mouth, or with mechanical resuscitator). The situation will dictate method of choice, e.g., contaminated face. Do not use mouth-to-mouth resuscitation when facial contamination exists. When appropriate and trained personnel are available, cardiopulmonary resuscitation (CPR) may be necessary.

4. An individual who has received a known agent exposure or who exhibits definite signs or symptoms of agent exposure shall be given an intramuscular injection immediately with the MARK I kit auto-injectors.

1. Some of the early symptoms of a vapor exposure may be rhinorrhea (runny nose) and/or tightness in the chest with shortness of breath (bronchial constriction).
2. Some of the early symptoms of a percutaneous exposure may be local muscular twitching or sweating at the area of exposure followed by nausea or vomiting.
3. Although myosis (pin-pointing of the pupils) may be an early sign of agent exposure, an injection shall not be administered when myosis is the only sign present. Instead, the individual shall be taken immediately to the medical facility for observation.
4. Injections using the MARK I kit injectors (or atropine only if directed by the local physician) may be repeated at 5 to 20 minute intervals if signs and symptoms are progressing until three series of injections have been administered. No more injections will be given unless directed by medical personnel. In addition, a record will maintained of all injections given.
5. Administer, in rapid succession, all three MARK I kit injectors (or atropine if directed by the local physician) in the case of SEVERE signs of agent exposure.

If indicated, CPR should be started immediately. Mouth-to-mouth resuscitation should be used when approved mask-bag or oxygen delivery systems are not available. Do not use mouth-to-mouth resuscitation when facial contamination exists.

CAUTION: Atropine does not act as a prophylactic and shall not be administered until an agent exposure has been ascertained.

Addendum C: Additional Information for Thickened GD

TRADE NAME AND SYNONYMS: Thickened GD, TGD

HAZARDOUS INGREDIENTS: K125 (acryloid copolymer, 5%) is used to thicken the GD. K125 is not known to be a hazardous material except in a finely divided, powder form.

PHYSICAL DATA: Essential the same as GD except for viscosity. The viscosity of TGD is approximately 1180 centistokes.

FIRE AND EXPLOSION DATA: Same as GD

HEALTH HAZARD DATA: Same as GD except for skin contact. For skin contact, don respiratory protective mask and remove contaminated clothing. Immediately scrape the TGD from the skin surface, then wash the contaminated surface with acetone. Administer Nerve Agent Antidote Kit, MARK I, only if local sweating and muscular twitching symptoms are observed. Seek medical attention IMMEDIATELY.

SPILL, LEAK AND DISPOSAL PROCEDURES: If spills or leaks of TGD occur, follow the same procedure as those for GD, but add the following step: Since TGD is not water soluble, dissolve the TGD in acetone prior to introducing any decontaminating solution. Containment of TGD is generally not necessary. Spilled TGD can be carefully scraped off the contaminated surface and placed in a drum with a fully removable head and a high density, polyethylene lining. The TGD can then be decontaminated after it has been dissolved in acetone, using the same procedures as for GD. Contaminated surfaces should be treated with acetone, then decontaminated using the same procedures as for GD.

SPECIAL PROTECTION INFORMATION: Same as GD.

SPECIAL PRECAUTIONS: Same as GD with the following addition. Handling the TGD requires careful observation of the "stringers" (elastic, thread-like attachments) formed when the agents are transferred or dispensed. These stringers must be broken cleanly before moving the contaminating device or dispensing device to another location, or unwanted contamination of a working surface will result.

TRANSPORTATION DATA: Same as GD.

APPENDIX 18:   Material Safety Data Sheet: Lethal Nerve Agent Tabun (GA)

Section I: General Information

MANUFACTURER'S NAME: Department of the Army

MANUFACTURER'S ADDRESS:
U.S. Army Armament, Munitions and Chemical Command
Chemical Research, Development and Engineering Center
ATTN: SMCCR-CMS-E
Aberdeen Proving Ground, MD 21010-5423

CAS REGISTRY NUMBER: None

CHEMICAL NAME: Ethyl N,N-dimethylphosphoramidocyanidate

TRADE NAME AND SYNONYMS:
- Ethyl dimethylplosphoramidocyanidate
- Dimethylaminoethoxy-cyanophosphine oxide
- Dimethylamidoethoxyphosphoryl cyanide
- Ethyldimethylaminocyanophosphonate
- Ethyl ester of dimethylphosphoroamidocyanidic acid
- Ethyl phosphorodimethylamidocyanidate
- GA
- EA1205
- Tabun

CHEMICAL FAMILY: Organophosphorous compound

FORMULA: C5 H11 N2 02 P

NFPA 704 SIGNAL:
- Health - 4
- Flammability - 2
- Reactivity - 1

Section II: Composition

Ingredients	Formula Name	Percentage by Weight	Airborne Exposure Limit (AEL)
GA	C5 H11 N2 02 P	100	0.0001 mg/m3

Section III: Physical Data

BOILING POINT DEG F (DEG C):   247.5 DEG C

VAPOR PRESSURE (mm hg):   0.07 @ 24 DEG C

VAPOR DENSITY (AIR=1):   5.6

SOLUBILITY IN WATER (g/100 g):   9.8 @ 25 DEG C/ 7.2 @ 20 DEG C

SPECIFIC GRAVITY (H20=1):   Not available

FREEZING (MELTING) POINT: -50 DEG C

AUTOIGNITION TEMPERATURE DEG F (DEG C):   Not available

VISCOSITY (CENTISTOKES):   2.18 @ 25 DEG C

PERCENTAGE VOLATILE BY VOLUME: 610 mg/m3 @ 25 DEG C

EVAPORATION RATE: Not available

APPEARANCE AND ODOR: Colorless to brown liquid. Faintly fruity; none when pure

Section IV: Fire and Explosion Data

FLASHPOINT: 78 DEG C

FLAMMABILITY LIMITS (% by volume):   Not available

EXTINGUISING MEDIA: Water, fog, foam, CO2 - Avoid using extinguishing methods that will cause splashing or spreading of the GA.

UNUSUAL FIRE & EXPLOSION HAZARDS: Fires involving this chemical may result in the formation of hydrogen cyanide.

SPECIAL FIRE FIGHTING PROCEDURES: All persons not engaged in extinguishing the fire should be immediately evacuated from the area. Fires involving GA should be contained to prevent contamination to uncontrolled areas. When responding to a fire alarm in buildings or areas containing agents, firefighting personnel should wear full firefighter protective clothing (without TAP clothing) during chemical agent firefighting and fire rescue operations.

Respiratory protection is required. Positive pressure, full facepiece, NIOSH-approved self- contained breathing apparatus (SCBA) will be worn where there is danger of oxygen deficiency and when directed by the fire chief or chemical accident/incident (CAI) operations officer. The M9 or M17 series mask may be worn in lieu of SCBA when there is no danger of oxygen deficiency. In cases where firefighters are responding to a chemical accident/incident for rescue/reconnaissance purposes vice firefighting, they will wear appropriate levels of protective clothing (see Section 8).

Section V: Health Hazard Data

AIRBORNE EXPOSURE LIMIT (AEL): The suggested permissible airborne exposure concentration for GA for an 8-hour workday or a 40 hour work week is an 8 hour time weight average (TWA) of 0.0001 mg/m3 (2 x 10-5 ppm). This value is based on the TWA of GA as proposed in the USaEHA Technical Guide 169, "Occupational Health Guildlines for the Evaluation and Control of Occupational Exposure to Nerve Agents, GA, GB, GD, and VX." To date, however, the Occupational Safety and Health Administration (OSHA) has not promulgated a permissible exposure concentration for GA.

EFFECTS OF OVEREXPOSURE: GA is an anticholinesterase agent similar in action to GB. Although only about half as toxic as GB by inhalation, GA in low concentrations is more irritating to the eyes than GB.

The number and severity of symptoms which appear are dependent on the quantity, and rate of entry of the nerve agent which is introduced into the body. (Very small skin dosages sometimes cause local sweating and tremors with few other effects.)

Individuals poisoned by GA display apaproximately the same sequence of symptoms regardless of the route by which the poison enters the body (whether by inhalation, absorption, or ingestion). These symptoms, in normal order of appearance, runny nose; tightness of chest; dimness of vision and pin pointing of the eye pupils; difficulty in breathing; drooling and excessive sweating; nausea; vomiting, cramps, and involuntary defecation and urination; twitching, jerking, and staggering; and headaches, confusion, drowsiness, coma, and convulsion. These symptoms are followed by cessation of breathing and death.

Onset Time of Symptoms: Symptoms appear much more slowly from skin dosage than from respiratory dosage. Although skin absorption great enough to cause death may occur in 1 to 2 minutes, death may be delayed for 1 to 2 hours. Respiratory lethal dosages kill in 1 to 10 minutes, and liquid in the eye kills almost as rapidly.

Median Lethal Dosage, Animals: LD50 (monkey, percutaneous) = 9.3 mg/kg (shaved skin); LCt50 (monkey, inhalation) = 187 mg-min/m3 (t = 10)

Median Lethal Dosage, Man: LCt50 (man, inhalation) = 135 mg-min/m3 (t = 0.5-2 min) at RMV (Respiratory Minute Volume) of 15 1/min; 200 mg-min/m3 at RMV of 10 1/min

GA is not listed by the International Agency for Research on Cancer (IARC), American Conference of Governmental Industrial Hygienists (ACGIH). Occupational Safety and Health Administration (OSHA), or National Toxicology Program (NTP) as a carcinogen.

EMERGENCY AND FIRST AID PROCEDURES:

- INHALATION: Hold breath until respiratory protective mask is donned. If severe signs of agent exposure appear (chest tightens, pupil construction, incoordination, etc.), immediately administer, in rapid succession, all three Nerve Agent Antidote Kit(s), Mark I injectors (or atropine if directed by the local physician). Injections using the Mark I kit injectors may be repeated at 5 to 20 minute intervals if signs and symptoms are progressing until three series of injections have been administered. No more injections will be given unless directed by medical personnel. In addition, a record will be maintained of all injections given. If breathing has stopped, give artificial respiration. Mouth-to-mouth resuscitation should be used when approved mask-bag or oxygen delivery systems are not available. Do not use mouth-to-mouth resuscitation when facial contamination exists. If breathing is difficult, administer oxygen. Seek medical attention IMMEDIATELY.
- EYE CONTACT: IMMEDIATELY flush eyes with water for 10-15 minutes then don respiratory protective mask. Although miosis (pinpointing of the pupils) may be an early sign of agent exposure, an injection will not be administered when miosis is the only sign present. Instead, the individual will be taken immediately to the medical treatment facility for observation.
- SKIN CONTACT: Don respiratory protection mask and remove contaminated clothing. Immediately wash contaminated skin with copious amounts of soap and water, 10 percent sodium carbonate solution, or 5 percent liquid household bleach. Rinse well with water to remove decontaminate. Administer an intramuscular injection with the MARK I kit injectors only if local sweating and muscular twitching symptoms are observed. Seek medical attention IMMEDIATELY.
- INGESTION: Do not induce vomiting. First symptoms are likely to be gastrointestinal. IMMEDIATELY administer 2 mg. intramuscular injection of the MARK I kit auto-injectors. Seek medical attention IMMEDIATELY.

Section VI: Reactivity Data

STABILITY: Stable

INCOMPATIBILITY: Not available

HAZARDOUS DECOMPOSITION: Decomposes within 6 months at 60 DEG C. Complete decomposition in 3-1/4 hours at 150 DEG C. May produce HCN. Oxides of nitrogen, oxides of phosphorus, carbon monoxide, and hydogen cyanide.

HAZARDOUS POLYMERIZATON: Not available

Section VII: Spill, Leak, and Disposal Procedures

STEPS TO BE TAKEN IN CASE MATERIAL IS RELEASED OR SPILLED: If leaks or spills occur, only personnel in full protective clothing (see Section VIII) will remain in area. In case of personnel contamination see Section V.

RECOMMENDED FIELD PROCEDURES: Spills must be contained by covering with vermiculite, diatomaceious earth, clay, fine sand, sponges, and paper or cloth towels. This containment is followed by treatment with copious amounts of aqueous Sodium Hydroxide solution (a minimum 10 wt percent). Scoop up all material and place in a fully removable head drum with a high density polyethylene liner. The decontamination solution must be treated with excess bleach to destroy the CN formed during hydrolysis. Cover the contents with additional bleach before affixing the drum head. After sealing the head, the exterior of the drum shall be decontaminated and then labeled IAW EPA and DOT regulations.

All leaking containers shall be overpacked with vermiculite placed between the interior and exterior containers. Decontaminate and label IAW EPA and DOT regulations. Dispose of the material IAW waste disposal methods provided below. Conduct general area monitoring with an approved monitor (see Section VIII) to confirm that the atmospheric concentrations do not exceed the airborne exposure limit (see Sections II and VIII).

If 10 wt percent Sodium Hydroxide is not available then the following decontaminants may be used instead and are listed in order of preference: Decontamination Solution No. 2 (DS2), Sodium Carbonate and Supertropical Bleach Slurry (STB).

RECOMMENDED LABORATORY PROCEDURES: A minimum of 56 grams of decon solution is required for each gram of GA. The decontamination solution is agitated while GA is added and the agitation is maintained for at least one hour. The resulting solution is allowed to react for 24 hours. At the end of 24 hours, the solution must be tritrated to a pH between 10 and 12. After completion of the 24 hour period, the decontamination solution must be treated with excess bleach (2.5 mole OCl-/mole GA) to destroy the CN formed during hydrolysis.

Scoop up all material and place in a fully removable head drum with a high density polyethylene liner. Cover the contents with additional bleach before affixing the drum head. All contaminated clothing will be placed in a fully removable head drum with a high density polyethylene liner. Cover the contents of the drum with decontaminating solution as above before affixing the drum head. After sealing the head, the exterior of the drum shall be decontaminated and then labeled IAW state, EPA and DOT regulations.

All leaking containers shall be overpacked with vermiculite placed between the interior and exterior containers. Decontaminate and label IAW State, EPA and DOT regulations. Conduct general area monitoring with an approved monitor (see Section 8) to confirm that the atmospheric concentrations do not exceed the airborne exposure limit (see Sections II and VIII).

WASTE DISPOSAL METHOD: Open pit burning or burying of GA or items containing GA is prohibited.

APPENDIX 19:   Agent T

Section I: General Information

MANUFACTURER'S ADDRESS:
U.S. ARMY CHEMICAL BIOLOGICAL DEFENSE COMMAND
EDGEWOOD RESEARCH DEVELOPMENT,
AND ENGINEERING CENTER (ERDEC)
ATTN: SCBRD-ODR-S
ABERDEEN PROVING GROUND, MD 20101-5423
Emergency telephone numbers: 0700-1630 EST: 410-671-4411/4414
After: 1630 EST: 410- 278-5201, Ask for Staff Duty Officer

CAS REGISTRY NUMBER: 63918-89-8

CHEMICAL NAME: Bis-(2-(2-chloroethylthio)ethyl) ether

ALTERNATE CHEMICAL NAMES:
Di (2- (2-chloroethylthio))ethyl ether
Di (2- (B-chloroethyl thio))ethyl ether

TRADE NAME AND SYNONYMS: T; Sulfur Mustard (Vesicant)

CHEMICAL FAMILY: Chlorinated Sulfur Compound

FORMULA/CHEMICAL STRUCTURE: C8H16Cl2OS2

NFPA 704 HAZARD SIGNAL:
Health - 4
Flammability - 1
Reactivity - 1
Special - 0

Section II: Hazardous Ingredients

Ingredients Name	Formula	Percentage by Weight	Airborne Exposure Limit (AEL)
TC8H16Cl2OS2100	None established	None established	

Section III: Physical Data

BOILING POINT: 120 C @ 0.02 torr; 174 C @ 2.0 torr

VAPOR PRESSURE (torr):   2.9 x E-5 @ 25 C (Calculated)

VAPOR DENSITY (AIR=1):   9.08 (Calculated)

SOLUBILITY IN WATER: Practically insoluble.

SPECIFIC GRAVITY (H2O=1):   1.2361 @ 25 C

FREEZING (MELTING) POINT: 9.6 - 9.9 C

VOLATILITY (mg/liter):   4.1 x E-4 @ 25 C (Calculated)

VISCOSITY (CENTISTOKE):   14.7 @ 25 C

EVAPORATION RATE: Very slow.

APPEARANCE AND ODOR: Yellow liquid with a garlic-like odor, similar to Mustard Agent.

Section IV: Fire and Explosion Data

FLASHPOINT (Method Used):   Unknown

FLAMMABILITY LIMITS: Unknown

EXTINGUISHING MEDIA: Water, fog, foam, CO2. Avoid use of extinguishing methods that will cause splashing or spreading of T.

SPECIAL FIRE FIGHTING PROCEDURES: All persons not engaged in extinguishing the fire should be evacuated immediately. Fires involving T should be contained to prevent contamination of uncontrolled areas. When responding to a fire alarm in buildings or areas containing agents, firefighting personnel should wear full firefighters protective clothing (Not TAP Clothing) during chemical agent firefighting and fire rescue operations. Respiratory protection is required. Positive pressure, full face piece, NIOSH approved self-contained breathing apparatus (SCBA) will be worn where there is danger of oxygen deficiency and when directed by the fire chief or chemical accident/incident (CAI)operations officer. In cases where firefighters are responding to a chemical accident/incident for rescue/reconnaissance purposes, they will wear appropriate levels of protective clothing (See Section VIII).

Do not breathe fumes. Skin contact with agents must be avoided always. Although the fire may destroy most of the agent, care must still be taken to assure the agent or contaminated liquids do not further contaminate other areas or sewers. Contact with the agent liquid or vapor can be fatal.

Section V: Health Hazard Data

AIRBORNE EXPOSURE LIMITS (AEL):   No detailed health hazard data on T is available. The following information is based upon the limited available information and the chemical similarity to Mustard (HD) Agent. Under no circumstances should any individual be intentionally exposed to any direct skin or eye contact.

T presently is not listed by the International Agency for Research on Cancer (IARC), National Toxicology Program (NTP), Occupational Safety and Health Administration (OSHA), or American Conference of Governmental Hygienist as a carcinogen. However, agent T should be treated as a suspect carcinogen due to its similarity to Mustard Agent (HD).

EFFECTS OF OVEREXPOSURE: T is a vesicant (blister agent) and alkylating agent producing cytotoxic action on the hematopoietic (blood forming) tissues, which are especially sensitive, much the same as for HD. The median lethal and incapacitating doses of T in man have not been established. The median lethal dosage (LCt50) of T in mice is 1650-2250 mg- min/m3, based upon a ten minute exposure time.

ACUTE PHYSIOLOGICAL ACTION OF T IS CLASSIFIED AS LOCAL AND SYSTEMIC.

ACUTE EFFECTS: T affects both the eyes and skin. Skin damage occurs after percutaneous absorption. Being lipid soluble, T can be absorbed into all organs. Skin penetration is rapid without skin irritation. Swelling (blisters) and reddening (erythema) of the skin occurs after a latency period of 4-24 hours following the exposure, depending on the degree of the exposure and individual sensitivity. The skin healing process is very slow. Tender skin, mucous membranes, and perspiration-covered skin is more sensitive to the effects of T. T's effect on the skin, however, is less than on the eyes. Severe exposure to the eyes produces severe necrotic damage and loss of eyesight. Exposure of the eyes to T vapors or aerosol produces lacrimation, photophobia, and inflammation of the cornea.

SYSTEMIC EFFECTS: Occurs primarily through inhalation and ingestion. The T vapor or aerosol is less toxic to the skin or eyes than the liquid form. When inhaled, the upper respiratory tract (nose, throat, tracheae) is inflamed after a few hours latency period, accompanied by sneezing, coughing and bronchitis, loss of appetite, diarrhea, fever, and apathy. Exposure to nearly lethal doses of T can produce injury to bone marrow, lymph nodes, and spleen as indicated by a drop in white blood cell (WBC) count and, therefore, results in increased susceptibility to local and systemic infections. Ingestion of T will produce severe stomach pains, vomiting, and bloody stools after a 15-20 minute latency period.

CHRONIC EXPOSURE: T can cause sensitization, chronic lung impairment (cough, shortness of breath, chest pain) and possibly cancer of the mouth, throat, respiratory tract and skin, and leukemia. Exposure to T may also cause birth defects.

EMERGENCY AND FIRST AID PROCEDURES:
- INHALATION: Hold breath until respiratory protective mask is donned. Remove from the source IMMEDIATELY. If breathing is difficult, administer oxygen. If breathing has stopped, give artificial respiration. Mouth-to-mouth resuscitation should be used when approved mask-bag or oxygen delivery systems are not avail-

able. Do not use mouth-to-mouth resuscitation when facial contamination exists. Seek medical attention IMMEDIATELY.

- EYE CONTACT: Speed in decontaminating the eyes is essential. Remove the person from the liquid source, flush the eyes immediately with water for at least 15 minutes by tilting the head to the side, pulling eyelids apart with fingers and pouring water slowly into the eyes. Do not cover eyes with bandages but, if necessary, protect eyes by means of dark or opaque goggles. Seek medical attention IMMEDIATELY.
- SKIN CONTACT: Remove the victim from the source and immediately decon skin and clothes by flushing with 5 percent sodium hypochlorite solution or liquid household bleach within one minute. Cut and remove contaminated clothing, flush affected areas again with decon. Wash skin area with soap and water. Seek medical attention IMMEDIATELY.
- INGESTION: Do not induce vomiting. Give victim milk to drink. Seek medical attention IMMEDIATELY.
- SECTION VI - REACTIVITY DATA
- STABILITY: Stable at ambient temperatures. Decomposition temperature is approximately 180 C. T is a persistent agent depending on pH and moisture.
- INCOMPATIBILITY: Unknown

HAZARDOUS DECOMPOSITION PRODUCTS: T will hydrolyze to form HCl and di-2-(2-hydroxy ethyl thio) ethyl ether.

HAZARDOUS POLYMERIZATION: Unknown

Section VII: Spill, Leak, and Disposal Procedures

STEPS TO BE TAKEN IN CASE MATERIAL IS RELEASED OR SPILLED: If spills or leaks of T occur only personnel in full protective clothing (see Section VIII) will be allowed in the area. See Section V for emergency and first aid procedures.

RECOMMENDED FIELD PROCEDURES: T should be contained using vermiculite, diatomaceous earth, clay, or fine sand and neutralized as soon as possible using copious amounts of alcoholic caustic, carbonate, or Decontaminating Solution, DS2. Caution must be exercised when using these decontaminates since acetylene will be given off. Household bleach can also be used if accompanied by stirring to allow contact. Scoop up all contaminated material and place in approved DOT containers. Cover the contents with additional decontaminant. All leaking containers will be over packed with vermiculite placed between the interior and exterior containers. Decontaminate the outside of the container and label according to DOT and EPA requirements. Dispose of according to waste procedures below. Dispose of decontaminate according to Federal, state, and local laws. Conduct general area monitoring with an approved monitor to confirm that the atmospheric concentrations do not exceed the airborne exposure limit (see Sections II and VIII).

WARNING: Never use dry High Test Hypochlorite (HTH) or Super Tropical Bleach (STB) since they will react violently with T and may burst into flames.

RECOMMENDED LABORATORY PROCEDURES: A minimum of 65 grams of decon solution per gram of T is allowed to agitate for a minimum of one hour. Agitation is not necessary following the first hour if a single phase is obtained. At the end of 24 hours, the resulting solution will be adjusted to a pH between 10 and 11. Test for presence of active chlorine by use of acidic potassium iodide solution to give free iodine color. Place 3 ml of the decontaminate in a test tube. Add several crystals of potassium iodine and swirl to dissolve. Add 3 ml of 50 wt. sulfuric acid: water and swirl. IMMEDIATE iodine color shows the presence of active chlorine. If negative, add additional 5.25 percent sodium hypochlorite solution to the decontamination solution, wait two hours, then test again for active chlorine. Continue procedure until positive chlorine is given by solution. Scoop up all material and place in approved DOT containers. Cover the contents with additional decontaminate as above. The exterior of the container will be decontaminated and labeled according to EPA and DOT regulations. All leaking containers will be over packed with vermiculite placed between the interior and exterior containers. Decontaminate and label according to EPA and DOT regulations. Dispose of the material according to waste disposal methods provided below. Dispose of decontaminate according to Federal, state and local regulations. Conduct general area monitoring with an approved monitor to confirm that the atmospheric concentrations do not exceed the airborne exposure limits (see Sections II and VIII).

A 10 wt.% calcium hypochlorite mixture may be substituted for sodium hypochlorite. Use 65 grams of decon per gram of T and continue the test as described for sodium hypochlorite.

Note: Surfaces contaminated with T, then rinse-decontaminated may evolve sufficient T vapor to produce a physiological response. T on laboratory glassware may be oxidized by it vigorous reaction with concentrated nitric acid.

WASTE DISPOSAL METHOD: All neutralized material should be collected, contained and thermally decomposed in EPA approved incinerators that will filter or scrub toxic by-products from effluent air before discharge to the atmosphere. Any contaminated materials or protective clothing should be decontaminated using HTH or bleach and analyzed to assure it is free of detectable contamination (3X) level. The clothing should then be sealed in plastic bags inside properly labeled drums and held for shipment back to the DA issue point.

Note: Some states define decontaminated surety material as a RCRA hazardous waste.

Section VIII: Special Protection Information

RESPIRATORY PROTECTION: CONCENTRATION RESPIRATORY PROTECTIVE EQUIPMENT

< 0.003 (mg/m3)NIOSH approved full face piece, chemical canister air-purifying, respirators or protective masks will be on hand for escape. (M9, M17, M40 series protective masks or other certified equivalent masks are acceptable for this, use with the M3 toxicological agent protective suit for dermal protection).

>0.003 or concentration unknown NIOSH approved pressure demand full face piece SCBA, suitable for use in unknown or high agent concentrations, with a protective ensemble. (See DA Pam 385-61 for examples.)

VENTILATION
- Local exhaust: Mandatory. Must be filtered or scrubbed to limit exit concentration to non-detectable level. Air emissions will meet Federal, state and local laws and regulations.
- Special: Chemical laboratory hoods will have an average inward face velocity of 100 linear feet per minute (1fpm) +/- 10% with the velocity at any point not deviating from the average face velocity by more than 20%. Existing laboratory hoods will have an inward face velocity of 150 lfpm +/- 20%. Laboratory hoods will be located such that cross drafts do not exceed 20% of the inward face velocity. A visual performance test utilizing smoke producing devices will be performed in the assessment of the inclosure's ability to contain T.
- Other: Recirculation of exhaust air from agent areas is prohibited. No connection between agent area and other areas through the ventilation system are permitted. Emergency backup power is necessary. Hoods should be tested semiannually or after modification or maintenance operations. Operations should be performed 20 centimeters inside hoods. Procedures should be developed for disposal of contaminated filters.

PROTECTIVE GLOVES: M3 and M4 Butyl Rubber, Norton, Chemical Protective Glove Set

EYE PROTECTION: As a minimum, chemical goggles will be worn. For splash hazards use goggles and face-shield.

OTHER PROTECTIVE EQUIPMENT: For laboratory operations, wear lab coats, gloves and have a mask readily accessible. In addition, daily clean smocks, foot covers, and head covers will be required when handling contaminated lab animals.

MONITORING: Real Time Analytical Platform (RTAP)
Real-time, low-level monitors (with alarm) are required for operations. In their absence, an Immediately Dangerous to Life and Health (IDLH) atmosphere must be presumed. Laboratory operations conducted in appropriately maintained and alarmed engineering controls require only periodic low-level monitoring.

Section IX: Special Precautions

PRECAUTIONS TO BE TAKEN IN HANDLING AND STORING: When handling agents, the buddy system will be incorporated. No smoking, eating, or drinking in areas containing agents are permitted. Containers should be periodically inspected for leaks, either visually or using a detector kit. Stringent control over all personnel handling agents must be exercised. Decontaminating equipment will be conveniently placed. Exits must be designed to permit rapid evacuation. Chemical showers, eye wash stations, and personal cleanliness facilities must be provided. Wash hands before meals and shower thoroughly with special attention given to hair, face, neck, and hands, using plenty of soap before leaving at the end of the workday.

OTHER PRECAUTIONS: T should be stored in containers made of glass for Research, Development Test and Evaluation (RDTE) quantities or one-ton steel containers for large quantities. Agents will be double contained in vapor and liquid tight containers when in storage or during transportation.

For additional information see "AR 385-61, The Army Toxic Chemical Agent Safety Program," "DA Pam 385-61, Toxic Chemical Agent Safety Standards," and "DA Pam 40-173, Occupational Health Guidelines for the Evaluation and Control of Occupational Exposure to Mustard H, HD, and HT."

Section X: Transportation Data

PROPER SHIPPING NAME: Poisonous liquids, n.o.s.

DOT HAZARDS CLASSIFICATION: 6.1, Packing Group I, Zone B

DOT LABEL: Poison

DOT MARKING: Poisonous liquids, n.o.s. Bis-(2-(2-chloroethylthio)ethyl) ether UN 2810, Inhalation Hazard

DOT PLACARD: POISON

EMERGENCY ACCIDENT PRECAUTIONS & PROCEDURES: See Sections IV, VII and VIII.

PRECAUTIONS TO BE TAKEN IN TRANSPORTATION: Motor vehicles will be placarded regardless of quantity. Drivers will be given full information regarding shipment and conditions in case of an emergency. AR 50-6 deals specifically with the shipment of chemical agents. Shipment of agents will be escorted according to AR 740-32.

While the Edgewood Research Development, and Engineering Center, Department of the Army believes that the data contained herein are factual and the opinions expressed are those of the experts regarding the results of the tests conducted, the data are not to be taken as a warranty or representation for which the Department of the Army or Edgewood Research Development, and Engineering Center assumes legal responsibility. They are offered solely for your consideration, investigation, and verification. Any use of these data and information must be determined by the user according to applicable Federal, state, and local laws and regulations.

APPENDIX 20: Decontaminating Agent, DS2

Section I: General Information

MANUFACTURER'S ADDRESS:
U.S. ARMY CHEMICAL BIOLOGICAL DEFENSE COMMAND
EDGEDWOOD RESEARCH DEVELOPMENT,
AND ENGINEERING CENTER (ERDEC)
ATTT: SCBRD-ODR-S
ABERDEEN PROVING GROUND, MD 20101-5423
Emergency telephone numbers: 0700-1630 EST: 410-671-4411/4414
After: 1630 EST: 410-278-5201, Ask for Staff Duty Officer

CAS Registry Numbers:
111-40-0 Diethylenetriamine
1310-73-2 Sodium Hydroxide
109-86-4 Ethylene Glycol Monomethyl Ether

CHEMICAL NAME: SYNONYMS:
Diethylenetriamine Bis (2-Aminoethyl) amine
DETA
N-(2-Aminoethyl)1,2-ethanediamine
2,2'-DiaminodiethylamineSodium
HydroxideCaustic sodaEthylene Glycol Monomethyl Ether2-Methoxyethanol
EGME
Methyl Cellosolve

TRADE NAME AND SYNONYMS:
Decontaminating Agent, DS2
DS2
Decon Agent DS2

CHEMICAL FAMILY: Mixture

CHEMICAL FORMULAS:
Diethylenetriamine
(NH2 CH2 CH2)2 NH
Sodium Hydroxide, NaOH
Ethylene Glycol Monomethyl Ether, CH3 O CH2 CH2 OH

NATIONAL STOCK NUMBERS (NSN):
Decontaminating Agent DS2, 1-1/3 quarts can, NSN: 6850-00-753-4827
Decontaminating Agent DS2, 5 gallon pails, NSN: 6850-00-753-4870
Decontaminating Apparatus, Portable, 14 liters, M13, NSN: 4230-01-133-4124
14 Liter Container, Fluid Filled, NSN: 6850-01-136-8888

NFPA 704 SIGNAL:
  Health - 3
  Flammability - 1
  Reactivity - 2
  Special - 0

Section II: Hazardous Ingredients

Ingredients	Percentages by Weight	Airborne Exposure Limits (AEL)
Diethylenetriamine	69-71%	4.2 mg/m3 (1 ppm) [skin]
Sodium Hydroxide	1.9-2.1%	2 mg/m3 [ceiling]
Ethylene Glycol Monomethyl Ether	26.9-29.1%	16 mg/m3 (5 ppm) [skin]

Section III: Physical Data

BOILING POINT: 193.3 C (380 F)

SPECIFIC GRAVITY (H20 = 1):   0.97 - 0.98

APPEARANCE AND ODOR: Clear amber solution with an ammonia-like odor.

VISCOSITY (CENTISTOKE):   9.9 @ 20 C

Section IV: Fire and Explosion Data

FLASHPOINT: (Method Used): The flashpoint of the mixture has been determined to be 168 F (75.5 C) by the closed cup method. The lowest flashing component of the mixture (ethylene glycol monomethyl ether) has a flashpoint of 115 F (46 C) by the closed cup method.

EXTINGUISHING MEDIA: Carbon dioxide, alcohol foam, water. Avoid use of extinguishing methods that will cause splashing or spreading of .

SPECIAL FIRE FIGHTING PROCEDURES: Incipient fires may be fought with a type BC fire extinguisher. If the fire cannot be controlled, contact the fire department. All persons not engaged in extinguishing the fire should be immediately evacuated from the area. Fires involving DS2 should be contained to prevent spreading to uncontrolled areas. When responding to fire alarms firefighters should wear full firefighter protective clothing during firefighting and rescue operations. Respiratory protection is required. Positive pressure, full face piece, NIOSH approved self-contained breathing apparatus (SCBA) will be worn where there is danger of oxygen deficiency and when directed by the fire chief.

Do not breathe fumes. Skin contact with DS2 must be avoided at all times. Care must be taken to avoid contamination of other areas and sewers. Contact with liquid or vapors can be fatal.

UNUSUAL FIRE AND EXPLOSION HAZARDS: Never mix or store acids, oxidizing agents, Super Tropical Bleach (STB) or High Test Hypochlorite (HTH) together with DS2; fire or explosion may result.

Section V: Health Hazard Data

THRESHOLD LIMIT VALUE: DS2 is made of two major components (EGME & DETA) with different toxicities and physical properties. The TLV of the mixture (calculated) is 5.2 mg/m3 as an 8-hour time weighted average (TWA). To date the Occupational Safety and Health Administration (OSHA) has not promulgated a permissible exposure limit for DS2 nor has the value proposed been officially adopted as a part of a special occupational safety and health standard for DS2 according to DOD 6055.1.

EFFECTS OF OVEREXPOSURE: No toxicity data is available on DS2, however, the toxicity of each component has been partially determined. DS2 is an alkali and direct contact will corrode tissue, e.g., skin, eye, respiratory mucosa or gastric mucosa. The effects exhibited depend on routes of exposure, dosage, and duration of exposure. Health effects can range from mild burns and primary irritation to corneal opacification, severe burns and esophageal strictures. Sufficient exposure to EGME, a major component of DS2, may cause central nervous system depression and liver damage. Although not definitely established in humans, reproductive effects (including teratogenesis) are also a major concern with this substance. The National Institute for Occupational Safety and Health (NIOSH) recommends that EGME be regarded in the workplace as having the potential to cause adverse reproductive effects in male and female workers. Appropriate controls must be instituted to minimize worker exposure to EGME. Exposure to high vapor concentrations of DS2 can cause nausea, vomiting, and respiratory irritation as acute effects. Repeated skin and respiratory exposures to DETA can cause skin sensitization and asthma.

EMERGENCY AND FIRST AID PROCEDURES:
- INHALATION: Remove to fresh air. If breathing is difficult, give oxygen If breathing has stopped, give artificial respiration. Mouth-to-mouth resuscitation should be used when approved mask-bag or oxygen delivery systems are not available. Do not use mouth-to-mouth resuscitation when facial contamination exists. Seek medical attention IMMEDIATELY.
- EYE CONTACT: Immediately flush the eyes with copious amounts of water for at least 15 minutes. Seek medical attention IMMEDIATELY.
- SKIN CONTACT: Flush away the DS2 from the skin with water until "soapiness" is no longer present. Seek medical attention IMMEDIATELY.
- INGESTION: If the patient is conscious, give as much milk or water as possible. Do not induce vomiting. Seek medical attention IMMEDIATELY.

Section VI: Reactivity Data

STABILITY: DS2 will deteriorate in air. Exposure of 48 hours or more to air will result in the formation of gelatin-like bodies on the surface of DS2 open.

INCOMPATIBILITY: DS2 is a corrosive material and because of its content, it is incompatible with some metals (e.g., cadmium, tin and zinc); some plastics (e.g., Lexan, cellulose acetate, polyvinyl chloride, Mylar, and acrylic); some paints; wool; leather; oxidizing materials (e.g., Super Tropical Bleach or High Test Hypochlorite); and acids.

HAZARDOUS DECOMPOSITION: Unknown

HAZARDOUS POLYMERIZATION: Unknown

Section VII: Spill, Leak, and Disposal Procedures

STEPS TO BE TAKEN IN CASE MATERIAL IS RELEASED OR SPILLED: Spills on porous surfaces (concrete, wood, etc.) should be cleaned and neutralized immediately. Otherwise, they will be absorbed and become an indefinite hazard. All spills must be contained, by covering with dry sodium bisulfate to neutralize and then absorbing them on vermiculite (NSN 5640-01-324-2664), clay or diatomaceous earth. Scoop up all material and any contaminated soil or substrate and place in a DOT approved container and label as corrosive according to EPA and DOT requirements. Leaking containers will be over packed with vermiculite between the interior and exterior containers. Decontaminate the exterior container and label according to DOT and EPA regulations. Dispose of contents according to procedures below. Dispose of decontaminate according to Federal, state, and local laws.

Approve containers include:
    7 gal NSN - 8110-00-254-5714
    27 gal NSN - 8110-00-082-2625
    57 gal NSN - 8110-00-082-2626

Other NSNs can also be used provided they meet the test requirements as specified in CFR Title 49.
    During spills provide adequate ventilation and remove any ignition source. During clean up, personnel should wear a full face respirator with an organic vapor cartridge effective against diethylenetriamine and methyl cellosolve, rubber gloves long enough to protect hands and arms, and a full length rubber apron. Contaminated clothing and shoes should be removed immediately and washed thoroughly with water before reuse. Avoid contact with leaking liquid or vapor. All wash water should have pH measured. All material with a pH less than 2.0 or greater than 12.5 is hazardous waste with an EPA number of D002.

WASTE DISPOSAL METHOD: Disposal methods for waste DS2 and accumulated spill cleanup residues must comply with RCRA, state, and local hazardous waste regulations and procedures. If the wastes are corrosive, they have the EPA Hazardous Waste Number of D002. This number should be used when the waste is manifested, to permit the use of off-site hazardous waste disposal facilities. For disposal of excess stocks of pure DS2, coordinate with the Defense Reutilization and Marketing Office (DRMO).

Disposal methods at overseas military installations must be according to the laws of the host country.

SECTION VIII: SPECIAL PROTECTION INFORMATION

RESPIRATORY PROTECTION: CONCENTRATION RESPIRATORY PROTECTIVE EQUIPMENT < 5.2 (as mixture i.e., 3.7 mg/m3 DETA and 1.5 mg/m3 EGME)NIOSH approved full face piece respirators with an organic vapor canister will be on hand for escape. (M9, M17, M40 series gas masks are acceptable for this purpose. Other masks certified as equivalent may be used.) NIOSH approved escape type SCBA can also be used >5.2 or concentration unknownNIOSH approved full face piece pressure demand SCBA. Or NIOSH approved full-face piece positive pressure, supplied-air respirators with auxiliary SCBA

Note: For military personnel engaged in training scenarios the M9, M17 or M40 series masks are acceptable. Filter elements and canisters should be changed after each use with DS2.

VENTILATION: Local exhaust - Necessary if TLV (TWA) exceeded.

PROTECTIVE GLOVES: Butyl Rubber

EYE PROTECTION: As a minimum, chemical goggles will be worn. For splash hazards use goggles and face shield.

OTHER PROTECTIVE EQUIPMENT: Hooded chemical-resistant clothing (i.e., overalls & long sleeve jacket, or one or two-piece chemical splash suit) and chemical resistant boots. Military personnel will use standard issue equipment during training operations.

Secction IX: Special Precautions

PRECAUTIONS TO BE TAKEN IN HANDLING AND STORING: In handling, the buddy system will be incorporated. No smoking, eating, and drinking in areas containing chemicals are permitted. Containers should be periodically inspected for leaks Stringent control over all personnel practices must be exercised. Exits must be designed to permit rapid evacuation. Chemical showers, eyewash stations and personal cleanliness facilities must be provided. Wash hands before meals and each worker will shower thoroughly with special attention given to hair, face, neck, and hands, using plenty of soap and water before leaving at the end of the workday.

OTHER PRECAUTIONS: Avoid extreme temperatures (e.g., 160 F or higher) during storage.

Section X: Transportaion Data

PROPER SHIPPING NAME: Caustic Alkali Liquids, n.o.s.

DOT HAZARD CLASSIFICATION: 8, Packing Group II

DOT LABEL: Corrosive

DOT MARKING: Caustic Alkàli Liquids, n.o.s. (Diethylenetriamine, Ethylene Glycol Monomethyl Ether, Sodium Hydroxide) UN 1719

DOT PLACARD: Corrosive

EMERGENCY ACCIDENT PRECAUTIONS & PROCEDURES: See Sections IV, VII, and VIII.

PRECAUTIONS TO BE TAKEN IN TRANSPORTATION: Shipping "on-deck" or under-deck" is permitted in cargo and passenger vessels subject to the requirements of 49 CFR 176.63 (b) and ©. MSDS for DS2 will be placed with all shipments. DS2 is limited to 5 gallons per package when shipped by cargo aircraft. Bulk packaging of DS2 (1 1/3 quart, 5 gallons, and M13 Portable Decontaminating Apparatus) are not authorized for shipment on passenger carrying aircraft or rail cars. Shipment on passenger carrying aircraft or rail-car is permitted in 1 quart packages. DS2 will be packed and shipped according to 49 CFR 173.202. Packaging exceptions can be found in 49 CFR 173.154.

While the Edgewood Research Development, and Engineering Center, Department of the Army believes that the data contained herein are factual and the opinions expressed are those of the experts regarding the results of the tests conducted, the data are not to be taken as a warranty or representation for which the Department of the Army or Edgewood Research Development and Engineering Center assume legal responsibility. They are offered solely for your consideration, investigation, and verification. Any use of these data and information must be determined by the user to be according to applicable Federal, state, and local laws and regulations.

APPENDIX 21: NATO Handbook on Medical Aspects of NBC Defensive Operations AMedP-6(B)

Part II: Biological Annex C Potentiona Biological Agents Operational Data Charts

Table C-I. Bacteria

Serial	Disease	Likely methods of dissemination	Transmissibility man to man	Infectivity	Incubation* time	Duration of illness	Lethality	Persistance	Vaccination	Antimicrobial therapy	Antisera
a	b	c	d	e	f	g	h	i	j	k	l
1	(Inhalation) Anthrax	Spores in aerosols	No	Moderate	1–6 days	3–5 days	High	Spores are highly stable	Yes	Little effect	Experimental
2	Brucellosis	1. Aerosol 2. Sabotage (food supply)	No	High	Days to months	Weeks to years	Low	Long persistence in wet soil & food	Yes	Moderately effective	No
3	Cholera	1. Sabotage (food/ water supply) 2. Aerosol	Negligible	Low	1–5 days	1 or more weeks	Moderate to high	Unstable in aerosols & pure water. More so in poluted water	Yes	Moderately effective	No
4	Melioidosis	Aerosol	Negligible	High	Days to years	4–20 days	Variable	Stable	None	Moderately effective	No
5	(Pneumonic) Plague	1. Aerosol 2. Infected vectors	High	High	2–3 days	1–2 days	Very high	Less important because of high transmissibility	Yes	Moderately effective	No
6	Tularemia	Aerosol	No	High	2–10 days	2 or more weeks	Moderate if untreated	Not very stable	Yes	Effective	No
7	Typhoid Fever	1. Sabotage (food/ water supply) 2. Aerosol	Negligible	Moderate	7–21 days	Several weeks	Moderate if untreated		Yes	Moderately effective	No

*Incubation applies to infectious disease. With toxins, it's application refers to the period between exposure and appearance of the symptoms and signs of poisoning.

Table C-II. *Rickettsiae*

Serial	Disease	Likely methods of dissemination	Transmissibility man to man	Infectivity	Incubation* time	Duration of illness	Lethality	Persistance	Vaccination	Antimicrobial therapy	Antisera
a	b	c	d	e	f	g	h	i	j	k	l
8	Epidemic Typhus	1. Aerosol 2. Infected vectors	No	High	6–16 days	Weeks to months	High	Not very stable	No	Effective	No
9	Q-Fever	1. Aerosol 2. Sabotage (food supply)	No	High	1–20 days	2 days to 2 weeks	Very low	Stable	Yes	Effective	No
10	Rocky Mountain Spotted Fever	1. Aerosol 2. Infected vectors	No	High	3–10 days	2 weeks to months	High	Not very stable	No	Effective	No
11	Scrub Typhus	1. Aerosol 2. Infected vector	No	High	4–15 days	Up to 16 days	Low	Not very stable	No	Effective	No

*Incubation applies to infectious disease. With toxins, it's application refers to the period between exposure and appearance of the symptoms and signs of poisoning.

Table C-III. *Chlamydia*

Serial	Disease	Methods of Dissemination	Transmissibility Man-to-Man	Infectivity	Incubation	Duration	Lethality	Persistance	Vaccination
12	Psittacosis	Aerosol	Negligible	Moderate	4/15 days	Weeks	Very Low	Stable	No
13	Coccidioido-mycosis	Aerosol	No	High	1-2 weeks	Weeks	Low	Stable	No
14	Histoplas-mosis	Aerosol	No	High	1-2 weeks	Weeks	Low	Persistence in Soil	No

Incubation applies to infectious disease. With toxins, it's application refers to the period between exposure and appearance of the symptoms and signs of poisoning.

Table C-IV. Viruses

Serial	Disease	Likely methods of dissemination	Transmissibility man to man	Infectivity	Incubation* time	Duration of illness	Lethality	Persistance	Vaccination	Antimicrobial therapy	Antisera
a	b	c	d	e	f	g	h	i	j	k	l
15	Chikun-Gunya Fever	Aerosol	None	High	2–6 days	2 weeks	Very low	Relatively stable	Experimental	Not effective	No
16	Crimean-Congo Hemorrhagic Fever	Aerosol	Moderate	High	3–12 days	Days to weeks	High	Relatively stable	Experimental (Bulgaria)	Effective	Yes (Bulgaria only)
17	Dengue Fever	Aerosol	None	High	3–5 days	Days to weeks	Low	Relatively unstable	Experimental	Not effective	No
18	Eastern Equine Encephalitis	Aerosol	None	High	5–15 days	1–3 weeks	High	Relatively unstable	Yes	Not effective	No
19	Ebola Fever	Aerosol	Moderate	High	7–9 days	5–15 days	High	Relatively unstable	No	Not effective	No
20	Korean Hemorrhagic Fever (Hanizan)	Aerosol	None	High	4–42 days	Days to weeks	Moderate	Relatively stable	Experimental	Effective	No
21	Lassa Fever	Aerosol	Low to moderate	High	10–14 days	1–4 weeks	Unknown	Relatively stable	No	Effective	Experimental
22	Omak Hemorrhagic Fever	1. Aerosol 2. Water	Negligible	High	3–7 days	7–10 days	Low	Relatively unstable	Experimental	Not effective	No
23	Rift Valley Fever	1. Aerosol 2. Infected vectors	Low	High	2–5 days	Days to weeks	Low	Relatively stable	Yes	Effective	No
24	Russian Spring-Summer Encephalitis	1. Aerosol 2. Milk	None	High	8–14 days	Days to months	Moderate	Relatively unstable	Yes	Not effective	Yes
25	Smallpox	Aerosol	High	High	10–17 days	1–2 weeks	High	Stable	Yes	Not effective	Yes
26	Western Equine Encephalitis	Aerosol	No	High	1–20 days	1–3 weeks	Low	Relatively unstable	Yes	Not effective	No
27	Venezuelan Equine Encephalitis	1. Aerosol 2. Infected vectors	Low	High	1–5 days	Days to weeks	Low	Relatively unstable	Yes	Not effective	No
28	Yellow Fever	Aerosol	None	High	3–5 days	1–2 weeks	High	Relatively unstable	Yes	Not effective	No

*Incubation applies to infectious disease. With toxins, it's application refers to the period between exposure and appearance of the symptoms and signs of poisoning.

Table C-V. Toxins

Serial	Disease	Likely methods of dissemination	Transmissibility man to man	Infectivity	Incubation* time	Duration of illness	Lethality	Persistance	Vaccination	Antimicrobial therapy	Antisera
a	b	c	d	e	f	g	h	i	j	k	l
29	Botulinum Toxin	1. Sabotage (food/ water supply) 2. Aerosol	No		Variable (hours to days)	24–72 hours Months if lethal	High	Stable	Yes	Not effective	Yes
30	Clostridium Perfringens Toxins	1. Sabotage 2. Aerosol	No		8–12 hourss	24 hours	Low	Stable	No	Not effective	No
31	Trichothecene Mycotoxins	1. Aerosol 2. Sabotage	No		Hours	Hours	High	Stable	No	Not effective	No
32	Palytoxin	1. Aerosol 2. Sabotage	No		Minutes	Minutes	High	Stable	No	Not effective	No
33	Ricin	Aerosol	No		Hours	Days	High	Stable	Under development	Not effective	No
34	Saxitoxin	1. Sabotage 2. Aerosol	No		Minutes to hours	Minutes to days	High	Stable	No	Not effective	No
35	Staphylococcai enterotoxin B	1. Aerosol 2. Sabotage	No		1–6 hours	Days to weeks	Low	Stable	Under development	Not effective	No
36	Tetrodotoxin	1. Sabotage 2. Aerosol	No		Minutes to hours	Minutes to days	High	Stable	No	Not effective	No

*Incubation applies to infectious disease. With toxins, it's application refers to the period between exposure and appearance of the symptoms and signs of poisoning.

Bibliography

Aiken County Emergency Services; Emergency Preparedness Family Planning Guide. Aiken, SC (undated).

Aiken County Emergency Management After-Action Report; Graniteville Train Wreck. Aiken, SC. January. Aiken County Hazardous Materials Team exclusive Web site, at http://www.Hazmat-Team.com, http://www.hazmatteam.com/ March 29, 2006.

Aiken County Sheriff's Office After-Action Report; Graniteville Train Wreck. Aiken, SC. January 2000.

Batts, Denish Watson. *S.C. Town Still Suffers a Year after Train Derailment.* Hampton Roads, VA: *Virginia Pilot,* January 7, 2006.

Biosafety in Microbiological and Biomedical Laboratories – Fourth Edition. U.S. Department of Health and Human Services, Public Health Services, CDC (Centers for Disease Control and Prevention). Washington, D.C.: U.S. Government Printing Office, 1999.

Boyd, Jean Clark, *Pictorial Timeline of Graniteville, South Carolina 1845–1996. Horse Creek Historical Society, Warrenville, South Carolina.* Second Printing, June 28, 2004.

Building a Systems Approach for Health and Medical Response to Acts of NBC Terrorism. Office of Emergency Preparedness (OEP), U.S. Department of Health and Human Services (HHS), 1996.

Cashman, John R., *Emergency Response to Chemical and Biological Agents.* Boca Raton, FL: Lewis Publishers, 2000.

Cashman, John R., *Interviews with 32 Victims at the Graniteville Train Wreck Site on January 6, 2005,* done from June 10 to June 17, 2006.

Center for Domestic Preparedness (DVD). U.S. Department of Homeland Security, http://www.cptraining.com Anniston, Alabama. December 8, 2004.

Chemical Accident Contamination Control. Department of the Army Field Manual FM 3-21, 1978.

Chemical/Biological Incident Handbook. Director of Central Intelligence for the Intelligence Committee on Terrorism, and the Community Counterterrorism Board. Washington, D.C., 1995

Chemical Operations, Principles, and Fundamentals. Marine Corps War-Fighting Publication (MCWP) 3-3.7.1. Washington, D.C.: U.S. Marine Corps.

Code of Federal Regulations Title 49-Transportation (Parts 178-199). Washington, D.C.: U.S. Government Printing Office, 1997.

Collision of Norfolk Southern Freight Train 192 With Standing Norfolk Southern Local Train P22 With Subsequent Hazardous Materials Release at Graniteville, South Carolina January 6, 2005, Railroad Accident Report NTSB/RAR-05/04. Washington, D.C.: National Transportation Safety Board.

Columbine High School Shootings (DVD),. Jefferson County Sheriff's Office, Colorado. April 20, 1999.

Combating Terrorism: Chemical and Biological Medical Supplies Are Poorly Managed. Washington, D.C.: U.S. Government Accounting Office (AIMD-00-36, Oct. 29), 1999.

Combating Terrorism: Observations on the threat of Chemical and Biological Terrorism. Washington, D.C.: U. . Government Accounting Office (T-NSIAD-00-50, Oct. 20) 1999.

Combating Terrorism: Need for Comprehensive Threat and Risk Assessments of Chemical and Biological Attack. Washington, D.C.: U.S. Government Accounting Office (NSIAD-99-163, Sept. 7) 1999.

Combating Terrorism: Analysis of Federal Counterterrorist Exercise. Washington, D.C.: U.S. Government Accouting Office (NSIAD-99-157BR, June 25) 1999.

Combating Terrorism: Observations on Growth in Federal Programs. Washington, D.C.: U. S. Government Accounting Office (GAO/T-NSIAD-99181, June 9) 1999.

Combating Terrorism: Analysis of Potential Emergency Response Equipment and Sustainment Costs. Washington, D.C.: U.S. Government Accounting Office (NSIAD-99-151, June 9) 1999.

Combating Terrorism: Use of National Guard Response Teams is Unclear. Washington, D.C.: U.S. Government Accounting Office (NSIAD-99-110. May 21) 1999.

Combating Terrorism: Issues in Managing Counterterrorist Programs: Statement of Norman J. Rabkin, Director; National Security Preparedness Issues, National Security and International Affairs Division. Washington, D.C.: U.S. Government Accounting Office (T-NSIAD-00-145, April 6) 2000.

Combating Terrorism: Issues to Be Resolved to Improve Counter-terrorist Operations. Washington, D.C.: U.S. Government Accounting Office (NSIAD-99-135, May 13) 1999.

Combating Terrorism: Observations on Biological Terrorism and Public Health Initiatives. Washington, D.C.: U.S. Government Accounting Office (T-NSIAD-99-112, March 16) 1999.

Combating Terrorism: Observations on Federal Spending to Combat Terrorism. Washington, D.C.: U.S. Government Accounting Office (T-NSIAD/GGD-99-107 March 11) 1999.

Combating Terrorism: Opportunities to Improve Domestic Preparedness Program Focus and Efficiency. Washington, D.C.: U.S. Government Accounting Office (NSIAD-99-3, Nov. 12) 1998.

Combating Terrorism: Observations on the Nunn-Lugar-Domenici Domestic Preparedness Program. Washington, D.C.: U.S. Government Accounting Office (T-NSIAD-99-16, Oct. 2) 1998.

Combating Terrorism: Observations of Cross-cutting Issues. Washington, D.C.: U. S. Government Accounting Office (T-NSIAD-98-164, April 23) 1998.

Combating Terrorism: Threat and Risk Assessments Can Help Prioritize and Target Program Investments. Washington, D.C.: U.S. Government Accounting Office (NSIAD-98-74, April 9) 1998..

Combating Terrorism: Spending on Government-wide Programs Requires Better Management and Coordination. Washington, D.C.: U.S. Government Accounting Office (NSIAD-9839, Dec. 1) 1997.

Combating Terrorism (Federal Agencies' Efforts to Implement National Policy and Strategy). Washington, D.C.: U.S. General Accounting Office (Report Number GAO/NSIAD – 97 – 254. September 26). 1997.

Combating Terrorism: Chemical and Biological Medical Supplies Are Poorly Managed. Washington, D.C.: U.S. General Accounting Office (Report Number GAO/HEHS/AIMD – 00 – 36. October 29) 1999.

Combating Terrorism (Federal Agencies' Efforts to Implement National Policy and Strategy). Washington, D.C.: U.S. General Accounting Office (Report Number GAO/NSIAD-97-254. September) 1997.

Combating Terrorism: Observations on the Threat of Chemical and Biological Terrorism. Washington, D.C.: U.S. General Accounting Office (Report Number GAO/T – NSIAD – 00 - 50)

Competencies for EMS Personnel Responding to Hazardous Materials Incidents - NFPA 473. Quincy, MA. National Fire Protection Associations, 1992.

Cutting Edge – A History of Fort Detrick, Maryland. (4th ed., October), 2000.

Defense against Toxin Weapons. U.S. Army Medical Research Institute of Infectious Diseases. Franz, David R. DVM, PhD. Fort Detrick, Maryland, 1997.

Department of Health and Human Services Health and Medical Services Support Plan for the Federal Response Acts of Chemical/Biological (C/B) Terrorism. Washington, D.C.: Office of Emergency Preparedness (OEP), of the Department of Health and Human Services (HHS), undated.

Domestic Preparedness Program in the Defense Against Weapons Of Mass Destruction. (Washington, D.C.: Department of Defense Report To Congress) 1997.

Emergency Response to Terrorism-Job Aid. U.S. Department of Justice, Office of Justice Programs, Bureau of Justice Assistance, and Federal Emergency Management Agency, United States Fire Administration - National Fire Academy, FEMA/USFA (May) 2000,

Emergency Response To Terrorism Self-Study. U.S. Department of Justice, Office of Justice Programs, Bureau of Justice Assistance, and Federal Emergency Management Agency, United States Fire Administration – National Fire Academy, FEMA/USFA (ERT:SS) 1997.

Evacuation Behavior in Response to the Graniteville, South Carolina Chlorine Spill; Quick Response Report. Department of Geography and Department of Journalism. Columbia, SC: University of South Carolina, 2005.

Federal Response Plan. Washington, D.C. 1992.

Field Behavior of NBC Agents. Department of the Army Field Manual FM 3-6. 1986.

Graniteville Train Wreck (DVD). Aiken, SC: Aiken County Hazardous Materials Team (January) 2005.

Graniteville Train Wreck (DVD). Aiken, SC: Aiken County Sheriff's Office (June 10) 2005.

Graniteville-Vaucluse-Warrenville Fire Department After-Action Report; Graniteville Train Wreck. Aiken, SC (January) 2000.

Guide to Managing An Emergency Service Infection Control Program. United States Fire Administration, FA-112 (January) 2002.

Guidelines for Mass Casualty Decontamination During a Terrorist Chemical Agent Incident. Prepared by U.S. Army Soldier and Biological Chemical Command (SBCCOM). Aberdeen Proving Ground, MD (January) 2000.

Hazardous Materials Emergency Response Planning Guide. The National Response Team of the National Oil and Hazardous Substances Contingency Plan. Washington, D.C., 1987.

Health Service Support in a Nuclear, Biological, and Chemical Environment. Department of the Army Field Manual FM8-10-7, 1993.

JTF-CS Media Outreach Information (DVD), Joint Task Force Civil Support. Fort Monroe, VA.

Liquid Splash-Protective Suits for Hazardous Chemical Emergencies – NFPA 1992. Quincy, MA: National Fire Protection Association, 1994.

Marshall, Stephen M.; Fedele, Paul D.; Lake, William A. *Guidelines for Incident Commander's Use of Firefighter Protective Ensemble (FFPE) with Self-Contained Breathing Apparatus (SCBA) for Rescue Operations During a Terrorist Chemical Agent Incident.* Prepared by U.S. Army Soldier and Biological Chemical Command (SBCCOM) and the Domestic Preparedness Chemical Team with the assistance of firefighters from Montgomery County Fire and Rescue Services (MD), and Baltimore County Fire Department (MD). Commander, U.S. Army, ECBC, Attn.: SSB-REN-HD-DI, Building E5307, Hanlon Road, APG, MD 21010-5424. August, 1999.

Medical Management Guidelines for Acute Chemical Exposures. U.S. Department of Human Services, Public Health Service, Agency for Toxic Substance and Disease Registry (August 1) 1992.

Medical Management of Biological Casualties Handbook – Fourth Edition. Fort Detrick, MD: U.S. Army Medical Research Institute of Infectious Diseases (February) 2001.

Medical Management of Biological Casualties – Satellite Broadcast/Student Booklet. Fort Detrick, MD: U.S. Army Medical Research Institute of Infectious Diseases (February) 2001.

Medical Management of Chemical Casualties Handbook – Second Edition. Aberdeen Proving Ground, MD: Medical Research Institute of Chemical Defense, 1995.

Medical Management and Treatment in Biological Warfare Operations, (Field Manual). FM 8-284/ NAVMED P-5042/AFMAN (l) 44-156/MCRP 4-ll.1C. 2000.

Medical Management of Chemical Casualties, MMCC Supplemental Training Materials v.3.00 (DVD). U.S. Army Medical Research Institute of Chemical Defense, Aberdeen Proving Ground, MD (January) 2002.

Metropolitan Medical Strike Team Operational System Description. Washington, D.C.: Metropolitan Washington Council of Governments, and the United States Public Health Service, Office of Emergency Preparedness, with the U.S. Department of Health and Human Services, 1996.

NATO Handbook on The Medical Aspects of NBC Defensive Operations: Part II – Biological AMedP-6(B). Department of the Army Field Manual (FM 8-9) 1996.

NATO Handbook on the Medical Aspects of NBC Defensive Operations: Part III – Chemical AMedP-6(B). Department of the Army Field Manual (FM 8-9) 1996.

NBC Decontamination. Department of the Army Field Manual (FM 3-5) 1993.

NBC Field Handbook. Department of the Army Field Manual (FM 3-7) 1994.

NBC Protection. Department of the Army Field Manual (FM 3-4) 1992.

Nidiffer, N. J. *Un-Natural Disaster: Stories of Survival after Graniteville Tragedy, Harbor House.* Augusta, GA, 2005.

NIOSH Pocket Guide To Chemical Hazards. U.S. Department of Health and Human Services, Public Health Service, Centers for Disease Control, and the National Institute for Occupational Health and Safety. Washington, D.C.: Superintendent of Documents, U.S. Government Printing Office, 1997.

Norfolk Southern Graniteville Derailment – Final Information Update as of January21,2005 (Homes, Schools, Traffic, Crash Site Activities, Environmental, Hospitals, Law Enforcement Fire Service, Family Pets, Government Services, Local Assistance Center, Aiken County Summary Court, Aiken County 211). U.S. E.P.A. Region 4, Southwest; March 12, 2006.

North American Emergency Response Guidebook (A Guidebook for First Responders During the Initial Phase of a Hazardous Materials/Dangerous Goods Incident). Washington, D.C.: U.S. Department of Transportation/Transport Canada/the Secretariat of Communications and Transportation for Mexico, 2000.

Nuclear, Biological, and Chemical Defense Operations. Marine Corps Warfighting Publication (MCWP) 3-37 MAGTF. Washington, D.C.: Department of the Navy, Headquarters United States Marine Corps (September) 1998.

Nuclear, Biological, and Chemical Defense Training. U.S. Marine Corps Order MCO 3400.3, Fleet Marine Force Manual, Washington, D.C.

Nuclear Contamination Avoidance. U.S. Marine Corps Fleet Marine Force Manual 11-18, Washington, D.C.

Occupational Safety and Health Guidance Manual for Hazardous Waste Site Activities. National Institute for Occupational Safety and Health (NIOSH), Occupational and Health Administration (OSHA), the U.S. Coast Guard (USCG), and the U.S. Environmental Protection Agency (EPA). Washington, D.C.: U.S. Department of Health and Human Services (HHS), Public Health Service (PHS), Centers for Disease Control (CDC), and the National Institute for Occupational Safety and Control (NIOSH), 1985.

Potential Military Chemical/Biological Agents and Compounds. Department of the Army Field Manual (FM 3-9) 1990.

Presidential Decision Directive – 39 (PDD-30), U.S. Policy on Counterterrorism. Washington, D.C.

Presidential Decision Directive – 62 (PDD – 62), Protection Against Unconventional Threats to the Homeland Proceedings of the Seminar on Responding to the Consequences of Chemical and Biological Terrorism Professional Competence of Responders to Hazardous Materials Incidents - NFPA 472. Quincy, MA: National *Protective Clothing for Emergency Medical Operations - NFPA 1999.* Quincy, MA. National Fire Protection *Public Health Consequences from Hazardous Substances Acutely Released During Rail Transit – South Carolina, 2005; Selected States, 1999–2004, MMWR Weekly.* Available online at http://www.cdc.gov/mmwr/review/mmwrhtml/mm5403a2.htm January 28, 2005.

Recommended Practice For Responding To Hazardous Materials Incidents – NFPA 471. Quincy, MA. National Fire Protection Association, 1992.

Second Annual Report of the Advisory to Assess Domestic Response Capabilities for Terrorism Involving Weapons of Mass Destruction (WMD): II Toward a National Strategy for Combating Terrorism (also called "the Gilmore Commission"). Report Number GAO/TSIAD – 00 – 145. Washington, D.C.: U.S. Government Accounting Office (December 15) 2000.

Situation Reports #1 Through #13 from the Forward State Emergency Operation Center (via ECV). State Emergency Operations Center, from January 6 through January 13, 2005, West Columbia, South Carolina.

Special Congressional Commission on Terrorism Report. Washington, D.C.: National Commission on Terrorism. June 5, 2000.

Standing Operating Procedure for Obtaining, Shipping, Receipt and Storage of Biomedical Samples. U.S. Army Medical Research Institute of Chemical Defense (USAMRICD), Aberdeen Proving Ground, MD. Undated.

Suit Smart (CD). Kappler Protective Apparel & Fabrics. Guntersville, AL 1997.

Support Function Protective Clothing for Hazardous Chemicals Operations – NFPA 1993. Quincy, MA. *Technical Bulletin – Assay Techniques for Detection of Exposure To Sulfur Mustard, Cholinesterase Inhibitors, Terrorism Incident Annex to the Federal Response Plan.* Washington, D.C. 1995.

Train Derailment and Chlorine Spill (DVD). Graniteville-Vaucluse-Warrenville Volunteer Fire Department. Graniteville, SC (January 6) 2005,

Treatment of Chemical Agent Casualties and Conventional Military Chemical Injuries. Department of the Army Field Manual, FM8-285.

USAMRMC Products Portfolio. U.S. Army Medical Research and Material Command, Fort Detrick, MD (May) 2004.

U.S. Navy Shipboard Chemical-Hazard Assessment Guide (C-HAG). 1990.

U.S. Policy on Counterterrorism (Presidential Decision Directive 39). 1995.

Vapor - Protective Suits for Hazardous Chemical Emergencies – NFPA 1991. Quincy, MA: National Fire Protection Association, 1994.

Index